MERCURY

Outboard
1965-89 REPAIR MANUAL
2-40 HORSEPOWER, 1 AND 2 CYLINDER

MW00838270

SELOC

Managing Partners	Dean F. Morgantini, S.A.E.
	Barry L. Beck
Executive Editor	Kevin M. G. Maher, A.S.E.
Production Specialists	Melinda Possinger
	Ronald Webb
Authors	Joan and Clarence Coles

Manufactured in USA
© 1994 Seloc Publishing
104 Willowbrook Lane
West Chester, PA 19382
ISBN 13: 978-0-89330-012-8
ISBN 10: 0-89330-012-8
11th Printing 9876543210

MARINE TECHNICIAN TRAINING

INDUSTRY SUPPORTED PROGRAMS
OUTBOARD, STERNDRIVE & PERSONAL WATERCRAFT

- *Dyno Testing* • *Boat & Trailer Rigging* • *Electrical & Fuel System Diagnostics*
- *Powerhead, Lower Unit & Drive Rebuilds* • *Powertrim & Tilt Rebuilds*
- *Instrument & Accessories Installation*

For information regarding housing, financial aid and employment opportunities in the marine industry, contact us today:

CALL TOLL FREE
1-800-528-7995

TRAIN IN SUNNY FLORIDA!

An Accredited Institution

SM

Name
Address
City   State   Zip
Phone

MARINE MECHANICS INSTITUTE
A Division of CTI

MEMBER NMMA

9751 Delegates Drive • Orlando, Florida 32837
2844 W. Deer Valley Rd. • Phoenix, AZ 85027

FINANCIAL ASSISTANCE AVAILABLE FOR THOSE WHO QUALIFY!

SAFETY NOTICE

Proper service and repair procedures are vital to the safe, reliable operation of all marine engines, as well as the personal safety of those performing repairs. This manual outlines procedures for servicing and repairing outboards using safe, effective methods. The procedures contain many NOTES, CAUTIONS and WARNINGS which should be followed, along with standard procedures, to eliminate the possibility of personal injury or improper service which could damage the vessel or compromise its safety.

It is important to note that repair procedures and techniques, tools and parts for servicing marine engines, as well as the skill and experience of the individual performing the work, vary widely. It is not possible to anticipate all of the conceivable ways or conditions under which these engines may be serviced, or to provide cautions as to all possible hazards that may result. Standard and accepted safety precautions and equipment should be used during cutting, grinding, chiseling, prying, or any other process that can cause material removal or projectiles.

Some procedures require the use of tools specially designed for a specific purpose. Before substituting another tool or procedure, you must be completely satisfied that neither your personal safety, nor the performance of the marine engine, will be compromised.

Although information in this manual is based on industry sources and is complete as possible at the time of publication, the possibility exists that some vehicle manufacturers made later changes which could not be included here. While striving for total accuracy, Seloc® cannot assume responsibility for any errors, changes or omissions that may occur in the compilation of this data.

PART NUMBERS

Part numbers listed in this reference are not recommendations by Seloc® for any product brand name. They are references that can be used with interchange manuals and aftermarket supplier catalogs to locate each brand supplier's discrete part number.

SPECIAL TOOLS

Special tools are recommended by the marine manufacturer to perform a specific task. Use has been kept to a minimum, but, where absolutely necessary, they are referred to in the text by the part number of the tool manufacturer. These tools can be purchased, under the appropriate part number, from your local dealer or regional distributor, or an equivalent tool can be purchased locally from a tool supplier or parts outlet. Before substituting any tool for the one recommended, read the SAFETY NOTICE at the top of this page.

ACKNOWLEDGMENTS

Seloc® expresses sincere appreciation to Mercury Marine for their assistance in the production of this manual.

ALL RIGHTS RESERVED

No part of this publication may be reproduced, transmitted or stored in any form or by any means, electronic or mechanical, including photocopy, recording, or by information storage or retrieval system, without prior written permission from the publisher.

This is a comprehensive tune-up and repair manual for Mercury outboards manufactured between 1965 and 1991. Competition, high-performance, and commercial units (including aftermarket equipment), are not covered. The book has been designed and written for the professional mechanic, the do-it-yourselfer, and the student developing his mechanical skills.

Professional Mechanics will find it to be an additional tool for use in their daily work on outboard units because of the many special techniques described.

Boating enthusiasts interested in performing their own work and in keeping their unit operating in the most efficient manner will find the step-by-step illustrated procedures used throughout the manual extremely valuable. In fact, many have said this book almost equals an experienced mechanic looking over their shoulder giving them advice.

Students and Instructors have found the chapters divided into practical areas of interest and work. Technical trade schools, from Florida to Michigan and west to California, as well as the U.S. Navy and Coast Guard, have adopted these manuals as a standard classroom text.

Troubleshooting sections have been included in many chapters to assist the individual performing the work in quickly and accurately isolating problems to a specific area without unnecessary expense and time-consuming work. As an added aid and one of the unique features of this book, many worn parts are illustrated to identify and clarify when an item should be replaced.

Illustrations and procedural steps are so closely related and identified with matching numbers that, in most cases, captions are not used. Exploded drawings show internal parts and their interrelationship with the major component.

TABLE OF CONTENTS

1
SAFETY

1-1 INTRODUCTION

In order to protect the investment for the boat and outboard, they must be cared for properly while being used and when out of the water. Always store the boat with the bow higher than the stern and be sure to remove the transom drain plug and the inner hull drain plugs. If any type of cover is used to protect the boat, be sure to allow for some movement of air through the hull. Proper ventilation will assure evaporation of any condensation that may form due to changes in temperature and humidity.

1-2 CLEANING, WAXING, AND POLISHING

Any boat should be washed with clear water after each use to remove surface dirt and any salt deposits from use in salt water. Regular rinsing will extend the time between waxing and polishing. It will also give you "pride of ownership", by having a sharp looking piece of equipment. Elbow grease, a mild detergent, and a brush will be required to remove stubborn dirt, oil, and other unsightly deposits.

Stay away from harsh abrasives or strong chemical cleaners. A white buffing compound can be used to restore the original gloss to a scratched, dull, or faded area. The finish of your boat should be thoroughly cleaned, buffed, and polished at least once each season. Take care when buffing or polishing with a marine cleaner not to overheat the surface you are working, because you will burn it.

1-3 CONTROLLING CORROSION

Since man first started out on the water, corrosion on his craft has been his enemy. The first form was merely rot in the wood and then it was rust, followed by other forms of destructive corrosion in the more modern materials. One defense against corrosion is to use similar metals throughout the boat. Even though this is difficult to do in designing a new boat, particularly the undersides, similar metals should be used whenever and wherever possible.

A second defense against corrosion is to insulate dissimilar metals. This can be done by using an exterior coating of Sea Skin or by insulating them with plastic or rubber gaskets.

A clean boat, properly tuned outboard unit, and attention to sensible safety practices are what make: "The worst day fishin' better than the best day workin'."

Zinc installation also used as the trim tab. The tab assists the helmsperson to maintain a true course without "fighting" the wheel.

Using Zinc

The proper amount of zinc attached to a boat is extremely important. The use of too much zinc can cause wood burning by placing the metals close together and they become "hot". On the other hand, using too small a zinc plate will cause more rapid deterioration of the metal you are trying to protect. If in doubt, consider the fact that it is far better to replace the zincs than to replace planking or other expensive metal parts from having an excess of zinc.

When installing zinc plates, there are two routes available. One is to install many

Accessory zinc installation on the boat transom to provide additional corrosion protection.

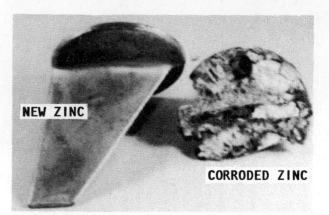

A new trim tab zinc, left, and a corroded zinc, right. An excellent example of the inexpensive zinc saving more costly parts of the outboard unit.

different zincs on all metal parts and thus run the risk of wood burning. Another route, is to use one large zinc on the transom of the boat and then connect this zinc to every underwater metal part through internal bonding. Of the two choices, the one zinc on the transom is the better way to go.

Small outboard engines have a zinc plate attached to the cavitation plate. Therefore, the zinc remains with the engine at all times.

1-4 PROPELLERS

As you know, the propeller is actually what moves the boat through the water. This is how it is done. The propeller operates in water in much the manner as a wood screw does in wood. The propeller "bites" into the water as it rotates. Water passes between the blades and out to the rear in the shape of a cone. The propeller "biting" through the water in much the same manner as a wood auger is what propels the boat.

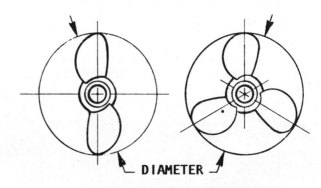

Diameter and pitch are the two basic dimensions of a propeller. The diameter is measured across the circumference of a circle scribed by the propeller blades, as shown.

Diameter and Pitch

Only two dimensions of the propeller are of real interest to the boat owner: the diameter and the pitch. These two dimensions are stamped on the propeller hub and always appear in the same order: the diameter first and then the pitch. For instance, the number 15-19 stamped on the hub, would mean the propeller had a diameter of 15 inches with a pitch of 19.

The diameter is the measured distance from the tip of one blade to the tip of the other as shown in the accompanying illustration.

The pitch of a propeller is the angle at which the blades are attached to the hub. This figure is expressed in inches of water travel for each revolution of the propeller. In our example of a 15-19 propeller, the propeller should travel 19 inches through the water each time it revolves. If the propeller action was perfect and there was no slippage, then the pitch multiplied by the propeller rpms would be the boat speed.

Most outboard manufacturers equip their units with a standard propeller with a diameter and pitch they consider to be best suited to the engine and the boat. Such a propeller allows the engine to run as near to the rated rpm and horsepower (at full throttle) as possible for the boat design.

The blade area of the propeller determines its load-carrying capacity. A two-blade propeller is used for high-speed running under very light loads.

A four-blade propeller is installed in boats intended to operate at low speeds under very heavy loads such as tugs, barges, or large houseboats. The three-blade propeller is the happy medium covering the wide range between the high performance units and the load carrying workhorses.

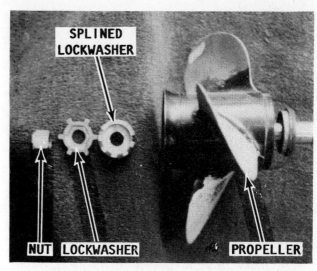

Typical attaching hardware for a propeller.

Propeller Selection

There is no standard propeller that will do the proper job in very many cases. The list of sizes and weights of boats is almost endless. This fact coupled with the many boat-engine combinations makes the propeller selection for a specific purpose a difficult job. In fact, in many cases the propeller is changed after a few test runs. Proper selection is aided through the use of charts set up for various engines and boats. These charts should be studied and understood when buying a propeller. However, bear in mind, the charts are based on average boats with average loads, therefore, it may be necessary to make a change in size or pitch, in order to obtain the desired results for the hull design or load condition.

Propellers are available with a wide range of pitch. Remember, a low pitch takes a smaller bite of the water than the high pitch propeller. This means the low pitch propeller will travel less distance through the water per revolution. The low

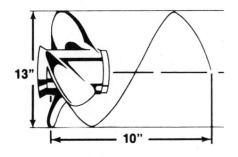

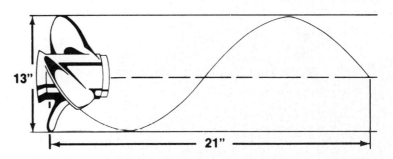

Diagram to explain the pitch dimension of a propeller. The pitch is the theoretical distance a propeller would travel through water if there were no friction.

pitch will require less horsepower and will allow the engine to run faster.

All engine manufacturers design their units to operate with full throttle at, or slightly above, the rated rpm. If you run your engine at the rated rpm, you will increase spark plug life, receive better fuel economy, and obtain the best performance from your boat and engine. Therefore, take time to make the proper propeller selection for the rated rpm of your engine at full throttle with what you consider to be an average load. Your boat will then be correctly balanced between engine and propeller throughout the entire speed range.

A reliable tachometer must be used to measure engine speed at full throttle to ensure the engine will achieve full horsepower and operate efficiently and safely. To test for the correct propeller, make your run in a body of smooth water with the lower unit in forward gear at full throttle. If the reading is above the manufacturer's recommended operating range, you must try propellers of greater pitch, until you find the one that allows the engine to operate continually within the recommended full throttle range.

If the engine is unable to deliver top performance and you feel it is properly tuned, then the propeller may not be to blame. Operating conditions have a marked effect on performance. For instance, an engine will lose rpm when run in very cold water. It will also lose rpm when run in salt water as compared with fresh water. A hot, low-barometer day will also cause your engine to lose power.

Cavitation

Cavitation is the forming of voids in the water just ahead of the propeller blades. Marine propulsion designers are constantly fighting the battle against the formation of these voids due to excessive blade tip speed and engine wear. The voids may be filled with air or water vapor, or they may actually be a partial vacuum. Cavitation may be caused by installing a piece of equipment too close to the lower unit, such as the knot indicator pickup, depth sounder, or bait tank pickup.

Vibration

Your propeller should be checked regularly to be sure all blades are in good condition. If any of the blades become bent or nicked, this condition will set up vibrations in the drive unit and the motor. If the vibration becomes very serious it will cause a loss of power, efficiency, and boat performance. If the vibration is allowed to continue over a period of time it can have a damaging effect on many of the operating parts.

Vibration in boats can never be completely eliminated, but it can be reduced by keeping all parts in good working condition and through proper maintenance and lubrication. Vibration can also be reduced in

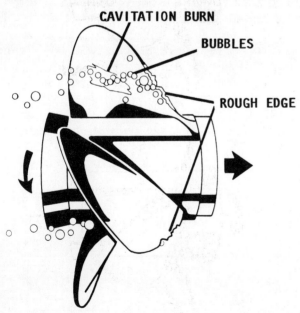

Cavitation (air bubbles) formed at the propeller. Manufacturers are constantly fighting this problem, as explained in the text.

Example of a damaged propeller. This unit should have been replaced long before this amount of damage was sustained.

some cases by increasing the number of blades. For this reason, many racers use two-blade props and luxury cruisers have four- and five-blade props installed.

Shock Absorbers

The shock absorber in the propeller plays a very important role in protecting the shafting, gears, and engine against the shock of a blow, should the propeller strike an underwater object. The shock absorber allows the propeller to stop rotating at the instant of impact while the power train continues turning.

How much impact the propeller is able to withstand, before causing the shock absorber to slip, is calculated to be more than the force needed to propel the boat, but less than the amount that could damage any part of the power train. Under normal propulsion loads of moving the boat through the water, the hub will not slip. However, it will slip if the propeller strikes an object with a force that would be great enough to stop any part of the power train.

If the power train was to absorb an impact great enough to stop rotation, even

for an instant, something would have to give and be damaged. If a propeller is subjected to repeated striking of underwater objects, it would eventually slip on its clutch hub under normal loads. If the propeller should start to slip, a new shock absorber/cushion hub would have to be installed.

Propeller Rake

If a propeller blade is examined on a cut extending directly through the center of the hub, and if the blade is set vertical to the propeller hub, as shown in the accompanying illustration, the propeller is said to have a zero degree ($0°$) rake. As the blade slants back, the rake increases. Standard propellers have a rake angle from $0°$ to $15°$.

A higher rake angle generally improves propeller performance in a cavitating or ventilating situation. On lighter, faster boats, higher rake often will increase performance by holding the bow of the boat higher.

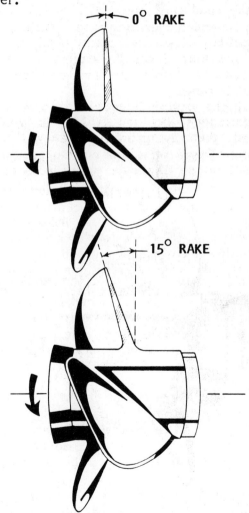

Rubber hub removed from the propeller because the hub was slipping in the propeller.

Illustration depicting the rake of a propeller, as explained in the text.

CONSTANT PITCH

PROGRESSIVE PITCH

Comparison of a constant and progressive pitch propeller. Notice how the pitch of the progressive pitch propeller, right, changes to give the blade more thrust and therefore, the boat more speed.

Progressive Pitch

Progressive pitch is a blade design innovation that improves performance when forward and rotational speed is high and/or the propeller breaks the surface of the water.

Progressive pitch starts low at the leading edge and progressively increases to the trailing edge, as shown in the accompanying illustration. The average pitch over the entire blade is the number assigned to that propeller. In the illustration of the progressive pitch, the average pitch assigned to the propeller would be 21.

Cupping

If the propeller is cast with an edge curl inward on the trailing edge, the blade is said to have a cup. In most cases, cupped blades improve performance. The cup helps the blades to **"HOLD"** and not break loose, when operating in a cavitating or ventilating situation.

The cup has the effect of adding to the propeller pitch. Cupping usually will reduce full-throttle engine speed about 150 to 300 rpm below the same pitch propeller without a cup to the blade. A propeller repair shop is able to increase or decrease the cup on the blades. This change, as explained, will alter engine rpm to meet specific operating demands. Cups are rapidly becoming standard on propellers.

In order for a cup to be the most effective, the cup should be completely concave (hollowed) and finished with a sharp corner. If the cup has any convex rounding, the effectiveness of the cup will be reduced.

Rotation

Propellers are manufactured as right-hand rotation (RH), and as left-hand rotation (LH). The standard propeller for outboard units is RH rotation.

A right-hand propeller can easily be identified by observing it as shown in the accompanying illustration. Observe how the blade of the right-hand propeller slants from the lower left to upper right. The left-hand propeller slants in the opposite direction, from lower right to upper left.

When the RH propeller is observed rotating from astern the boat, it will be rotating clockwise when the engine is in forward gear. The left-hand propeller will rotate counterclockwise.

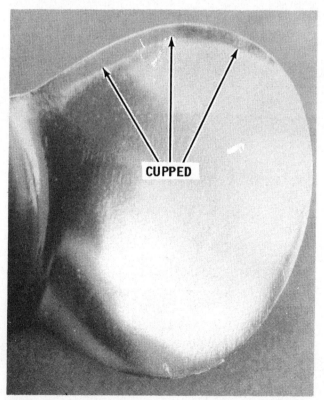

Propeller with a "cupped" leading edge. "Cupping" gives the propeller a better "hold" in the water.

U.S. Coast Guard plate affixed to all new boats. When the blanks are filled in, the plate will indicate the Coast Guard's recommendations for persons, gear, and horsepower to ensure safe operation of the boat. These recommendations should not be exceeded, as explained in the text.

1-5 LOADING

In order to receive maximum enjoyment, with safety and performance, from your boat, take care not to exceed the load capacity given by the manufacturer. A plate attached to the hull indicates the U.S. Coast Guard capacity information in pounds for persons and gear. If the plate states the maximum person capacity to be 750 pounds and you assume each person to weigh an average of 150 lbs., then the boat could carry five persons safely. If you add another 250 lbs. for motor and gear, and the maximum weight capacity for persons and gear is 1,000 lbs. or more, then the five persons and gear would be within the limit.

Try to load the boat evenly port and starboard. If you place more weight on one side than on the other, the boat will list to the heavy side and make steering difficult. You will also get better performance by placing heavy supplies aft of the center to keep the bow light for more efficient planing.

Clarification

Much confusion arises from the terms, certification, requirements, approval, regulations, etc. Perhaps the following may clarify a couple of these points.

1- The Coast Guard does not approve boats in the same manner as they "Approve" life jackets. The Coast Guard applies a formula to inform the public of what is safe for a particular craft.

2- If a boat has to meet a particular regulation, it must have a Coast Guard certification plate. The public has been led to believe this indicates approval of the Coast Guard. Not so.

3- The certification plate means a willingness of the manufacturer to meet the Coast Guard regulations for that particular craft. The manufacturer may recall a boat if it fails to meet the Coast Guard requirements.

4- The Coast Guard certification plate, see accompanying illustration, may or may not be metal. The plate is a regulation for the manufacturer. It is only a warning plate and the public does not have to adhere to the restrictions set forth on it. Again, the plate sets forth information as to the Coast Guard's opinion for safety on that particular boat.

5- Coast Guard Approved equipment is equipment which has been approved by the Commandant of the U.S. Coast Guard and has been determined to be in compliance with Coast Guard specifications and regulations relating to the materials, construction, and performance of such equipment.

1-6 HORSEPOWER

The maximum horsepower engine for each individual boat should not be increased by any significant amount without checking requirements from the Coast Guard in the local area. The Coast Guard determines horsepower requirements based on the length, beam, and depth of the hull. **TAKE CARE NOT** to exceed the maximum horsepower listed on the plate or the warranty, and possibly the insurance, on the boat may become void.

1-7 FLOTATION

If the boat is less than 20 ft. overall, a Coast Guard or BIA (Boating Industry of America), now changed to NMMA (National Marine Manufacturers Association) requirement is that the boat must have buoyant material built into the hull (usually foam) to keep it from sinking if it should become swamped. Coast Guard requirements are mandatory but the NMMA is voluntary.

"Kept from sinking" is defined as the ability of the flotation material to keep the boat from sinking when filled with water and with passengers clinging to the hull. One restriction is that the total weight of the motor, passengers, and equipment aboard does not exceed the maximum load capacity listed on the plate.

Life Preservers —Personal Flotation Devices (PFDs)

The Coast Guard requires at least one Coast Guard approved life-saving device be carried on board all motorboats for each person on board. Devices approved are identified by a tag indicating Coast Guard approval. Such devices may be life preservers, buoyant vests, ring buoys, or buoyant cushions. Cushions used for seating are serviceable if air cannot be squeezed out of it. Once air is released when the cushion is squeezed, it is no longer fit as a flotation device. New foam cushions dipped in a rubberized material are almost indestructable.

Life preservers have been classified by the Coast Guard into five type categories. All PFDs presently acceptable on recreational boats fall into one of these five designations. All PFDs **MUST** be U.S. Coast Guard approved, in good and serviceable condition, and of an appropriate size for the persons who intend to wear them. Wearable PFDs **MUST** be readily accessible and throwable devices **MUST** be immediately available for use.

Type I PFD has the greatest required buoyancy and is designed to turn most

UNCONSCIOUS persons in the water from a face down position to a vertical or slightly backward position. The adult size device provides a minimum buoyancy of 22 pounds and the child size provides a minimum buoyancy of 11 pounds. The Type I PFD provides the greatest protection to its wearer and is most effective for all waters and conditions.

Type II PFD is designed to turn its wearer in a vertical or slightly backward position in the water. The turning action is not as pronounced as with a Type I. The device will not turn as many different type persons under the same conditions as the Type I. An adult size device provides a minimum buoyancy of 15½ pounds, the medium child size provides a minimum of 11 pounds, and the infant and small child sizes provide a minimum buoyancy of 7 pounds.

Type III PFD is designed to permit the wearer to place himself (herself) in a vertical or slightly backward position. The Type III device has the same buoyancy as the Type II PFD but it has little or no turning ability. Many of the Type III PFD are designed to be particularly useful when water skiing, sailing, hunting, fishing, or engaging in other water sports. Several of this type will also provide increased hypothermia protection.

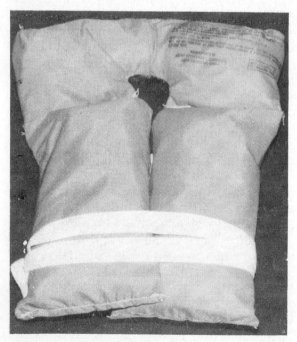

*Type I PFD Coast Guard approved life jacket. This type flotation device provides the greatest amount of buoyancy. **NEVER** use them for cushions or other purposes.*

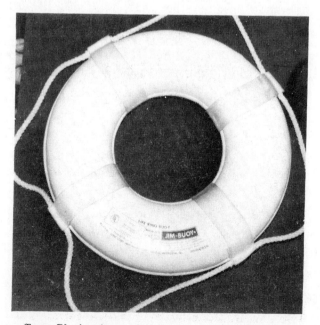

Type IV ring buoy also designed to be thrown to a person in the water. On ocean cruisers, this type device usually has a weighted pole with flag and light attached to the buoy.

Type IV PFD is designed to be thrown to a person in the water and grasped and held by the user until rescued. It is **NOT** designed to be worn. The most common Type IV PFD is a ring buoy or a buoyant cushion.

Type V PFD is any PFD approved for restricted use.

Coast Guard regulations state, in general terms, that on all boats less than 16 ft. overall, one Type I, II, III, or IV device shall be carried on board for each person in the boat. On boats over 26 ft., one Type I, II, or III device shall be carried on board for each person in the boat **plus** one Type IV device.

It is an accepted fact that most boating people own life preservers, but too few actually wear them. There is little or no excuse for not wearing one because the modern comfortable designs available today do not subtract from an individual's boating pleasure. Make a life jacket available to your crew and advise each member to wear it. If you are a crew member ask your skipper to issue you one, especially when boating in rough weather, cold water, or when running at high speed. Naturally, a life jacket should be a must for non-swimmers any time they are out on the water in a boat.

1-8 ANCHORS

One of the most important pieces of equipment in the boat next to the power plant is the ground tackle carried. The engine makes the boat go and the anchor and its line are what hold it in place when the boat is not secured to a dock or on the beach.

The anchor must be of suitable size, type, and weight to give the skipper peace of mind when his boat is at anchor. Under certain conditions, a second, smaller, lighter anchor may help to keep the boat in a favorable position during a non-emergency daytime situation.

In order for the anchor to hold properly, a piece of chain must be attached to the anchor and then the nylon anchor line attached to the chain. The amount of chain should equal or exceed the length of the boat. Such a piece of chain will ensure that the anchor stock will lay in an approximate horizontal position and permit the flutes to dig into the bottom and hold.

1-9 BOATING ACCIDENT REPORTS

New federal and state regulations require an accident report to be filed with the nearest State boating authority within 48

A Type IV PFD cushion device intended to be thrown to a person in the water. If air can be squeezed out of the cushion, it is no longer fit for service as a PFD.

Moisture-protected flares should be carried on board for use as a distress signal.

hours if a person is lost, disappears, or is injured to the degree of needing medical treatment beyond first aid.

Accidents involving only property or equipment damage **MUST** be reported within 10 days if the damage is in excess of $200. Some states require reporting of accidents with property damage less than $200 or total boat loss.

A **$500 PENALTY** may be assessed for failure to submit the report.

WORD OF ADVICE

Take time to make a copy of the report to keep for your records or for the insurance company. Once the report is filed, the Coast Guard will not give out a copy, even to the person who filed the report.

The report must give details of the accident and include:

1- The date, time, and exact location of the occurrence.

2- The name of each person who died, was lost, or injured.

3- The number and name of the vessel.

4- The names and addresses of the owner and operator.

If the operator cannot file the report for any reason, each person on board **MUST** notify the authorities, or determine that the report has been filed.

Where the fish are bitin' is where the action is, even on a cold rainy day. Gear in the boat for safety and pleasure add to the day's fun on the water.

2
TUNING

2-1 INTRODUCTION

The efficiency, reliability, fuel economy and enjoyment available from engine performance are all directly dependent on having it tuned properly. The importance of performing service work in the sequence detailed in this chapter cannot be over emphasized. Before making any adjustments, check the specifications in the Appendix. **NEVER** rely on memory when making critical adjustments.

Before beginning to tune any engine, check to be sure the engine has satisfactory compression. An engine with worn or bro-

ken piston rings, burned pistons, or scored cylinder walls, cannot be made to perform properly no matter how much time and expense is spent on the tune-up. Poor compression must be corrected or the tune-up will not give the desired results.

A practical maintenance program that is followed throughout the year, is one of the best methods of ensuring the engine will give satisfactory performance at any time.

With his "best friend", this man is off to a good day on the water with a properly tuned outboard to get him to where the "big ones" are hidin'.

Damaged piston, probably caused by inaccurate fuel mixture, or improper timing.

BURNED AREA

The extent of the engine tune-up is usually dependent on the time lapse since the last service. A complete tune-up of the entire engine would entail almost all of the work outlined in this manual. A logical sequence of steps will be presented in general terms. If additional information or detailed service work is required, the chapter containing the instructions will be referenced.

Each year higher compression ratios are built into modern outboard engines and the electrical systems become more complex, especially with electronic (capacitor discharge) units. Therefore, the need for reliable, authoritative, and detailed instructions becomes more critical. The information in this chapter and the referenced chapters fulfill that requirement.

2-2 TUNE-UP SEQUENCE

During a major tune-up, a definite sequence of service work should be followed to return the engine to the maximum performance desired. This type of work should not be confused with attempting to locate problem areas of "why" the engine is not performing satisfactorily. This work is classified as "trouble shooting". In many cases, these two areas will overlap, because many times a minor or major tune-up will correct the malfunction and return the system to normal operation.

The following list is a suggested sequence of tasks to perform during the tune-up service work. The tasks are merely listed here. Generally procedures are given in subsequent sections of this chapter. For more detailed instructions, see the referenced chapter.

1- Perform a compression check of each cylinder. See Chapter 8.

2- Inspect the spark plugs to determine their condition. Test for adequate spark at the plug. See Chapter 5.

3- Start the engine in a body of water and check the water flow through the engine. See Chapter 9.

4- Check the gear oil in the lower unit. See Chapter 9.

5- Check the carburetor adjustments and the need for an overhaul. See Chapter 4.

6- Check the fuel pump for adequate performance and delivery. See Chapter 4.

7- Make a general inspection of the ignition system. See Chapter 5.

8- Test the starter motor and the solenoid, if so equipped. See Chapter 7.

9- Check the internal wiring.

10- Check the timing and synchronization. See Chapter 6.

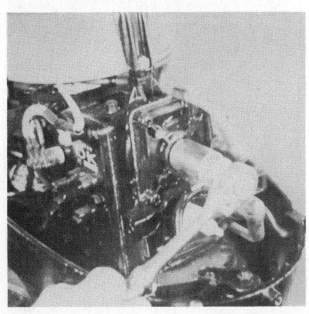

Removing the spark plugs for inspection. Worn plugs are one of the major contributing factors to poor engine performance.

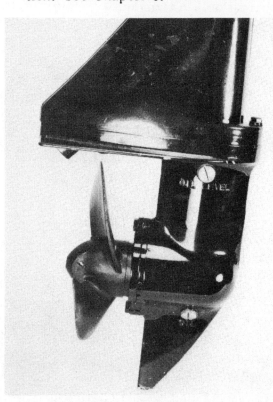

"OIL LEVEL" and "OIL" are cast into the lower unit housing near the two openings. Both screws are removed and lubricant added through the lower opening.

2-3 COMPRESSION CHECK

A compression check is extremely important, because an engine with low or uneven compression between cylinders **CANNOT** be tuned to operate satisfactorily. Therefore, it is essential that any compression problem be corrected before proceeding with the tune-up procedure. See Chapter 8.

If the powerhead shows any indication of overheating, such as discolored or scorched paint, inspect the cylinders visually thru the transfer ports for possible scoring. It is possible for a cylinder with satisfactory compression to be scored slightly. Also, check the water pump. The overheating condition may be caused by a faulty water pump.

Checking Compression

Remove the spark plug wires. **ALWAYS** grasp the molded cap and pull it loose with a twisting motion to prevent damage to the connection. Remove the spark plugs and keep them in **ORDER** by cylinder for evaluation later. Ground the spark plug leads to the engine to render the ignition system inoperative while performing the compression check.

A compression check should be taken in each cylinder before spending time and money on tune-up work. Without adequate compression, efforts in other areas to regain powerhead performance will be wasted.

Insert a compression gauge into the No. 1, top, spark plug opening. Crank the engine with the starter thru at least 4 complete strokes with the throttle at the wide-open position, to obtain the highest possible reading. Record the reading. Repeat the test and record the compression for each cylinder. A variation between cylinders is far more important than the actual readings. A variation of more than 15 psi (103 kPa), between cylinders indicates the lower compression cylinder is defective. The problem may be worn, broken, or sticking piston rings, scored pistons or worn cylinders.

Use of an engine cleaner will help to free stuck rings and to disolve accumulated carbon. Follow the directions on the can.

2-4 SPARK PLUG INSPECTION

Inspect each spark plug for badly worn electrodes, glazed, broken, blistered, or lead fouled insulators. Replace all of the plugs, if one shows signs of excessive wear.

Make an evaluation of the cylinder performance by comparing the spark condition with those shown in Chapter 5. Check each

*Damaged spark plugs. Notice the broken electrode on the left plug. The broken part **MUST** be found and removed before returning the powerhead to service.*

*Today, numerous type spark plugs are available for service. **ALWAYS** check with the local marine dealer to be sure the proper plug is purchased for the unit being serviced.*

spark plug to be sure they are all of the same manufacturer and have the same heat range rating.

Inspect the threads in the spark plug opening of the block, and clean the threads before installing the plug.

When purchasing new spark plugs, **AL-WAYS** ask the marine dealer if there has been a spark plug change for the engine being serviced.

Crank the engine through several revolutions to blow out any material which might have become dislodged during cleaning.

Install the spark plugs and tighten them to a torque value of 17 ft-lbs (23 Nm). **ALWAYS** use a new gasket and wipe the seats in the block clean. The gasket must be fully compressed on clean seats to complete the heat transfer process and to provide a gas tight seal in the cylinder. If the torque value is too high, the heat will dissipate too rapidly. Conversely, if the torque value is too low, heat will not dissipate fast enough.

2-5 IGNITION SYSTEM

Four, yes, four different ignition systems are used on the outboard engines covered in this manual. If the engine performance is less than expected, and the ignition is diagnosed as the problem area, refer to Chapter 6 for detailed service procedures. The various types are clearly identified and cross-referenced in the Appendix. Once the Type system for the engine being serviced is known, the work can proceed smoothly. To properly time and synchronize the ignition system with the fuel system, see Chapter 5.

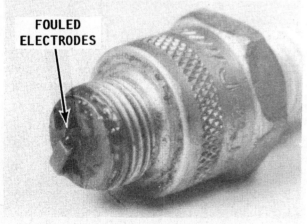

A fouled spark plug. The condition of this plug indicates problems in the cylinder which should be corrected.

Breaker Points

SOME GOOD WORDS: High primary voltage in Thunderbolt ignition systems (referenced as Type I and Type II in the Appendix), will darken and roughen the breaker points within a short period. This is not cause for alarm. Normally points in this condition would not operate satisfactorily in the conventional magneto, but they will give good service in the Thunderbolt systems. Therefore, **DO NOT** replace the points in a Type I or Type II Thunderbolt system unless an obvious malfunction exists, or the contacts are loose or burned. Rough or discolored contact surfaces are **NOT** sufficient reason for replacement. The cam follower will usually have worn away by the time the points have become unsatisfactory for efficient service.

Check the resistance across the contacts. If the test indicates zero resistance, the points are serviceable. A slight resistance across the points will affect idle operation. A high resistance may cause the ignition system to malfunction and loss of spark. Therefore, if any resistance across the points is indicated, the point set should be replaced.

2-6 TIMING AND SYNCHRONIZING

Correct timing and synchronization are essential to efficient engine operation. An engine may be in apparent excellent mechanical condition, but perform poorly, un-

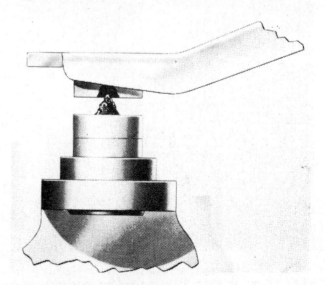

Worn ignition points are a common problem area with units having a distributor with points.

less the timing and synchronization have been adjusted precisely, according to the Specifications in the Appendix. To time and synchronize the engine, see Chapter 7.

2-7 BATTERY CHECK

Inspect and service the battery, cables and connections. Check for signs of corrosion. Inspect the battery case for cracks or bulges, dirt, acid, and electrolyte leakage. Check the electrolyte level in each cell.

Fill each cell to the proper level with distilled water or water passed thru a demineralizer.

Clean the top of the battery. The top of a 12-volt battery should be kept especially clean of acid film and dirt, because of the high voltage between the battery terminals. For best results, first wash the battery with a diluted ammonia or baking soda solution to neutralize any acid present. Flush the solution off the battery with clean water. Keep the vent plugs tight to prevent the neutralizing solution or water from entering the cells.

Check to be sure the battery is fastened securely in position. The hold-down device should be tight enough to prevent any movement of the battery in the holder, but not so tight as to place a strain on the battery case.

If the battery posts or cable terminals are corroded, the cables should be cleaned separately with a baking soda solution and a wire brush. Apply a thin coating of Multipurpose Lubricant to the posts and cable clamps before making the connections. The lubricant will help to prevent corrosion.

If the battery has remained under-charged, check for high resistance in the charging circuit. If the battery appears to be using too much water, the battery may be defective, or it may be too small for the job.

Jumper Cables

If booster batteries are used for starting an engine the jumper cables must be connected correctly and in the proper sequence to prevent damage to either battery, or diodes in the circuit.

*The fuel and ignition systems on any powerhead **MUST** be properly synchronized before maximum performance can be obtained from the unit.*

A check of the electrolyte in the battery should be a regular task on the maintenance schedule on any boat.

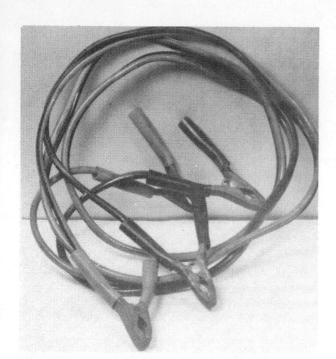

Common set of jumper cables for use with a second battery to crank and start the engine. EXTREME care should be exercised when using a second battery, as explained in the text.

ALWAYS connect a cable from the positive terminals of the dead battery to the positive terminal of the good battery **FIRST**. **NEXT**, connect one end of the other cable to the negative terminals of the good battery and the other end of the **ENGINE** for a good ground. By making the ground connection on the engine, if there is an arc when you make the connection it will not be near the battery. An arc near the battery could cause an explosion, destroying the battery and causing serious personal injury.

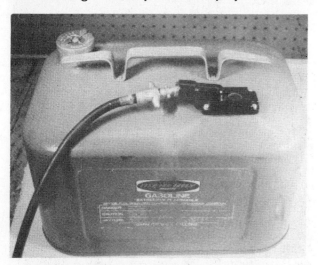

An ideal fuel tank and fuel line arrangement for an outboard unit. The tank should be kept clean and well secured in the boat. The quick-disconnect device affords easy removal for filling and safety.

DISCONNECT the battery ground cable before replacing an alternator or before connecting any type of meter to the alternator.

If it is necessary to use a fast-charger on a dead battery, **ALWAYS** disconnect one of the boat cables from the battery first, to prevent burning out the diodes in the circuit.

NEVER use a fast charger as a booster to start the engine because the diodes will be **DAMAGED**.

Alternator Charging

When the battery is partially discharged, the ammeter should change from discharge to charge between 800 to 1000 rpm for all models. If the battery is fully-charged, the rpm will be a little higher.

2-8 CARBURETOR ADJUSTMENT

Fuel and Fuel Tanks

Take time to check the fuel tank and all of the fuel lines, fittings, couplings, valves, flexible tank fill and vent. Turn on the fuel supply valve at the tank. If the gas was not drained at the end of the previous season,

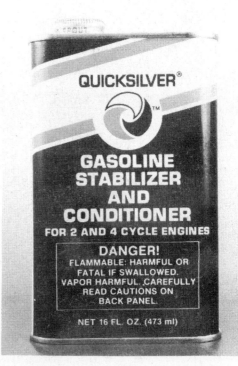

Quicksilver Gasoline Stabilizer and Conditioner may be used to keep the gasoline in the tank fresh. Such an additive will prevent the fuel from "souring" for up to twelve months.

make a careful inspection for gum formation. When gasoline is allowed to stand for long periods of time, particularly in the presence of copper, gummy deposits form. This gum can clog the filters, lines, and passageway in the carburetor.

If the condition of the fuel is in doubt, drain, clean, and fill the tank with fresh fuel.

Check other than a Mercury fuel tank for the following:

1- Adequate air vent in the fuel cap.

2- Fuel line of sufficient size, should be 5/16" to 3/8" (8 mm to 9.5 mm).

3- Filter on the end of the pickup is too small or is clogged.

4- Fuel pickup tube is too small.

High-speed Adjustment

The high-speed jet is fixed at the factory and is **NOT** adjustable. However, larger or smaller jets may be installed for different elevations. These jet sizes are clearly identified in the Appendix.

Idle Adjustment
Integral Fuel Pump Type Carburetor Only

The idle mixture and idle speed are set at the factory. Due to local conditions, it may be necessary to adjust the carburetor while the engine is running in a test tank or with the boat in a body of water. For

Filters used with the side-bowl carburetor. The two on the left are obsolete and should be replaced with the new type on the right.

maximum performance, the idle mixture and the idle rpm should be adjusted under actual operating conditions.

Refer to the idle adjustment procedures on the carburetor being serviced for the correct idle setting.

Start the engine and allow it to warm to operating temperature.

CAUTION: Water must circulate through the lower unit to the engine any time the engine is run to prevent damage to the water pump in the lower unit. Just five seconds without water will damage the water pump.

NEVER, AGAIN NEVER, operate the engine at high speed with a flush device attached. The engine, operating at high speed with such a device attached, would **RUN-A-WAY** from lack of load on the propeller, causing extensive damage.

With the engine running in forward gear, slowly turn the idle mixture screw **COUNTERCLOCKWISE** until the affected cylin-

The tank and fuel line can be easily removed for draining and flushing to remove stale fuel or deposits.

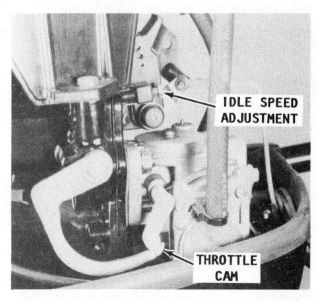

Side view of the powerhead showing the idle speed adjustment and the throttle cam.

ders start to load up or fire unevenly, due to an over-rich mixture. Slowly turn the idle mixture screw **CLOCKWISE** until the cylinders fire evenly and engine rpm increases. Continue to slowly turn the screw **CLOCKWISE** until too lean a mixture is obtained and the rpms fall off and the engine begins to misfire. Now, set the idle mixture screw one-quarter (1/4) turn out (counterclockwise) from the lean-out position. This adjustment will result in an approximate true setting. A too-lean setting is a major cause of hard starting a cold engine. It is better to have the adjustment on the rich side rather than on the lean side. Stating it another way, do not make the adjustment any leaner than necessary to obtain a smooth idle. When working on units equipped with the Mikuni side-draft carburetor (referenced "D" and "E" in the Appendix), the adjustment controls the amount of air instead of fuel.

If the engine hesitates during acceleration after adjusting the idle mixture, the mixture is too lean. Enrich the mixture slightly, by turning the adjustment screw inward until the engine accelerates correctly.

Loosen the locknut and adjust the idle stop screw on the stop bracket until the engine idles the recommended rpm in forward gear. Tighten the locknut to hold the adjustment.

Repairs and Adjustments

For detailed procedures to disassemble, clean, assemble, and adjust the carburetor, see the appropriate section in Chapter 4 for the carburetor type on the engine being serviced.

Typical separate fuel pumps installed on the power-heads covered in this manual.

2-9 FUEL PUMPS

Many times, a defective fuel pump diaphragm is mistakingly diagnosed as a problem in the ignition system. The most common problem is a tiny pin-hole in the diaphragm. Such a small hole will permit gas to enter the crankcase and wet foul the spark plug at idle-speed. During high-speed operation, gas quantity is limited, the plug is not foul and will therefore fire in a satisfactory manner.

If the fuel pump fails to perform properly, an insufficient fuel supply will be delivered to the carburetor. This lack of fuel will cause the engine to run lean, lose rpm or cause piston scoring.

Tune-up Task

Remove the fuel filter on the carburetor. Wash the parts in solvent and then dry them with compressed air. Install the clean element. A fuel pump pressure test should be made any time the engine fails to perform satisfactorily at high speed.

NEVER use liquid Neoprene on fuel line fittings. Always use Permatex when making fuel line connections. Permatex is available at almost all marine and hardware stores.

To service the fuel pump, see Chapter 4.

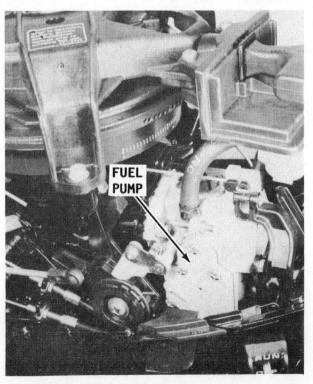

The fuel pump on late model powerheads is an integral part of the carburetor.

2-10 CRANKING MOTOR AND SOLENOID

Cranking Motor Test

Check to be sure the battery has a 70-ampere rating and is fully charged. Would you believe, many cranking motors are needlessly disassembled, when the battery is actually the culprit.

Lubricate the pinion gear and screw shaft with No. 10 oil.

Connect one lead of a voltmeter to the positive terminal of the cranking motor. Connect the other meter lead to a good ground on the engine. Check the battery voltage under load by turning the ignition switch to the **START** position and observing the voltmeter reading.

If the reading is 9-1/2 volts or greater, and the cranking motor fails to operate, repair or replace the cranking motor. See Chapter 7.

Solenoid Test

A magneto analyzer is required for this test. Turn the selector switch of the magneto analyzer to position No. 2 (distributor resistance). Clip the small red and black leads together. Turn the meter adjustment knob for Scale No. 2 until the meter pointer aligns with the set position on the left side of the **OK** block on Scale No. 2. Separate the red and black leads. Connect the small red test lead to one large terminal of the solenoid. Connect the small black test lead to the other large terminal.

NEVER connect the battery leads to the large terminals of the solenoid, or the meter will be damaged.

Using battery jumper leads, connect the positive lead from the positive terminal of the battery to the the small **"S"** terminal of the solenoid. Connect the negative lead to the negative battery terminal and the **"I"** terminal of the solenoid. If the meter pointer hand moves into the **OK** block, the solenoid is serviceable. If the pointer fails to reach the **OK** block, the solenoid must be replaced.

2-11 INTERNAL WIRING HARNESS

Check the internal wiring harness if problems have been encountered with any of the electrical components. Check for frayed or chafed insulation and/or loose connections between wires and terminal connections.

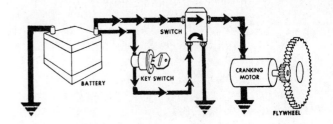

Simple functional diagram of a cranking circuit.

Check the harness connector for signs of corrosion. Inspect the electrical "prongs" to be sure they are not bent or broken. If the harness shows any evidence of the foregoing problems, the problem must be corrected before proceeding with any harness testing.

Verify that the "prongs" of the harness connector are clean and free of corrosion. Convince yourself that a good electrical connection is being made between the harness connector and the remote control harness.

Short Test (See the Wiring Diagram in the Appendix)

Disconnect the internal wiring harness from the electrical components. Use a magneto analyzer, set on Scale No. 3 and check for continuity between any of the wires in the harness. Use Scale No. 3 and check for continuity between any wire and a good ground. If continuity exists, the harness **MUST** be repaired or replaced.

Resistance Test (See the Wiring Diagram in the Appendix.)

Use a magneto analyzer, set on Scale No. 2. Clip the small red and black leads together. Turn the meter adjustment knob for Scale No. 2 until the meter pointer aligns with the set position on the left side of the **"OK"** block on Scale No. 2. Separate

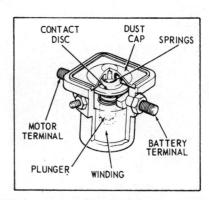

Exploded view of a cranking motor solenoid.

Worn water pump impeller, unfit for service.

the small red and black leads. Use the Wiring Diagram in the Appendix, and check each wire for resistance between the harness connection and the terminal ends. If resistance exists (meter reading outside the **"OK"** block) the harness **MUST** be repaired or replaced.

2-12 WATER PUMP CHECK

FIRST A GOOD WORD: The water pump **MUST** be in very good condition for the engine to deliver satisfactory service.

Flush attachment and garden hose attached to a small horsepower unit to clean the engine with fresh water. This arrangment may also be used while operating the engine at IDLE speeds only while making adjustments.

The pump performs an extremely important function by supplying enough water to properly cool the engine. Therefore, in most cases, it is advisable to replace the complete water pump assembly at least once a year, or anytime the lower unit is disassembled for service.

Sometimes during adjustment procedures, it is necessary to run the engine with a flush device attached to the lower unit. **NEVER** operate the engine over 1000 rpm with a flush device attached, because the engine may **"RUNAWAY"** due to the no-load condition on the propeller. A "run-a-way" engine could be severely damaged.

As the name implies, the flush device is primarily used to flush the engine after use in salt water or contaminated fresh water. Regular use of the flush device will prevent salt or silt deposits from accumulating in the water passage-way. During and immediately after flushing, keep the motor in an upright position until all of the water has drained from the drive shaft housing. This will prevent water from entering the power head by way of the drive shaft housing and the exhaust ports, during the flush. It will also prevent residual water from being trapped in the drive shaft housing and other passageways.

Most of the engines covered in this manual have water exhaust ports which deliver a tattle-tale stream of water, if the water pump is functioning properly during engine operation. Water pressure at the cylinder block should be checked if an overheating condition is detected or suspected.

To test the water pump, the lower unit **MUST** be placed in a test tank or the boat moved into a body of water. The pump must

Major parts included in a water pump kit, available at the local marine dealer at modest cost.

now work to supply a volume to the engine. A tattle-tale stream of water should be visible from the ports.

Lack of adequate water supply from the water pump thru the engine will cause any number of powerhead failures, such as stuck rings, scored cylinder walls, burned pistons, etc.

To service the water pump, see Chapter 9.

2-13 PROPELLER

Check the propeller blades for nicks, cracks, or bent condition. If the propeller is damaged, the local marine dealer can make repairs or send it out to a shop specializing in such work.

Remove the propeller and the thrust hub. Check the propeller shaft seal to be sure it is not leaking. Check the area just forward of the seal to be sure a fish line is not wrapped around the shaft.

Operation At Recommended RPM

Check with the local marine dealer, or a propeller shop for the recommended size and pitch for a particular size engine, boat, and intended operation. The correct propeller should be installed on the engine to enable operation at the upper end of the factory recommended rpm.

2-14 LOWER UNIT

NEVER remove the vent or filler plugs when the lower unit is hot. Expanded lubricant would be released through the plug hole. Check the lubricant level after the unit has been allowed to cool. Add only

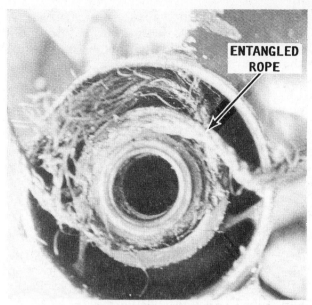

This rope became entangled behind the propeller around the propeller shaft. The propeller should be removed periodically and this area checked for foreign material.

Super-Duty Gear Lubricant. NEVER use regular automotive-type grease in the lower unit, because it expands and foams too much. Outboard lower units do not have provisions to accommodate such expansion.

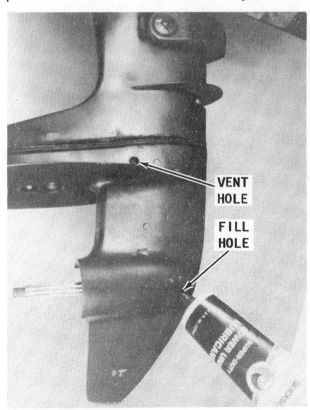

Damage was caused to this unit when the propeller struck an underwater object. If the propeller should suffer this much abuse, the propeller shaft should be carefully checked.

Filling the lower unit with new lubricant. Notice the unit is filled through the lower plug, but the upper plug MUST be removed to allow trapped air to escape.

If the lubricant appears milky brown, or if large amounts of lubricant must be added to bring the lubricant up to the full mark, a thorough check should be made to determine the cause of the loss.

Draining Lower Unit

Remove the **FILL** plug from the lower end of the gear housing on the port side and the **VENT** plug just above the anti-cavitation plate.

Filling Lower Unit

Position the drive unit approximately vertical and without a list to either port or starboard. Insert the lubricant tube into the **FILL/DRAIN** hole at the bottom plug hole, and inject lubricant until the excess begins to come out the **VENT** hole. Install the **VENT** plug first then replace the **FILL** plug with **NEW** gaskets. Check to be sure the gaskets are properly positioned to prevent water from entering the housing.

For detailed lower unit service procedures, see Chapter 9. For lower unit lubrication capacities, see the Appendix.

2-15 BOAT TESTING

Operation of the outboard unit, mounted on a boat with some type of load, is the ultimate test. Failure of the power unit or the boat under actual movement through the water may be detected much more quickly than operating the power unit in a test tank.

Hook and Rocker

Before testing the boat, check the boat bottom carefully for marine growth or evidence of a "hook" or a "rocker" in the bottom. Either one of these conditions will greatly reduce performance.

Performance

Mount the motor on the boat. Install the remote control cables (if used), and check for proper adjustment.

Make an effort to test the boat with what might be considered an average gross load. The boat should ride on an even keel, without a list to port or starboard. Adjust the motor tilt angle, if necessary, to permit the boat to ride slightly higher than the stern. If heavy supplies are stowed aft of the center, the bow will be light and the boat will "plane" more efficiently. For this test the boat must be operated in a body of water.

If the motor is equipped with an adjustable trim tab, the tab should be adjusted to permit boat steerage in either direction with equal ease.

Check the engine rpm at full throttle. The rpm should be within the Specifications in the Appendix. If the rpm is not within specified range, a propeller change may be in order. A higher pitch propeller will decrease rpm, and a lower pitch propeller will increase rpm.

For maximum low speed engine performance, the idle mixture and the idle rpm should be readjusted under actual operating conditions.

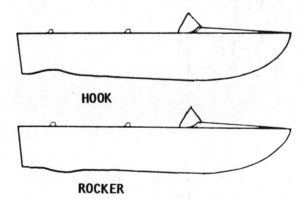

Boat performance will be drastically impaired, if the bottom is damaged by a dent (hook) or bulge (rocker).

3
MAINTENANCE

3-1 INTRODUCTION

GOOD WORDS: The authors estimate 75% of engine repair work can be directly or indirectly attributed to lack of proper care for the engine. This is especially true of care during the off-season period. There is no way on this green earth for a mechanical engine, particularly an outboard motor, to be left sitting idle for an extended period of time, say for six months, and then be ready for instant satisfactory service.

Imagine, if you will, leaving your automobile for six months, and then expecting to turn the key, have it roar to life, and be able to drive off in the same manner as a daily occurrence.

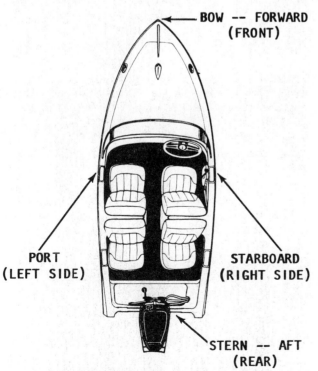

BOW -- FORWARD (FRONT)

PORT (LEFT SIDE)

STARBOARD (RIGHT SIDE)

STERN -- AFT (REAR)

Common terminology used throughout the world for reference designation on boats of all sizes. These are the terms used in this book.

It is critical for an outboard engine to be run at least once a month, preferably, in the water, but if this is not possible, then a flush attachment **MUST** be connected to the lower unit.

CAUTION: Water must circulate through the lower unit to the engine any time the engine is run to prevent damage to the water pump in the lower unit. Just five seconds without water will damage the water pump.

NEVER, AGAIN NEVER, operate the engine at high speed with a flush device attached. The engine, operating at high speed with such a device attached, would **RUNAWAY** from lack of load on the propeller, causing extensive damage.

. At the same time, the shift mechanism should be operated through the full range several times and the steering operated from hard-over to hard-over.

Only through a regular maintenance program can the owner expect to receive long life and satisfactory performance at minimum cost.

The material presented in this chapter is divided into five general areas.

1- General information every boat owner should know.

2- Maintenance tasks that should be performed periodically to keep the boat operating at minimum cost.

3- Care necessary to maintain the appearance of the boat and to give the owner that "Pride of Ownership" look.

4- Winter storage practices to minimize damage during the off-season when the boat is not in use.

In nautical terms, the front of the boat is the **bow**; the rear is the **stern**; the right side, when facing forward, is the **starboard** side; and the left side is the **port** side. All directional references in this manual use this terminology. Therefore, the direction from which an item is viewed is of no consequence, because **starboard** and **port** **NEVER** change no matter where the individual is located.

3-2 OUTBOARD SERIAL NUMBERS

The engine serial numbers are the manufacturer's key to engine changes. These numbers identify the year of manufacture, the qualified horsepower rating, and the parts book identification. If any correspondence or parts are required, the engine serial number **MUST** be used or proper identification is not possible. The accompanying illustration will be very helpful in locating the engine identification tag for the various models.

ONE MORE WORD:

The serial number establishes the year in which the engine was produced and not necessarily the year of first installation.

Two serial number locations are used on each of the outboard units covered in this manual. The most accessible location is on the serial/instruction plate on the swivel bracket. The other location is on the engine cylinder block.

3-3 LUBRICATION - COMPLETE UNIT

As with every type mechanical invention with moving parts, lubrication plays a prominent role in operation, enjoyment, and longevity of the unit.

If an outboard unit is operated in salt water the frequency of applying lubricant to

fittings is usually cut in half for the same fitting if the unit is used in fresh water. The few minutes involved in moving around the outboard applying lubricant and at the same time making a visual inspection of its general condition will pay in rich rewards with years of continued service.

It is not uncommon to see outboard units well over 20-years of age moving a boat through the water as if the unit had recently been purchased from the current line of models. An inquiry with the proud owner will undoubtedly reveal his main credit for its performance to be regular periodic maintenance.

The following chart can be used as a guide to periodic maintenance while the outboard is being used during the season. The pictures numbers reference in the chart are keyed to the illustrations in this section.

In addition to the normal lubrication listed in the lubrication chart, the prudent owner will inspect and make checks on a regular basis as listed in the accompanying chart.

Serial numbers on the identification plate of a late model outboard unit covered in this manual. This is a standard location for the plate although some units may have it elswhere.

The electrical system should be checked at the beginning of each season. Check all connections to be sure they are clean, secure, and free of corrosion. Inspect the wiring for any type of damage.

LUBRICATION POINT/FREQUENCY CHART

DESCRIPTION	LUBRICANT	FREQUENCY FRESH WATER	FREQUENCY SALT WATER
Ride-Guide steering cable	Multipurpose Lubricant	Every 60 days	Every 30 days
Throttle linkage and throttle cable			
Upper shift shaft			
Lube fitting in power trim cylinder			
Reverse lock lever			
Shift linkage and shift cable			
Steering link rod pivot points	SAE 30W engine oil		
Clamp screws	Anti-Corrosion Lubricant		
Propeller shaft	Perfect Seal	Once in season	Every 60 days
Gear Housing	Super-Duty Lubricant	After 10 days then ea. 30 days	After 10 days then ea. 30 days
Tilt lock lever	Multipurpose Lubricant	Every 60 days	Every 30 days
Cranking motor pinion gear	SAE 10W Engine oil	Once ea. season	Once ea. season

3-4 PRE-SEASON PREPARATION

Satisfactory performance and maximum enjoyment can be realized if a little time is spent in preparing the outboard unit for service at the beginning of the season. Assuming the unit has been properly stored, as outlined in Section 3-12, a minimum amount of work is required to prepare the unit for use.

The following steps outline an adequate and logical sequence of tasks to be performed before using the outboard the first time in a new season.

1- Lubricate the outboard according to the manufacturer's recommendations. Refer to the lubrication chart on this page. Remove, clean, inspect, adjust, and install the spark plugs with new gaskets (if they require gaskets). Make a thorough check of the ignition system. This check should include: the points, coil, condenser, stator assembly, condition of the wiring, and the battery electrolyte level and charge.

2- If a built-in fuel tank is installed, take time to check the gasoline tank and all of the fuel lines, fittings, couplings, valves, flexible tank fill and vent. Turn on the fuel

INSPECTION AND CHECKS

Item or Area to be Checked	Every 30 Days	Every 60 Days	Once In Season	Twice In Season
Lubricant level in lower unit	X			
Lubricant level in Power Trim/Tilt		X		
Clean battery terminals				X
Spark plugs & leads			X	
Clean fuel filter/s			X	
Fuel filter on oil injection unit		X		
Fuel filter at fuel tank			X	
All fuel lines and connections				X
Complete outboard -- damaged parts			X	
Breaker points			X	
Propeller for damage		X		
Clean outboard -- touch-up paint			X	

SPECIAL WORDS

Time for inspection, checks, and maintenance can almost be cut in half, if the unit is operated in salt water.

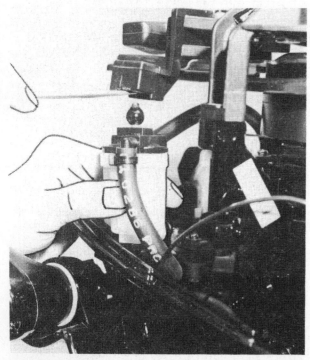

A late model clear plastic fuel filter may be remov-ed ("popped off") with a screwdriver and installed without the use of any tools. With such easy access to the filter, the chances of "contaminated" fuel reaching the carburetor are drastically reduced.

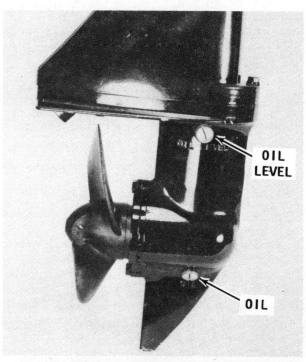

The oil level (vent) plug should be removed periodi-cally and a piece of wire, or similar material, inserted to determine the amount of lubricant in the lower unit. The oil level should be kept close to the hole and the lubricant show no sign of water (milky color).

supply valve at the tank. If the fuel was not drained at the end of the previous season, make a careful inspection for gum formation. If a six-gallon fuel tank is used, take the same action. When gasoline is allowed to stand for long periods of time, particularly in the presence of copper, gummy deposits form. This gum can clog the filters, lines, and passageways in the carburetor. See Chapter 4, Fuel System Service.

3- Check the oil level in the lower unit by first removing the vent screw on the port side just above the anti-cavitation plate. Insert a short piece of wire into the hole and check the level. Fill the lower unit according to procedures outlined in Section 3-11.

UNITS WITH OIL INJECTION

4- First, remove the front cover of the unit by simultaneously pushing in on the cutaway tabs located on both sides of the cover, and at the same time pulling the cover away from the unit. Check to be sure the fuel drain plug is tight. Replace the front cover by aligning the cover openings on both sides of the unit, and then pushing in on the cover until it snaps into place.

Next, fill the oil tank with 2-cycle outboard oil with a BIA rating of TC-W. Tighten the fill cap securely.

Remove any plugs in the fuel lines, and then connect the hoses to the fuel tank and the powerhead. Remember, the squeeze bulb **MUST** be in the hose between the oil injection unit and the fuel pump on the powerhead.

Connect the low oil warning wire harness to the battery. Connect the **RED** lead to the positive battery terminal and the **BLACK** lead to the negative battery terminal.

Check to be sure the low oil warning system is functioning correctly. First, verify the tank is full of oil, and then the fill cap is tightened securely. Now, turn the oil injection unit upside down. This position will allow the float to activate the horn.

If the horn sounds, immediately turn the unit rightside up and position it in the mounting bracket. Secure it in place with the strap and Velcro material.

If the horn does not sound, check the 0.5 amp fuse in the fuse holder of the positive battery lead. Check both the battery connections and the charge condition of the battery.

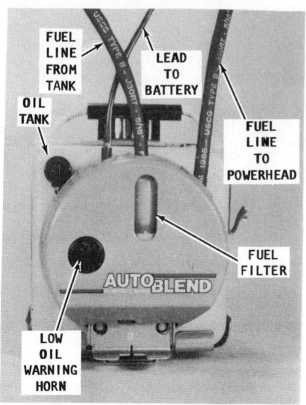

Exterior view of an Autoblend oil injection unit showing routing of the hoses, electrical lead, and location of the warning horn and the fuel filter.

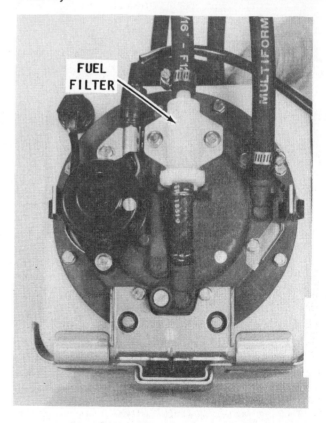

Autoblend oil injection unit with the plastic cover removed, ready for a pre-season check or maintenance.

GOOD WORDS

The manufacturer recommends the fuel filter be replaced at the start of each season or at least once a year. The manufacturer also recommends oil be added to the fuel tank at the ratio of 50:1 for the first 6-gallons of fuel used after the unit is brought out of storage. The oil in the fuel tank plus the 50:1 oil mixture in the oil injection unit will deliver a mixture of 25:1 to the powerhead. This ratio will **ENSURE** adequate lubrication of moving parts which have been drained of oil during the storage period.

ALL UNITS

5- Close all water drains. Check and replace any defective water hoses. Check to be sure the connections do not leak. Replace any spring-type hose clamps, if they have lost their tension, or if they have distorted the water hose, with band-type clamps.

6- The engine can be run with the lower unit in water to flush it. If this is not

*Flushing with a flush attachment connected to the lower unit. The powerhead should **NEVER** be run above idle speed with this type device attached.*

practical, a flush attachment may be used. This unit is attached to the water pick-up in the lower unit. Attach a garden hose, turn on the water, allow the water to flow into the engine for awhile, and then run the engine.

CAUTION: Water must circulate through the lower unit to the engine any time the engine is run to prevent damage to the water pump in the lower unit. Just five seconds without water will damage the water pump.

Check the exhaust outlet for water discharge. Check for leaks. Check operation of the thermostat.

7- Check the electrolyte level in the battery and the voltage for a full charge. Clean and inspect the battery terminals and cable connections. **TAKE TIME** to check the polarity, if a new battery is being installed. Cover the cable connections with grease or special protective compound as a prevention to corrosion formation. Check all electrical wiring and grounding circuits.

8- Check all electrical parts on the engine and lower portions of the hull to be sure they are not of a type that could cause ignition of an explosive atmosphere. Rubber caps help keep spark insulators clean and reduce the possibility of arcing. Starters, generators, distributors, alternators, electric fuel pumps, voltage regulators, and high-tension wiring harnesses should be of a marine type that cannot cause an explosive mixture to ignite.

ONE FINAL WORD

Before putting the boat in the water, **TAKE TIME** to **VERIFY** the drain plugs are installed. Countless number of boating excursions have had a very sad beginning because the boat was eased into the water only to have the boat begin to fill with the "wet stuff" from the river, lake, reservoir, etc.

3-5 FIBERGLASS HULLS

Fiberglass-reinforced plastic hulls are tough, durable, and highly resistant to impact. However, like any other material they can be damaged. One of the advantages of this type of construction is the relative ease with which it may be repaired. Because of its break characteristics, and the simple techniques used in restoration, these hulls have gained popularity throughout the world. From the most congested urban marina, to isolated lakes in wilderness areas, to the severe cold of far off northern seas, and in sunny tropic remote rivers of primative islands or continents, fiberglass boats can be found performing their daily task with a minimum of maintenance.

A fiberglass hull has almost no internal stresses. Therefore, when the hull is broken or stove-in, it retains its true form. It will not dent to take an out-of-shape set. When the hull sustains a severe blow, the impact will be either absorbed by deflection of the laminated panel or the blow will result in a definite, localized break. In addition to hull damage, bulkheads, stringers, and other stiffening structures attached to the hull may also be affected and therefore, should be checked. Repairs are usually confined to the general area of the rupture.

3-6 BELOW WATERLINE SERVICE

A foul bottom can seriously affect boat performance. This is one reason why racers, large and small, both powerboat and sail, are constantly giving attention to the condition of the hull below the waterline.

In areas where marine growth is prevalent, a coating of vinyl, anti-fouling bottom paint should be applied. If growth has developed on the bottom, it can be removed with a solution of muriatic acid applied with a brush or swab and then rinsed with clear water. **ALWAYS** use rubber gloves when working with muriatic acid and **TAKE EXTRA CARE** to keep it away from your face and hands. The **FUMES ARE TOXIC.** Therefore, work in a well-ventilated area, or if outside, keep your face on the windward side of the work.

Barnacles have a nasty habit of making their home on the bottom of boats which have not been treated with anti-fouling paint. Actually they will not harm the fiberglass hull, but can develop into a major nuisance.

If barnacles or other crustaceans have attached themselves to the hull, extra work will be required to bring the bottom back to a satisfactory condition. First, if practical, put the boat into a body of fresh water and allow it to remain for a few days. A large percentage of the growth can be removed in this manner. If this remedy is not possible, wash the bottom thoroughly with a high-

pressure fresh water source and use a scraper. Small particles of hard shell may still hold fast. These can be removed with sandpaper.

3-7 SUBMERGED ENGINE SERVICE

A submerged engine is always the result of an unforeseen accident. Once the engine is recovered, special care and service procedures **MUST** be closely followed in order to return the unit to satisfactory performance.

NEVER, again we say **NEVER** allow an engine that has been submerged to stand more than a couple hours before following the procedures outlined in this section and making every effort to get it running. Such delay will result in serious internal damage. If all efforts fail and the engine cannot be started after the following procedures have been performed, the engine should be disassembled, cleaned, assembled, using new gaskets, seals, and O-rings, and then started as soon as possible.

Submerged engine treatment is divided into three unique problem areas: Submersion in salt water; submerged engine while running; and a submerged engine in fresh water, including special instructions.

The most critical of these three circumstances is the engine submerged in salt water, with submersion while running a close second.

Salt Water Submersion

NEVER attempt to start the engine after it has been recovered. This action will only result in additional parts being damag-

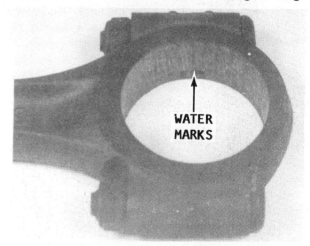

Damaged rod and rod cap unfit for further service. The needle bearing water marks shown were caused from water entering the powerhead.

ed and the cost of restoring the engine increased considerably. If the engine was submerged in salt water the complete unit **MUST** be disassembled, cleaned, and assembled with new gaskets, **O**-rings, and seals. The corrosive effect of salt water can only be eliminated by the complete job being properly performed.

Submerged While Running
Special Instructions

If the engine was running when it was submerged, the chances of internal engine damage is greatly increased. After the engine has been recovered, remove the spark plugs to prevent compression in the cylinders. Make an attempt to rotate the crankshaft with the rewind starter or the flywheel. On larger horsepower engines without a rewind starter, use a socket wrench on the flywheel nut to rotate the crankshaft. If the attempt fails, the chances of serious internal damage, such as: bent connecting rod, bent crankshaft, or damaged cylinder, is greatly increased. If the crankshaft cannot be rotated, the powerhead must be completely disassembled.

CRITICAL WORDS

Never attempt to start powerhead that has been submerged. If there is water in the cylinder, the piston will not be able to compress the liquid. The result will most likely be a bent connecting rod.

Submerged Engine — Fresh Water

SPECIAL WORD: As an aid to performing the restoration work, the following steps are numbered and should be followed in sequence. However, illustrations are not included with the procedural steps because the work involved is general in nature.

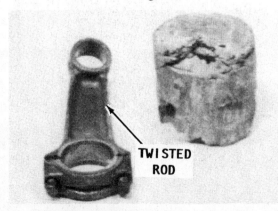

TWISTED ROD

Damaged rod and piston caused when the powerhead was submerged while running.

1- Recover the engine as quickly as possible.

2- Remove the cowl and the spark plugs.

3- Remove the carburetor float bowl cover, or the bowl.

4- Flush the outside of the engine with fresh water to remove silt, mud, sand, weeds, and other debris. **DO NOT** attempt to start the engine if sand has entered the powerhead. Such action will only result in serious damage to powerhead components. Sand in the powerhead means the unit must be disassembled.

CRITICAL WORDS

Never attempt to start powerhead that has been submerged. If there is water in the cylinder, the piston will not be able to compress the liquid. The result will most likely be a bent connecting rod.

5- Remove as much water as possible from the powerhead. Most of the water can be eliminated by first holding the engine in a horizontal position with the spark plug holes **DOWN**, and then cranking the powerhead with the rewind starter or with a socket wrench on the flywheel nut. Rotate the crankshaft through at least 10 complete revolutions. If you are satisfied there is no water in the cylinders, proceed with Step 6 to remove moisture.

6- Alcohol will absorb moisture. Therefore, pour alcohol into the carburetor throat and again crank the powerhead.

7- Rotate the outboard in the horizontal position until the spark plug openings are facing **UPWARD**. Pour alcohol into the spark plug openings and again rotate the crankshaft.

8- Rotate the outboard in the horizontal position until the spark plug openings are again facing **DOWN**. Pour engine oil into the carburetor throat and, at the same time, rotate the crankshaft to distribute oil throughout the crankcase.

9- Rotate the outboard in the horizontal position until the spark plug holes are again facing **UPWARD**. Pour approximately one teaspoon of engine oil into each spark plug opening. Rotate the crankshaft to distribute the oil in the cylinders.

10- Install and connect the spark plugs.

11- Install the carburetor float bowl cover, or the bowl.

12- Obtain **FRESH** fuel and attempt to start the engine. If the powerhead will start, allow it to run for approximately an hour to eliminate any unwanted moisture remaining in the powerhead.

CAUTION: Water must circulate through the lower unit to the engine any time the engine is run to prevent damage to the water pump in the lower unit. Just five seconds without water will damage the water pump.

13- If the powerhead fails to start, determine the cause, electrical or fuel, correct the problem, and again attempt to get it running. **NEVER** allow a powerhead to remain unstarted for more than a couple hours without following the procedures in this section and attempting to start it. If attempts to start the powerhead fail, the unit should be disassembled, cleaned, assembled, using new gaskets, seals, and O-rings, just as **SOON** as possible.

3-8 PROPELLER SERVICE

The propeller should be checked regularly to be sure all the blades are in good condition. If any of the blades become bent or nicked, this condition will set up vibrations in the motor. Remove and inspect the propeller. Use a file to trim nicks and burrs. **TAKE CARE** not to remove any more material than is absolutely necessary. For a complete check, take the propeller to your marine dealer where the proper equip-

ment and knowledgeable mechanics are available to perform a proper job at modest cost.

Inspect the propeller shaft to be sure it is still true and not bent. If the shaft is not perfectly true, it should be replaced.

Install the thrust hub. Coat the propeller shaft splines with Perfect Seal No. 4, and the rest of the shaft with a good grade of anti-corrosion lubricant. Install the propeller, and then the splined washer, tab washer, and propeller nut.

Position a block of wood between the propeller and the anti-cavitation tab to keep the propeller from turning. Tighten the propeller nut to a torque value of 35-45 ft. lbs (47.6-61.2 Nm). Adjust the nut to fit the tab lock space. Bend three of the tab washer tabs into the spline washer using a punch and hammer. The tabs will prevent the nut from backing out.

3-9 POWER TRIM/TILT

Check the power trim/tilt system for proper operation.

Check the oil level in the pump reservoir and be sure the vent screw is left **OPEN.**

Check to be sure all connections are secure. Check and adjust the shift and throttle cables. Lubricate all external lubrication points with Multipurpose Lubricant. Check and clean the water intake opening.

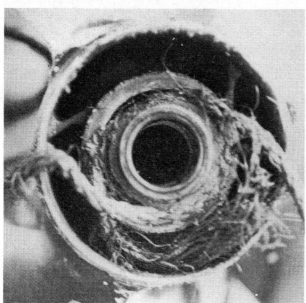

Excellent view of rope and fish line entangled behind the propeller. Entangled fish line can actually cut through the seals allowing water to enter and oil to escape from the lower unit.

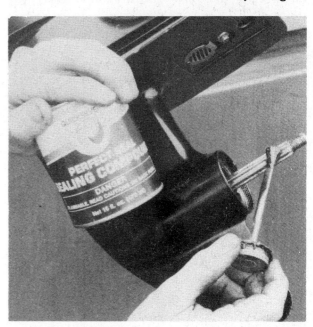

Applying Perfect Seal compound to the propeller shaft. This compound should be used each and every time the propeller is removed to prevent the propeller from "freezing" onto the propeller shaft.

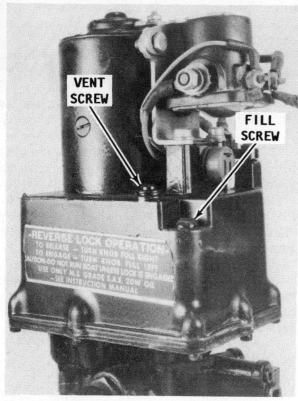

Overall exterior view of an early model Power Trim/Tilt unit. Use ONLY the manufacturer's recommended oil.

Trim Tabs

Check the trim tab and the anodic heads. Replace them, if necessary. The trim tab must make a good ground inside the lower unit. Therefore, the trim tab and the cavity **MUST NOT** be painted. In addition to trimming the boat, the trim tab acts as a zinc electrode to prevent electrolysis from acting on more expensive parts. It is normal for the tab to show signs of erosion. The tabs are inexpensive and should be replaced frequently.

A new trim tab (left), and a badly deteriorated tab (right). Actually, such extensive erosion of the tab suggests a possible electrolysis problem.

Clean the exterior surface of the unit thoroughly. Inspect the finish for damage or corrosion. Clean any damaged or corroded areas, and then apply primer and matching paint.

Check the entire unit for loose, damaged, or missing parts.

3-10 INSIDE THE BOAT

The following points may be lubricated with Quicksilver Multipurpose Lubricant:

a- Ride-Guide steering cable end next to the hand nut. **DO NOT** over-lubricate the cable.

b- Steering arm pivot socket.

c- Exposed shaft of the cable passing through the cable guide tube.

d- Steering link rod to the steering cable.

3-11 LOWER UNIT

Draining Lower Unit

Remove the **FILL** plug from the lower end of the gear housing on the port side and the **VENT** plug just above the anti-cavitation plate.

NEVER remove the vent or filler plugs when the drive unit is hot. Expanded lubricant would be released through the plug hole. Check the lubricant level after the unit has been allowed to cool. Add only Super-Duty Gear Lubricant. **NEVER** use regular automotive-type grease in the lower unit because it expands and foams too much. Lower units do not have provisions to accommodate such expansion.

Replacing the anode on a Model 2.2 lower unit is a simple task. The anode is "sacrificed" to corrosion and thus saves more expensive parts.

If the lubricant appears milky brown, or if large amounts of lubricant must be added to bring the lubricant up to the full mark, a thorough check should be made to determine the cause of the loss.

Filling Lower Unit

Position the drive unit approximately vertical and without a list to either port or starboard. Insert the lubricant tube into the **FILL/DRAIN** hole at the bottom plug hole, and inject lubricant until the excess begins to come out the **VENT** hole. Install the **VENT** and **FILL** plugs with **NEW** gaskets. Check to be sure the gaskets are properly positioned to prevent water from entering the housing.

See the Appendix for lower unit capacities.

3-12 WINTER STORAGE

Taking extra time to store the boat properly at the end of each season, will increase the chances of satisfactory service at the next season. **REMEMBER**, idleness is the greatest enemy of an outboard motor. The unit should be run on a monthly basis. The boat steering and shifting mechanism should also be worked through complete cycles several times each month. The owner who spends a small amount of time involved in such maintenance will be rewarded by satisfactory performance, and greatly reduced maintenance expense for parts and labor.

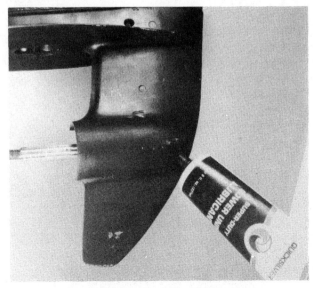

Filling the lower unit with the manufacturer's approved gear lubricant. The vent screw should always be removed prior to filling to allow trapped air to escape.

Proper storage involves adequate protection of the unit from physical damage, rust, corrosion, and dirt.

The following steps provide an adequate maintenance program for storing the unit at the end of a season.

1- Remove the cowl. Start the engine and allow it to warm to operating temperature.

CAUTION: Water must circulate through the lower unit to the engine any time the engine is run to prevent damage to the water pump in the lower unit. Just five seconds without water will damage the water pump.

Disconnect the fuel line from the engine and allow the unit to run at **LOW** rpm and, at the same time, inject about 4 ounces of Quicksilver Storage Seal through each carburetor throat. Allow the engine to run until it shuts down from lack of fuel, indicating the caburetor/s are dry of fuel.

2- Drain the fuel tank and the fuel lines. Pour approximately one quart (0.96 liters) of benzol (benzine) into the fuel tank, and then rinse the tank and pickup filter with the benzol. Drain the tank. Store the fuel tank in a cool dry area with the vent **OPEN** to allow air to circulate through the tank. **DO NOT** store the fuel tank on bare concrete. Place the tank to allow air to circulate around it.

3- Clean the carburetor fuel filter/s with benzol, see Chapter 4, Carburetor Repair Section.

4- Drain, and then fill the lower unit with Super-Duty Lower Unit Gear Lubricant, as outlined in Section 3-11.

5- Lubricate the throttle and shift linkage. Lubricate the swivel pin and the tilt

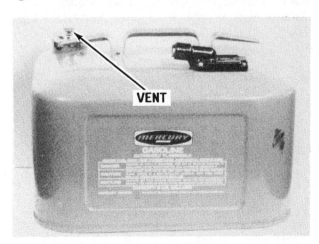

The fuel tank should be drained and stored in a cool dry area with the vent open to allow air to circulate through the tank during the off-season.

tube with Multipurpose Lubricant, or equivalent.

Clean the outboard unit thoroughly. Coat the powerhead with Corrosion and Rust Preventative spray. Install the cowl, and then apply a thin film of fresh engine oil to all painted surfaces.

Remove the propeller. Apply Perfect Seal or a waterproof sealer to the propeller shaft splines, and then install the propeller back in position.

STORAGE FOR UNITS WITH OIL INJECTION

Proper storage procedures are **CRITICAL** to ensure efficient operation when the unit is again placed in service.

First, disconnect the battery leads from the battery.

Next, disconnect and plug the fuel lines at the fuel tank and powerhead.

Now, drain all fuel from the oil injection unit. Remove the front cover of the unit by simultaneously pushing in on the cutaway tabs located on both sides of the cover, and at the same time pulling the cover away from the unit. Remove the drain plug and allow at least 5-minutes for all fuel to drain from the pump. Install the drain plug and tighten it securely.

CRITICAL WORDS

All fuel **MUST** be drained from the oil injection fuel "pump". The percentage of alcohol in modern fuels seems to increase each year. This alcohol in the fuel is a definite enemy of the diaphragm in the "pump". Therefore, if any fuel is left in the

Manufacturer recommended lubricants and additives will not only keep the unit within the limits of the warranty, but will be a major contributing factor to dependable performance and reduced maintenance cost.

"pump" during storage the diaphragm will most likely be damaged.

Install the front cover by aligning the cover openings on both sides of the unit, and then pushing in on the cover until it snaps into place.

Oil may remain in the oil injection tank during storage without any harmful effects.

FINAL WORDS: Be sure all drain holes in the gear housing are open and free of obstruction. Check to be sure the **FLUSH** plug has been removed to allow all water to drain. Trapped water could freeze, expand, and cause expensive castings to crack.

ALWAYS store the outboard unit off the boat with the lower unit below the powerhead to prevent any water from being trapped inside.

BATTERY STORAGE

Remove the batteries from the boat and keep them charged during the storage period. Clean the batteries thoroughly of any dirt or corrosion, and then charge them to full specific gravity reading. After they are fully charged, store them in a clean cool dry place where they will not be damaged or knocked over.

NEVER store the battery with anything on top of it or cover the battery in such a manner as to prevent air from circulating around the fillercaps. All batteries, both new and old, will discharge during periods of storage, more so if they are hot than if they remain cool. Therefore, the electrolyte level and the specific gravity should be checked at regular intervals. A drop in the specific gravity reading is cause to charge them back to a full reading.

In cold climates, **EXERCISE CARE** in selecting the battery storage area. A fully-charged battery will freeze at about 60 degrees below zero. A discharged battery, almost dead, will have ice forming at about 19 degrees above zero.

ALWAYS remove the drain plug and position the boat with the bow higher than the stern. This will allow any rain water and melted snow to drain from the boat and prevent "trailer sinking". This term is used to describe a boat that has filled with rain water and ruined the interior, because the plug was not removed or the bow was not high enough to allow the water to drain properly.

4
FUEL

4-1 INTRODUCTION

The carburetion and ignition principles of two-cycle engine operation **MUST** be understood in order to perform a proper tune-up on an outboard motor.

If you have any doubts concerning your understanding of two-cycle engine operation, it would be best to study the Introduction section in the first portion of Chapter 8, before tackling any work on the fuel system.

The fuel system includes the fuel tank, fuel pump, fuel filters, carburetor, connecting lines, with a squeeze bulb, and the associated parts to connect it all together. Regular maintenance of the fuel system to obtain maximum performance, is limited to changing the fuel filter at regular intervals and using fresh fuel.

If a sudden increase in gas consumption is noticed, or if the engine does not perform properly, a carburetor overhaul, including boil-out, or replacement of the fuel pump may be required.

4-2 GENERAL CARBURETION INFORMATION

The carburetor is merely a metering device for mixing fuel and air in the proper proportions for efficient engine operation. At idle speed, an outboard engine requires a mixture of about 8 parts air to 1 part fuel. At high speed or under heavy duty service, the mixture may change to as much as 12 parts air to 1 part fuel.

Float Systems

A small chamber in the carburetor serves as a fuel reservoir. A float valve admits fuel into the reservoir to replace the fuel consumed by the engine. If the carburetor has more than one reservoir, the fuel

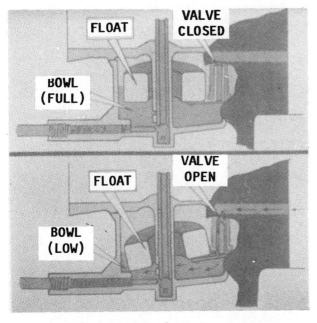

Fuel flow principle of a modern carburetor.

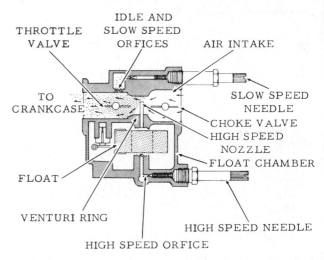

Fuel flow through the venturi, showing principle and related parts controlling intake and outflow.

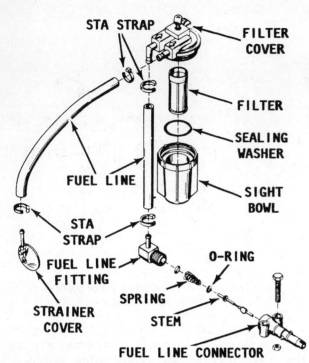

Exploded illustration of a late model fuel filter with a clear sight bowl. Principle parts are identified.

level in each reservoir (chamber) is controlled by identical float systems.

Fuel level in each chamber is extremely critical and must be maintained accurately. Accuracy is obtained through proper adjustment of the float/s. This adjustment will provide a balanced metering of fuel to each cylinder at all speeds.

Following the fuel through its course, from the fuel tank to the combustion chamber of the cylinder, will provide an appreci-

ation of exactly what is taking place. In order to start the engine, the fuel must be moved from the tank to the carburetor by a squeeze bulb installed in the fuel line. This action is necessary because the fuel pump does not have sufficient pressure to draw fuel from the tank during cranking before the engine starts.

After the engine starts, the fuel passes through the pump to the carburetor. All systems have some type of filter installed somewhere in the line between the tank and the carburetor. Many units have a filter as an integral part of the carburetor.

At the carburetor, the fuel passes through the inlet passage to the needle and seat, and then into the float chamber (reservoir). A float in the chamber rides up and down on the surface of the fuel. After fuel enters the chamber and the level rises to a predetermined point, a tang on the float closes the inlet needle and the flow entering the chamber is cutoff. When fuel leaves the chamber as the engine operates, the fuel level drops and the float tang allows the inlet needle to move off its seat and fuel once again enters the chamber. In this manner a constant reservoir of fuel is maintained in the chamber to satisfy the demands of the engine at all speeds.

A fuel chamber vent hole is located near the top of the carburetor body to permit atmospheric pressure to act against the fuel in each chamber. This pressure assures an adequate fuel supply to the various operating systems of the engine.

Late model clear fuel filter held in place with a snap-on fitting on the rewind hand starter.

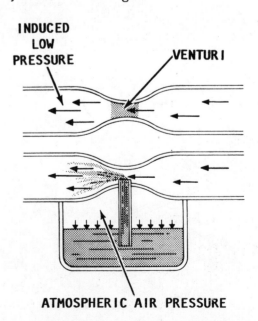

Air flow principle of a modern carburetor.

Air/Fuel Mixture

A suction effect is created each time the piston moves upward in the cylinder. This suction draws air through the throat of the carburetor. A restriction in the throat, called a venturi, controls air velocity and has the effect of reducing air pressure at this point.

The difference in air pressures at the throat and in the fuel chamber, causes the fuel to be pushed out of metering jets extending down into the fuel chamber. When the fuel leaves the jets, it mixes with the air passing through the venturi. This fuel/air mixture should then be in the proper proportion for burning in the cylinder/s for maximum engine performance.

In order to obtain the proper air/fuel mixture for all engine speeds, high and low speed jets are provided. These jets have adjustable needle valves which are used to compensate for changing atmospheric conditions. In almost all cases, the high-speed circuit has fixed high-speed jets that are not adjustable.

Engine operation at sea level compared with performance at high altitudes is quite noticeable. A jet/altitude chart is provided in the Appendix for operation from sea level to above 7500 ft.

A throttle valve controls the flow of air/fuel mixture drawn into the combustion

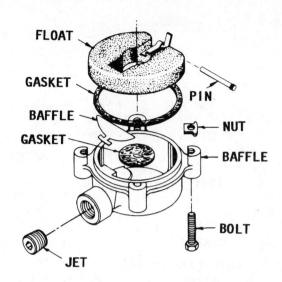

Exploded view of a single float system.

chambers. A cold powerhead requires a richer fuel mixture to start and during the brief period it is warming to normal operating temperature. A choke valve is placed ahead of the metering jets and venturi. As this valve begins to close, the volume of air intake is reduced, thus enriching the mixture entering the cylinders.

When this choke valve is closed, a **VERY** rich fuel mixture is drawn into the cylinders.

The throat of the carburetor is usually referred to as the "barrel". Carburetors with single, double, or four barrels have individual metering jets, needle valves, throttle and choke plates, for each barrel. Single and two barrel carburetors are fed by a single float and chamber.

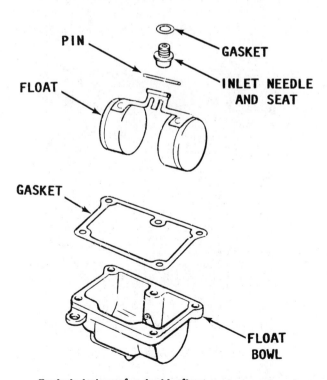

Exploded view of a double float system carburetor.

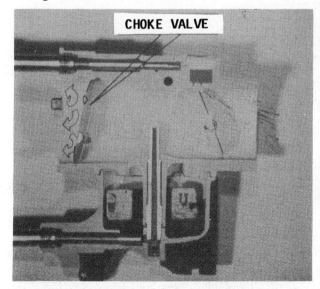

Choke valve location in the carburetor venturi. The choke valve on most carburetors covered in this manual is located in front of the venturi.

4-3 FUEL SYSTEM

The fuel system includes the fuel tank, fuel pump, fuel filters, carburetor, connecting lines with a squeeze bulb, and the associated parts to connect it all together. Regular maintenance of the fuel system to obtain maximum performance, is limited to changing the fuel filter at regular intervals and using fresh fuel. Even with the high price of fuel, removing gasoline that has been standing unused over a long period of time, is still the easiest and least expensive preventive maintenance possible. In most cases this old gas, even with some oil mixed with it, can be used without harmful effects in an automobile using regular gasoline.

If a sudden increase in gas consumption is noticed, or if the engine does not perform properly, a carburetor overhaul, including boil-out, or replacement of the fuel pump may be required.

LEADED GASOLINE AND GASOHOL

In the United States, the Environmental Protection Agency (EPA) has slated a proposed national phase-out of leaded fuel, "Regular" gasoline, by 1988. Lead in gasoline boosts the octane rating (energy). Therefore, if the lead is removed, it must be replaced with another agent. Unknown to the general public, many refineries are adding alcohol in an effort to hold the octane rating.

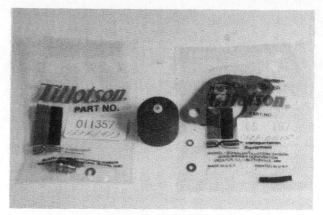
Major parts found in carburetor repair kits.

Alcohol in gasoline can have a deteriorating effect on certain fuel system parts. Seals can swell, pump check valves can swell, diaphragms distort, and other rubber or neoprene composition parts in the fuel system can be affected.

Since 1980, the manufacturer has made every effort to use materials that will resist the alcohol being added to fuels.

Fuels containing alcohol will slowly absorb moisture from the air. Once the moisture content in the fuel exceeds about 1%,

Damaged piston, possibly caused by insufficient oil mixed with the fuel; using too-low an octane fuel; or using fuel that has "soured" (stood too long without a preservative added).

Quicksilver Gasoline Stabilizer and Conditioner may be used to prevent the fuel from "souring" for up to twelve full months.

it will separate from the fuel taking the alcohol with it. This water/alcohol mixture will settle to the bottom of the fuel tank. The engine will fail to operate. Therefore, storage of this type of gasoline for use in marine engines is not recommended for more than just a few days.

One temporary, but aggravating, solution to increase the octane of "Unleaded" fuel is to purchase some aviation fuel from the local airport. Add about 10 to 15 percent of the tank's capacity to the unleaded fuel.

REMOVING FUEL FROM THE SYSTEM

For many years there has been the widespread belief that simply shutting off the fuel at the tank and then running the engine until it stops is the proper procedure before storing the engine for any length of time. Right? **WRONG.**

It is **NOT** possible to remove all of the fuel in the carburetor by operating the engine until it stops. Some fuel is trapped in the float chamber and other passages and in the line leading to the carburetor. The **ONLY** guaranteed method of removing **ALL** of the fuel is to take the time to remove the carburetor, and drain the fuel.

If the engine is operated with the fuel supply shut off until it stops the fuel and oil mixture inside the engine is removed, leav-

ing bearings, pistons, rings, and other parts with little protective lubricant, during long periods of storage.

Proper procedure involves: Shutting off the fuel supply at the tank; disconnecting the fuel line at the tank; operating the engine until it begins to run **ROUGH**; then stopping the engine, which will leave some fuel/oil mixture inside; and finally removing and draining the carburetor. By disconnecting the fuel supply, all **SMALL** passages are cleared of fuel even though some fuel is left in the carburetor. A light oil should be put in the combustion chamber as instructed in the Owner's Manual. On some model carburetors the high-speed jet plug can be removed to drain the fuel from the carburetor.

For short periods of storage, simply running the carburetor dry may help prevent severe gum and varnish from forming in the carburetor. This is especially true during hot weather.

4-4 TROUBLESHOOTING

The following paragraphs provide an orderly sequence of tests to pinpoint problems in the system. It is very rare for the carburetor by itself to cause failure of the engine to start.

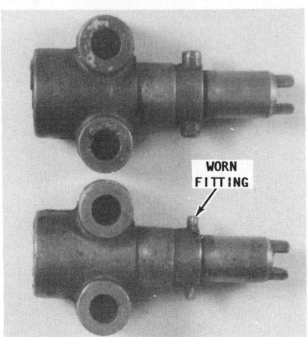

Comparison of a new (top) and worn (bottom) male fuel connector. The pins on the bottom connector are worn -- smaller and tapered, therefore, the connector will fail to maintain adequate fuel flow.

Female portion of the quick disconnect fitting ready to be mated with the male ʰortion on the powerhead.

FUEL PROBLEMS

Many times fuel system troubles are caused by a plugged fuel filter, a defective fuel pump, or by a leak in the line from the fuel tank to the fuel pump. A defective choke may also cause problems. **WOULD YOU BELIEVE,** a majority of starting troubles which are traced to the fuel system are the result of an empty fuel tank or aged "sour" fuel.

"SOUR" FUEL

Under average conditions (temperate climates), fuel will begin to breakdown in about four months. A gummy substance forms in the bottom of the fuel tank and in other areas. The filter screen between the tank and the carburetor and small passages in the carburetor will become clogged. The gasoline will begin to give off an odor similar to rotten eggs. Such a condition can cause the owner much frustration, time in cleaning components, and the expense of replacement or overhaul parts for the carburetor.

Even with the high price of fuel, removing gasoline that has been standing unused over a long period of time is still the easiest and least expensive preventative maintenance possible. In most cases, this old gas can be used without harmful effects in an automobile using regular gasoline.

The gasoline preservative additive Quicksilver Gasoline Stabilizer and Conditioner, shown below, will keep the fuel "fresh" for up to twelve months. If this particular product is not available in your area, other similar additives are produced under various trade names.

Fouled spark plug, possibly caused by the operator's habit of overchoking or a malfunction holding the choke closed. Either of these conditions delivered a too-rich fuel mixture to the cylinder.

The choke plate on the Model 2.2 swings down to restrict the flow of air causing a "rich" mixture until the powerhead reaches operating temperature. At that time, the choke lever must be manually pushed downward in an arc.

Choke Problems

When the engine is hot, the fuel system can cause starting problems. After a hot engine is shut down, the temperature inside the fuel bowl may rise to 200°F and cause the fuel to actually boil. All carburetors are vented to allow this pressure to escape to the atmosphere. However, some of the fuel may percolate over the high-speed nozzle.

If the choke should stick in the open position, the engine will be hard to start. If the choke should stick in the closed position, the engine will flood making it very difficult to start.

In order for this raw fuel to vaporize enough to burn, considerable air must be added to lean out the mixture. Therefore, the only remedy is to remove the spark plug/s; ground the leads; turn the engine

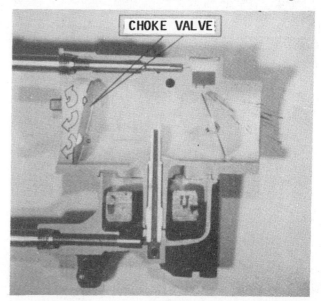

Choke valve location in the carburetor venturi. The choke valve on most carburetors covered in this manual is located in front of the venturi.

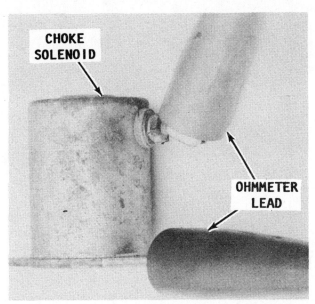

A choke solenoid may be tested with an ohmmeter. If the meter indicates continuity, the solenoid is satisfactory for further service. If continuity is not indicated, the unit cannot be repaired, it must be replaced.

over about 10 times; clean the plugs; install the plugs again; and start the engine.

If the needle valve and seat assembly is leaking, an excessive amount of fuel may enter the intake manifold in the following manner: After the engine is shut down, the pressure left in the fuel line will force fuel past the leaking needle valve. This extra fuel will raise the level in the fuel bowl and cause fuel to overflow into the intake manifold.

A continuous overflow of fuel into the intake manifold may be due to a sticking inlet needle or to a defective float which

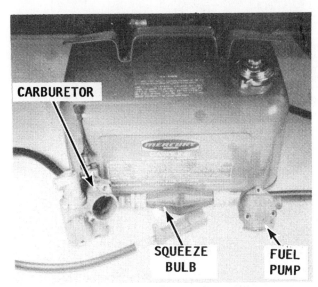

Major parts of a complete outboard motor fuel system from the tank to the carburetor.

would cause an extra high level of fuel in the bowl and overflow into the intake manifold.

FUEL PUMP TEST

First, These Words

On many units, the fuel pump is an integral part of the carburetor. On other units the fuel pump is a separate piece of equipment. The fuel pump on Models 3.5, 3.6, and 4.0, **CANNOT** be serviced. Therefore, if troubleshooting indicates the fuel pump to be at fault on these models, the fuel pump **MUST** be replaced with a new unit.

CAUTION Gasoline will be flowing in the engine area during this test. Therefore, guard against fire by grounding the high-tension wire to prevent it from sparking.

The high tension wire between the coil and the distributor can be grounded by either pulling it out of the coil and grounding it, or by connecting a jumper wire from the primary (distributor) side of the ignition coil

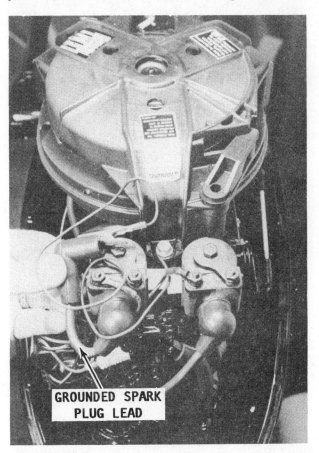

*Grounding the spark plug leads to the powerhead in preparation to making fuel flow tests. The grounding is **NECESSARY** to prevent a spark from igniting fuel being handled in the open.*

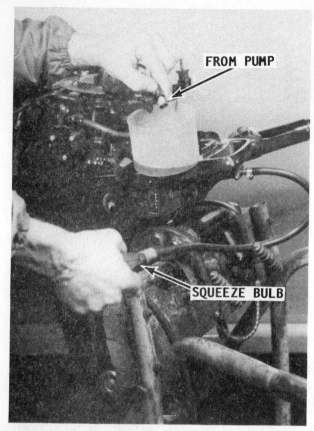

Testing the fuel pickup in the fuel tank **AND** operation of the squeeze bulb by observing fuel flow from the line disconnected at the fuel pump and discharged into a suitable container.

Working the squeeze bulb and observing the fuel flow from the line diconnected at the carburetor and discharged into a suitable container. This verifies fuel flow through the fuel pump. (The two photographs in this column were taken with a smaller powerhead than those covered in this manual. However, the procedure is the same.)

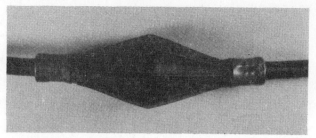

Common squeeze bulb used with outboard engine fuel systems.

to a good ground. An alternate safety method, and perhaps a better one, is to ground each spark plug lead. Disconnect the fuel line at the carburetor. Place a suitable container over the end of the fuel line to catch the fuel discharged. Now, squeeze the primer bulb and observe if there is satisfactory flow of fuel from the line.

If there is no fuel discharged from the line, the check valve in the squeeze bulb may be defective, or there may be a break or obstruction in the fuel line.

If there is a good fuel flow, then crank the engine. If the fuel pump is operating properly, a healthy stream of fuel should pulse out of the line.

Continue cranking the engine and catching the fuel for about 15 pulses to determine if the amount of fuel decreases with each

If tests indicate a satisfactory fuel flow to the carburetor, but adequate fuel quantity is not reaching the cylinders, then the carburetor **MUST** be removed and serviced.

pulse or maintains a constant amount. A decrease in the discharge indicates a restriction in the line. If the fuel line is plugged, the fuel stream may stop. If there is fuel in the fuel tank but no fuel flows out of the fuel line while the engine is being cranked, the problem may be in one of four areas:

1- The line from the fuel pump to the carburetor may be plugged as already mentioned.

2- The fuel pump may be defective.

3- The line from the fuel tank to the fuel pump may be plugged; the line may be leaking air; or the squeeze bulb may be defective.

4- If the engine does not start even though there is adequate fuel flow from the fuel line, the fuel filter in the carburetor inlet may be plugged or the fuel inlet needle valve and the seat may be gummed together and prevent adequate fuel flow.

FUEL LINE TEST

On most installations, the fuel line is provided with quick-disconnect fittings at

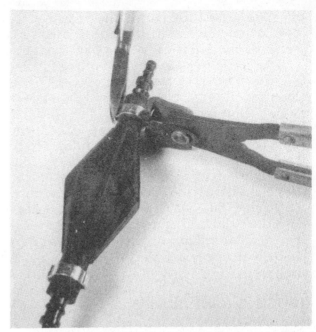

Using the proper tools to install a clamp around the squeeze bulb check valve.

the tank and at the engine. If there is reason to believe the problem is at the quick-disconnects, the hose ends should be replaced as an assembly. For a small addi-

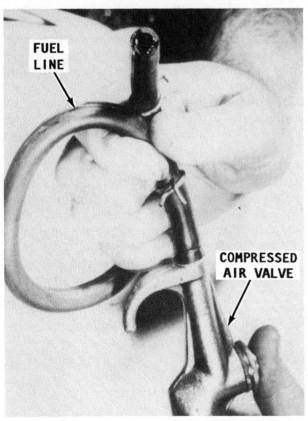

FUEL LINE

COMPRESSED AIR VALVE

Many times, restrictions such as foreign material may be cleared from the fuel line using compressed air. Use CARE to be sure the open end of the hose is pointing clear to avoid personal injury to the eyes.

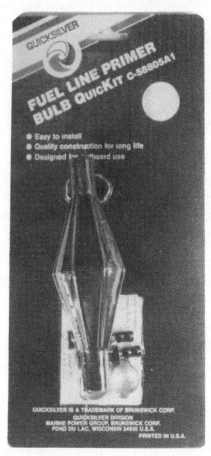

A replacement squeeze bulb kit includes parts necessary to return this section of the fuel line to service.

tional expense, the entire fuel line can be replaced and eliminate this entire area as a problem source for many future seasons.

The primer squeeze bulb can be replaced in a short time. First, cut the hose line as close to the old bulb as possible. Slide a small clamp over the end of the fuel line from the tank. Next, install the **SMALL** end of the check valve assembly into this side of the fuel line. The check valve always goes towards the fuel tank. Place a large clamp over the end of the check valve assembly. Use Primer Bulb Adhesive when the connections are made. Tighten the clamps. Repeat the procedure with the other side of the bulb assembly and the line leading to the engine.

ROUGH ENGINE IDLE

If an engine does not idle smoothly, the most reasonable approach to the problem is

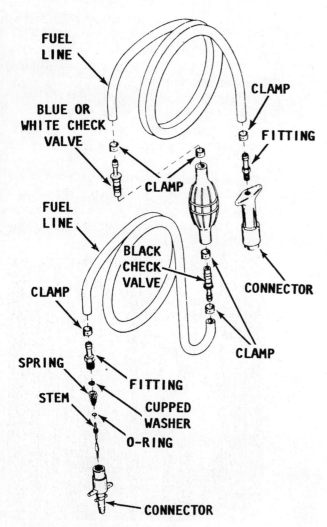

Exploded drawing of a typical fuel line and primer bulb with major parts identified.

to perform a tune-up to eliminate such areas as: defective points; faulty spark plugs; and timing out of adjustment.

Other problems that can prevent an engine from running smoothly include: An air leak in the intake manifold; uneven compression between the cylinders; and sticky or broken reeds.

Of course any problem in the carburetor affecting the air/fuel mixture will also prevent the engine from operating smoothly at idle speed. These problems usually include: Too high a fuel level in the bowl; a heavy float; leaking needle valve and seat; defective automatic choke; and improper adjustments for idle mixture or idle speed.

EXCESSIVE FUEL CONSUMPTION

Excessive fuel consumption can be the result of any one of three conditions, or a combination of all three.

1- Inefficient engine operation.
2- Faulty condition of the hull, including excessive marine growth.
3- Poor boating habits of the operator.

If the fuel consumption suddenly increases over what could be considered normal, then the cause can probably be attributed to the engine or boat and not the operator.

Marine growth on the hull can have a very marked effect on boat performance. This is why sail boats always try to have a haul-out as close to race time as possible. While you are checking the bottom take note of the propeller condition. A bent blade or other damage will definitely cause poor boat performance.

If the hull and propeller are in good shape, then check the fuel system for possible leaks. Check the line between the fuel pump and the carburetor while the engine is running and the line between the fuel tank and the pump when the engine is not running. A leak between the tank and the pump many times will not appear when the engine is operating, because the suction created by the pump drawing fuel will not allow the fuel to leak. Once the engine is turned off and the suction no longer exists, fuel may begin to leak.

If a minor tune-up has been performed and the spark plugs, points, and timing are properly adjusted, then the problem most likely is in the carburetor and an overhaul is in order. Check the needle valve and seat for leaking. Use extra care when making any adjustments affecting the fuel consumption, such as the float level or automatic choke.

ENGINE SURGE

If the engine operates as if the load on the boat is being constantly increased and decreased, even though an attempt is being made to hold a constant engine speed, the problem can most likely be attributed to the fuel pump, or a restriction in the fuel line between the tank and the carburetor.

Operational description and service procedures for the fuel pump are given in Section 4-14.

4-5 CARBURETOR MODELS

Eight, yes eight different carburetors are used on the outboard powerheads covered in this manual. Procedures for each carburetor is presented in a separate section of this chapter. To determine which carburetor is installed on the outboard being serviced, check the Tune-up Specifications in the Appendix, under engine model and manufactured year. The carburetor model is designated by a letter in the third to last column. The carburetor identification used in the Appendix and throughout this book is as follows:

A Side bowl and back drag carburetors.
B Round bowl -- single float carburetor with integral fuel pump.
C Center round bowl -- single float --

side-draft carburetor.
D Mikuni rectangular bowl -- double float carburetor.
E Mikuni round bowl -- single float carburetor.
F Tillotson rectangular bowl -- double float carburetor with fuel pump.
G Walbro round bowl -- single float carburetor with fuel pump.
H Round bowl -- Stamped "F" --single float carburetor with "Keikhin" integral fuel pump.

Complete detailed service procedures for each carburetor are presented in separate sections.

4-6 SIDE BOWL AND BACK DRAG CARBURETORS REFERENCED "A" IN THE APPENDIX

This section provides complete detailed procedures for removal, disassembly, cleaning and inspecting, assembling including bench adjustments, installation, and operating adjustments for the side bowl and back drag carburetor, referenced "A" in the Tune-up Specifications (third to last column), in the Appendix.

This "A" carburetor was used on a wide range of 1- and 2-cylinder powerheads over the years.

Make an attempt to keep the work area organized and to cover parts after they have been cleaned to prevent foreign matter from entering passageways or adhering to critical parts.

Example of a damaged propeller, probably caused by striking an underwater object. Such a unit will restrict performance of the boat and the outboard unit.

The fuel pump being removed from the intake transfer port of a 35hp powerhead.

To service the separate fuel pump, see Section 4-14, this chapter. For synchronizing adjustments with the ignition system, see Chapter 6.

REMOVAL AND DISASSEMBLING

FIRST, THESE WORDS

Good shop practice dictates a carburetor repair kit be purchased and new parts be installed any time the carburetor is disassembled.

1- Remove the battery leads from the battery terminals. Using the quick-disconnect fitting, detach the fuel line from the engine or from the fuel tank.

2- Remove the hood assembly from the engine. Remove the choke and throttle linkage to the carburetor.

3- Remove the fuel line from the carburetor. This may be accomplished by either one of two methods. One is to remove the line from the strainer cover. The other is to remove the strainer cover bolt, and then lift the cover from the carburetor as shown in the accompanying illustration.

4- Remove the nuts securing the carburetor to the crankcase. Lift the carburetor from the engine.

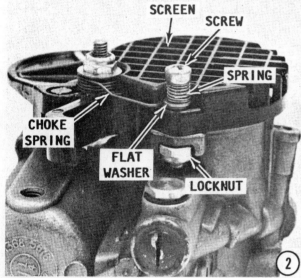

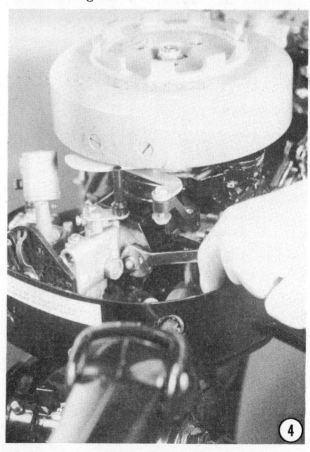

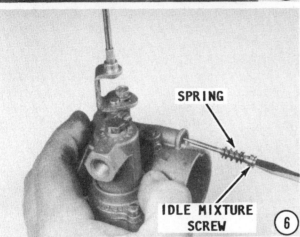

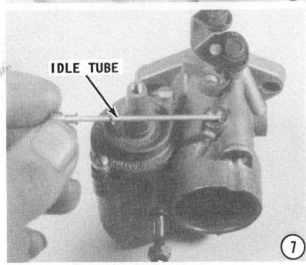

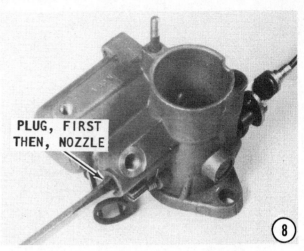

5- Remove the 3/8" cap screw securing the strainer cover to the carburetor, if it was not removed during carburetor removal. **OBSERVE** the gasket under the screw as it is removed. Keep them together to ensure the gasket is installed during assembling. Remove the strainer cover, and then the gasket inside the strainer cover. Remove the gas fuel line fitting and the base gasket.

6- Remove the idle screw and spring from the carburetor.

7- Carefully remove the idle tube from the top of the carburetor. This tube extends down inside the main discharge nozzle. A new tube gasket should **ALWAYS** be used. Therefore, **DISCARD** the old gasket.

8- Remove the discharge screw plug from the bottom side of the carburetor. Use the **PROPER** size screwdriver and remove the main discharge nozzle.

9- Remove the 7/16" brass hex head plug and gasket from the front of the carburetor and the other plug from the bottom. Do not

High-speed jet and gasket for the Type "A" carburetor.

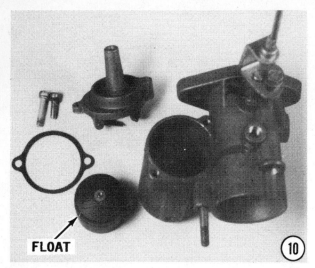

FLOAT   (10)

Filters used with the side-bowl Type "A" carburetor. The two on the left are obsolete and should be replaced with the new type on the far right.

remove the tube unless absolutely necessary. Only check to make sure it is tight. Use the **PROPER** size screwdriver and remove the main fuel jet and gasket.

10- Remove the float assembly by removing the two screws from the top of the carburetor. Lift the float assembly from the carburetor body, and then remove the gasket.

11- Turn the float cover upside down and notice the assembly has two levers. Remove the top lever pin and hinge back the other lever. Now, remove the inlet needle from the needle seat. Use the proper size socket and remove the needle seat. This seat has a standard right-hand thread. Reach into the body with a small punch and gently remove the gasket.

12- To remove the Welch plug on the side of the carburetor, use a sharp punch to puncture the center of the plug, and then pry out the plug A new Welch plug is **ONLY** available in a carburetor overhaul kit.

A GOOD WORD: Further disassembly of the carburetor is not necessary.

CLEANING AND INSPECTING

NEVER dip rubber parts, plastic parts, diaphragms, or pump plungers in carburetor cleaner. These parts should be cleaned **ONLY** in solvent, and then blown dry with compressed air.

Place all metal parts in a screen-type tray and dip them in carburetor cleaner until they appear completely clean, then blow them dry with compressed air.

Blow out all passages in the castings with compressed air. Check all parts and passages to be sure they are not clogged or contain any deposits. **NEVER** use a piece of wire or any type of pointed instrument to clean drilled passages or calibrated holes in a carburetor.

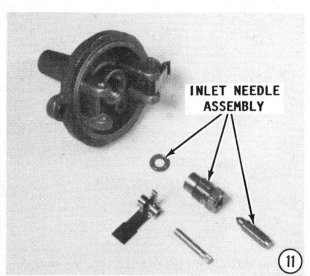

**INLET NEEDLE
ASSEMBLY**   (11)

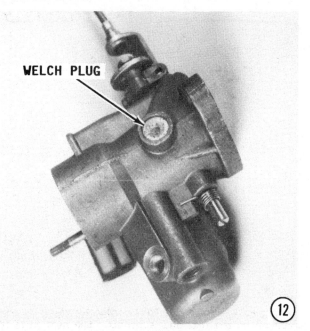

WELCH PLUG   (12)

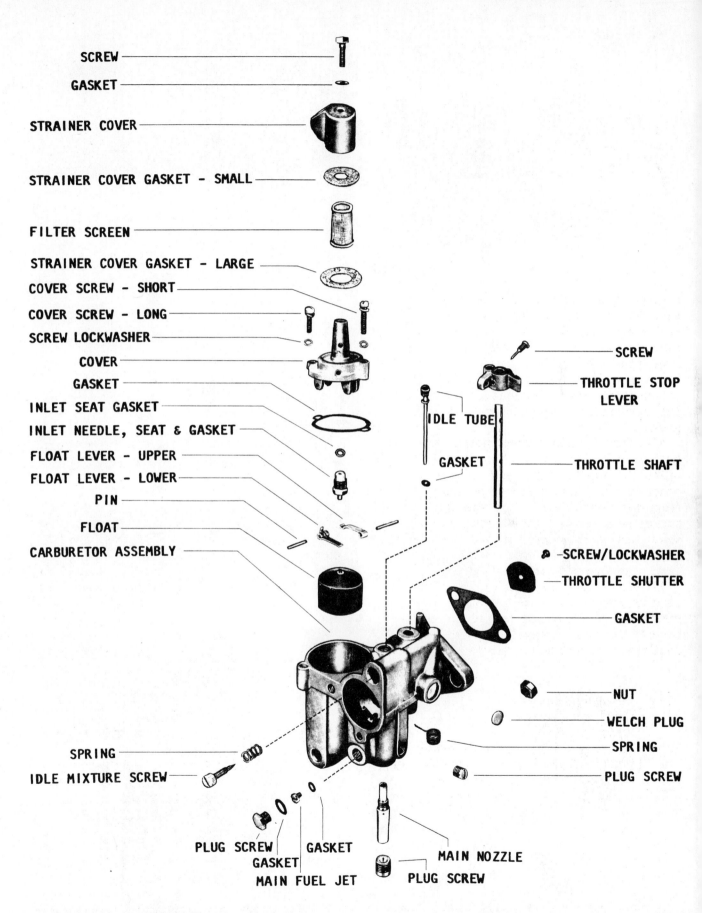

SCREW

GASKET

STRAINER COVER

STRAINER COVER GASKET - SMALL

FILTER SCREEN

STRAINER COVER GASKET - LARGE

COVER SCREW - SHORT

COVER SCREW - LONG

SCREW LOCKWASHER

COVER

GASKET

INLET SEAT GASKET

INLET NEEDLE, SEAT & GASKET

FLOAT LEVER - UPPER

FLOAT LEVER - LOWER

PIN

FLOAT

CARBURETOR ASSEMBLY

IDLE TUBE

GASKET

SCREW

THROTTLE STOP LEVER

THROTTLE SHAFT

SCREW/LOCKWASHER

THROTTLE SHUTTER

GASKET

NUT

WELCH PLUG

SPRING

PLUG SCREW

SPRING

IDLE MIXTURE SCREW

PLUG SCREW

GASKET

GASKET

MAIN FUEL JET

MAIN NOZZLE

PLUG SCREW

Exploded view of a side bowl carburetor showing arrangement of major parts. This carburetor is identified as an "A" carburetor in the text and Appendix. See next page for the "back drag" carburetor, also identified in the text and appendix as an "A" carburetor because it is so similar in appearance, operation, and for service procedures.

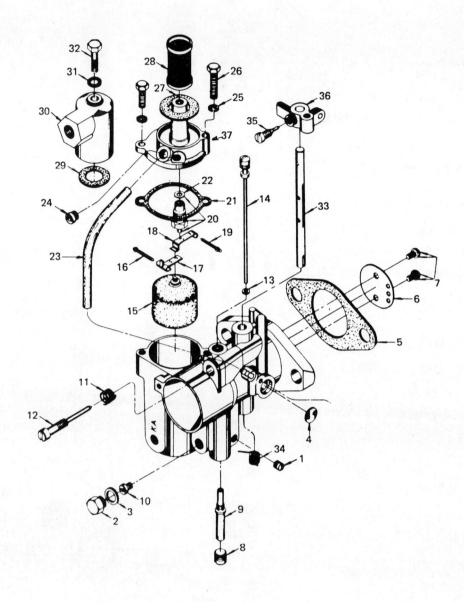

1- PLUG		
2- MAIN FUEL JET PLUG	14- IDLE TUBE	26- FLOAT COVER BOLT
3- GASKET	15- FLOAT	27- GASKET
4- WELCH PLUG	16- LOWER FLOAT LEVER PIN	28- FILTER SCREEN
5- GASKET	17- LOWER FLOAT LEVER	29- GASKET
6- THROTTLE SHUTTER	18- UPPER FLOAT LEVER	30- FUEL INLET COVER
7- THROTTLE SHUTTER SCREW	19- UPPER FLOAT LEVER PIN	31- GASKET
8- PLUG	20- INLET NEEDLE & SEAT	32- INLET COVER SCREW
9- MAIN FUEL NOZZLE	21- GASKET	33- THROTTLE SHAFT
10- MAIN FUEL JET	22- GASKET	34- SPRING
11- SPRING	23- BACK DRAG TUBE	35- THROTTLE STOP LEVER
12- IDLE MIXTURE SCREW	24- BACK DRAG AIR JET	36- THROTTLE STOP LEVER
13- GASKET	25- LOCKWASHER	37- FLOAT BOWL COVER

 Exploded view of the side bowl "back drag" carburetor and like the carburetor on the previous page, identified as an "A" carburetor in the text and appendix. This carburetor has an additional circuit which lowers the atmospheric pressure in the float bowl to increase fuel economy at certain mid-range rpm. Major parts are identified.

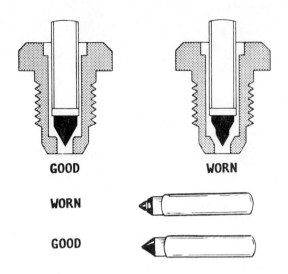

Needle and seat arrangement on the carburetor covered in this section, showing a worn and new needle for comparison.

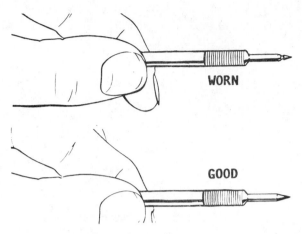

Carburetor idle mixture adjustment needles. The top needle is worn and unfit for service. The bottom needle is new.

Move the throttle shaft back-and-forth to check for wear. If the shaft appears to be too loose, replace the complete throttle body because individual replacement parts are **NOT** available.

Inspect the main body, airhorn, and venturi cluster gasket surfaces for cracks and burrs which might cause a leak. Check the float for deterioration. Check to be sure the float spring has not been stretched. If any part of the float is damaged, the unit must be replaced. Check the float arm

needle contacting surface and replace the float if this surface has a groove worn in it.

Inspect the tapered section of the idle adjusting needles and replace any that have developed a groove.

Most of the parts that should be replaced during a carburetor overhaul are included in overhaul kits available from your local marine dealer. One of these kits will contain a matched fuel inlet needle and seat. This combination should be replaced each time the carburetor is disassembled as a precaution against leakage.

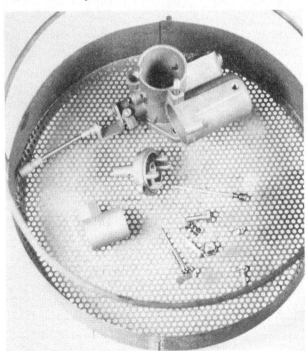

*All rubber and plastic parts **MUST** be removed before carburetor parts are placed in a basket to be submerged in carburetor cleaner.*

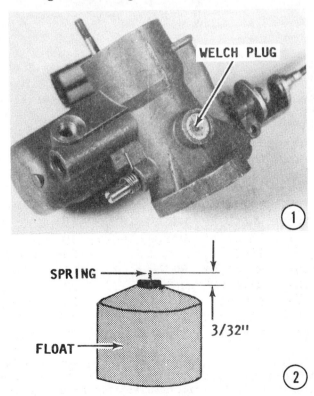

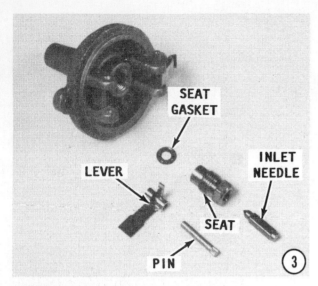

ASSEMBLING
CARBURETOR "A"

1- If the Welch plug was removed, insert a new plug in position, and then tap it into place. Seal the outside edge of the plug with Gasketcinch, or equivalent.

2- Check the spring on the top of the float. If it does not extend out 3/32" (2.40 mm) the float **MUST** be replaced. Install the float onto the float pin, and then slide the float into the carburetor body.

3- Insert a new needle seat gasket into place. Thread the inlet seat into the body and tighten the seat with the proper size socket to a torque value of 60 in.-lbs (6.78Nm). Discharge a drop of oil into the center of the seat, and then insert the inlet needle into the seat. Hinge over the lever that was not removed on top of the inlet needle. Install the other lever on top of the lever in place, and then install the hinge pin.

Float Lever Adjustment
4- Turn the float bowl cover upside down. Measure the distance from the face of the shoulder to the secondary lever. This

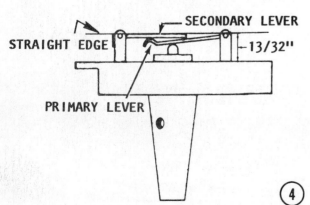

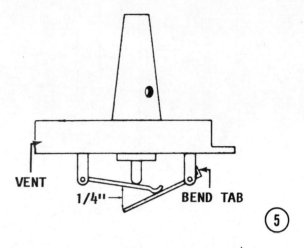

measurement should be 13/32" ± 1/64" (10.32mm ± 0.40mm). **CAREFULLY** bend the primary lever as required to obtain the correct measurement.

Float Drop Adjustment
5- Turn the float bowl cover upright. Check to be sure the needle moves freely on the actuating primary lever and that it is not sticking in the seat. Hold the bowl cover upright and measure the distance between the primary and secondary levers. This distance should be 1/4" (6.35mm). **CAREFULLY** bend the secondary lever stop tang to obtain the proper measurement.

6- Place a **NEW** gasket onto the float bowl. Then place the float into the float chamber. The vent hole in the cover **MUST** be installed toward the carburetor mounting flange. Invert the carburetor and check the float for free movement.

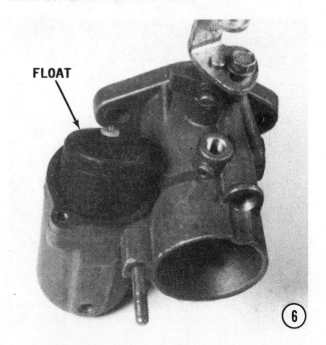

(7)

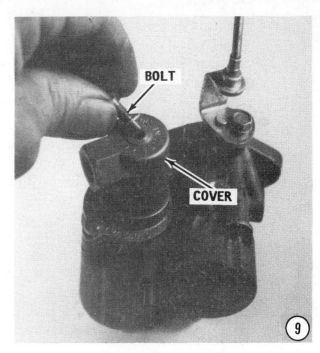

BOLT

COVER

(9)

7- Place the float bowl cover over the float. Install the two screws and lockwashers, and tighten them alternately.

8- Position the large gasket over the tower of the float bowl cover. Install the filter screen. Insert a **NEW** gasket inside the strainer cover. Place the strainer cover over the float bowl cover. **ONE WORD:** If the strainer cover was left attached to the fuel line during disassembly, and is therefore, still on the engine, then bypass the next instruction. The cover will be installed in Step 14. **BE SURE** the strainer cover is setting squarely on the float bowl cover, because if it is not positioned properly, the

tower assembly on the float bowl cover will be broken when the strainer bolt is tightened.

9- Slide a **NEW** gasket onto the strainer bolt, and then install the bolt into the float bowl cover.

10- Thread the main nozzle into the bottom of the carburetor, and then tighten it securely using the **PROPER** size screwdriver. Install the nozzle plug. Use a wood toothpick or Mercury special tool (high-speed jet).

GOOD WORD

Main fuel (high-speed) jet size recommendations are intended as a guide. If any change in size is to be made, check the Jet Size/Elevation Chart in the Appendix.

COVER

FILTER

(8)

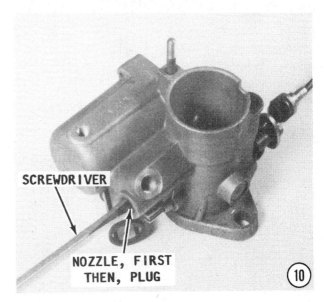

SCREWDRIVER

NOZZLE, FIRST
THEN, PLUG

(10)

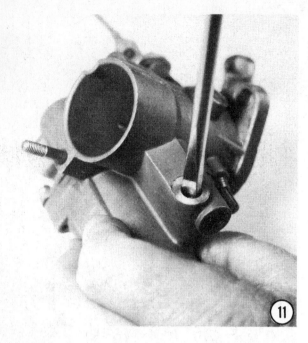

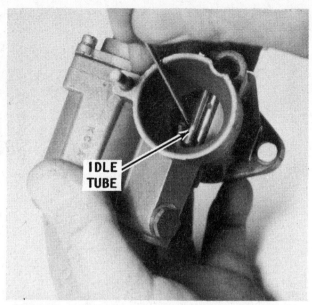

The idle restriction tube MUST contact the front of the venturi tube.

11- Slide a **NEW** gasket onto the fixed high-speed jet. (On some models, this gasket is not used.) Use the **PROPER** size screwdriver and install the jet into the carburetor body. Position a **NEW** gasket onto the 7/16" brass plug, and then install the plug into the carburetor body.

12- Slide a **NEW** gasket onto the idle restriction tube. (On some models, this gasket is not used.) Thread the tube into the top of the carburetor and tighten it securely. When properly installed, idle restriction tube **MUST** touch front of venturi tube.

13- Position the spring over the idle adjusting screw, and then **SLOWLY** thread it into the carburetor body, until you can feel

it seat. Now, as a preliminary adjustment, back the screw out 1 to 1½ full turns. Check the throttle shutters to be sure they do not bend in the carburetor venturi.

INSTALLATION
CARBURETOR "A"

14- Install a **NEW** gasket onto the intake manifold. Position the carburetor onto the intake manifold and at the same time, check to be sure the cam on the carburetor is in front of the cam on the magneto plate. Connect the fuel line to the carburetor strainer cover. If the strainer cover was not installed in Step 8, then place the cover over the tower assembly of the carburetor.

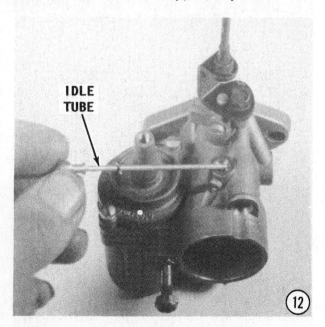

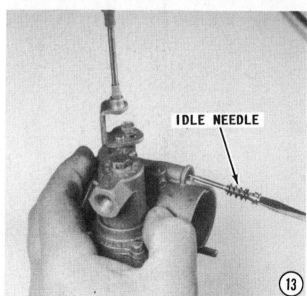

CHECK TO BE SURE the strainer cover is setting squarely on the float bowl cover, because if it is not positioned properly, the tower assembly on the float bowl cover will be broken when the strainer bolt is tightened. Connect the manual choke to the choke rod.

15- Install the throttle pickup bracket and the throttle pickup lever with the mounting screw. Check to be sure the cam on the carburetor is in front of the cam on the magneto plate.

Connect the throttle and choke linkage.

Synchronizing

To synchronize the fuel and ignition systems, see Chapter 6.

ADJUSTMENTS

FIRST A WORD: Before fine carburetor adjustments can be properly made, the following conditions must exist:

a. The regular engine-propeller combination must be used.

b. The power unit must be in forward gear.

c. The lower unit must be in the water.

d. The engine must be warmed to normal operating temperature.

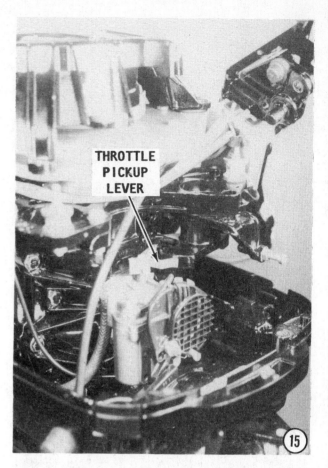

THROTTLE PICKUP LEVER

15

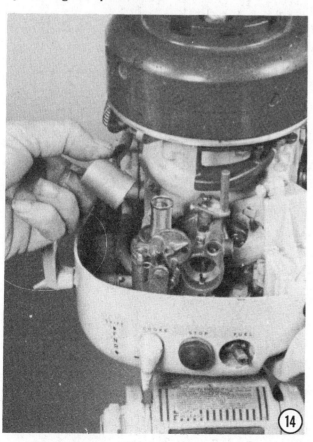

14

IDLE SPEED ADJUSTING SCREW

16

Typical location of the idle speed adjustment screw on a late model powerhead with a Type "A" carburetor.

Idle-Speed Adjustment

16- After the engine has been warmed to operating temperature, turn the idle speed adjusting screw on the stop bracket until the engine idles at approximately 650 rpm in forward gear.

Idle Mixture Adjustment

17- Turn the adjusting screw counter-clockwise for the 3.5, 3.6, 3.9, 60, and 110 hp engine models, and clockwise for all others, until the engine fires evenly and rpm begins to increase. Continue turning the adjusting screw until the mixture is so lean that the rpm begin to drop and the engine begins to misfire. Set the adjusting screw halfway between the rich and lean points.

ADVICE

It is better to have the mixture set slightly on the rich side, rather than too lean.

High-Speed Adjustment

The main metering high-speed jet is not adjustable. If the engine is to be operated at elevations above 4000 ft., replace the main metering jet as indicated in the Jet Size/Elevation Chart in the Appendix.

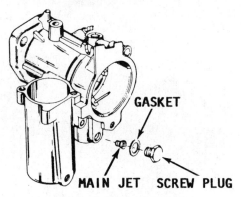

Access to the main jet of carburetor "A" is from outside the carburetor. This permits quick and easy changes for varying altitudes.

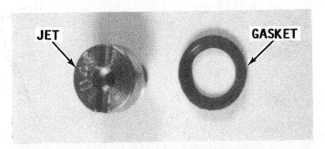

The main (high-speed) jet is not adjustable, but it is available in varying sizes. The jet may be replaced if the engine is to be operated at different elevations.

FUEL INLET HOSE

4-7 INTEGRAL FUEL PUMP CARBURETOR — REFERENCED "B" IN THE APPENDIX

This section provides complete detailed procedures for removal, disassembly, cleaning and inspecting, assembling including bench adjustments, installation, and operating adjustments for the integral fuel pump carburetor, referenced "B" in the Tune-up Specifications (third to last column), in the Appendix. The fuel pump is an intergral part of the carburetor. To synchronize the fuel and ignition systems, see Chapter 6.

This new carburetor has an integral fuel pump, which should be overhauled every time the carburetor is disassembled.

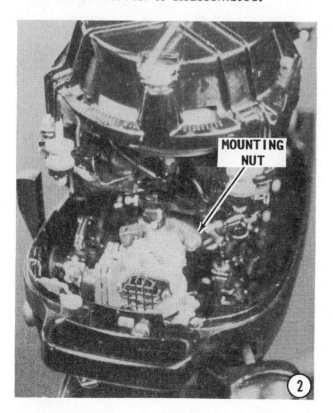

MOUNTING NUT

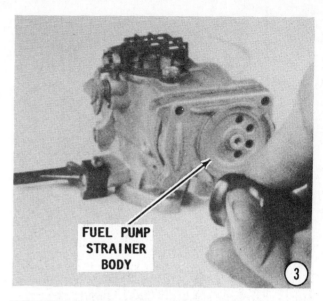

FUEL PUMP STRAINER BODY

REMOVAL AND DISASSEMBLING

1- Remove the battery leads from the battery terminals. Remove the hood assembly. Disconnect the fuel line at the fuel

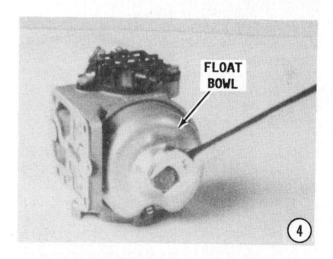

FLOAT BOWL

MAIN FUEL JET

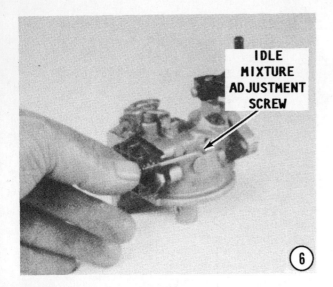

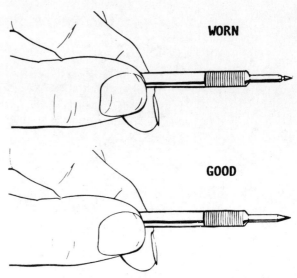

Carburetor idle mixture adjustment needles. The top needle is worn and unfit for service. The bottom needle is new.

tank. Disconnect the choke cable from the choke lever. Remove the cap screw and spacer securing the choke cable to the carburetor. Remove the fuel line from the inlet cover. An alternative is to remove the screw securing the inlet cover to the carburetor and leave the fuel line attached.

2- Remove the two nuts attaching the carburetor to the manifold, and then remove the carburetor.

3- Remove the four screws holding the fuel pump strainer body to the carburetor. Remove the gaskets and diaphragm. Remove the fuel pump body and gaskets.

4- Remove the bolt securing the float bowl to the carburetor casting. **OBSERVE** and **REMEMBER** there is a gasket under the bolt and one between the float bowl and the casting. Withdraw the float retaining pin, and then lift off the float assembly. Lift out the inlet needle valve and spring.

DO NOT attempt to remove the needle valve seat. This seat is pressed into the carburetor body.

5- Remove the main fuel (high-speed) jet. A gasket is not used under this jet. **DO NOT** attempt to remove the main nozzle even though it has a screwdriver slot. The boost venturi is very difficult to install if the main nozzle has been removed.

6- Remove the idle mixture adjusting screw and spring.

7- Remove the plug screw, and then unscrew the idle tube. Slide the gasket free of the idle tube.

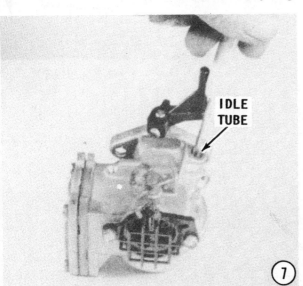

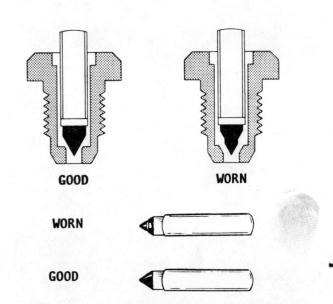

Needle and seat arrangement on the carburetor covered in this section, showing a worn and new needle for comparison.

CLEANING AND INSPECTING

NEVER dip rubber parts, plastic parts, diaphragms, or pump plungers in carburetor cleaner. These parts should be cleaned **ONLY** in solvent, and then blown dry with compressed air.

Place all of the metal parts in a screen-type tray and dip them in carburetor cleaner until they appear completely clean, then blow them dry with compressed air.

Blow out all of the passages in the castings with compressed air. Check all of the parts and passages to be sure they are not clogged or contain any deposits.

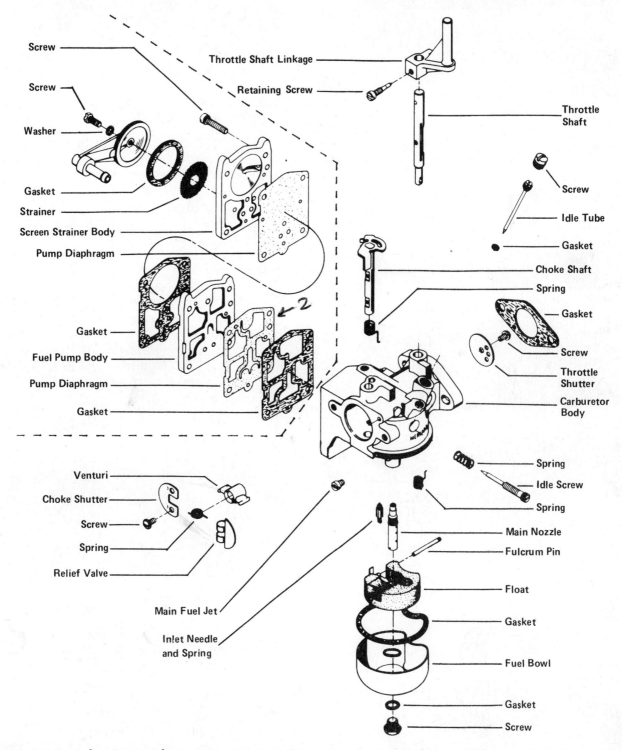

Exploded view of an integral fuel pump carburetor. This carburetor is identified as a "B" carburetor in the text and appendix. Major parts are identified. Fuel pump parts are to the left and above the dotted line.

MAIN
FUEL JET

Inspect the main body, airhorn, and venturi cluster gasket surfaces for cracks and burrs which might cause a leak. Check the float for deterioration. If a hollow float is used, check to be sure it does not contain any fluid. Check to be sure the float spring has not been stretched. If any part of the float is damaged, the unit must be replaced. Check the float arm needle contacting surface and replace the float if this surface has a groove worn in it.

Inspect the tapered section of the idle adjusting needles and replace any that have developed a groove.

Most of the parts that should be replaced during a carburetor overhaul are included in an overhaul kit available from your local marine dealer.

Check the jet sizes with a drill of the proper size. **ALWAYS** hold the drill in a pin vise to avoid enlarging the jet orifice. Refer to the Carburetor Jet Size/Elevation Chart in the Appendix for the proper size for your engine, carburetor, and anticipated elevation of operation.

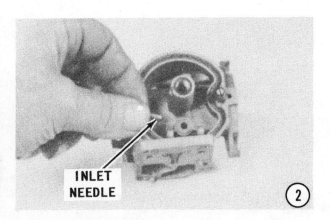

INLET
NEEDLE

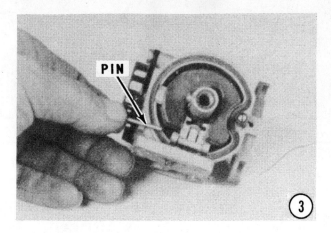

PIN

ASSEMBLING CARBURETOR "B"

1- Install the main fuel (high-speed) jet. As mentioned during removal, a gasket is not used under this jet.

2- Install a **NEW** inlet valve needle and spring to reduce the chances of a leak. Install a **NEW** float bowl gasket.

3- Install the float, and then insert the float retaining pin to secure the float in place.

Float Level Adjustment

4- Hold the carburetor as shown, and measure the distance to the bottom edge of the float. This measurement should be 1/4" $\pm$ 1/64" (6.35 $\pm$ 0.40mm). **CAREFULLY** bend the float needle actuating lever to obtain the correct measurement.

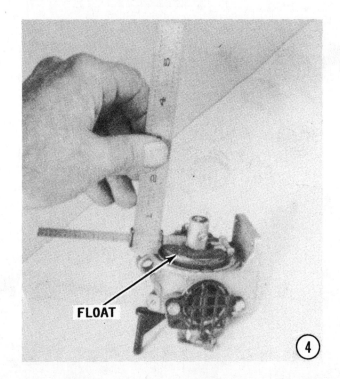

FLOAT

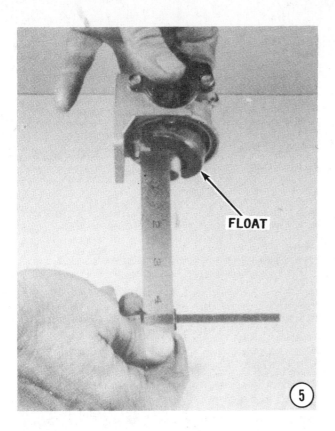

FLOAT

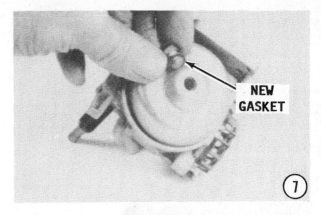

NEW GASKET

(7)

NEW GASKET

(8)

Float Drop Adjustment

5- Hold the carburetor upside down, as shown, to allow the float to drop to its lowest point. Measure the distance from the bottom of the float to the top of the main fuel (high-speed) jet. This distance should be from 1/64" to 1/32" (0.40 to 0.80mm). CAREFULLY bend the float tang to obtain the correct measurement.

6- Check to be sure the float bowl gasket is in place properly. Position the float bowl gasket on the carburetor casting. Install the float bowl.

7- Slide a NEW gasket onto the retaining bolt, and then install the bolt onto the float bowl cover.

8- Position a NEW fuel pump gasket onto the carburetor casting taking care to index it over the alignment dowel. Install a NEW valve diaphragm. OBSERVE the three valve flaps, two for inlet control and one for outlet.

9- Install the valve body with the alignment dowel entering the hole in the casting properly.

10- Position a NEW gasket on the valve body. Install a NEW pump diaphragm.

11- Install the fuel pump strainer body and secure it in place with the four retaining screws. Tighten the screws EVENLY and

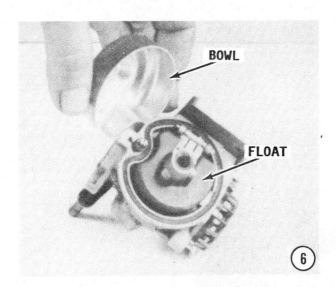

BOWL

FLOAT

(6)

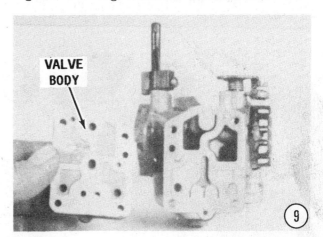

VALVE BODY

(9)

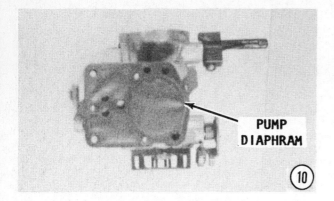

PUMP DIAPHRAM

(10)

FUEL PUMP STRAINER BODY

(11)

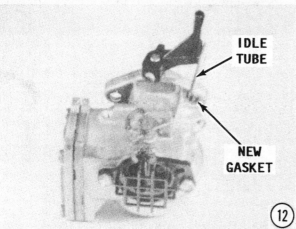

IDLE TUBE

NEW GASKET

(12)

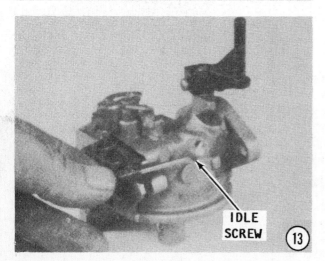

IDLE SCREW

(13)

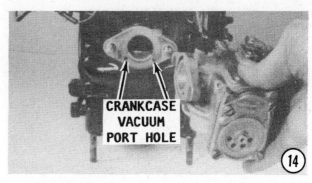

CRANKCASE VACUUM PORT HOLE

(14)

a little-at-a-time. Install the filter screen and a **NEW** gasket.

12- Slide a **NEW** gasket onto the idle tube and then thread it into place. Install the plug screw over the idle tube.

13- Slowly thread the idle mixture adjusting screw into the carburetor body until you can feel it seat. **DO NOT** tighten the screw or you will damage the tip. Now, as a preliminary adjustment, back it out 1-1/4 turns.

14- Place a **NEW** flange gasket in position on the intake manifold. Check to be sure the two crankcase vacuum port holes are aligned with the holes in the casting. If this gasket is not installed **PROPERLY**, the fuel pump will **NOT** function.

INSTALLATION CARBURETOR "B"

15- Place the carburetor in position on the intake manifold. Install and tighten the two carburetor retaining nuts alternately to the torque value given in the Appendix. Install the fuel pump inlet cover. Slide a **NEW** lockwasher onto the retaining screw, and then install and tighten the screw. If the fuel line was removed from the inlet cover, install the hose and tighten the hose clamps.

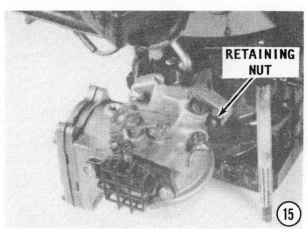

RETAINING NUT

(15)

16- Install the cap screw and spacer securing the choke cable to the carburetor. Connect the choke cable to the lever. Connect the fuel line to the tank. Activate the fuel line squeeze bulb several times. Check delivery of fuel to the carburetor and the lines and their fittings for possible leaks. Connect the battery leads.

Synchronizing

To synchronize the fuel and ignition systems, see Chapter 6.

ADJUSTMENTS

FIRST A WORD: Before fine carburetor adjustments can be properly made, the following conditions must exist:

a. The correct engine-propeller combination must be used.

b. The power unit must be in forward gear.

c. The lower unit must be in the water.

d. The engine must be warmed to normal operating temperature.

Idle Mixture Adjustment

17- After the above conditions have been met, including the engine run until it has reached operating temperature, set the idle mixture screw 1-1/2 turns open from a lightly seated position. Now, with the engine running, **SLOWLY** turn the idle mixture screw counterclockwise until the affected cylinders start to load up or begin to fire unevenly, due to an over-rich mixture. **SLOWLY** turn the idle mixture screw clockwise until the cylinders fire evenly and the engine rpm increase. Continue turning the screw clockwise until the engine rpm drop off and the engine begins to misfire. Now, turn the idle mixture screw **COUNTER-CLOCKWISE** halfway between lean and rich position. Favor the rich side.

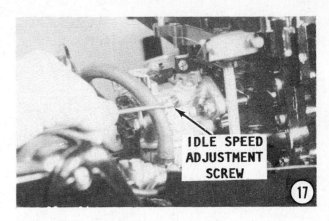

SOME ADVICE: Do not adjust to a leaner position than necessary. It is better to have the mixture set slightly on the rich side, rather than too lean. Too lean a mixture is often the cause of hard starting.

MORE ADVICE: If the engine hesitates during acceleration after adjusting the idle mixture, the mixture is set too lean and should be changed to the richer side until engine acceleration is smooth.

Idle Speed Adjustment

18- After the conditions listed at the beginning of this ADJUSTMENT section have been met, and the idle mixture adjustment has been properly made, as described in the previous step, then adjust the idle speed stop screw on the stop bracket until the engine idles at the recommended rpm given in the Tune-up Specifications in the Appendix. Continue running the engine in forward gear at the recommended wide open throttle range (WOT) to clear the engine, and then recheck the idle speed.

4-8 CENTER ROUND BOWL CARBURETOR REFERENCED "C" IN APPENDIX

This section provides complete detailed procedures for removal, disassembly, cleaning and inspecting, assembling including bench adjustments, installation, and operating adjustments for the square bowl carburetor, referenced "C" in the Tune-up Specifications (third to last column), in the Appendix. To service the separte fuel pump, see Section 4-14, this chapter. For adjustments with the ignition system, see Chapter 6.

REMOVAL

1- Remove the air box cover and the air box from the carburetor.

2- Close the gas shutoff valve to the carburetor to prevent gas from draining out of the gas tank when the lines are disconnected at the carburetor.

3- Disconnect the choke cable from the mounting bracket. Disconnect the cable end from the carburetor. Close the fuel line shutoff valve. Cut the tie strap on the fuel line and remove the fuel line from the carburetor.

4- Pry the remote idle lever from the idle mixture screw. The lever will pop off the screw.

5- Remove the attaching nuts, and lift the carburetor from the intake manifold.

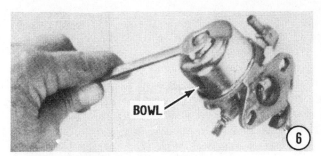

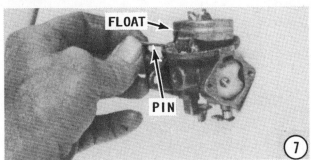

DISASSEMBLING CARBURETOR "C"

6- Turn the carburetor upside down and remove the center bolt securing the bowl to the body. This bolt also serves as a fixed jet to meter fuel.

7- Push the hinge pin out of the float lever, and then lift off the float and inlet needle. Determine if the float has been leaking, by shaking it and listening for the movement of fuel inside. A leaky float **MUST** be replaced.

8- Remove the inlet needle seat.

9- Now remove the carburetor nozzle.

10- Remove the idle mixture screw from the side of the carburetor.

11- Remove the fuel inlet hose fitting.

12- Check the filter screen inside the hole. Remove the screen. The factory does not recommend using a filter in the carburetor.

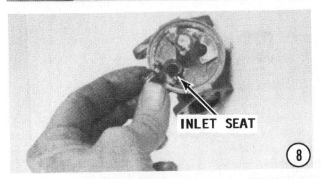

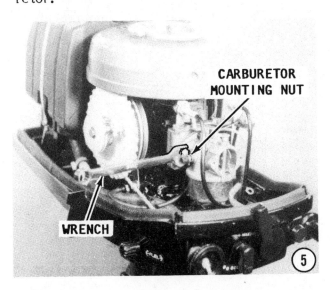

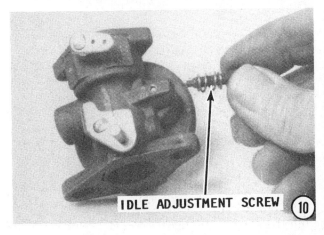

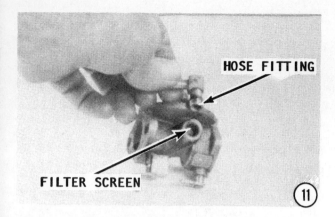

A carburetor removed from a submerged powerhead. Note the corrosion in the float bowl and deterioration of the fixed jet.

CLEANING AND INSPECTING

NEVER dip rubber parts, plastic parts, diaphragms, or pump plungers in carburetor cleaner. These parts should be cleaned **ONLY** in solvent, and then blown dry with compressed air.

Place all of the metal parts in a screen-type tray and dip them in carburetor cleaner until they appear completely clean, then blow them dry with compressed air.

Blow out all of the passages in the castings with compressed air. Check all of the parts and passages to be sure they are not clogged or contain any deposits. **NEVER** use a piece of wire or any type of pointed instrument to clean drilled passages or calibrated holes in a carburetor.

Move the throttle shaft back-and-forth to check for wear. If the shaft appears to be too loose, replace the complete throttle body because individual replacement parts are **NOT** available.

Inspect the main body, airhorn, and venturi cluster gasket surfaces for cracks and burrs which might cause a leak. Check the float for deterioration. Check to be sure the float spring has not been stretched. If any part of the float is damaged, the unit must be replaced. Check the float arm

needle contacting surface and replace the float if this surface has a groove worn in it.

Inspect the tapered section of the idle adjusting needles and replace any that have developed a groove.

Most of the parts that should be replaced during a carburetor overhaul are included in an overhaul kit available from your local marine dealer. This kit will also contain a matched fuel inlet needle and seat. This combination should be replaced each time the carburetor is disassembled as a precaution against leakage.

Check the jet sizes with a drill of the proper size. **ALWAYS** hold the drill in a pin vise to avoid enlarging the jet orifice. Refer to the Carburetor Jet Size/Elevation Chart in the Appendix for the proper size jet for your engine, carburetor, and anticipated elevation of operation.

Examine the throttle shaft for wear and the throttle shutter plates for damage.

ASSEMBLING
CARBURETOR "C"

1- The factory does not recommend replacing the fuel filter. Install the fuel hose fitting.

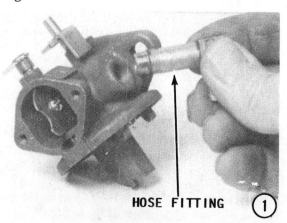

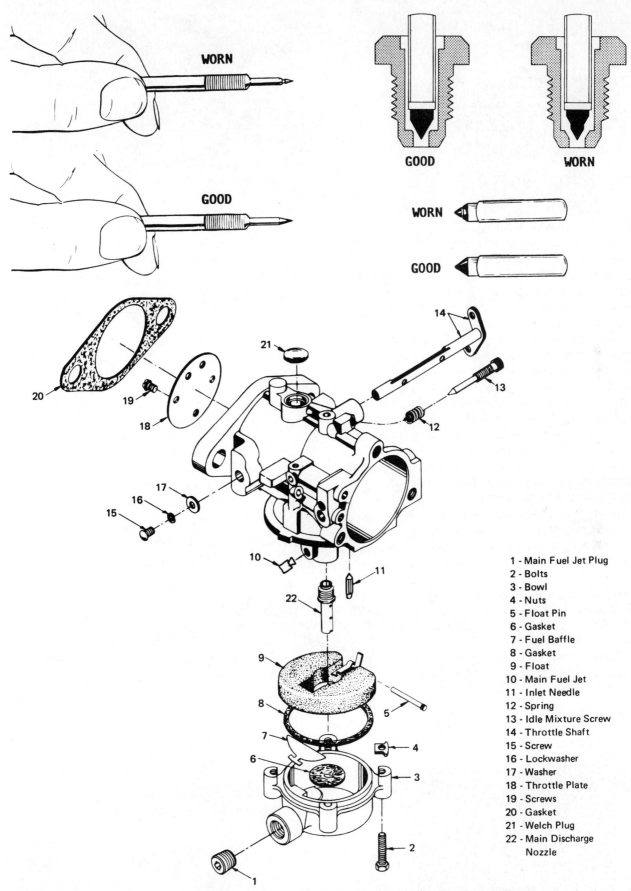

1 - Main Fuel Jet Plug
2 - Bolts
3 - Bowl
4 - Nuts
5 - Float Pin
6 - Gasket
7 - Fuel Baffle
8 - Gasket
9 - Float
10 - Main Fuel Jet
11 - Inlet Needle
12 - Spring
13 - Idle Mixture Screw
14 - Throttle Shaft
15 - Screw
16 - Lockwasher
17 - Washer
18 - Throttle Plate
19 - Screws
20 - Gasket
21 - Welch Plug
22 - Main Discharge
       Nozzle

Exploded view of a center round bowl carburetor. This carburetor is identified as a "C" carburetor in the text and appendix. Major parts are identified. The upper illustrations compare new carburetor needles with ones unfit for further service.

NOZZLE

②

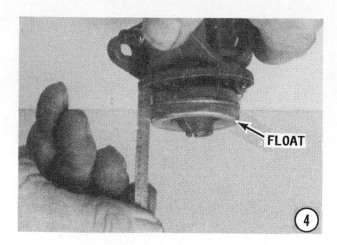

FLOAT

④

2- Install the nozzle. Install needle seat into the body.

3- Attach the inlet needle, with its spring, to the float. Lower the float into position and at the same time guide the inlet needle into its seat. Insert the hinge pin through the float tab.

4- Turn the bowl upside down and allow the float to drop. Measure the distance from the base of the bowl to the bottom of the float. This measurement should be 5/64" to 7/64" (1.98 to 2.78mm). **CARE-FULLY** bend the float tab to obtain the proper measurement.

5- Hold the carburetor upside down and place a **NEW** bowl gasket into the recess of the body. **OBSERVE** the flat area of the bowl. This area **MUST** be positioned over the hinge pin area. Mate the bowl to the carburetor body and secure it in place with the jet bolt retaining screw.

6- Slide the spring onto the idle mixture screw. **SLOWLY** thread the screw into the the carburetor body until you can just feel it bottom in the seat. **NEVER** use force when installing a jet in any carburetor. Now, as a preliminary adjustment, back the mixture screw out 1-1/2 turns from the seated position.

INSTALLATION
CARBURETOR "C"

7- Place a **NEW** gasket in position on the intake manifold. Install and secure the carburetor in place with the two attaching nuts.

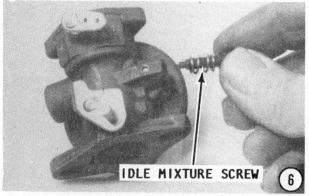

FLAT SPOT   BOWL

⑤

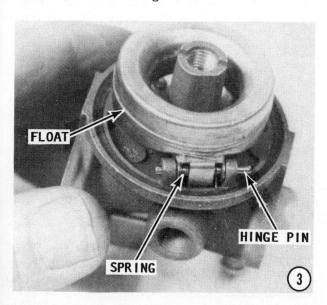

FLOAT

HINGE PIN

SPRING

③

IDLE MIXTURE SCREW

⑥

Synchronizing

To synchronize the fuel and ignition systems, see Chapter 6.

ADJUSTMENTS

FIRST A WORD: Before fine carburetor adjustments can be properly made, the following conditions must exist:

a. The regular engine-propeller combination must be used.

b. The power unit must be in forward gear.

c. The lower unit must be in the water.

d. The engine must be warmed to normal operating temperature.

High-Speed Adjustment

The high-speed jet is set at the factory and is **NOT** adjustable. However, to compensate for engine operation at various altitudes, refer to the Carburetor Jet Size-Elevation Chart in the Appendix. Adjust

8- Connect the fuel hose to the carburetor fuel line fitting. Attach the tie strap to the fuel line. Open the fuel line shutoff valve and allow fuel to reach the carburetor. Check the fuel connections for possible leaks. **DO NOT** install the remote idle lever onto the idle mixture screw until the idle mixture has been properly adjusted with the engine operating, as described in the paragraphs under **Adjustments.**

9- Connect the choke linkage to the carburetor.

10- Install the airbox using Loctite, or equivalent, on the attaching screw threads. Use a **NEW** gasket and install the airbox cover.

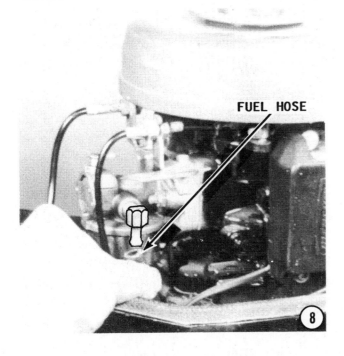

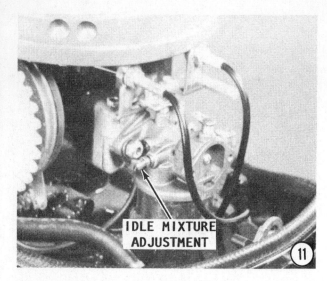

idle speed rpm according to the recommendations given in the Specifications in the Appendix.

Idle Mixture Adjustment

11- The engine will respond very slowly to changes in idle mixture adjustment when operating at a slow rpm. Therefore, set the idle mixture adjustment with the engine running at approximately 1000 rpm. At this speed, the response to adjustment will be more rapid. After the a, b, c, and d, conditions listed above have been met, including the engine run until it has reached operating temperature, then proceed with the idle mixture adjustment.

SLOWLY turn the idle mixture screw counterclockwise until the engine begins to load up or fires unevenly due to the over-rich mixture. **SLOWLY** turn the idle mixture screw clockwise until the highest engine rpm and smoothest operation is obtained.

With the engine still in forward gear, rotate the twist grip on the tiller handle and return the engine to the slowest idle, then determine if the engine runs smooth at slowest idle. If necessary, again adjust the idle mixture screw at slowest speed for smoothest performance. Allow at least 15 seconds for the engine to respond to adjustment.

SOME ADVICE

Do not adjust to a leaner position than necessary. It is better to have the mixture set slightly on the rich side, rather than too lean. Too lean a mixture is often the cause of hard starting.

MORE ADVICE

If the engine hesitates during acceleration, after adjusting the idle mixture, the mixture is set too lean and should be changed to the richer side (screw turned counterclockwise) until engine acceleration is smooth.

Idle Adjustment

12- After the conditions listed at the beginning of this ADJUSTMENT section have been met, and the idle mixture adjustment has been completed, and without disturbing the idle mixture setting, push the remote idle lever onto the idle mixture screw with the lever positioned at 10 o'clock. Push the lever onto the screw securely.

*Exterior view of a Mercury Gnat engine. The vent screw **MUST** be opened to allow air to enter the tank and prevent a vacumn condition in the system.*

4-9 MIKUNI RECTANGULAR BOWL SIDE-DRAFT CARBURETOR REFERENCED "D" IN APPENDIX

This section provides complete detailed procedures for removal, disassembly, cleaning and inspecting, assembling including bench adjustments, and installation for the rectangular bowl carburetor referenced "D" in the Tune-up Specifications (third to last column) in the Appendix. If a fuel pump is installed on the powerhead, it is a disposable type and cannot be serviced. For adjustments with the ignition system, see Chapter 6.

REMOVAL

1- Close the fuel valve between the fuel tank and the carburetor. Disconnect the fuel line at the carburetor. Remove the jamnut and cam follower at the carburetor. **DO NOT** allow the shaft to rotate. Remove the two carburetor flange mounting nuts. Lift the carburetor free of the powerhead.

DISASSEMBLING CARBURETOR "D"

2- Remove the idle mixture adjustment screw and spring.
3- Remove the idle speed adjustment screw and spring.
4- Remove the mixing chamber cover and throttle valve spring.
5- Remove the throttle valve assembly.
6- Disassemble the throttle valve assembly consisting of the throttle valve, jet needle with an E-ring on the fourth groove, retainer, and throttle valve shaft.

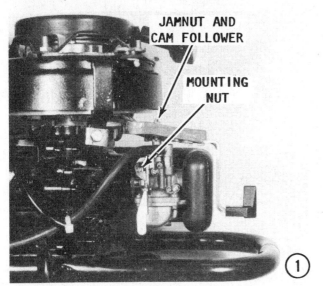
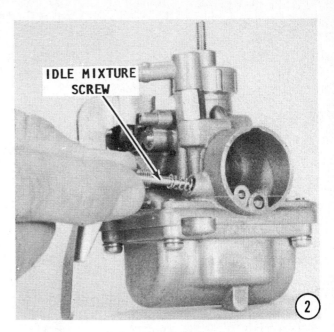
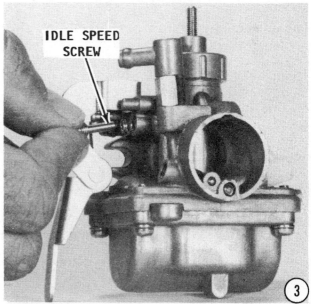
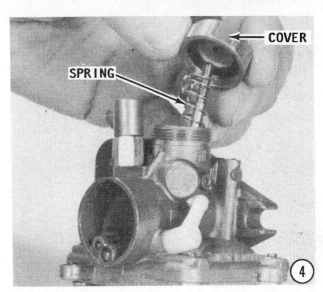

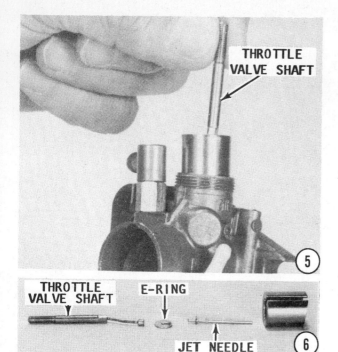

THROTTLE VALVE SHAFT

⑤

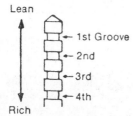

THROTTLE VALVE SHAFT · E-RING · JET NEEDLE

⑥

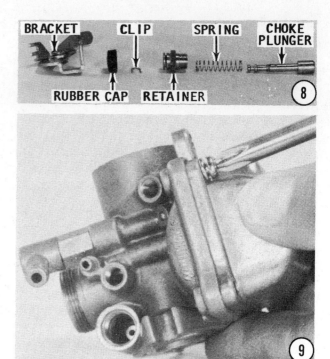

BRACKET · CLIP · SPRING · CHOKE PLUNGER

RUBBER CAP · RETAINER

⑧

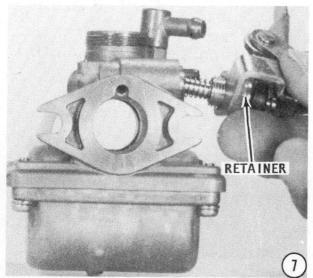

⑨

Lean

↑

1st Groove
2nd
3rd
4th

Rich

Jet needle E-ring grooves are provided to permit changes in air/fuel mixture. For leaner mixture, raise the ring; to enrich the mixture, lower the ring.

GOOD WORDS

The **E-ring** must be installed in the correct groove from which it is removed. If the ring is lowered, the carburetor will cause the engine to operate rich at mid-range. Raising the **E-ring** will cause the engine to operate too lean.

RETAINER

⑦

7- Rotate the choke assembly retainer **COUNTERCLOCKWISE**, and then remove the choke assembly.

8- Remove the U-shaped clip, and then slide the choke lever, rubber cap, retainer, and retainer spring free of the choke plunger.

9- Remove the four screws securing the float bowl to the carburetor body. Lift the bowl free.

10- Hold the float with one hand and remove the float pin with the other hand.

11- Use the proper size socket and remove the inlet needle seat nut. Shake the

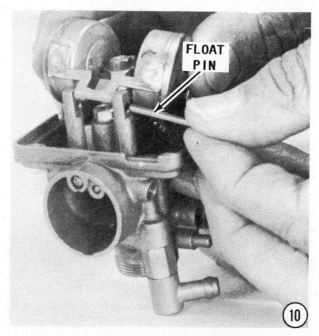

FLOAT PIN

⑩

1-   CARBURETOR BODY
2-   THROTTLE VALVE
3-   JET NEEDLE
4-   E-RING
5-   RETAINER
6-   POST
7-   SPRING
8-   GASKET
9-   MIXING CHAMBER COVER
10-   GASKET
11-   INLET NEEDLE & SEAT
12-   FLOAT PIN
13-   FLOAT
14-   PILOT JET (not removable – some models)

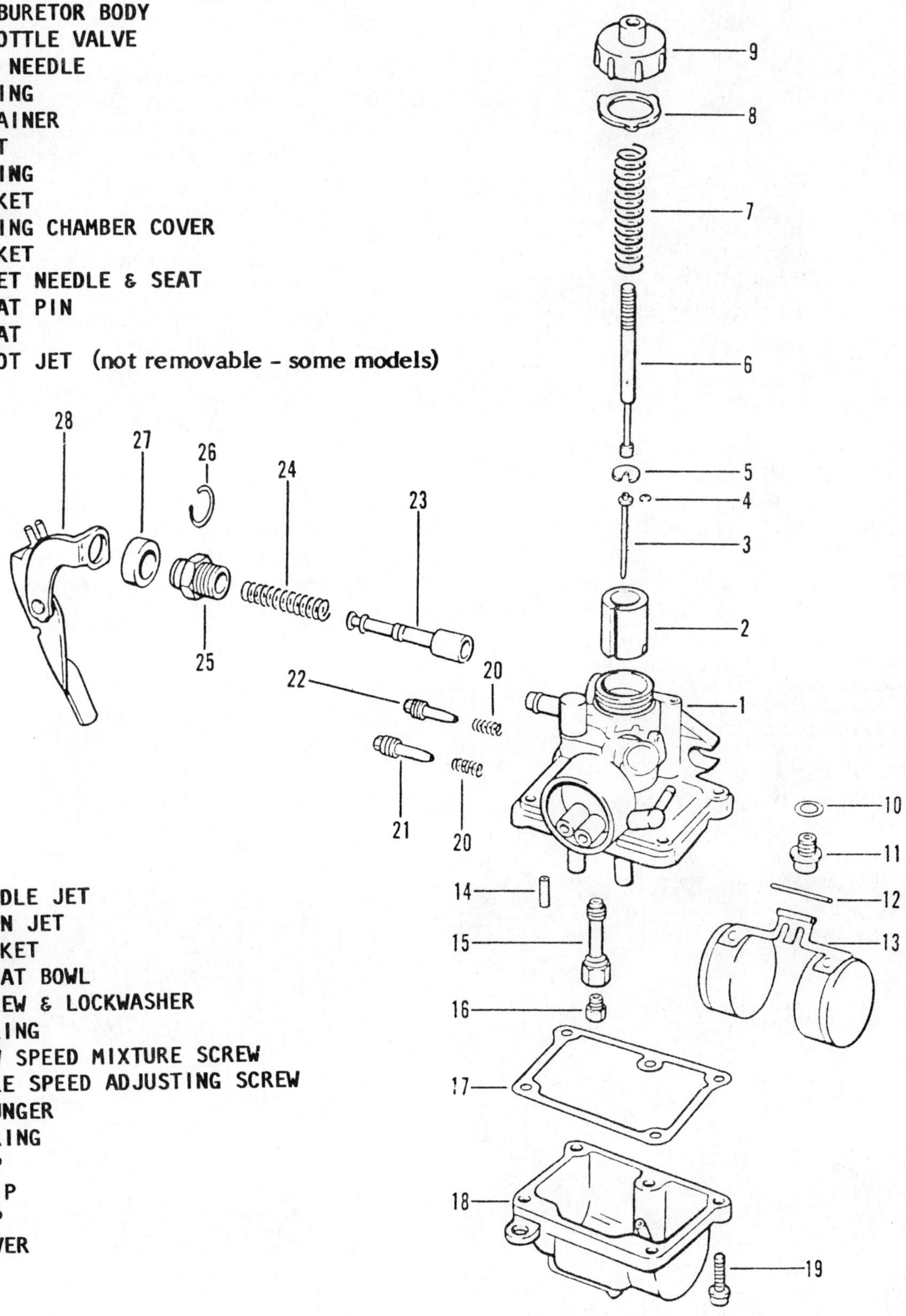

15-   NEEDLE JET
16-   MAIN JET
17-   GASKET
18-   FLOAT BOWL
19-   SCREW & LOCKWASHER
20-   SPRING
21-   LOW SPEED MIXTURE SCREW
22-   IDLE SPEED ADJUSTING SCREW
23-   PLUNGER
24-   SPRING
25-   CAP
26-   CLIP
27-   CAP
28-   LEVER

Exploded view of a Mikuni rectangular bowl side-draft carburetor. This carburetor is identified as a "D" carburetor in the text and appendix. Major parts are identified.

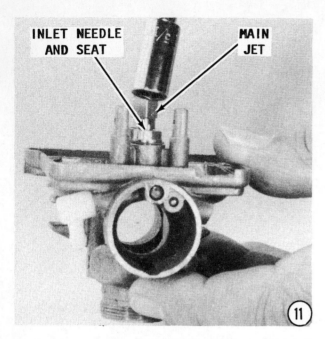

carburetor and the inlet needle will fall free. Remove the main jet.

CLEANING AND INSPECTING

NEVER dip rubber parts, plastic parts, diaphragms, or pump plungers in carburetor cleaner. These parts should be cleaned **ONLY** in solvent, and then blown dry with compressed air.

Place all the metal parts in a screen-type tray and dip them in carburetor cleaner until they appear completely clean -- free of gum and varnish which accumulates from stale fuel. Blow the parts dry with compressed air.

Blow out all passageways in the castings with compressed air. Check all of the parts and passages to be sure they are not clogged or contain any deposits. **NEVER** use a piece of wire or any type of pointed instrument to clean drilled passages or calibrated holes in a carburetor.

Carefully inspect the casting for cracks, stripped threads, or plugs for any sign of leakage. Inspect the float hinge in the hinge pin area for wear and the float for any sign of leakage.

Examine the inlet needle for wear and if there is any evidence of wear, the inlet needle **MUST** be replaced.

ALWAYS replace any and all worn parts.

A carburetor service kit is available at modest cost from the local marine dealer. The kit will contain all necessary parts to perform the usual carburetor overhaul work.

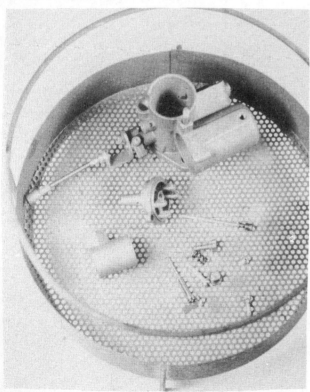

Carburetor parts in a basket ready to be submerged in carburetor cleaner.

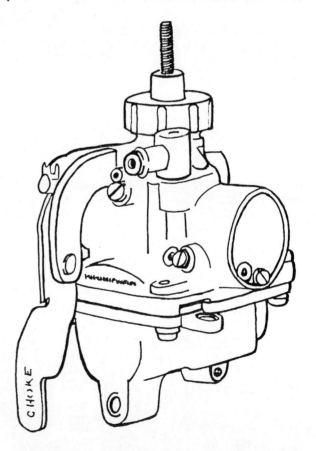

The exterior appearance of the rectangular float bowl carburetor may differ between two designs. Operation and procedures for service are identical.

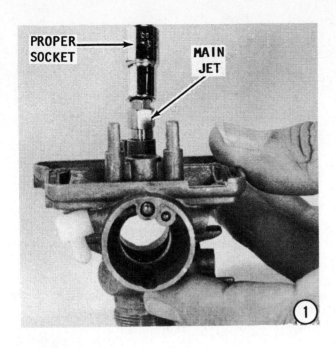

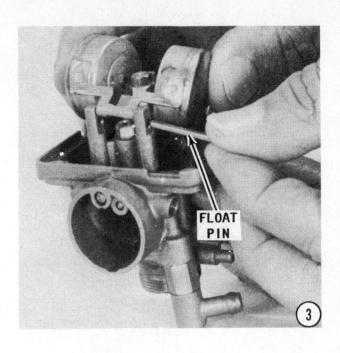

ASSEMBLING CARBURETOR "D"

1- Use the proper size socket and install the main jet. Thread the inlet needle seat into the carburetor body.

2- Slide the inlet needle into place in the seat.

3- Align the float in place and then slide the float pin through to secure the float.

Float Adjustment

4- Hold the carburetor on its side, as shown. Move the float until it just barely touches the top of the spring loaded inlet needle pin.

SPECIAL WORDS

The weight of the float can compress the inlet needle spring pin by its own weight. Therefore, make the adjustment when the float and inlet needle **BARELY** make contact. Measure the distance between the carburetor body surface and the bottom of the float, as shown. **CAREFULLY** adjust the tab until this measurement is 7/8" (22.23mm).

5- Place a **NEW** bowl gasket in position on the carburetor. Mate the bowl to the surface of the carburetor body, and then secure the bowl with the four attaching screws.

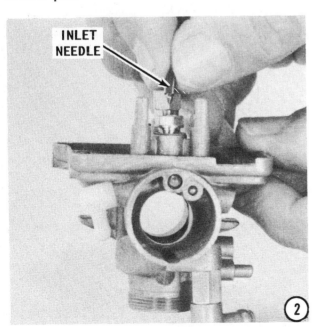

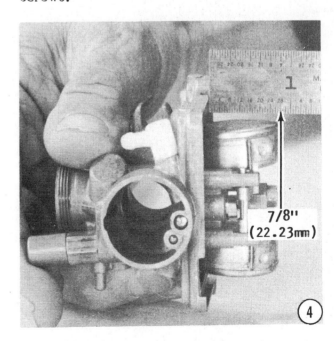

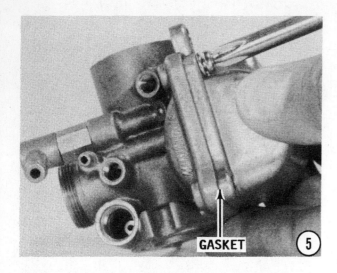

GASKET ⑤

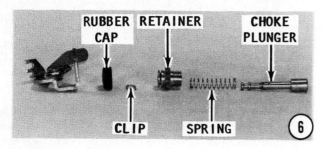

RUBBER CAP   RETAINER   CHOKE PLUNGER

CLIP   SPRING   ⑥

6- Slide the spring, retainer, rubber cap, and choke lever onto the choke plunger. Secure the parts together with the **U**-shaped clip.

7- Slide the choke assembly shaft into the carburetor body. Compress the spring with pressure on the choke bracket, and then thread the retainer into the carburetor body.

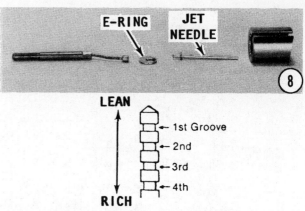

E-RING   JET NEEDLE   ⑧

LEAN

1st Groove
2nd
3rd
4th

RICH

Jet needle E-ring grooves are provided to permit changes in air/fuel mixture. For leaner mixture, raise the ring; to enrich the mixture, lower the ring.

8- Assemble the throttle valve parts in the order shown. **BE SURE** the E-ring is installed into the same groove on the jet needle from which it was removed.

GOOD WORDS

The **E**-ring must be installed in the correct groove from which it was removed. If the ring is lowered, the carburetor will cause the engine to operate rich at mid-range. Raising the **E**-ring will cause the engine to operate too lean. Install the assembled throttle valve into the carburetor mixing chamber.

9- Slide the assembled throttle valve unit into the carburetor body.

10- Slide the spring onto the shaft, and then thread the mixing chamber cover into place with the shaft extended through the hole in the cap.

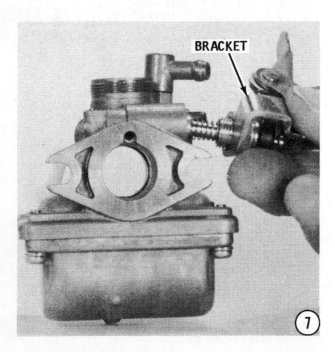

BRACKET   ⑦

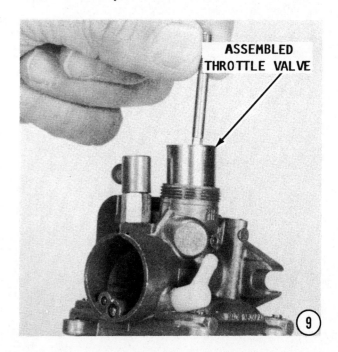

ASSEMBLED THROTTLE VALVE   ⑨

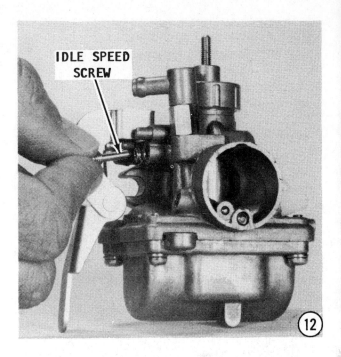

11- Slide the spring onto the idle mixture adjusting screw shaft, and then thread the screw into the carburetor body. Tighten the screw until it **BARELY** seats, and then back it out two (2) turns as a preliminary rough adjustment.

12- Slide the spring onto the idle speed adjusting screw shaft, and then thread the screw into the carburetor body. Tighten the screw until it **BARELY** seats, and then tighten two (2) more complete turns as a preliminary rough adjustment.

Timing and Synchronizing
See appropriate section in Chapter 6.

INSTALLATION CARBURETOR "D"

13- Place a **NEW** gasket in position on the powerhead. Install the carburetor and secure it in place with the two mounting nuts. Install the cam follower to the throttle lever. Connect the fuel line to the carburetor. See appropriate section in Chapter 6 for timing and synchronizing.

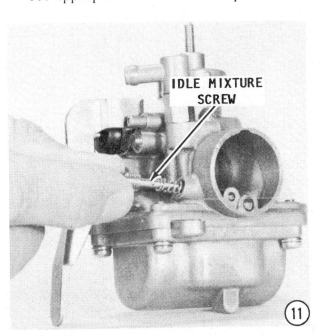

4-10 MIKUNI ROUND BOWL SIDE DRAFT CARBURETOR REFERENCED "E" IN APPENDIX

This section provides complete detailed procedures for removal, disassembly, cleaning and inspecting, assembling including bench adjustments, and installation for the Mikuni round bowl side draft carburetor, referenced "E" in the Tune-up Specifications (third to last column) in the Appendix. If a fuel pump is installed on the powerhead, it is a disposable type and cannot be serviced. For adjustments with the ignition system, see Chapter 6.

REMOVAL AND DISASSEMBLING

1- Remove the screws securing the knobs at the end of the choke and throttle levers. Remove the two screws securing the rectangular intake cover to the carburetor body. Lift the cover free.

SPECIAL WORDS

It is not necessary to disconnect the wires attached to the cover. Simply move the cover to one side out of the way.

2- Close the fuel valve between the fuel tank and the carburetor. Squeeze the wire-type hose clamp on the fuel line enough for the hose to slip free of the inlet fuel fitting.

3- Loosen the clamp screw and remove the carburetor from the powerhead.

4- Work the carburetor-to-powerhead seal out of the recess in the carburetor throat. Discard the seal.

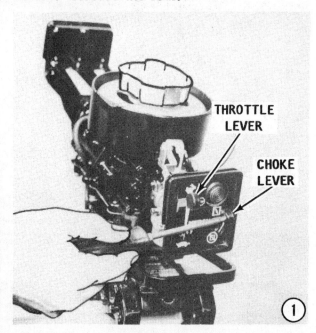

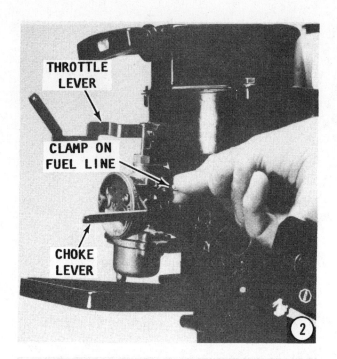

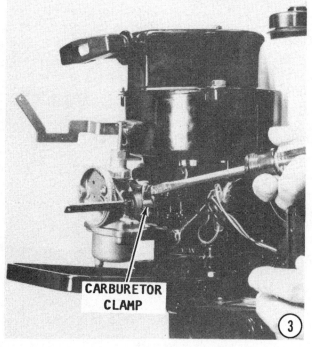

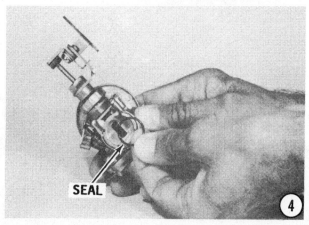

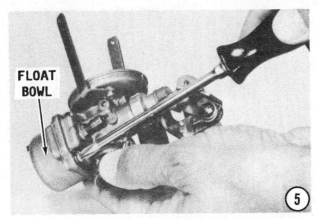

FLOAT BOWL

5

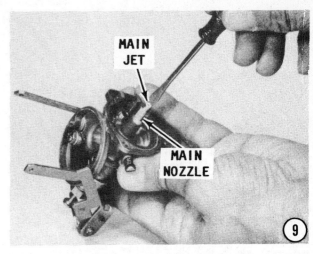

MAIN JET

MAIN NOZZLE

9

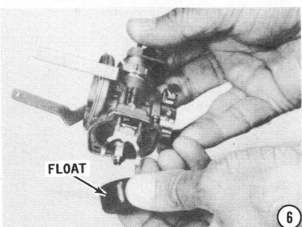

FLOAT

6

5- Back out the two Phillips head screws securing the float bowl to the carburetor.

6- Remove the float from the carburetor.

7- Slide the hinge pin free, and then remove the float hinge and the inlet needle.

8- Lift off and discard the float bowl gasket.

9- Unscrew the main jet from the center of the main nozzle.

10- Use the proper size wrench and remove the main nozzle.

11- Rotate the idle speed screw clockwise and **COUNT** the number of turns

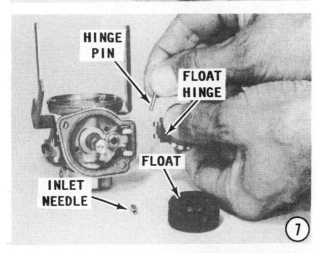

HINGE PIN

FLOAT HINGE

FLOAT

INLET NEEDLE

7

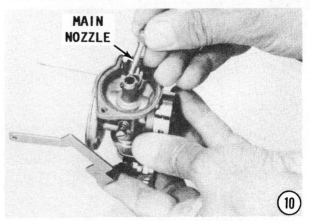

MAIN NOZZLE

10

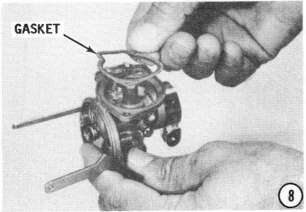

GASKET

8

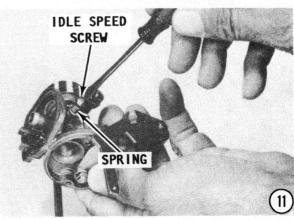

IDLE SPEED SCREW

SPRING

11

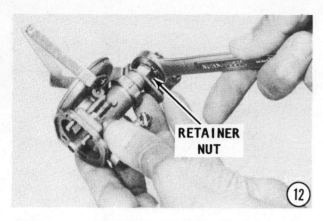

necessary to seat it **LIGHTLY**. After the number of turns has been noted and recorded somewhere, back out the idle speed screw. Slide the spring free of the screw.

12- Loosen the retainer nut with the proper size wrench. Rotate the nut several turns **COUNTERCLOCKWISE** but **DO NOT** remove the nut.

13- Unscrew the mixing chamber cover with a pair of pliers. Once the cover is free, lift off the throttle valve assembly. Disconnect the throttle lever from the bracket.

14- Compress the spring in the throttle valve assembly to allow the throttle cable end to clear the recess in the base of the throttle valve and to slide down the slot.

15- Disassemble the throttle valve consisting of the throttle valve, spring, jet needle (with an E-clip on the second groove), jet retainer, and throttle cable end.

GOOD WORDS

It is not necessary to remove the E-clip from the jet needle, unless replacement is required or if the powerhead is to be operated at a significantly different elevation.

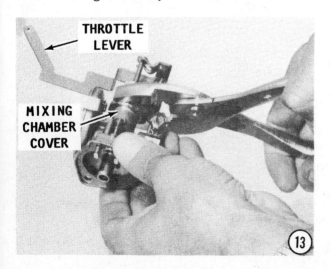

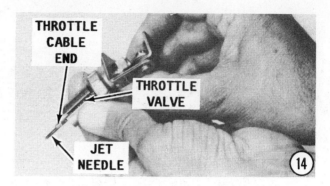

CLEANING AND INSPECTING

NEVER dip rubber parts, plastic parts, diaphragms, or pump plungers in carburetor cleaner, because they tend to absorb liquid and expand. These parts should be cleaned **ONLY** in solvent, and then blown dry with compressed air immediately.

Place all of the metal parts in a screen-type tray and dip them in carburetor cleaner until they appear completely clean -- free of gum and varnish which accumulates from stale fuel. Blow the parts dry with compressed air.

Blow out all of the passageways in the carburetor with compressed air. Check all of the parts and passageways to be sure they are clear and not clogged with any deposits.

NEVER use a piece of wire or any type of pointed instrument to clean drilled passages or calibrated holes in a carburetor.

Carefully inpsect the casting for cracks, stripped threads, or plugs for any sign of leakage. Inspect the float hinge in the hinge pin area for wear and the float for any sign of leakage.

Examine the inlet needle for wear and if there is any evidence of wear, the needle **MUST** be replaced.

Always replace any worn or damaged parts.

A carburetor service kit is available at modest cost from the local marine dealer. This kit contains all necessary parts to perform the usual carburetor overhaul work.

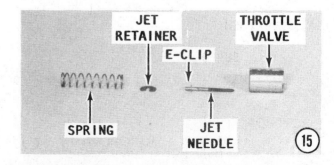

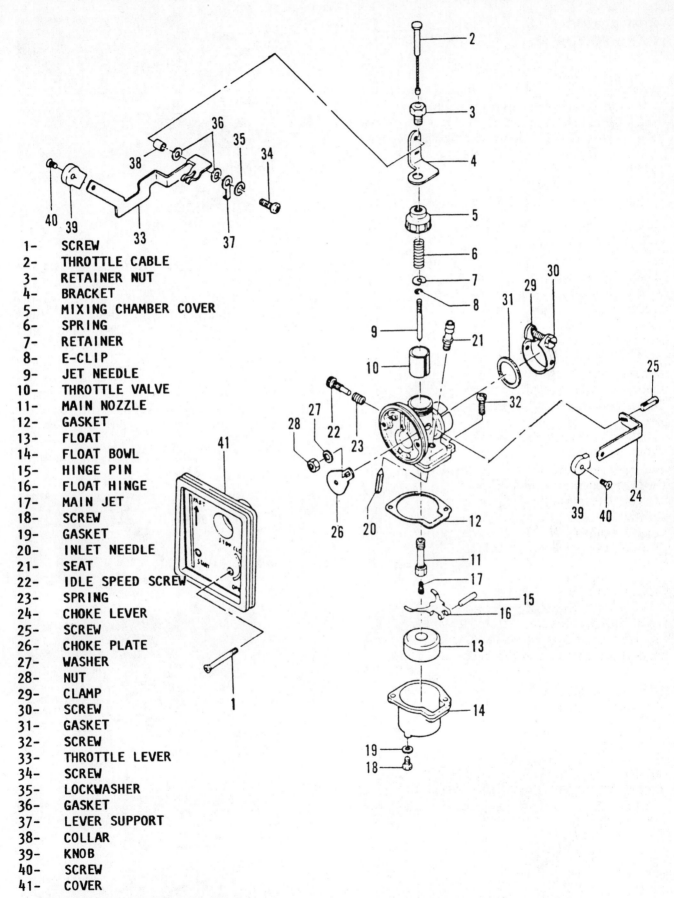

1- SCREW
2- THROTTLE CABLE
3- RETAINER NUT
4- BRACKET
5- MIXING CHAMBER COVER
6- SPRING
7- RETAINER
8- E-CLIP
9- JET NEEDLE
10- THROTTLE VALVE
11- MAIN NOZZLE
12- GASKET
13- FLOAT
14- FLOAT BOWL
15- HINGE PIN
16- FLOAT HINGE
17- MAIN JET
18- SCREW
19- GASKET
20- INLET NEEDLE
21- SEAT
22- IDLE SPEED SCREW
23- SPRING
24- CHOKE LEVER
25- SCREW
26- CHOKE PLATE
27- WASHER
28- NUT
29- CLAMP
30- SCREW
31- GASKET
32- SCREW
33- THROTTLE LEVER
34- SCREW
35- LOCKWASHER
36- GASKET
37- LEVER SUPPORT
38- COLLAR
39- KNOB
40- SCREW
41- COVER

Exploded view of a Mikuni round bowl side-draft carburetor. This carburetor is identified as an "E" carburetor in the text and appendix. Major parts are identified.

ASSEMBLING CARBURETOR "E"

SPECIAL WORDS

The E-clip **MUST** be installed into the same groove from which it was removed. If the clip is lowered, the carburetor will allow the powerhead to operate "rich". Raising the E-clip will cause the powerhead to operate "lean".

1- Assemble the throttle valve components in the following order: Insert the E-clip end of the jet needle into the throttle valve (the end with the recess for the throttle cable end). Place the needle retainer into the throttle valve over the E-clip and align the retainer slot with the slot in the throttle valve.

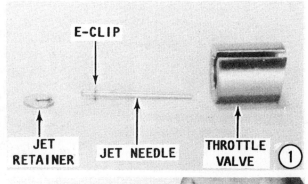

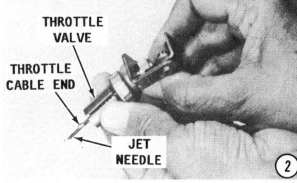

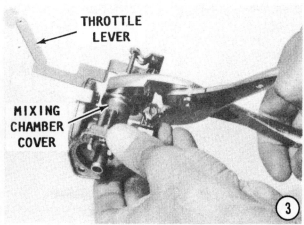

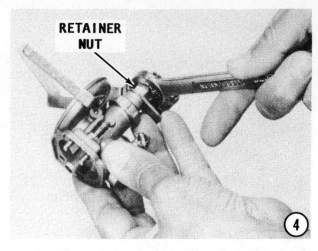

2- Thread the spring over the end of the throttle cable and insert the cable into the retainer end of the throttle valve. Compress the spring and at the same time guide the cable end through the slot until the end locks into place in the recess. Position the assembled throttle valve in such a manner to permit the slot to slide over the alignment pin while the throttle valve is lowered into the carburetor. Attach the throttle lever to the bracket on top of the throttle valve assembly.

3- Carefully tighten the mixing chamber cover with a pair of pliers.

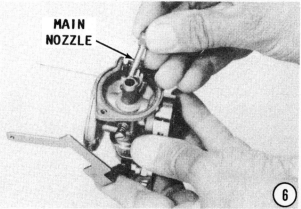

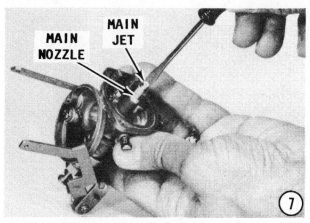

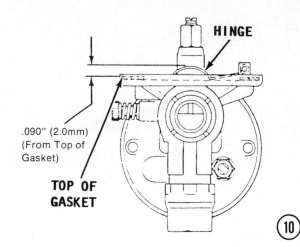

4- Position the bracket to allow the throttle to just clear the front of the carburetor, and then tighten the retainer nut.

5- Slide the spring onto the idle speed screw, and then start to thread the screw into the carburetor body. **SLOWLY** rotate the screw until it seats **LIGHTLY**. From this position, back out the screw the same number of complete turns as noted during removal (Step 11). If the count was lost, back the screw out two turns as a preliminary adjustment. A fine idle adjustment will be made later in Step 17.

6- Thread the main nozzle into the carburetor body and tighten it just "snug" with the proper size wrench. **DO NOT** overtighten the nozzle.

7- Install the main jet into the main nozzle and tighten it just "snug" with a screwdriver.

8- Position a new float bowl gasket in place, and then install the inlet needle.

9- Hold the carburetor body in a perfect upright position on a firm surface. Set the hinge in position, and then slide the hinge pin into place through the hinge.

10- Check the float hinge adjustment. The vertical distance between the top of the hinge and the top of the gasket should be .090" (2.0mm). **CAREFULLY** bend the hinge, as necessary, to achieve the required measurement.

11- Install the float.

12- Place the float bowl in position on the carburetor body, and then secure it with the two Phillips head screws.

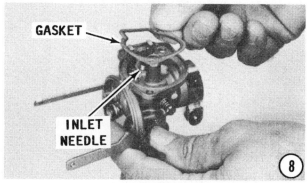

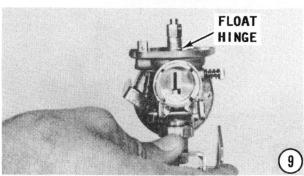

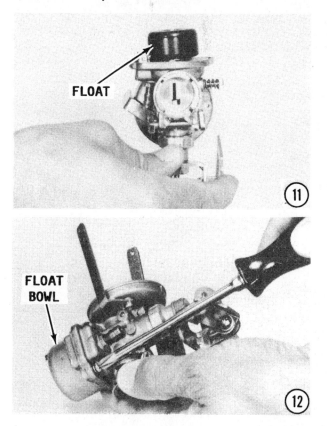

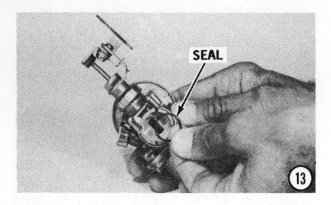

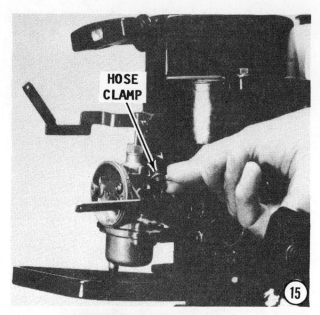

INSTALLATION
CARBURETOR "E"

13- Position a new carburetor-to-power-head seal in the recess of the carburetor throat.

14- Slide the carburetor onto the crank-case cover and secure it in place with the clamp screw.

15- Connect the fuel line to the fuel inlet fitting. Open the fuel valve between the fuel tank and the carburetor.

16- Install the front cover and the knobs onto the choke and throttle levers.

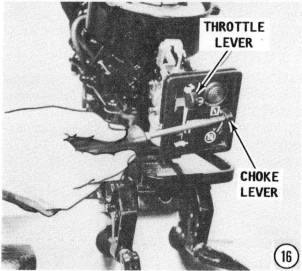

Idle Speed Adjustment

17- Connect a tachometer to the power-head. Start the engine and allow it to warm to operating temperature. Adjust the throttle lever to the lowest speed, and then adjust the idle speed screw until the engine idles at 900-1000 rpm.

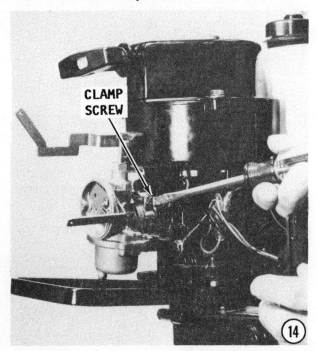

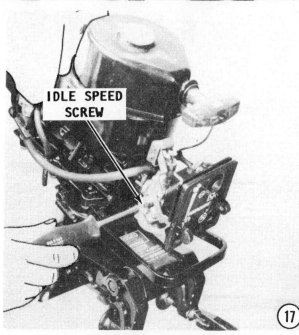

4-11 TILLOTSON "BC" CARBURETOR RECTANGULAR BOWL — DOUBLE FLOAT WITH INTEGRAL FUEL PUMP REFERENCED "F" IN APPENDIX

This section provides complete detailed procedures for removal, disassembly, cleaning and inspecting, assembling including bench adjustments, and installation for the Tillotson "BC", rectangular bowl, double float carburetor with integral fuel pump. This carburetor is referenced "F" in the Tune-up Specifications (third to last column) in the Appendix. For adjustments with the ignition system, see Chapter 6.

Positive identification of the carburetor can be made by the embossed letters on the mounting flange, as shown in the accompanying illustration.

If troubleshooting indicates the fuel pump to be the cause of problems, the pump may be serviced without removing the carburetor from the powerhead.

REMOVAL AND DISASSEMBLING

1- Snip the Sta-strap on the fuel line to the carburetor. Disconnect the fuel line. Loosen, but **DO NOT** remove the choke securing nut holding the choke rod to the choke arm on top of the carburetor. Move the rod to one side. Remove the two nuts securing the carburetor to the adaptor. Lift the carburetor free of the powerhead.

2- Remove the center screw securing the cover to the carburetor. Remove the cover, then lift out the strainer and gasket. Discard the gasket. (A new one will be provided in a carburetor repair kit.)

3- Slowly tighten the low speed mixture

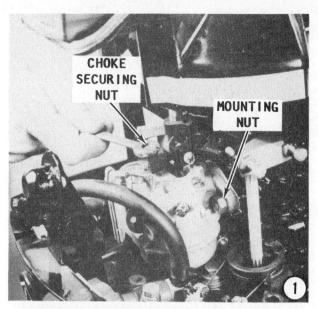

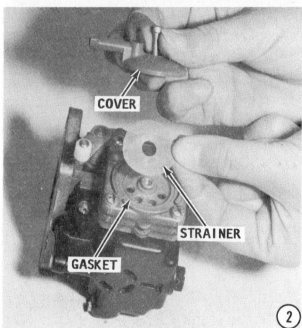

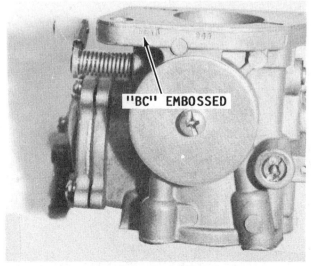

Exterior view of the Type "F" carburetor.

MAIN JET

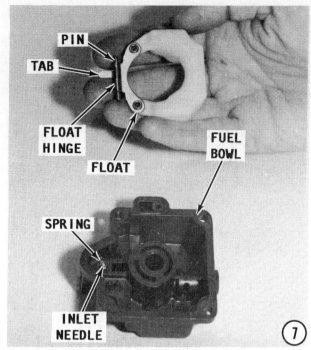

PIN

TAB

FLOAT HINGE

FLOAT

FUEL BOWL

SPRING

INLET NEEDLE

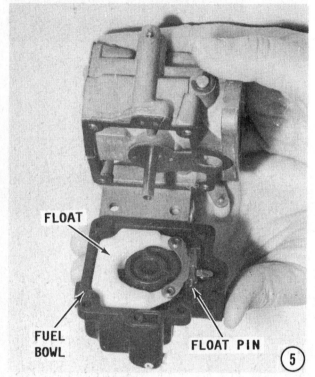

FLOAT

FUEL BOWL

FLOAT PIN

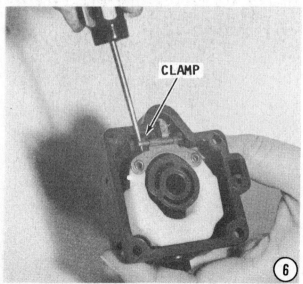

CLAMP

screw and **COUNT** the number of turns necessary to seat the screw **LIGHTLY**. This count will be most helpful during assembling. Back out the low speed mixture screw. **TAKE CARE** to not lose the tiny spring on the shaft of the screw.   If this spring remains in the carburetor body, remove and save the spring.

4- Unscrew the main jet from the fuel bowl. Discard the gasket under the screw head.

5- Remove the four bolts securing the fuel bowl to the carburetor body. Separate the bowl from the body. Discard the gasket.

6- Pry out the float pin from the built-in clamps in the fuel bowl.

7- Lift the fuel float, the hinge, and the hinge pin clear of the float bowl. Release the tiny wire spring from the tab on the float hinge.

8- Remove the spring and inlet needle from the recess in the fuel bowl.

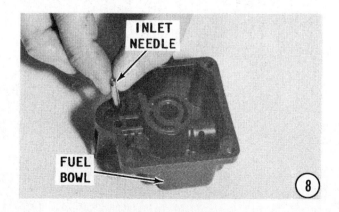

INLET NEEDLE

FUEL BOWL

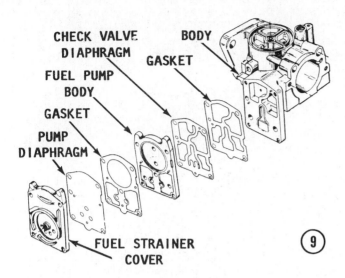

CHECK VALVE
DIAPHRAGM
BODY
FUEL PUMP
BODY
GASKET
GASKET
PUMP
DIAPHRAGM
FUEL STRAINER
COVER

⑨

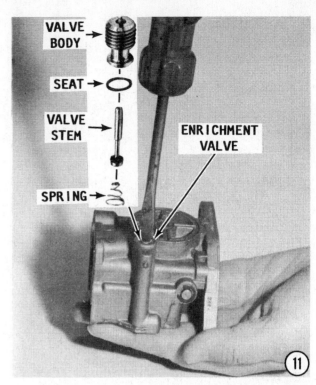

VALVE
BODY
SEAT
VALVE
STEM
ENRICHMENT
VALVE
SPRING

⑪

DISASSEMBLING FUEL PUMP

The manufacturer recommends the pump be serviced anytime the carburetor is overhauled. Actually, if troubleshooting the fuel system indicates the fuel pump to be at fault, the pump may be serviced without removing the carburetor from the powerhead.

9- Remove the four screws securing the fuel strainer cover to the carburetor body. Remove the cover and then the following parts in order, as shown in the accompanying illustration: pump diaphragm, gasket, fuel pump body, check valve diaphragm, and another gasket. Notice how the outlet and inlet check valves are a part of the check valve diaphragm.

10- Back out the center screw, and then lift off the mixing chamber cover. Remove and discard the gasket from the recess in the carburetor body.

11- Remove the enrichment valve. Exercise **CARE** to not lose any of the small associated parts. After the valve body is removed, the valve body seat, which resembles a washer or gasket, and valve stem may be lifted out. Remove the spring and notice how the small end is facing **UP**.

CLEANING AND INSPECTING

NEVER dip rubber parts, plastic parts, diaphragms, or pump plungers in carburetor cleaner. These parts should be cleaned **ONLY** in solvent, and then blown dry with compressed air.

Place all the metal parts in a screen-type tray and dip them in carburetor cleaner until they appear completely clean -- free

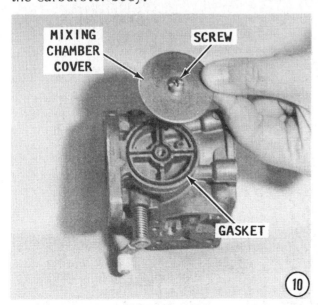

MIXING
CHAMBER
COVER
SCREW
GASKET

⑩

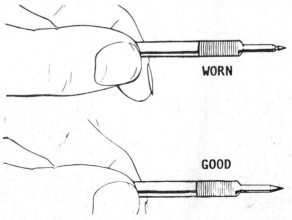

WORN

GOOD

Carburetor idle adjustment screws. The top screw is worn and unfit for service. The bottom screw is new.

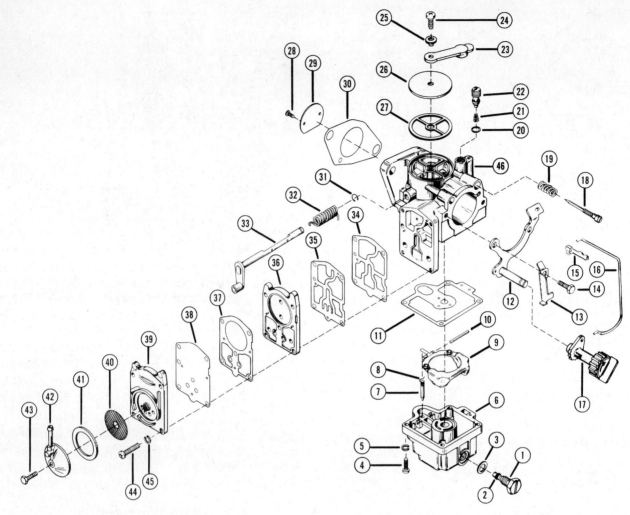

1 - Main (High Speed) Jet	13 - Spring - Choke Detent	25 - Shoulder Washer
2 - "O" Ring (Seal)	14 - Bolt (2 Req'd)	26 - Mixing Chamger Cover
3 - Gasket	15 - Clip	27 - Gasket
4 - Screw (4 Required)	16 - Rod, choke	28 - Screw with Lockwasher
5 - Lockwasher (4 Required)	17 - Knob, choke	29 - Throttle Shutter Plate
6 - Fuel Bowl	18 - Low Speed Mixture Screw	30 - Flange Gasket (Carb. Mounting)
7 - Fuel Inlet Needle	19 - Spring	31 - Retaining Clip
8 - Clip	20 - Seat (Enrichment Valve)	32 - Spring
9 - Float	21 - Spring	33 - Throttle Shaft
10 - Float Pin	22 - Enrichment Valve	34 - Gasket
11 - Gasket (Float Bowl)	23 - Rubber Seal	35 - Check Valve Diaphragm
12 - Plate	24 - Screw	36 - Fuel Pump Body

37 - Gasket
38 - Pump Diaphragm
39 - Fuel Strainer Body
40 - Fuel Strainer
41 - Gasket
42 - Fuel Strainer Cover
43 - Screw
44 - Screw (4 Req'd)
45 - Lockwasher (4 Req'd)
46 - Carburetor Body

Exploded view of a Tillotson "BC" rectangular bowl -- double float carburetor with integral fuel pump. This carburetor is identified as an "F" carburetor in the text and appendix. Major parts are identified.

of gum and varnish which accumulates from stale fuel. Blow the parts dry with compressed air.

Blow out all passageways in the castings with compressed air. Check all of the parts and passages to be sure they are not clogged or contain any deposits. **NEVER** use a piece of wire or any type of pointed instrument to clean drilled passages or calibrated holes in a carburetor.

Make a thorough inspection of the fuel pump diaphragm for the tinest pin hole. If

one is discovered, the hole will only get bigger. Therefore, the diaphragm must be replaced in order to obtain full performance from the powerhead.

Carefully inspect the casting for cracks, stripped threads, or plugs for any sign of leakage. Inspect the float hinge in the hinge pin area for wear and the float for any sign of leakage.

Examine the inlet needle for wear and if there is any evidence of wear, the inlet needle **MUST** be replaced.

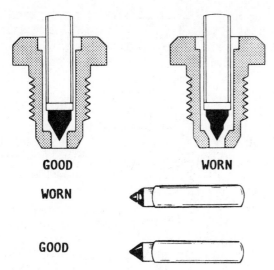

Line drawing to compare a worn and new needle and seat arrangement. The worn needle would have to be replaced for full carburetor efficiency.

ALWAYS replace any and all worn parts.

A carburetor service kit is available at modest cost from the local marine dealer. The kit will contain all necessary parts to perform the usual carburetor overhaul work.

ASSEMBLING CARBURETOR "F"

1- Slide the enrichment valve spring into the carburetor body with the **LARGE** end going in first. Insert the valve stem with the "plunger" end going in first. The spring will allow the stem to move up and down.

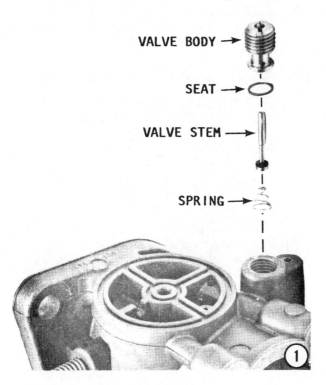

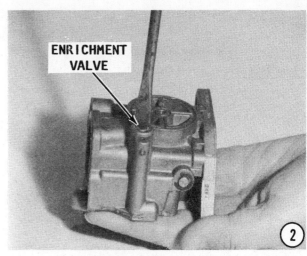

Set the valve seat in place and then thread the enrichment body into place.

2- Tighten the valve body just **"SNUG"**. The body is made of brass and can be easily damaged by overtightening.

3- Place a **NEW** gasket in position on the mixing chamber. The gasket will fit properly in only way way. Install the mixing chamber cover and secure it in place with the center screw.

4- Slide the inlet needle into its recess in the fuel bowl. Place the small spring on top of the needle.

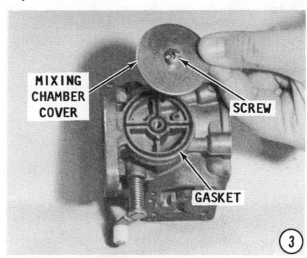

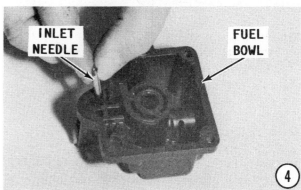

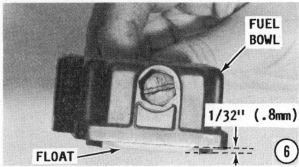

5- Hook the spring over the tab on the float hinge. Lower the fuel float into the bowl and snap the pin into place onto the built-in clamps of the bowl. Place a **NEW** gasket over the main jet, and then thread the jet into the carburetor body. Tighten the jet just **"SNUG"**.

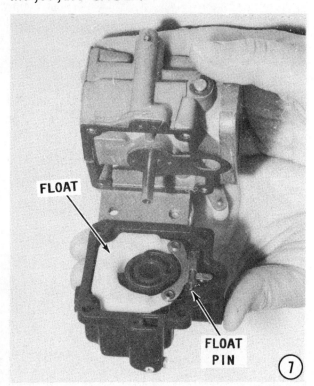

6- Invert the fuel bowl allowing the fuel float to hang free. Measure the distance between the base of the fuel bowl and the "bottom" surface of the float -- "bottom" as the float hangs, as shown. This measurement should be as close to 1/32" (.8mm) as possible. To make an adjustment, remove the float and bend the float tab ever so **SLIGHTLY** to increase or decrease the measurement. Just a "whisker" of bend on the tab will move the float and change the measurement.

7- Position a **NEW** gasket onto the float bowl mating surface, and then bring the bowl and carburetor body together. Secure the bowl in place with the four attaching screws.

8- Slide the tiny spring onto the low speed mixture screw and then thread the screw into the carburetor body. Seat the screw **LIGHTLY**, and then back it out the same number of turns counted in Step 3 of disassembling. If the count was lost, back the screw out from the lightly seated position 1-1/4 turns as a preliminary bench adjustment.

ASSEMBLING FUEL PUMP

9- Assemble the various part of the fuel pump to the carburetor body in the order shown in the accompanying illustration. Refer to the illustration to ensure the proper gasket is installed in the correct location or the pump will not function properly, perhaps not at all. First, the gasket, then the check valve diaphragm, the fuel pump body, another gasket, the pump diaphragm, and finally the fuel strainer cover. Secure the cover in place with the four screws and lockwashers.

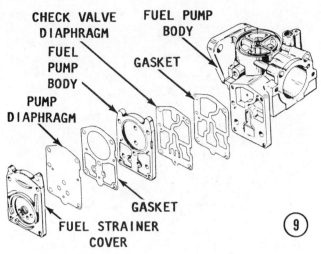

CHECK VALVE DIAPHRAGM
FUEL PUMP BODY
FUEL PUMP BODY
GASKET
PUMP DIAPHRAGM
GASKET
FUEL STRAINER COVER

9

10- After the cover is secured, place the fuel strainer into its recess, place a **NEW** gasket in place, and then install the fuel strainer cover. Secure the cover with the center screw.

11- Position a **NEW** gasket in place over the two adaptor studs. Mount the carburetor to the adaptor and secure it in place with the two nuts. Tighten the nuts securely. Connect the choke rod to the choke arm. Slide the choke rod under the retainer. Tighten the choke securing nut. Connect the fuel line to the carburetor using a **NEW** Sta-strap.

12- Position the outlet on the fuel strainer cover to ensure a minimum of 1/4" (7mm) clearance between the throttle cam, when the cam is at the throttle wide open position, and the fuel line. If the throttle cam contacts the fuel line, the throttle

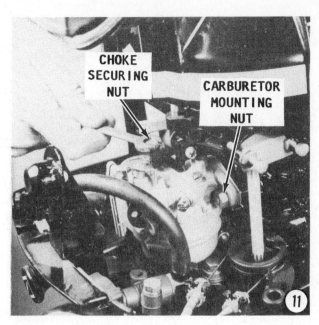

CHOKE SECURING NUT
CARBURETOR MOUNTING NUT

11

could remain open when an attempt is made to reduce power. This could result in a very **HAZARDOUS** condition.

ADJUSTMENTS

FIRST A WORD: Before fine carburetor adjustments can be properly made, the following conditions must exist:

a. The regular engine-propeller combination must be used.

b. The power unit must be in forward gear.

c. The lower unit must be in the water.

d. The engine must be warmed to normal operating temperature.

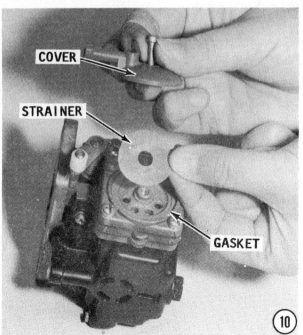

COVER
STRAINER
GASKET

10

CAM
FUEL STRAINER COVER

12

The high speed jet on the "F" carburetor is easily accessible. This location permits quick changes for operation at different altitudes.

Idle Speed Adjustment

13- After the engine has been warmed to operating temperature, turn the idle speed adjusting screw on the stop bracket until the engine idles at approximately 700-800 rpm in forward gear in a test tank or 600-700 on a boat.

Idle Mixture Adjustment

14- Turn the adjusting screw counter clockwise until the engine fires unevenly from the mixture being too rich. Now, begin to turn the screw clockwise and rpm will begin to increase. Continue turning the adjusting screw until the mixture is so lean that the rpm begins to drop and the engine begins to misfire. Set the adjusting screw halfway between the rich and lean points.

ADVICE

It is better to have the mixture set slightly on the rich side, rather than too lean.

High-Speed Adjustment

The main metering high speed jet is not adjustable. If the engine is to be operated at elevations above 4000 ft. (1219m), replace the main metering jet as indicated in the Jet Size/Elevation Chart in the Appendix.

4-12 WALBRO WM CARBURETOR ROUND BOWL — DOUBLE FLOAT WITH INTEGRAL FUEL PUMP REFERENCED "G" IN THE APPENDIX

INTRODUCTION

This section provides complete detailed procedures for removal, disassembly, cleaning and inspecting, assembling including bench adjustments, and installation for the Walbro WM series carburetor -- round bowl -- double float with an integral fuel pump. This carburetor is referenced "G" in the Tune-up Specifications (third to last column) in the Appendix.

Positive identification of the carburetor can be made by the embossed letters on the mounting flange, as shown in the accompanying illustration.

If troubleshooting indicates the fuel pump to be the cause of problems, the pump may be serviced without removing the carburetor from the powerhead.

For adjustments with the ignition system, see Chapter 6.

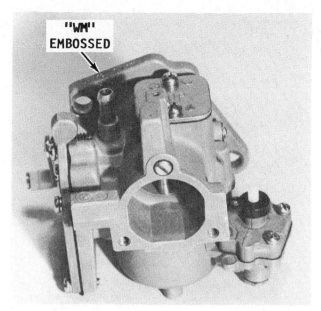

*Exterior view of the Walbro carburetor covered in this section and identified as **"G"** in the text and appendix. The letters "WM" are embossed on the mounting flange.*

REMOVAL AND DISASSEMBLING

1- Using a pair of pliers, remove the Circlip securing the choke knob in place.

2- Remove the choke knob by pulling it straight out. Unsnap the idle wire from the primer bracket. Remove the two bolts securing the primer bracket to the carburetor. The primer bracket will come free with the bracket. Using a pair of snips, cut the Sta-strap from the fuel hose to the carburetor. Pull the fuel line free of the fitting. Remove the two carburetor mounting bolts and lift the carburetor off the powerhead.

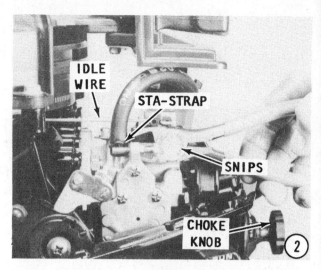

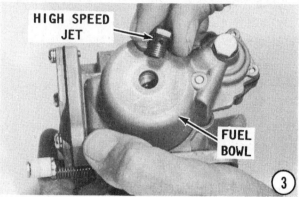

3- Remove the high speed jet from the bottom of the carburetor. The jet also serves as a retaining bolt for the fuel bowl.

4- Lift the fuel bowl off the carburetor. The primer system is attached to the bowl and therefore, will come with it.

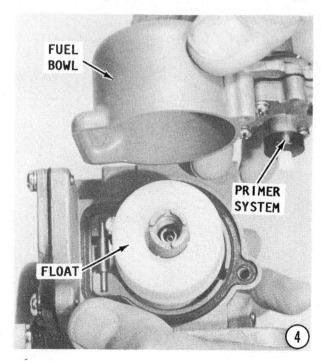

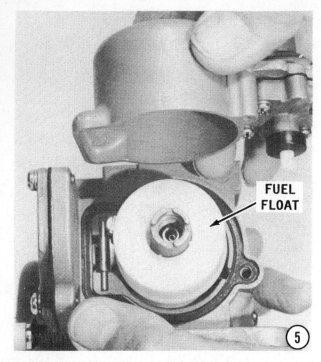

FUEL FLOAT

(5)

5- Lift out the fuel float.

6- Back out the screw securing the hinge pin in place. Pull the hinge pin, and then lift the hinge out of the carburetor body.

7- Remove the inlet needle from its position under the hinge. Remove the carburetor bowl gasket from its recess.

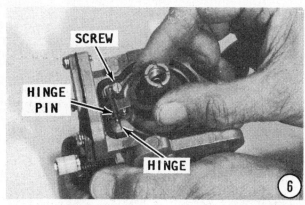

SCREW

HINGE PIN

HINGE

(6)

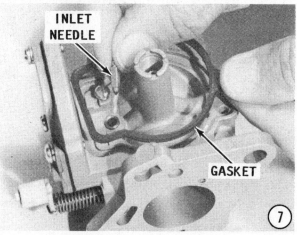

INLET NEEDLE

GASKET

(7)

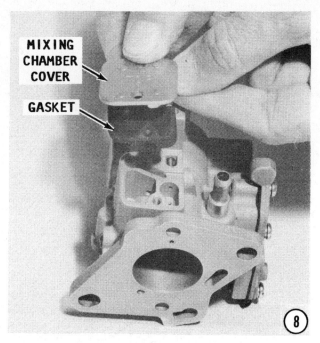

MIXING CHAMBER COVER

GASKET

(8)

8- Remove the two screws securing the mixing chamber cover to the carburetor body. Remove the cover, and then the gasket.

9- Remove the five screws securing the fuel pump cover in place. Remove in order, the cover, gasket, and then the fuel pump check valve diaphragm.

ADVICE

Count the number of turns as the low speed mixture screw is removed, as an aid during assembling to making a rough bench adjustment.

10- Back out and count the number of turns required to remove the low speed mixture screw. The spring will come with the screw.

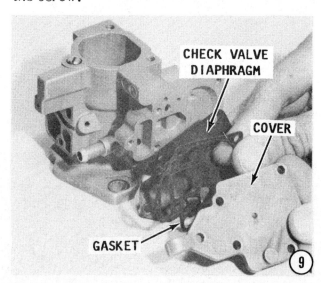

CHECK VALVE DIAPHRAGM

COVER

GASKET

(9)

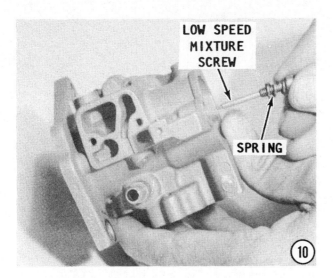

11- Remove the four screws securing the primer system cover to the float bowl flange. Gently lift the cover and the primer diaphragm, gasket, and spring will then be free, as shown. A seal is installed in the cover.

12- Remove the base plug from the underside of the float bowl flange (primer system housing). **TAKE CARE** not to lose the spring from the shaft of the plug or the tiny check ball. Remove the washer.

CLEANING AND INSPECTING

NEVER dip rubber parts, plastic parts, diaphragms, or pump plungers in carburetor cleaner. These parts should be cleaned **ONLY** in solvent, and then blown dry with compressed air.

Place all the metal parts in a screen-type tray and dip them in carburetor cleaner until they appear completely clean -- free of gum and varnish which accumulates from stale fuel. Blow the parts dry with compressed air.

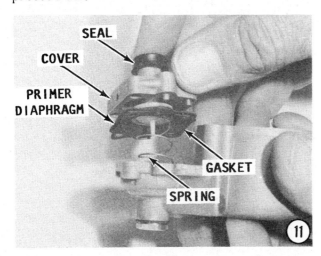

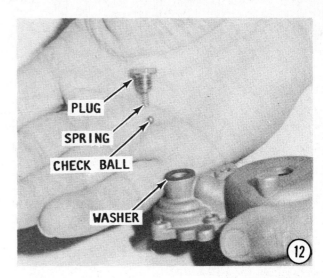

Blow out all passageways in the castings with compressed air. Check all of the parts and passages to be sure they are not clogged or contain any deposits. **NEVER** use a piece of wire or any type of pointed instrument to clean drilled passages or calibrated holes in a carburetor.

Carefully inspect the casting for cracks, stripped threads, or plugs for any sign of leakage. Inspect the float hinge in the hinge pin area for wear and the float for any sign of leakage.

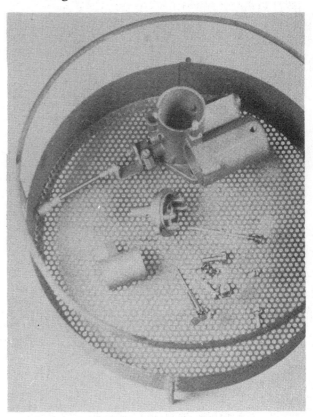

*All rubber and plastic parts **MUST** be removed before other carburetor parts are placed in a basket to be submerged in carburetor cleaner.*

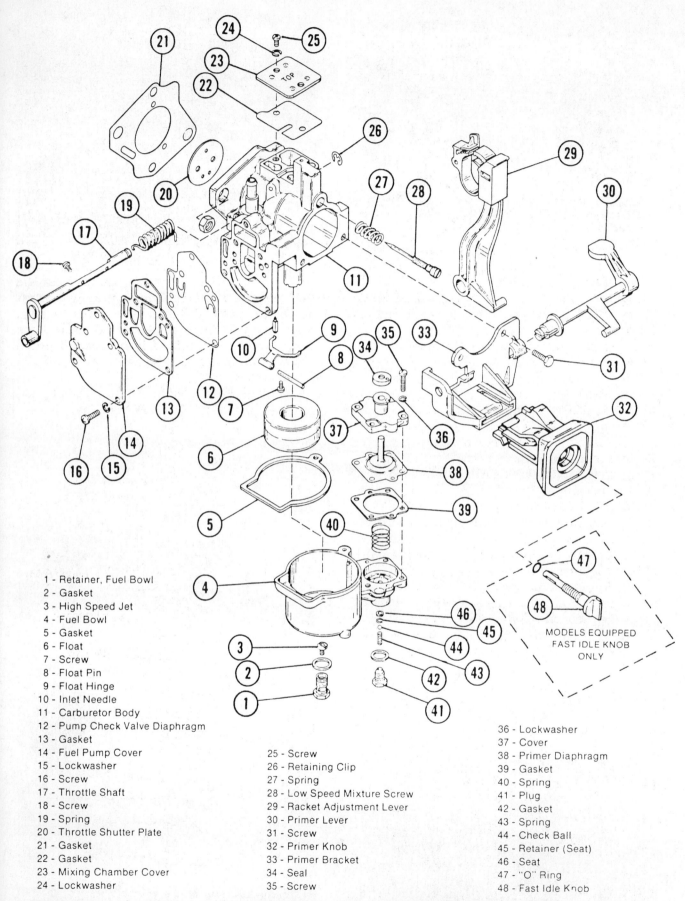

1 - Retainer, Fuel Bowl
2 - Gasket
3 - High Speed Jet
4 - Fuel Bowl
5 - Gasket
6 - Float
7 - Screw
8 - Float Pin
9 - Float Hinge
10 - Inlet Needle
11 - Carburetor Body
12 - Pump Check Valve Diaphragm
13 - Gasket
14 - Fuel Pump Cover
15 - Lockwasher
16 - Screw
17 - Throttle Shaft
18 - Screw
19 - Spring
20 - Throttle Shutter Plate
21 - Gasket
22 - Gasket
23 - Mixing Chamber Cover
24 - Lockwasher

25 - Screw
26 - Retaining Clip
27 - Spring
28 - Low Speed Mixture Screw
29 - Racket Adjustment Lever
30 - Primer Lever
31 - Screw
32 - Primer Knob
33 - Primer Bracket
34 - Seal
35 - Screw

36 - Lockwasher
37 - Cover
38 - Primer Diaphragm
39 - Gasket
40 - Spring
41 - Plug
42 - Gasket
43 - Spring
44 - Check Ball
45 - Retainer (Seat)
46 - Seat
47 - "O" Ring
48 - Fast Idle Knob

MODELS EQUIPPED
FAST IDLE KNOB
ONLY

Exploded view of a Walbro "WM" round bowl -- single float carburetor with integral fuel pump. This carburetor is identified as a "G" carburetor in the text and appendix. Major parts are identified.

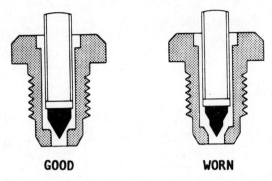

GOOD   **WORN**

Line drawing to compare a worn and new needle and seat arrangement. The worn needle would have to be replaced for full carburetor efficiency.

Examine the inlet needle for wear and if there is any evidence of wear, the inlet needle **MUST** be replaced.

ALWAYS replace any and all worn parts.

A carburetor service kit is available at modest cost from the local marine dealer. The kit will contain all necessary parts to perform the usual carburetor overhaul work.

ASSEMBLING CARBURETOR "G"

1- Drop the tiny check ball, and then the little spring into the cavity of the primer system housing. Set a **NEW** washer in place on the housing. Thread the plug into place and tighten it securely.

2- Turn the carburetor body over and set the spring in place in the primer housing. Next, position a **NEW** gasket in place on the housing. Set the primer diaphragm on top of the spring with the holes in the diaphragm for the mounting screws roughly aligned with the holes in the primer housing. Slip the seal over the lip on the cover Now, compress the spring with the primer system

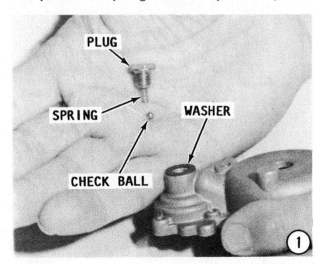

cover allowing the shaft of the primer diaphragm to pass up through the hole in the seal and the holes in the diaphragm to align perfectly with the holes in the housing. Secure the cover in place with the four attaching screws and lockwashers. Tighten the screws securely.

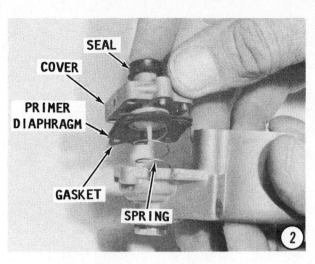

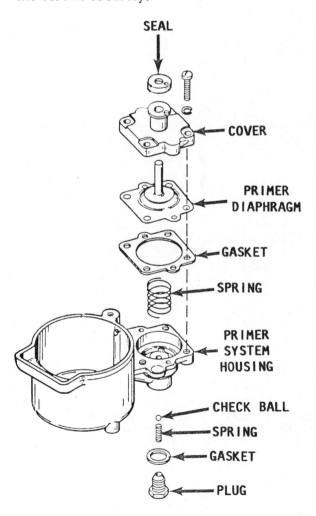

Exploded drawing of the primer system of the "G" carburetor, with major parts identified.

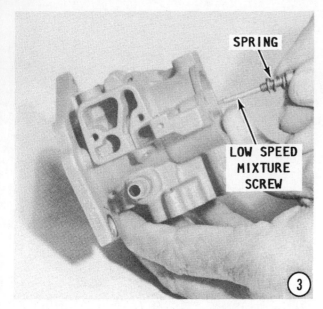

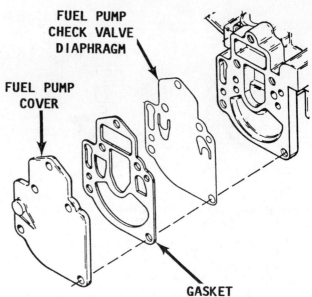

Exploded drawing of the integral fuel pump of the "G" carburetor to show arrangement of diaphragm and gasket.

3- Slide the spring onto the shaft of the low speed mixture screw. Thread the screw into the carburetor housing and seat it **LIGHTLY. NEVER** tighten the screw, because such action would damage the tip. From the lightly seated position, back the screw out the same number of turns recorded during disassembly in Step 10. If the number of turns was lost, back the screw out 1-1/4 turns as a preliminary bench adjustment. This adjustment will allow the powerhead to be started. Final adjustment will be made in Step 15.

4- Place the fuel pump check valve diaphragm onto the carburetor body with the holes for the mounting screws aligned with the holes in the body. Position a **NEW** gasket on top of the diaphragm with the mounting holes aligned. Now, place the fuel pump cover onto the carburetor body and secure it with the five attaching screws and lockwashers. tighten the screws securely.

5- Position a **NEW** gasket in place on the mixing chamber. Install the mixing chamber cover onto the carburetor and secure it with the two screws. Tighten the screws securely.

6- Position a **NEW** carburetor bowl gasket into place on the carburetor body. Slide the inlet needle into its hole in the body.

7- Slide the hinge pin through the hinge. Next, install the hinge and secure it in place with the screw. Tighten the screw just **"SNUG"**.

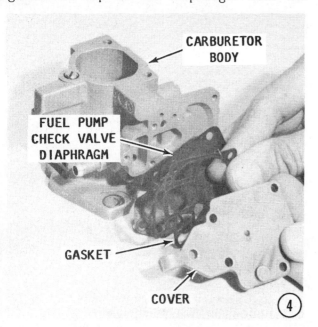

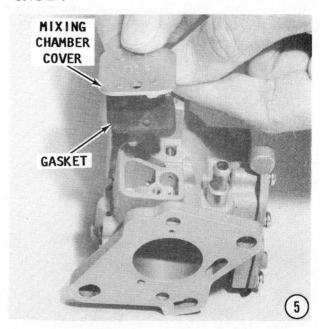

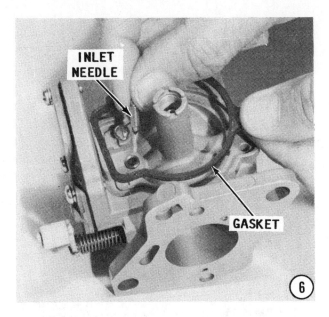

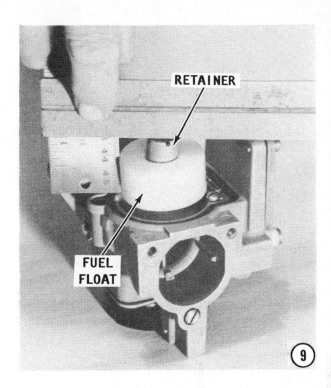

Check to be sure the hinge will move without binding.

8- Slide the float down over the shaft and onto the hinge.

9- Check the float level as shown in the accompanying illustration. With the carbu-

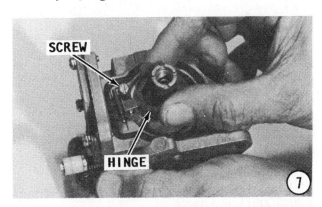

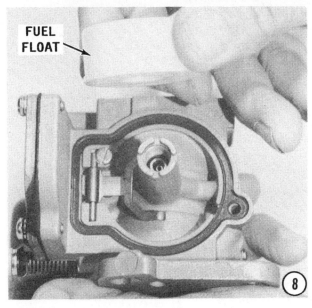

retor in the position shown, the distance from the lower edge of the float and the top of the retainer, should be 1" (25.4mm). If necessary, adjust the float level by ever so **CAREFULLY** bending the hinge **SLIGHTLY**. Just a "whisker" of change in the hinge will move the float.

10- Cover the float with the float bowl and primer system housing.

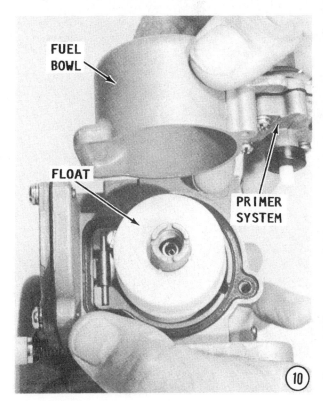

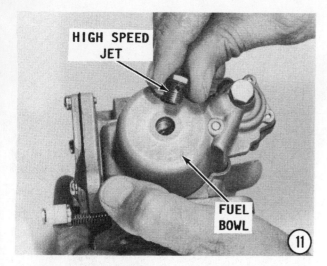

11- Slide a **NEW** gasket onto the high speed jet. Install the high speed jet. This jet also secures the bowl and primer system to the carburetor body.

12- Position a **NEW** carburetor gasket onto the powerhead. Install the carburetor to the powerhead and secure it in place with the two nuts. Tighten the nuts to the torque value listed in the appendix. Connect the fuel hose and secure it with a **NEW** Sta-strap. Install the small primer bracket to the side of the carburetor with the two bolts. Snap the idle wire into place on the ratchet adjustment lever.

13- Position the choke knob in place through the opening in the lower cowl. Install the Circlip securing the choke knob.

Mount the outboard unit in a test tank or connect a flush attachment to the lower unit.

FIRST A WORD: Before fine carburetor adjustments can be properly made, the following conditions must exist:

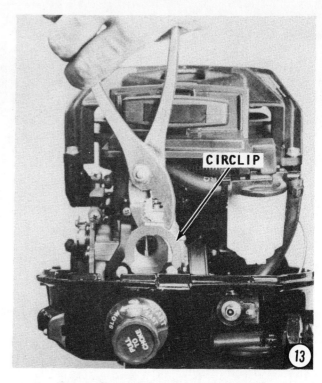

a- The regular engine-propeller combination must be used.

b- The power unit must be in forward gear.

c- The lower unit must be in the water.

d- The engine must be warmed to normal operating temperature.

Idle-Speed Adjustment

14- After the engine has been warmed to operating temperature, turn the idle speed adjusting screw on the stop bracket until the engine idles at approximately 700-800 rpm in forward gear in a test tank or 600-700 on a boat.

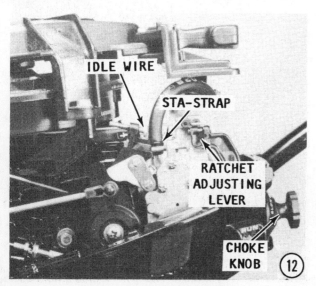

Idle Mixture Adjustment

15- Turn the adjusting screw counter clockwise until the engine fires unevenly from the mixture being too rich. Now, begin to turn the screw clockwise and rpm will begin to increase. Continue turning the adjusting screw until the mixture is so lean that the rpm begins to drop and the engine begins to misfire. Set the adjusting screw halfway between the rich and lean points.

ADVICE

It is better to have the mixture set slightly on the rich side, rather than too lean.

High-Speed Adjustment

The main metering high-speed jet is not adjustable. If the engine is to be operated at elevations above 4000 ft. (1219m), replace the main metering jet as indicated in the Jet Size/Elevation Chart in the Appendix.

16- Stop the engine. Loosen the hex nut or screw on the throttle cam. Notice the two different designs of cam followers, as shown in Illustration 16 and 16A. Adjust the cam until the center of the roller on the throttle shaft aligns with the straight line at the end of the throttle cam. (The throttle cam is elongated behind the hex attaching screw). Tighten the attaching hex nut or screw.

17- Start the engine and then retard the throttle to the **SLOW** position. Shift the unit into **NEUTRAL**, and move the **Primer/-Fast/Idle** knob into the center detent. Manually adjust the ratchet until the rpm ranges between 1400 - 1700 rpm.

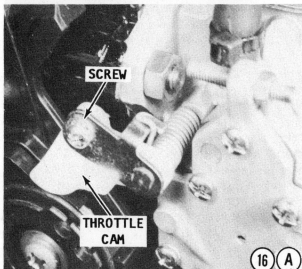

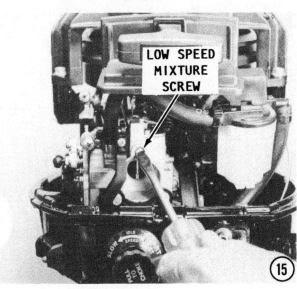

4-13 ROUND BOWL
SINGLE FLOAT CARBURETOR
STAMPED "F"
WITH "KEIKHIN" INTEGRAL FUEL PUMP
REFERENCED "H" IN APPENDIX

INTRODUCTION

This section provides complete detailed procedures for removal, disassembly, cleaning and inspecting, assembling, including bench adjustments, and installation for the Round Bowl -- Single Float Carburetor Stamped "F" with "Keikhin" Integral Fuel Pump. This carburetor is referenced **"H"** in the Tune-up Specifications (third to last column) in the Appendix.

If troubleshooting indicates the fuel pump to be the cause of problems, the pump may be serviced without removing the carburetor from the powerhead.

For adjustments with the ignition system, see Chapter 6.

REMOVAL

1- Squeeze the wire clamp around the fuel inlet line and pull the line from the pump. Plug the fuel line to prevent leakage.

2- Loosen, but do not remove, the screw retaining the throttle wire in the barrel clamp. Pull the throttle wire **DOWNWARD**, to clear the barrel clamp. Pry the choke link from the plastic retainer on the carburetor linkage.

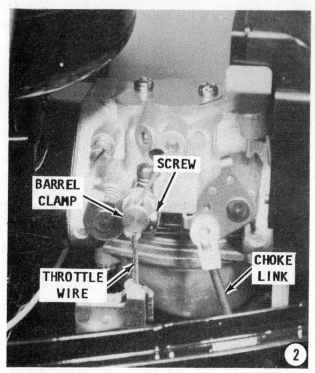

3- Remove the two screws securing the baffle cover to the carburetor.

4- Loosen the two carburetor retaining bolts. These two long bolts extend through the baffle bracket, the body of the carburetor and into the block.

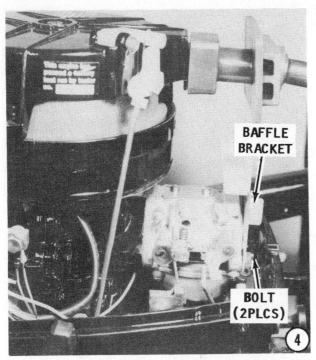

5- Lift away the baffle bracket and carburetor with integral fuel pump still attached.

Remove and discard the carburetor mounting gasket.

DISASSEMBLING

6- Remove the pilot screw and spring. The number of turns out from a lightly seated position will be given during assembling.

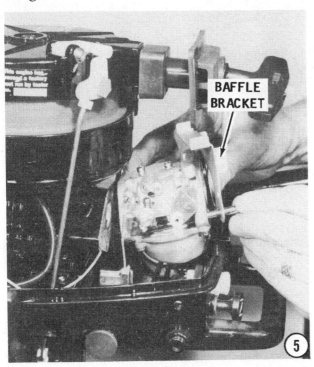

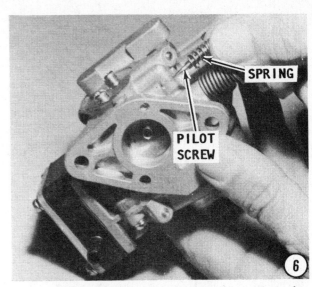

7- Remove the two screws securing the top cover and lift off the cover. Gently pry the two rubber plugs out with an awl. Remove the oval air jet cover and the round bypass cover.

8- Remove the four screws securing the fuel pump to the carburetor body. Disassemble the fuel pump components in the

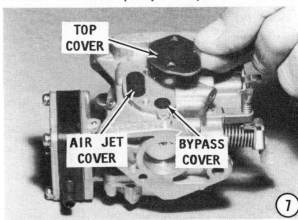

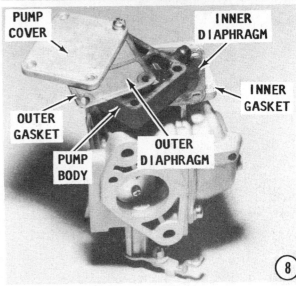

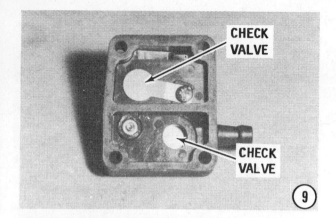

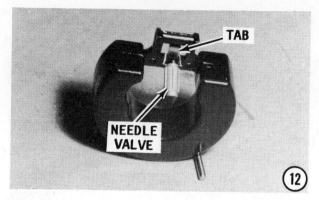

following order: the pump cover, the outer gasket, the outer diaphragm, the pump body, the inner diaphragm, and finally the inner gasket.

9- Remove the screws from both sides of the fuel pump body. Remove the check valves.

10- Remove the four screws securing the float bowl cover in place. Lift off the float bowl. Remove and discard the rubber sealing ring.

11- Remove the small Phillips screw securing the float hinge to the mounting posts. Lift out the float, the hinge pin and the needle valve. Slide the hinge pin free of the float.

12- Slide the wire attaching the needle valve to the float free of the tab.

13- Use the proper size slotted screwdriver and remove the main jet, then unscrew the main nozzle from beneath the

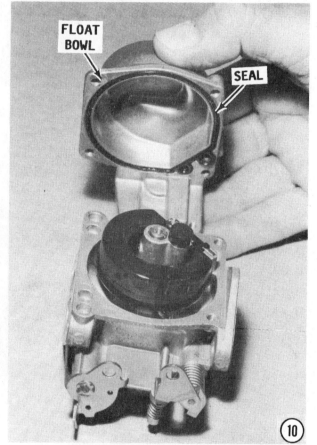

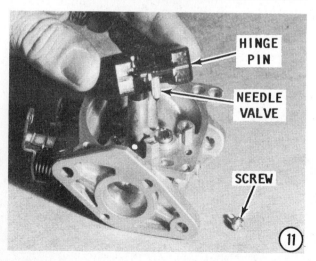

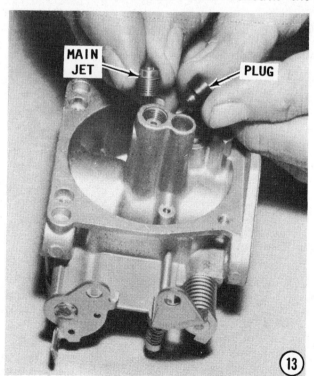

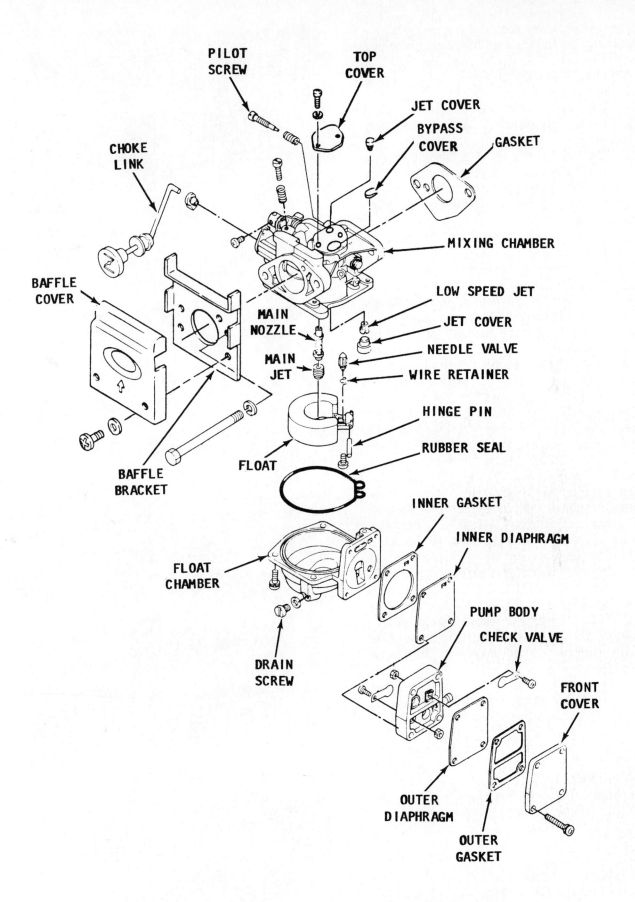

Exploded drawing of a Type "H" carburetor, with a Keikhin integral fuel pump. Major parts are identified.

main jet. Remove the plug, and then unscrew the pilot jet located beneath the plug.

CLEANING AND INSPECTING

Inspect the check valves in the fuel
pump for varnish build up as well as any
deformity.

NEVER dip rubber parts, plastic parts,
diaphragms, or pump plungers in carburetor
cleaner. These parts should be cleaned
ONLY in solvent, and then blown dry with
compressed air.

Place all metal parts in a screen-type
tray and dip them in carburetor cleaner
until they appear completely clean, then
blow them dry with compressed air.

Blow out all passages in the castings
with compressed air. Check all parts and
passages to be sure they are not clogged or
contain any deposits. **NEVER** use a piece of
wire or any type of pointed instrument to
clean drilled passages or calibrated holes in
a carburetor.

Move the throttle shaft back and forth
to check for wear. If the shaft appears to
be too loose, replace the complete throttle
body because individual replacement parts
are **NOT** available.

Inspect the main body, airhorn, and venturi cluster gasket surfaces for cracks and
burrs which might cause a leak. Check the
float for deterioration. Check to be sure
the float spring has not been stretched. If
any part of the float is damaged, the unit
must be replaced. Check the float arm
needle contacting surface and replace the
float if this surface has a groove worn in it.

Inspect the tapered section of the idle
adjusting needles and replace any that have
developed a groove.

As previously mentioned, most of the
parts which should be replaced during a
carburetor overhaul are included in overhaul
kits available from your local marine dealer.
One of these kits will contain a matched
fuel inlet needle and seat. This combination
should be replaced each time the carburetor
is disassembled as a precaution against leakage.

ASSEMBLING

1- Install the main nozzle into the center hole and tighten it snugly. Install the
main jet on top of the nozzle and tighten it
snugly also. Install the pilot jet into the

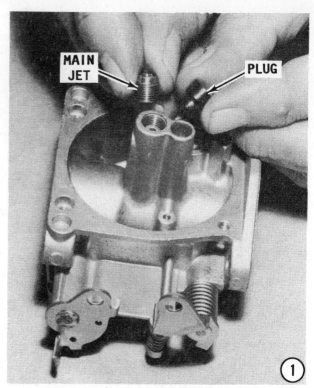

center hole and then the plug. Tighten the
jet and the plug securely.

2- Slide the wire attached to the needle
valve onto the float tab.

3- Guide the hinge pin through the float
hinge. Lower the float and needle assembly

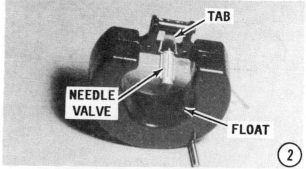

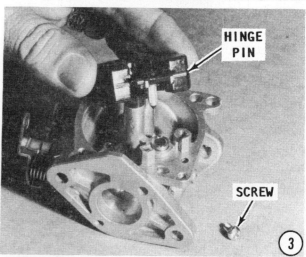

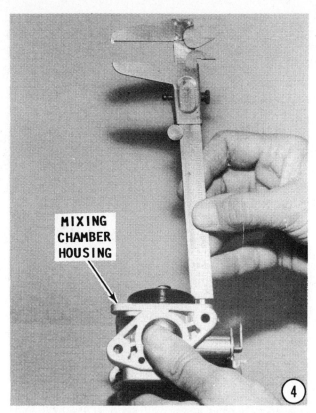

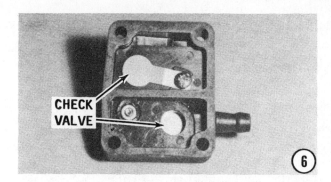

down into the float chamber and guide the needle valve into the needle seat. Check to be sure the hinge pin indexes into the mounting posts. Secure the pin in place with the small Phillips screw.

Float Adjustment

4- Invert the carburetor and allow the float to rest on the needle valve. Measure the distance between the top of the float and the mixing chamber housing. This distance should be 0.47 - 0.53" (12 - 14mm).

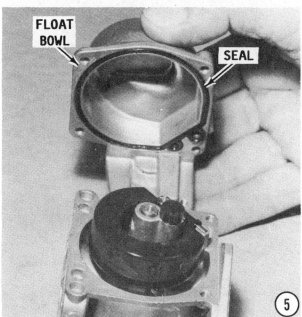

This dimension, with the carburetor inverted, places the "lower" surface of the float parallel to the carburetor body. If the dimension is not within the limits listed, the needle valve must be replaced.

CAREFULLY bend the float arm, as required, to obtain a satisfactory measurement. Install the float.

5- Insert the rubber sealing ring into the groove in the float bowl. Install the float bowl and secure it in place with the four Phillips screws.

6- Place the check valves, one at a time, in position on both sides of the fuel pump body. Secure each valve with the attaching screw.

7- Assemble the fuel pump components onto the carburetor body in the following order: the inner gasket, the inner diaphragm, the pump body, the outer diaphragm, the outer gasket, and finally the pump cover. Check to be sure all the parts are properly aligned with the mounting holes. Secure it all in place with the four attaching screws.

8- Install the oval air jet cover and the round bypass cover in their proper recesses. Place the top cover over them, no gasket is

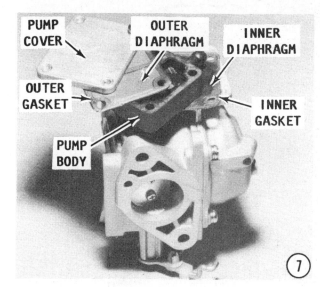

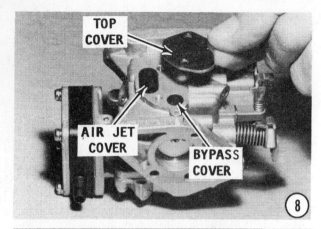

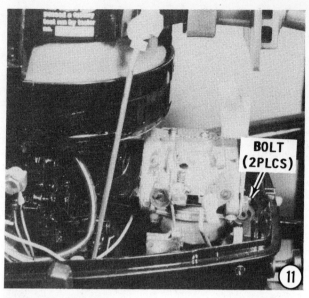

INSTALLATION

10- Slide the two long carburetor retaining bolts through the baffle bracket, carburetor body and mounting gasket.

11- Start the bolts, and then tighten them, to a torque value of 5.8 ft lbs (8Nm).

12- Secure the baffle cover to the baffle bracket using two screws and washers. Tighten the screws securely.

used. Install and tighten the two attaching screws.

9- Slide the spring over the pilot screw, and then install the screw. Tighten the screw until it **BARELY** seats. Back the pilot screw out 1-1/2.

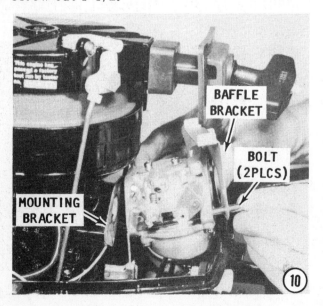

13- Snap the choke link into the plastic fitting on the starboard side of the carburetor.

Check the action of the choke knob to be sure there is no evidence of binding.

Slide the throttle cable up into the barrel clamp from the bottom up, but do not secure the wire until the following adjustment has been made. Back off the vertical idle speed screw until it no longer contacts the throttle arm. Rotate the same screw clockwise until it barely makes contact with the throttle arm, then continue to rotate the screw two more turns to slightly open the throttle plate. Grasp the cable end and lightly pull up to take up the slack. Tighten the screw on the clamp to secure the cable in position.

14- Slide the inlet fuel hose onto the fitting on the carburetor. Squeeze the two ends of the wire clamp and bring the clamp up over the inlet fitting.

Mount the outboard unit in a test tank, on the boat in a body of water, or connect a flush attachment and hose to the lower unit. Connect a tachometer to the powerhead

FIRST A WORD: Before fine carburetor adjustments can be properly made, the following conditions must exist:

a- The regular engine-propeller combination must be used.

b- The power unit must be in forward gear.

c- The lower unit must be in the water.

d- The engine must be warmed to normal operating temperature.

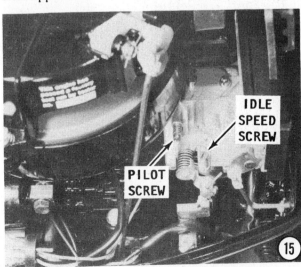

Idle-Speed Adjustment

15- After the engine has been warmed to operating temperature, turn the vertical idle speed adjusting screw on the stop bracket until the engine idles at approximately 900-1000 rpm in forward gear. Slowly rotate the horizontal pilot screw **COUNTER-CLOCKWISE** until the powerhead starts to run unevenly, then rotate the vertical idle speed screw until the rpm increases and the powerhead runs smoothly.

The manufacturer recommends the powerhead be adjusted on the rich side, because there is a possibility the powerhead may be adjusted according to the method described, and yet may experience hesitation upon acceleration. Setting the idle screw slightly on the rich side will eliminate this problem.

High-Speed Adjustment

The main metering high-speed jet is not adjustable, and there are no external high speed adjustments. If the engine is to be operated at elevations above 4000 ft. (1219m), replace the main metering jet as indicated in the Jet Size/Elevation Chart in the Appendix.

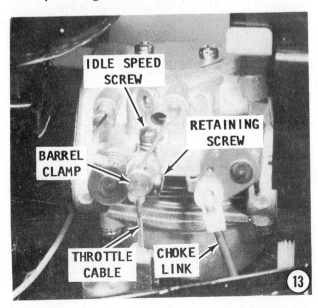

4-14 FUEL PUMP

FIRST, THESE WORDS

This section provides detailed instructions to service the fuel pump on powerheads equipped with a Carburetor "A" or "C" (as identified from the third to last column in the Tune-up Specifications in the Appendix). The fuel pump is usually installed on the transfer port/s.

The fuel pump on powerheads equipped with a Carburetor "B", "F", "G", and "H" is an integral part of the carburetor. Therefore, do not search for a separate fuel pump on these powerheads

Powerheads equipped with Carburetor "D" and "E" may not have a fuel pump of any type. Fuel is provided to the carburetor by gravity flow from the fuel tank atop the powerhead. If the powerhead does have a fuel pump, it is a disposable Mikuni pump and cannot be serviced.

THEORY OF OPERATION

The next few paragraphs briefly describe operation of the fuel pump used on Mercury

Mikuni fuel pump installed on the Model 3.5 powerhead. The pump is disposable and **CANNOT** *be serviced. Care* **MUST** *be taken to prevent the pump fitting from rubbing against the fuel tank. Adjust the 90° pulse fitting as required.*

Outboard engines. This description is followed by detailed procedures for testing the pressure, testing the volume, removing, and installing the fuel pump.

The fuel pump, installed on Mercury Outboard engines, is a diaphragm displacement type. The pump is attached to the cylinder bypass. Therefore, it is operated

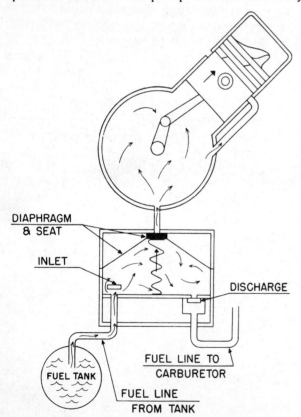

Simplified drawing of the fuel pump with the powerhead piston on the upward stroke. Notice the position of the diaphragm; the inlet disc is open; and the discharge disc is closed. The springs to preload the discs are not shown for clarity.

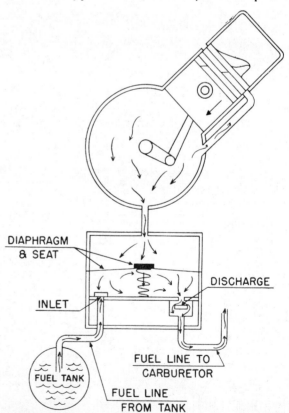

Drawing similar to the one to the left, with the powerhead piston on the downward stroke. Notice the position of the diaphragm; inlet disc is closed; and the discharge disc is open. Again, the springs to preload the discs are not shown for clarity.

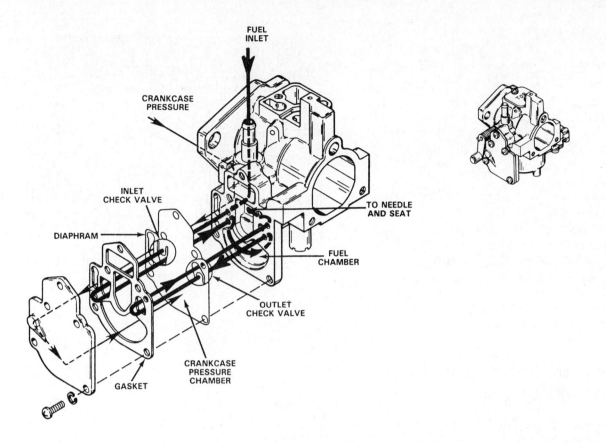

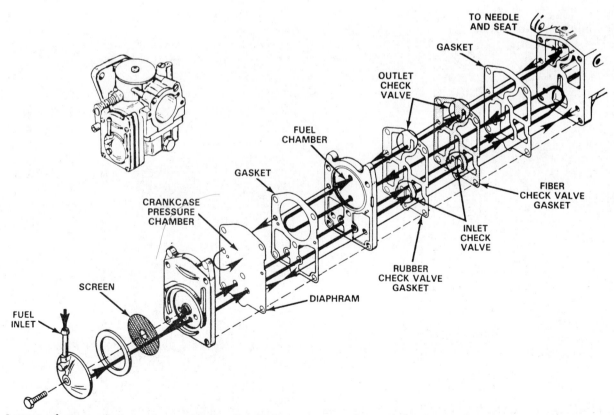

Layout of the various parts comprising the fuel pump on an integral fuel pump carburetor. The black arrows indicate fuel flow as pressure and vacuum from the crankcase move the diaphragm and check valves. The upper pump is used on carburetor identified in this manual as "G". The lower pump is part of carburetor "F". Identification can be made instantly by the pump cover.

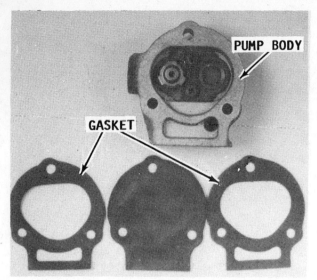

Major parts of a typical early model fuel pump.

by crankcase impulses. A hand-operated squeeze bulb is installed in the fuel line to fill the fuel pump and carburetor with fuel before the engine starts. After engine start, the pump is able to supply an adequate supply of fuel to the carburetor to meet engine demands under all speeds and conditions.

The pump consists of a diaphragm, two similar spring loaded disc valves, one for inlet (suction) and the other for outlet (discharge), and a small opening leading directly into the crankcase bypass. The suction and compression created, as the piston travels

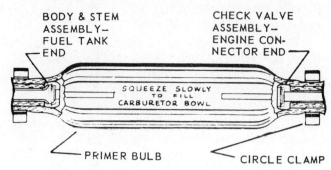

Major parts of a typical fuel line squeeze bulb, used to prime the system and deliver fuel to the carburetor until the engine is operating and the pump/s can deliver fuel on their own.

up and down in the cylinder, causes the diaphragm to flex.

As the piston moves upward, the diaphragm will flex inward displacing volume on its opposite side to create suction. This suction will draw liquid fuel in through the inlet disc valve.

When the piston moves downward, compression is created in the crankcase. This compression causes the diaphragm to flex in the opposite direction. This action causes the discharge valve disc to lift off its seat. Fuel is then forced through the discharge valve into the carburetor.

The pump has the capacity to lift fuel two feet and deliver approximately five gallons per hour at four pounds pressure psi.

Problems with the fuel pump are limited to possible leaks in the flexible neoprene suction lines; a punctured diaphragm; air leaks between sections of the pump assembly; or possibly from the disc valves not seating properly.

The pump is activated by one cylinder. If this cylinder indicates a wet fouled condition, as evidenced by a wet fouled spark plug, be sure to check the fuel pump diaphragm for possible puncture or leakage.

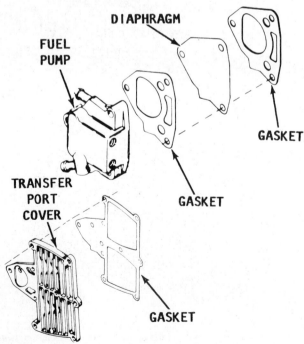

Fuel pump diaphragm and gasket arrangement. The pump is attached to the transfer port cover.

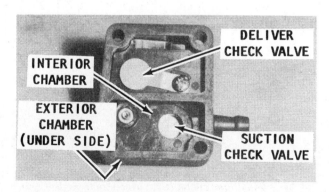

Integral fuel pump -- part of the carburetor -- with major parts and areas identified.

PUMP PRESSURE CHECK

First, these words: Lack of an adequate fuel supply will cause the engine to run lean, lose rpm, or cause piston scoring. If an integral fuel pump carburetor is installed, the fuel pressure **cannot** be checked.

With a multiple carburetor installation, fuel pressure at the top carburetor should be checked whenever insufficient fuel is suspected.

Fuel pressure should be checked if a fuel tank, other than the one supplied by the outboard unit's manufacturer, is being used. When the tank is checked, be sure the fuel cap has an adequate air vent. Verify that the fuel line from the tank is of sufficient size to accommodate the engine demands. An adequate size line would be one measuring from 5/16" to 3/8" (7.94 to 9.52mm) ID (inside diameter). Check the fuel filter on the end of the pickup in the fuel tank, to be sure it is not too small and that it is not clogged. Check the fuel pickup tube. The tube must be large enough to accommodate the fuel demands of the engine under all conditions. Be sure to check the filter at the carburetor. Sufficient quantities of fuel cannot pass through into the carburetor to meet engine demands if this screen becomes clogged.

To test: Install the fuel pressure gauge in the fuel line between the fuel pump and the carburetor. If multiple carburetors are installed, connect the gauge in the line to the top carburetor. Operate the engine at full throttle and check the pressure reading. The gauge should indicate at least 2 psi.

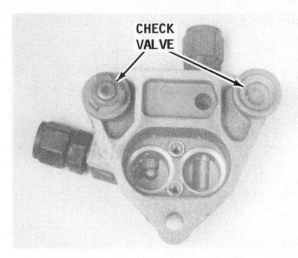

Typical fuel pump with the check valves removed. Notice how the valves face in opposite directions.

REMOVAL

Turn the fuel shut-off valve to the **OFF** position or disconnect the fuel line either at the fuel tank or at the engine.

TAKE CARE: In most cases the bolts attaching the pump to the engine also secure the pump together. Therefore, hold

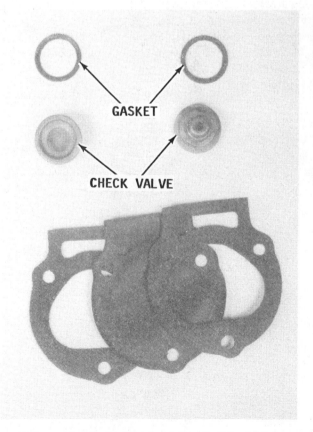

Major parts required to properly rebuild a fuel pump and return it to satisfactory service.

If the diaphragm in the fuel pump should rupture, an excessive amount of fuel would enter the cylinder and foul the spark plug, as shown.

the pump together with one hand and remove the attaching bolts with the other.

Remove the pump and lay it on a suitable work surface. Now **CAREFULLY** separate the parts and keep them in **ORDER** as an assist in assembling. As you remove the check valves **TAKE TIME** to **OBSERVE** and **REMEMBER** how each valve faces, because it **MUST** be installed in exactly the same manner, or the pump will not function.

CLEANING AND INSPECTING

Wash all parts thoroughly in solvent, and then blow them dry with compressed air. **USE CARE** when using compressed air on the check valves. **DO NOT** hold the nozzle too close because the check valve can be damaged from an excessive blast of air.

Inspect each part for wear and damage. Verify that the valve seats provide a flat contact area for the valve disc. Tighten all elbows and check valve connections firmly as they are replaced.

Test each check valve by blowing through it with your mouth. In one direction the valve should allow air to pass through. In the other direction, air should not pass through.

Check the diaphragm for pin holes by holding it up to the light. If pin holes are detected or if the diaphragm is not pliable, it **MUST** be replaced.

The fuel pump **MAY** be an integral part of the carburetor. If so, procedures for the pump are presented in the appropriate carburetor section of this chapter.

ASSEMBLING

Proper operation of the fuel pump is essential for maximum performance of the powerhead. Therefore, always use **NEW** gaskets.

NEVER use any type of sealer on fuel pump gaskets.

Place **NEW** check valve gaskets in position in their seats. Insert the check valve discs in their seats. The inlet check valve seat is identified by the protruding tip in the casting. The flat side of the check valve seats over this tip. The outlet check valve is set in opposite, with the flat end up. In this position the tension is against the valves.

Install the retainer on the check valves in the housing and secure it in place with the two retaining screws.

Place a **NEW** gasket on the pump body, then the neoprene diaphragm, and finally another **NEW** gasket. Mate the fuel pump cover to the body and hold it all together.

CAREFULLY place the fuel pump on the engine base. Install the retaining screws through the pump and into the engine block. Tighten the screws alternately to the torque value given in the Appendix.

Connect the fuel lines or turn the fuel valve to the **ON** position.

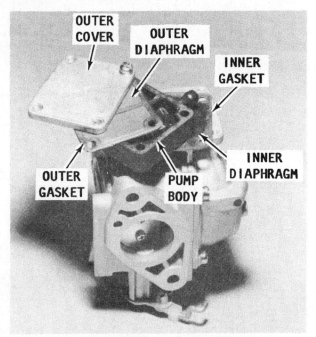

Type "H" carburetor with Keikhin integral fuel pump. The gaskets and diaphragms under the cover are exposed to indicate their order of installation.

Typical location of a separate fuel pump installed on the powerheads covered in this manual.

4-15 OIL INJECTION

DESCRIPTION

In 1986, an oil injection system became optional equipment for the 35 hp outboard units and in later years, the system became available for units as small as the 6hp. The system is produced by the manufacturer under the name "Auto Blend".

This oil injection system replaces the age-old method of manually mixing oil with the fuel for lubrication of internal moving parts in the powerhead.

"Auto Blend" is not considered an optional accessory by the manufacturer and is not recommended to be installed on units not equipped with the system from the factory.

The system consists of an oil reservoir (tank), oil screen, diaphragm vacuum-operated fuel "pump", low oil warning horn, fuel filter, and the necessary fittings and hoses for efficient operation. The reservoir and associated parts are supported in a bracket mounted on a bulkhead or the boat transom. The unit is secured in the bracket with a nylon strap and Velcro fastener. This arrangement provides quick and easy removal from the boat for refilling, testing, or for security reasons.

Oil Reservoir (Tank)

The tank is constructed of slightly transparent material and the quantity of oil can be determined by a quick glance at the tank. The tank has sight level lines in half quart (0.47L) increments. Total capacity is 3.5 quarts (3.3L). **ONLY** 2-cycle outboard oil with a BIA rating of TC-W should be used.

A screen installed in the tank filters the oil mixing with the fuel. Normally, this screen does not require service.

Fuel "Pump"

A positive displacement diaphragm vacuum-operated fuel "pump" is mounted on the front of the oil tank. (In the strict sense of the word, it is not a pump because it is dependent on operation from another source.) The "pump" mixes oil with the fuel and is operated under vacuum supplied by the fuel pump mounted on the powerhead. The "pump" is provided with a drain plug.

Low Oil Warning Horn

A warning horn will sound to indicate one of two conditions:

a- The level of oil in the tank is dangerously low.

b- The oil screen in the tank has become clogged.

The horn circuit is connected to a 12-volt battery through leads and a harness plug. The **RED** lead is connected to the positive terminal of the battery. This lead has a 0.5 amp fuse installed as protection against damage to more expensive parts in the circuit. The black lead is connected to the negative battery terminal.

The harness plug should **NOT** be connected until the tank is filled with oil to prevent the horn from automatically sounding.

Fuel Lines

A 5/16" (7.87 mm) I.D. hose is used to connect the fuel tank to the oil injection unit and also from the unit to the powerhead. The hose between the oil injection tank and the powerhead pump should never exceed 5 feet (1.50 meters). Clear plastic support tubing (not shown in the accompanying illustration) is usually used over the fuel lines and secured with standard hose clamps.

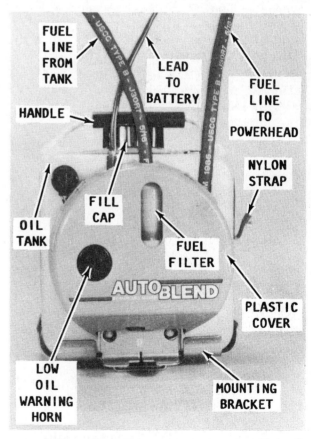

The Auto Blend unit ready to be secured in the bracket for service. Usually, short pieces of clear plastic support tubing (not shown), are placed over the end of the fuel lines to prevent kinking at the fittings.

These support tubes will prevent kinking at the fittings and subsequent restriction of fuel flow through the fuel lines.

The inlet fitting of the oil tank also serves as a fuel filter. With this arrangement, the fuel must pass through the filter before mixing with the oil in the tank. The filter is transparent, therefore, sediment can quickly be identified when it is present. The filter can be easily removed, cleaned, and installed, without any special tools.

A primer bulb must be installed between the oil injection unit and the fuel pump on the powerhead. **NEVER** connect the primer bulb between the fuel tank and the oil tank. This error could cause serious damage to the oil injection unit by providing excessive pressure.

OPERATION

While the powerhead is operating, the oil injection unit provides a variable fuel/oil mixture to the fuel pump in a ratio of 50:1 at full throttle. This is standard mixture for normal operation.

During the break-in period for a new or overhauled powerhead, oil should be added to the fuel tank in a ratio of 50:1. This ratio of oil in the fuel tank, added with the ratio of 50:1 from the oil injection unit will provide a mixture of 25:1 to the powerhead. This ratio is recommended by the manufacturer during the break-in period.

During normal operation, the level of oil in the tank will drop at a steady rate.

The unit must be positioned where the helmsperson may occasionally notice the decreasing amount in the tank. The decreasing oil level indicates the system is functioning properly and is supplying the correct proportions of fuel and oil to the powerhead.

TROUBLESHOOTING

Lack of Fuel

If the powerhead fails to operate properly and troubleshooting indicates a lack of fuel to the fuel pump, the problem may be blockage of fuel in the fuel lines, in the fuel passageway in the oil injection unit, or a clogged fuel filter.

First, check the fuel lines to be sure they are free of stress, kinks, and nothing is laying on them, i.e.: tackle box, bait tank, etc.

Next, check the filter at the oil injection unit. Because the filter is transparent, any foreign material may be quickly discovered.

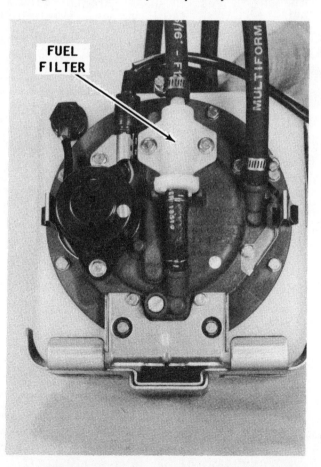

The fuel filter is transparent allowing visual inspection for foreign material. The filter may be removed and installed without the use of special tools.

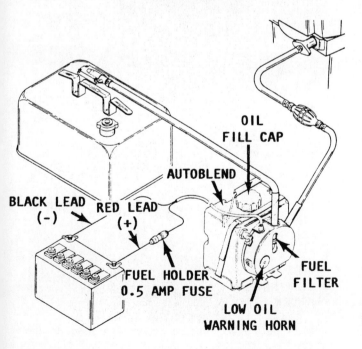

Functional diagram depicting complete hookup of the Auto Blend oil injection system.

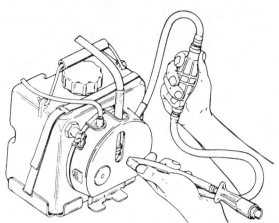

The outlet line and fitting may be back-flushed by kinking the hose and using the squeeze bulb, as explained in the text.

The filter may be removed, cleaned, and installed quickly, without the use of any special tools.

Finally, the fuel passageway in the unit may need to be back-flushed. This is accomplished by simply kinking the line and at the same time slowly squeezing the primer bulb, as shown in the accompanying illustration. The primer bulb should only be squeezed a few times to prevent building up excessive pressure in the unit.

Warning Horn Sounds

If the horn should sound during operation of the outboard unit, shut down the unit immediately and make a couple of quick checks.

First, check the level of oil in the tank and replenish as required.

Next, check the oil screen in the tank. If it is clogged with sediment, remove the filter, clean it, and install it back in the tank.

Warning Horn Does Not Sound
(When it should — low or no oil)

First, check the 0.5 amp fuse in the in-line holder of the positive lead to the battery.

Check the battery connections and then the charge condition of the battery. Correct the condition as necessary.

STORAGE

Proper storage procedures are **CRITICAL** to ensure efficient operation when the unit is again placed in service.

First, disconnect the battery leads from the battery.

Next, disconnect and plug the fuel lines at the fuel tank and powerhead.

Now, drain all fuel from the unit. Remove the front cover of the unit by simultaneously pushing in on the cutaway tabs located on both sides of the cover, and at the same time pulling the cover away from the unit. Remove the drain plug and allow at least 5-minutes for all fuel to drain from the pump. Install the drain plug and tighten it securely.

CRITICAL WORDS

All fuel **MUST** be drained from the oil injection fuel "pump". The percentage of alcohol in modern fuels seems to increase each year. This alcohol in the fuel is a definite enemy of the diaphragm in the "pump". Therefore, if any fuel is left in the "pump" during storage the diaphragm will most likely be damaged.

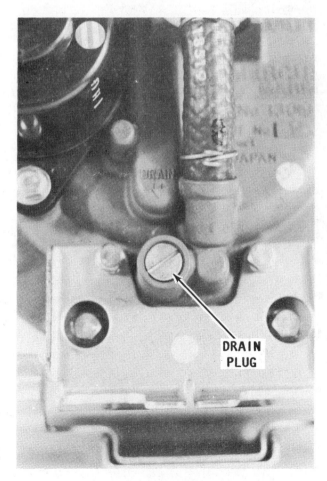

DRAIN PLUG

*All fuel **MUST** be drained prior to placing the unit in storage, to prevent damage to the "pump" diaphragm.*

Install the front cover by aligning the cover openings on both sides of the unit, and then pushing in on the cover until it snaps into place.

Oil may remain in the oil injection tank during storage without any harmful effects.

PREPARATION FOR USE

First, remove the front cover of the unit by simultaneously pushing in on the cutaway tabs located on both sides of the cover, and at the same time pulling the cover away from the unit. Check to be sure the fuel drain plug is tight. Replace the front cover by aligning the cover openings on both sides of the unit, and then pushing in on the cover until it snaps into place.

Next, fill the oil tank with 2-cycle outboard oil with a BIA rating of TC-W. Tighten the fill cap securely.

Remove any plugs in the fuel lines, and then connect the hoses to the fuel tank and the powerhead. Remember, the squeeze bulb **MUST** be in the hose between the oil injection unit and the fuel pump on the powerhead.

Connect the low oil warning wire harness to the battery. Connect the **RED** lead to the positive battery terminal and the **BLACK** lead to the negative battery terminal.

Check to be sure the low oil warning system is functioning correctly. First, verify the tank is full of oil, and then the fill cap is tightened securely. Now, turn the oil injection unit upside down. This position will allow the float to activate the horn.

If the horn sounds, immediately turn the unit rightside up and position it in the mounting bracket. Secure it in place with the strap and Velcro material.

If the horn does not sound, check the 0.5 amp fuse in the fuse holder of the positive battery lead. Check both the battery connections and the charge condition of the battery.

GOOD WORDS

The manufacturer recommends the fuel filter be replaced at the start of each season or at least once a year. The manufacturer also recommends oil be added to the fuel tank at the ratio of 50:1 for the first 6-gallons of fuel used after the unit is brought out of storage. The oil in the fuel tank plus the 50:1 oil mixture in the oil injection unit will deliver a mixture of 25:1 to the powerhead. This ratio will **ENSURE** adequate lubrication of moving parts which have been drained of oil during the storage period.

Maximum enjoyment can only be obtained if the boat and power unit are properly maintained and adequate storage is provided during the off-season.

5
IGNITION

5-1 INTRODUCTION

The less an outboard engine is operated, the more care it needs. Allowing an outboard engine to remain idle will do more harm than if it is used regularly. To maintain the engine in top shape and always ready for efficient operation at any time, the engine should be operating every 3 to 4 weeks throughout the year.

The carburetion and ignition principles of two-cycle engine operation **MUST** be understood in order to perform a proper tune-up on an outboard motor.

If you have any doubts concerning your understanding of two-cycle engine operation, it would be best to study the operation theory section in the first portion of Chapter 8, before tackling any work on the ignition system.

Four different types of ignition systems are used on the outboard units covered in this manual. Where similarities exist, the systems are grouped together in the same section. The first sections of this chapter will be devoted to an explanation of each ignition system and its theory of operation. The latter sections will provide troubleshooting and repair instructions for the systems. For timing and synchronizing procedures, see Chapter 6.

Outboard unit mounted on a stand ready for service work on the ignition system. A thick piece of wood clamped in a vise serves the same purpose as the stand.

*The carburetion and ignition system **MUST** be properly adjusted and synchronized for optimum powerhead performance.*

For convenience, each ignition system has been identified with a code numeral from I thru IV. These code numerals will be used throughout this chapter and are referenced in the Appendix. To determine the code numeral of the ignition system used on any particular powerhead, simply find the engine and appropriate model year in the Tune-up Specifications, then move across the table to the IGN TYPE column. The ignition system used on that engine will be identified. Once this numeral is determined, refer to the appropriate section in this chapter for detailed troubleshooting and service procedures. The identification code numerals are as follows:

Type I Phelon -- Flywheel -- Magneto with points.
Type II Thunderbolt -- Flywheel -- Phasemaker -- with points.
Type III Thunderbolt -- Flywheel -- CD -- pointless.
Type IV Thunderbolt -- Flywheel -- CD -- pointless -- coil per cylinder.

5-2 SPARK PLUG EVALUATION

Removal

Remove the spark plug wires by pulling and twisting on only the molded cap. **NEVER** pull on the wire or the connection inside the cap may become separated or the boot damaged. Remove the spark plugs and keep them in order. **TAKE CARE** not to tilt the socket as you remove the plug or the insulator may be cracked.

*Damaged spark plugs. Notice the broken electrode on the left plug. The missing part **MUST** be found and removed before returning the powerhead to service, to prevent serious damage to expensive internal parts.*

This spark plug has been operating too-cold, because it is rated with a too-low heat range for the powerhead.

Examine

Line the plugs in order of removal and carefully examine them to determine the firing conditions in each cylinder. If the side electrode is bent down onto the center electrode, the piston is traveling too far upward in the cylinder and striking the spark plug. Such damage indicates the piston pin or the rod bearing is worn excessively. In most cases, an engine overhaul is required to correct the condition. To verify the cause of the problem, turn the engine over by hand. As the piston moves to the full up position, push on the piston crown with a screwdriver inserted through the spark plug

Example of a center-fire spark plug operating under favorable conditions.

This spark plug is foul from operating with an overrich air/fuel mixture, possibly caused by an improper carburetor adjustment.

hole, and at the same time rock the flywheel back-and-forth. If any play in the piston is detected, the engine must be rebuilt.

Correct Color

A proper firing plug should be dry and powdery. Hard deposits inside the shell indicate too much oil is being mixed with the fuel. The most important evidence is the light gray to tan color of the porcelain, which is an indication this plug has been running at the correct temperature. This means the plug is one with the correct heat range and also that the air-fuel mixture is correct.

Rich Mixture

A black, sooty condition on both the spark plug shell and the porcelain is caused by an excessively rich air-fuel mixture, both at low and high speeds. The rich mixture lowers the combustion temperature so the spark plug does not run hot enough to burn off the deposits.

Deposits formed only on the shell is an indication the low-speed air-fuel mixture is too rich. At high speeds with the correct mixture, the temperature in the combustion chamber is high enough to burn off the deposits on the insulator.

Too Cool

A dark insulator, with very few deposits, indicates the plug is running too cool. This condition can be caused by low compression or by using a spark plug of an incorrect heat range. If this condition shows on only one plug it is most usually caused by low compression in that cylinder. If all of the plugs have this appearance, then it is probably due to the plugs having a too-low heat range.

Fouled

A fouled spark plug may be caused by the wet oily deposits on the insulator shorting the high-tension current to ground inside the shell. The condition may also be caused by ignition problems which prevent a high-tension pulse from being delivered to the spark plug.

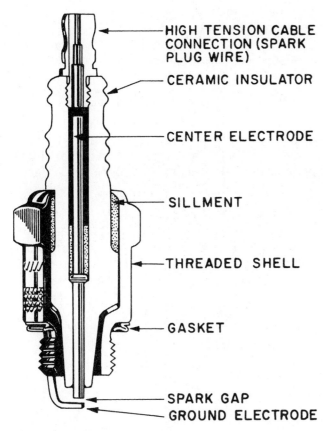

HIGH TENSION CABLE CONNECTION (SPARK PLUG WIRE)

CERAMIC INSULATOR

CENTER ELECTRODE

SILLMENT

THREADED SHELL

GASKET

SPARK GAP
GROUND ELECTRODE

Cutaway drawing of a typical spark plug with principle parts identified.

The spark plugs should be kept in order as they are removed from the powerhead to enable a proper diagnosis to be made of each cylinder operating condition.

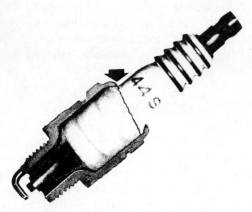

A crack in the porcelain is usually caused by removing or installing the plug using the wrong size wrench. Such damage will cause the spark to be grounded by jumping from the crack to the base of the plug.

Carbon Deposits

Heavy carbon-like deposits are an indication of excessive oil in the fuel. This condition may be the result of worn piston rings or excessive ring end gap.

Overheating

A dead white or gray insulator, which is generally blistered, is an indication of overheating and pre-ignition. The electrode gap wear rate will be more than normal and in the case of pre-ignition, will actually cause the electrodes to melt as shown in this illustration. Overheating and pre-ignition are usually caused by overadvanced timing, detonation from using too-low an octane

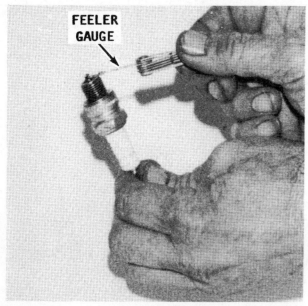

The spark plug gap should always be checked with a wire-type feeler gauge before installing new or used plugs.

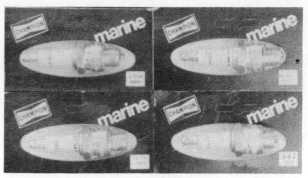

Today, numerous type spark plugs are available for service. ALWAYS check with the Specifications or the local marine shop to be sure the manufacturer has not initiated a late-change for the model being serviced.

rating fuel, an excessively lean air-fuel mixture, or problems in the cooling system.

Electrode Wear

Electrode wear results in a wide gap and if the electrode becomes carbonized it will form a high-resistance path for the spark to jump across. Such a condition will cause the engine to misfire during acceleration. If all of the plugs are in this condition, it can cause an increase in fuel consumption and very poor performance at high-speed operation. The solution is to replace the spark plugs with a rating in the proper heat range and gapped to specification.

Red rust-colored deposits on the entire firing end of a spark plug can be caused by water in the cylinder combustion chamber. This can be the first evidence of water entering the cylinders through the exhaust manifold because of an accumulation of scale or defective exhaust shutter. This condition **MUST** be corrected at the first opportunity. Refer to Chapter 8, Powerhead.

5-3 POLARITY CHECK

Coil polarity is extremely important for proper battery ignition system operation. If a coil is connected with reverse polarity,

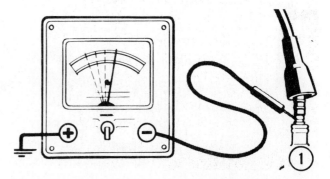

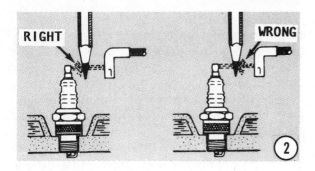

the spark plugs may demand from 30 to 40 percent more voltage to fire, or on most C.D. systems, there will be **NO** spark. Under such demand conditions, in a very short time the coil would be unable to supply enough voltage to fire the plugs. Any one of the following three methods may be used to quickly determine coil polarity.

1- The polarity of the coil can be checked using an ordinary D.C. voltmeter set on the maximum scale. Connect the positive lead to a good ground. With the engine running, momentarily touch the negative lead to a spark plug terminal. The needle should swing upscale. If the needle swings downscale, the polarity is reversed.

2- If a voltmeter is not available, a pencil may be used in the following manner: Disconnect a spark plug wire and hold the metal connector at the end of the cable about 1/4" (6.35mm) from the spark plug terminal. Now, insert an ordinary pencil tip between the terminal and the connector. Crank the engine with the ignition switch ON. If the spark feathers on the plug side and has a slight orange tinge, the polarity is correct. If the spark feathers on the cable connector side, the polarity is reversed.

3- The firing end of a used spark plug can give a clue to coil polarity. If the ground electrode is "dished", it may mean polarity is reversed.

5-4 WIRING HARNESS

CRITICAL WORDS: These next two paragraphs may well be the most important words in this chapter. Probably the No. 1 cause of electrical problems with outboard power plants is misuse of the wiring harness.

A wiring harness is used between the key switch and the engine. This harness seldom contains wire of sufficient size to allow connecting accessories. Therefore, anytime a new accessory is installed, **NEW** wiring should be used between the battery and the accessory. A separate fuse panel **MUST** be installed on the dash. To connect the fuse panel, use one red and one black No. 10 gauge wires from the battery. If a small amount of 12-volt current should be accidently attached to the magneto system, the coil may be damaged or **DESTROYED**. Such a mistake in wiring can easily happen if the source for the 12-volt accessory is taken from the key switch. Therefore, again let it be said, **NEVER** connect accessories through the key switch.

A damaged wiring lead or harness cannot be repaired, it ***MUST*** *be replaced.*

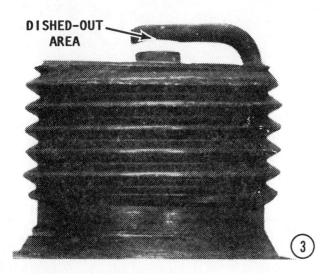

DISHED-OUT AREA

5-5 TYPE I IGNITION SYSTEM PHELON — FLYWHEEL — MAGNETO WITH POINTS

DESCRIPTION

This magneto system is identified as Type 1 in the Appendix.

READ AND BELIEVE. A battery installed to crank the engine **DOES NOT** mean the engine is equipped with a battery-type ignition system. A magneto system uses the battery only for cranking the engine. Once the engine is running, the battery has absolutely no affect on engine operation. Therefore, if the battery is low and fails to crank the engine properly for starting, the engine may be cranked manually, started, and operated.

The flywheel and distributor type magneto systems operate on the same principle.

A magneto system is a self-contained unit. The unit does not require assistance from an outside source for starting or continued operation. Therefore, as previously mentioned, if the battery is dead, the engine may be cranked manually and the engine started.

The flywheel-type magneto unit consists of an armature plate, and a permanent magnet built into the flywheel, or magnet rotor. The ignition coil, condenser and breaker points are mounted on the armature plate.

As the pole pieces of the magnet pass over the heels of the coil, a magnetic field is built up about the coil, causing a current to flow through the primary winding.

Now, at the proper time, the breaker points are separated by action of a cam, and the primary circuit is broken. When the circuit is broken, the flow of primary current stops and causes the magnetic field about the coil to break down instantly. At this precise moment, an electrical current of extremely high voltage is induced in the fine secondary windings of the coil. This high voltage is conducted to the spark plug where it jumps the gap between the points of the plug to ignite the compressed charge of air-fuel mixture in the cylinder.

A magneto type distributor unit operates similar to the flywheel type unit just described, except the distributor type has a distributor and is belt driven.

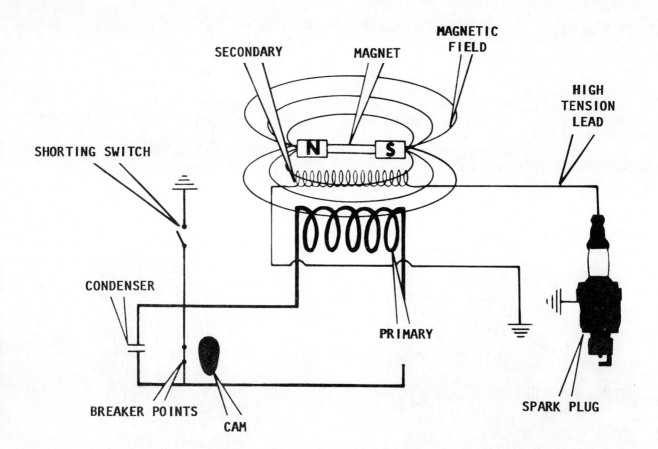

Schematic diagram of a magneto ignition system.

TROUBLESHOOTING
TYPE I IGNITION SYSTEM

Always attempt to proceed with the troubleshooting in an orderly manner. The shotgun approach will only result in wasted time, incorrect diagnosis, replacement of unnecessary parts, and frustration.

Begin the ignition system troubleshooting with the spark plug/s and continue through the system until the source of trouble is located.

Remember, a magneto system is a self-contained unit. Therefore, if the engine has a key switch and wire harness, remove them from the engine and then make a test for spark. If a good spark is obtained with these two items disconnected, but no spark is available at the plug when they are connected, then the trouble is in the harness or the key switch. If a test is made for spark at the plug with the harness and switch connected, check to be sure the key switch is turned to the **ON** position.

Key Switch

A magneto key switch operates in **RE-VERSE** of any other type key switch. When the key is moved to the **OFF** position, the circuit is **CLOSED** between the magneto and ground. For this reason, an automotive-type switch **MUST NEVER** be used, because the circuit would be opened and closed in reverse, and if 12-volts should reach the coil, the coil may be **DESTROYED**.

CRITICAL WORDS: These next two paragraphs may well be the most important words in this chapter. Misuse of the wiring harness is the most single cause of electrical problems with outboard power plants.

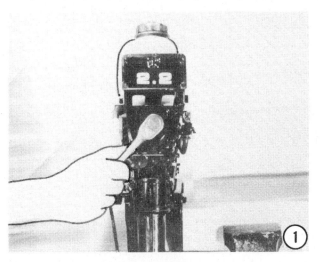

A wiring harness is used between the key switch and the engine. This harness seldom contains wire of sufficient size to allow connecting accessories. Therefore, anytime a new accessory is installed, **NEW** wiring should be used between the battery and the accessory. A separate fuse panel **MUST** be installed on the dash. To connect the fuse panel, use one red and one black No. 10 gauge wires from the battery. If a small amount of 12-volt current should be accidentally attached to the magneto system, the coil may be severely damaged or **DESTROYED**. Such a mistake in wiring can easily happen if the source for the 12-volt accessory is taken from the key switch. Therefore, again let it be said, **NEVER** connect accessories through the key switch.

Spark Plugs

1- Check the plug wires to be sure they are properly connected. Check the entire length of the wire/s from the plug/s to the magneto under the armature plate. If the wire is to be removed from the spark plug, **ALWAYS** use a pulling and twisting motion as a precaution against damaging the connection.

Attempt to remove the spark plug/s by hand. This is a rough test to determine if the plug is tightened properly. You should not be able to remove the plug without using the proper socket size tool. Remove the spark plug/s and keep them in order. Examine each plug and evaluate its condition as described in Section 5-2.

2- Use a spark tester and check for spark at each cylinder. If a spark tester is

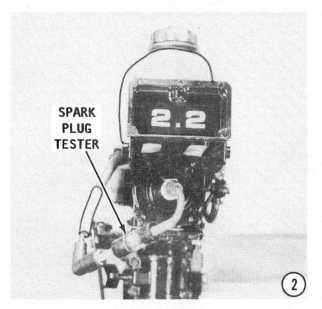

SPARK PLUG TESTER

not available, hold the plug wire about 1/4" (6.4mm) from the engine. Turn the flywheel with a pull starter or electrical starter and check for spark. A strong spark over a wide gap must be observed when testing in this manner, because under compression a strong spark is necessary in order to ignite the air-fuel mixture in the cylinder. This means it is possible to think you have a strong spark, when in reality the spark will be too weak when the plug is installed. If there is no spark, or if the spark is weak, the trouble is most likely under the flywheel in the magneto.

ONE MORE WORD: Each cylinder has its own ignition system in a flywheel-type ignition system. This means if a strong spark is observed on any one cylinder and not at another, only the weak system is at fault. However, it is always a good idea to check and service all systems while the flywheel is removed.

Compression

Before spending too much time and money attempting to trace a problem to the ignition system, a compression check of each cylinder should be made. If the cylinder does not have adequate compression, troubleshooting and attempted service of the ignition or fuel system will fail to give the desired results of satisfactory engine performance.

Remove the spark plug wire/s by pulling and twisting **ONLY** on the molded cap. **NEVER** pull on the wire because the con-

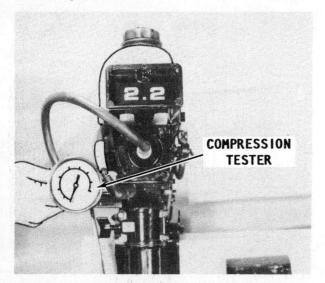

Making a compression check should become standard procedure before any major work or tune-up is perform-ed. Without adequate compression in each cylinder, any other work will have little affect on performance.

nection inside the cap may be separated or the boot may be damaged. Remove the spark plug/s. Insert a compression gauge into the cylinder spark plug hole. Crank the engine for several revolutions and note the compression reading. Repeat the procedure for each cylinder.

A variation in reading between the cylinders is far more important than the actual individual readings. If one cylinder varies more than 20 psi (138 kPa) from the other one, the cylinder may be scored, the rings frozen, or the piston burned. The power-heads covered in this manual, with the exception of the Model 2.2, do not use a cylinder head. Therefore, low compression in one cylinder **CANNOT** be attributed to a blown head gasket.

An acceptable pressure reading for a single cylinder powerhead is 110 psi (760 kPa) or more.

Condenser

In simple terms, a condenser is composed of two sheets of tin or aluminum foil laid one on top of the other, but separated by a sheet of insulating material such as waxed paper, etc. The sheets are rolled into a cylinder to conserve space and then inserted into a metal case for protection and to permit easy assembly.

The purpose of the condenser is to absorb or store the secondary current built up in the primary winding at the instant the breaker points are separated. By absorbing or storing this current the condenser prevents excessive arcing and the useful life of the breaker points is extended. The condenser also gives added force to the charge produced in the secondary winding as the condenser discharges.

Modern condensers seldom cause problems, therefore, it is not necessary to install

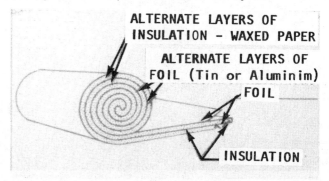

Rough sketch to illustrate how the waxed paper, aluminum foil, and insulation are rolled in a typical condenser.

a new one each time the points are replaced. However, if the points show evidence of arcing, the condenser may be at fault and should be replaced. A faulty condenser may not be detected without the use of special test equipment. The modest cost of a new condenser justifies its purchase and installation to eliminate this item as a source of trouble.

Breaker Points

The breaker points in an outboard motor are an extremely important part of the ignition system. A set of points may appear to be in good condition, but they may be the source of hard starting, misfiring, or poor engine performance. The rules and knowledge gained from association with 4-cycle engines does not necessarily apply to a 2-cycle engine. The points should be replaced every 100 hours of operation or at least once a year. **REMEMBER**, the less an outboard engine is operated, the more care it needs. Allowing an outboard engine to remain idle will do more harm than if it is used regularly.

A breaker point set consists of two points. One is attached to a stationary bracket and does not move. The other point is attached to a moveable mount. A spring is used to keep the points in contact with each other, except when they are separated by the action of a cam built into the flywheel or machined on the crankshaft. Both points are constructed with a steel base and a tungsten cap fused to the base.

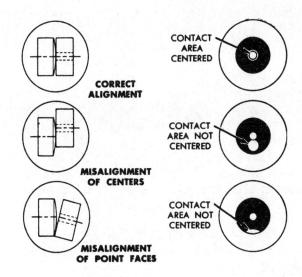

Before setting the breaker point gap, the points must be properly aligned (top). **ALWAYS** *bend the stationary point,* **NEVER** *the breaker lever. Attempting to adjust an old worn set of points is not practical, when compared with the modest cost of a new set, thus eliminating this area as a possible cause of trouble. If a worn set of points is to be retained for emergency use, both contact surfaces of the set should be refaced with a point file.*

To properly diagnose magneto (spark) problems, the theory of electricity flow must be understood. The flow of electricity through a wire may be compared with the flow of water through a pipe. Consider the voltage in the wire as the water pressure in the pipe and the amperes as the volume of water. Now, if the water pipe is broken, the water does not reach the end of the

Each set of points has its own condenser. Some powerhead models have the condenser mounted vertically, others horizontally.

A normal set of breaker points used in a magneto will show evidence of a shallow crater and build-up after a few hours of operation. The left set of points is considered normal and need not be replaced. The set on the right has been in service for more than 450 hours and should be replaced.

pipe. In a similar manner if the wire is broken the flow of electricity is broken. If the pipe springs a leak, the amount of water reaching the end of the pipe is reduced. Same with the wire. If the installation is defective or the wire becomes grounded, the amount of electricity (amperes) reaching the end of the wire is reduced.

Check the wiring carefully, inspect the points closely, and adjust them accurately according to the Specifications in the Appendix.

SERVICING
TYPE I IGNITION SYSTEM

General Information
Magnetos installed on outboard engines will usually operate over extremely long periods of time without requiring adjustment or repair. However, if ignition system problems are encountered, and the usual corrective actions such as replacement of spark plugs does not correct the problem, the magneto output should be checked to determine if the unit is functioning properly.

Magneto overhaul procedures may differ slightly on various outboard models, but the following general basic instructions will apply to all high speed flywheel type magnetos covered in this manual.

DISASSEMBLING
TYPE I IGNITION SYSTEM

1- Remove the cowl or enough of the engine cover to expose the flywheel. Disconnect the battery connections from the battery terminals.

2- Remove the nut securing the flywheel to the crankshaft. It may be necessary to use some type of flywheel strap to prevent the flywheel from turning as the nut is loosened.

3- Install the proper flywheel puller. **NEVER** attempt to use a puller which pulls

WOOD BLOCK

FLYWHEEL HOLDER

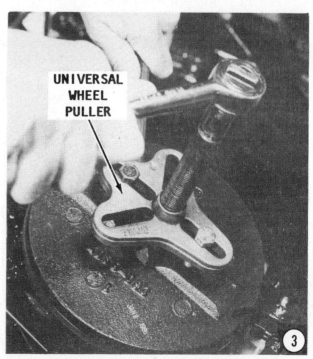

UNIVERSAL WHEEL PULLER

on the outside edge of the flywheel or the flywheel may be damaged. After the puller is installed, tighten the center screw onto the end of the crankshaft. Continue tightening the screw until the flywheel is released from the crankshaft. Remove the flywheel.

4- **STOP,** and carefully observe the layout of the magneto and associated wiring. Study how the magneto is assembled. Because there are so many different magneto installation arrangements, it is not possible to illustrate all of them, and if they were shown, you would not be able to identify the circuitry for the engine you are servicing. **TAKE TIME** to make notes of the wire routing. Observe how the heels of the laminated core with the coil attached is flush with the boss on the stator plate. These items must be replaced in their proper positions. You may elect to follow the practice of many professional mechanics by taking a series of photographs of the engine with the flywheel removed, one from the top, and a couple from the sides showing the wiring and arrangement of parts.

5- Remove the screw attaching the wires from the coil and condenser to one set of points. Remove the hold-down screw on

the points and condenser. Repeat the procedure for the other set of points.

6- Clean the surface of the stator plate where the points and condenser attach.

SPECIAL WORDS

If the coil and laminated core do need to be replaced, proceed directly to Cleaning & Inspecting and then begin assembling with Step 4.

Coil Replacement

On some models, the laminated core and coil can be removed from the stator plate and on others the core cannot. If the laminated core cannot be removed, bypass Step 7 and 8.

STATOR PLATE

Typical installation of non-removable laminated core. The core is riveted in place.

7- Bend down the tab in the center of the coil which protrudes from the center piece of the laminated core.

8- Remove the two hold-down screws securing the laminated core to the plate, and then lift the core free.

9- Disconnect the coil ground wire from the laminated core, and then pull the coil free of the core. If servicing a 2-cylinder unit, perform Steps 7, 8, and 9 for the second coil.

CLEANING AND INSPECTING

Inspect the flywheel for cracks or other damage, especially around the inside of the center hub. Check to be sure metal parts have not become attached to the magnets. Verify each magnet has good magnetism by using a screwdriver or other tool.

Throughly clean the inside taper of the flywheel and the taper on the crankshaft to prevent the flywheel from "walking" on the crankshaft during operation.

Check the top seal around the crankshaft to be sure no oil has been leaking onto the stator plate. If there is **ANY** evidence the seal has been leaking, it **MUST** be replaced. See Chapter 8.

Test the stator assembly to verify it is not loose. Attempt to lift each side of the plate. There should be little or no evidence of movement.

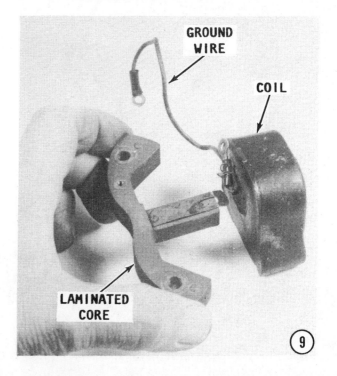

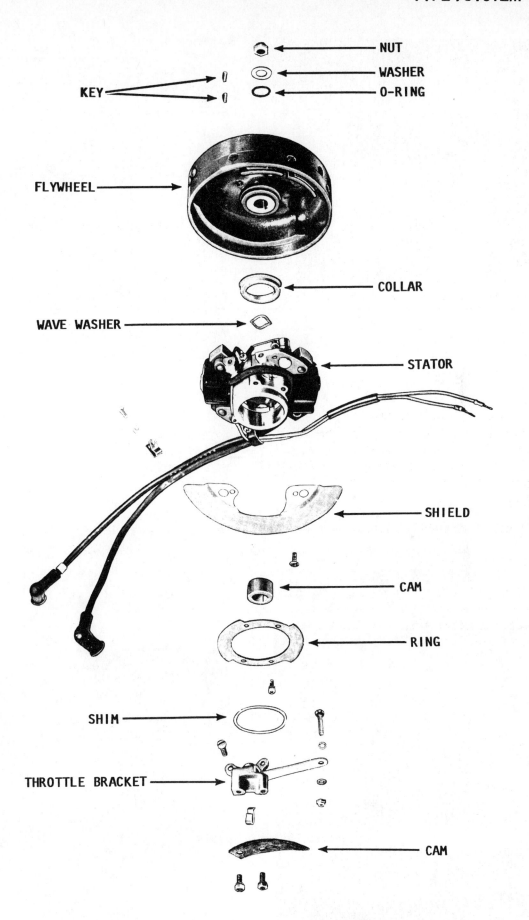

NUT

WASHER

O-RING

KEY

FLYWHEEL

COLLAR

WAVE WASHER

STATOR

SHIELD

CAM

RING

SHIM

THROTTLE BRACKET

CAM

Arrangement of principle parts for the Type I Phelon Magneto ignition system with points.

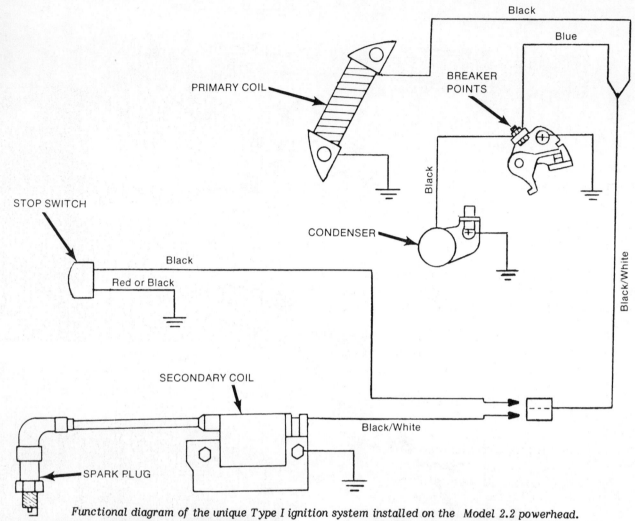

Functional diagram of the unique Type I ignition system installed on the Model 2.2 powerhead.

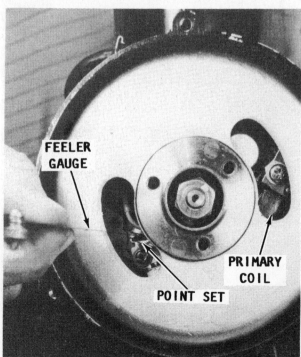

The point gap on a Model 2.2 powerhead may be checked without removing the flywheel.

After the flywheel is removed, components on the stator plate may be easily serviced.

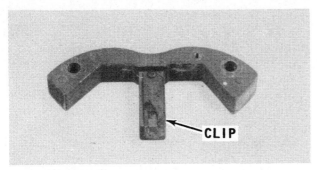

*Laminated core ready for service. If the clip is broken, the core **MUST** be replaced.*

ASSEMBLING
TYPE I IGNITION SYSTEM

1- Slide the center piece of the laminated core into the coil, as shown. Connect the wire between the coil and the laminated core.

2- Place the assembled laminated core and coil in position on the stator plate. Secure the assembly with the two hold-down screws.

3- Ensure the coil is bottomed onto the center piece of the laminated core, and then bend the tab upward to secure the coil in place. If servicing a 2-cylinder unit, repeat Steps 1, 2, and 3 for the second coil.

4- Insert the condenser into the grounding bracket, and then secure the bracket in place on the stator place with the attaching screw.

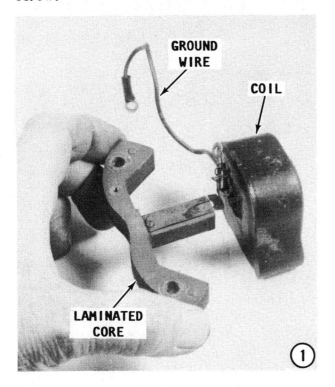

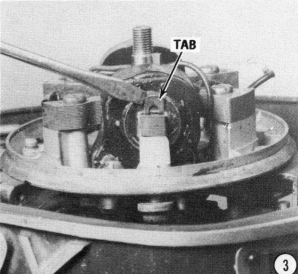

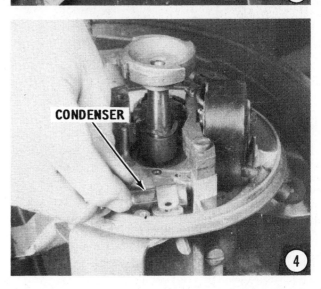

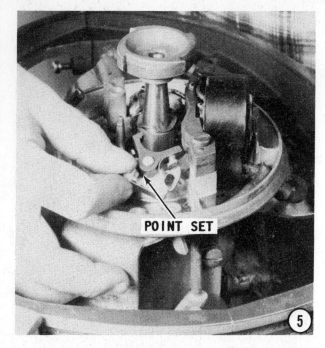

POINT SET

⑤

5- Place a **NEW** set of breaker points in position on the stator plate with the pin on the bottom side indexed into the hole in the plate. Secure the point set with the attaching screw. Do not tighten the screw at this time. Allow just a bit of slack to allow movement of the point set for adjustment later.

6- Insert the terminal block into the recess of the stator plate. Connect the condenser and coil wires to the outer terminal of the block. Check to be sure the flywheel will not make contact with the point set as it rotates. Verify the points are free to open and close.

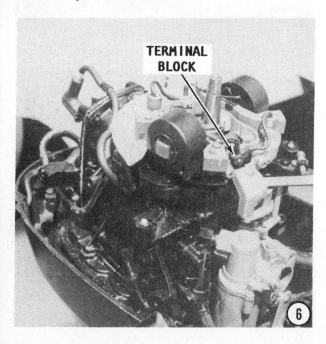

TERMINAL BLOCK

⑥

Point Adjustment

7- Temporarily install the collar onto the end of the crankshaft. Rotate the crankshaft **CLOCKWISE** and at the same time observe the cam on the crankshaft. Continue rotating the crankshaft until the rubbing block of the point set is at the high point of the cam. In this position, use a wire feeler gauge and set the points according to the Specifications in the Appendix.

CRITICAL WORDS

If the powerhead has been disassembled and both coil wires have been disconnected, the No. 1 piston must be at TDC when the point set for that cylinder is about to open. One very simple method of verifying the cylinder is at TDC, is to remove the spark plug, and then hold your thumb tightly over the opening. Now, with your other hand, rotate the crankshaft with the collar. As the piston moves upward, pressure will be felt against your thumb. If the piston moves passed TDC, the pressure will be reduced and replaced by a suction feeling on the thumb. In this manner, the position of the piston at TDC can be determined. A dial indicator or other arrangement may also be employed to determine when a piston is at TDC.

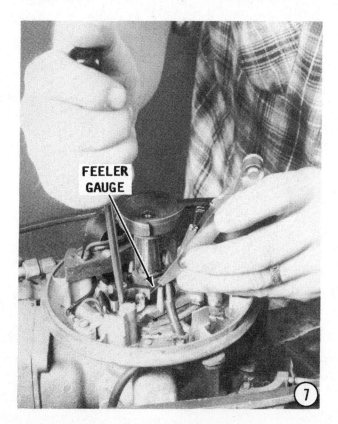

FEELER GAUGE

⑦

8- After the proper point gap has been obtained in Step 7, tighten the hold-down screw securely. Rotate the crankshaft a couple of complete revolutions, and then check the point gap adjustment.

If servicing a 2-cylinder unit, repeat Steps 7 and 8 for the second set of points.

Remove the collar from the crankshaft.

Flywheel Installation

When servicing the Model 200 and the 350, 1965-69, maximum spark **MUST** be set before the flywheel is installed. This procedure is covered in Chapter 6.

9- Check to be sure there are no metal parts stuck to the magnets. Place the key in the crankshaft keyway. Check to be sure the inside taper of the flywheel and the taper on the crankshaft are clean of dirt or oil, to prevent the flywheel from "walking" on the crankshaft during operation. Slide the flywheel down over the crankshaft with the keyway in the flywheel aligned with the key on the crankshaft.

On some models a flat washer with a boss is used. Install this washer with the boss on the washer facing **DOWN** toward the flywheel.

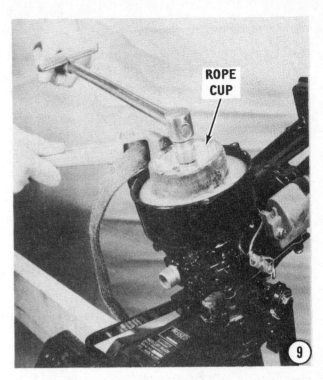

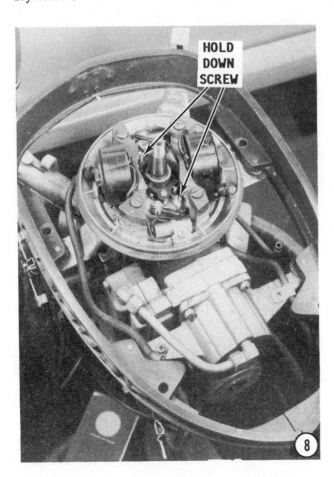

Rotate the flywheel clockwise and check to be sure the flywheel does not contact any part of the magneto or the wiring. Thread the flywheel nut onto the crankshaft and tighten it to the torque value given in the Appendix.

10- Check the gap on each spark plug according to the Specifications in the Appendix. Install the spark plugs and tighten them to a torque value of 20-1/2 ft lbs (27 Nm). Connect the battery leads to the battery terminals. For timing and synchronizing, see Chapter 6.

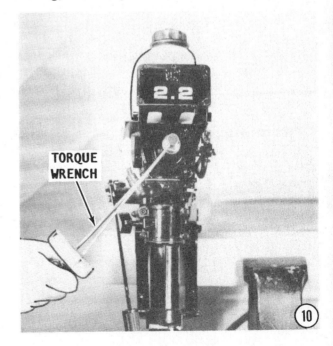

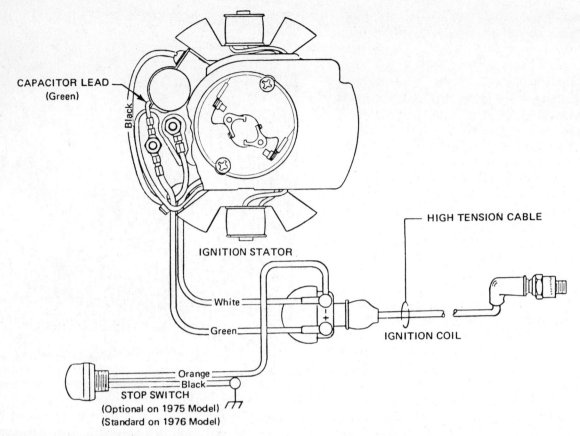

CAPACITOR LEAD —
(Green)

Black

IGNITION STATOR

HIGH TENSION CABLE

White

Green

IGNITION COIL

Orange
Black
STOP SWITCH
(Optional on 1975 Model)
(Standard on 1976 Model)

Functional diagram of the Type II ignition system with points and electrical module.

5-6 TYPE II IGNITION SYSTEM THUNDERBOLT -- FLYWHEEL PHASE MAKER -- WITH POINTS

DESCRIPTION

This ignition system is identified as Type II in the Specifications in the Appendix.

Basically, the Type II ignition system consists of a low-speed and high-speed generating coil. The maker contacts and generating coils are phased (rotate together for ignition timing change). This arrangement results in a very efficient ignition power source. The voltage generated by the coils is rectified and used to charge a capacitor. The capacitor is allowed to discharge into the primary of the ignition when the contacts close.

In this system, the current flows to the spark plug and the air-fuel mixture in the cylinder is ignited when the points **CLOSE**. This fact is in direct contrast to most other systems where the current reaches the spark plug when the points are open.

High resistance at the points is not a factor, since the capacitor voltage (primary) is approximately 1000 volts.

TROUBLESHOOTING TYPE II IGNITION SYSTEM

Always attempt to proceed with the troubleshooting in an orderly manner. The shotgun approach will only result in wasted time, incorrect diagnosis, replacement of unnecessary parts, and frustration. Begin the ignition system troubleshooting with the spark plug/s and continue through the system until the source of trouble is located.

Remember, the Phase Maker system is a self-contained unit. Therefore, if the engine has a key switch and wire harness, remove them from the engine and then make a test for spark. If a good spark is obtained with these two items disconnected, but no spark is available at the plug when they are connected, then the trouble is in the harness or the key switch. If a test is made for spark at the plug with the harness and switch connected, check to be sure the key switch is turned to the **ON** position.

Key Switch

A marine-type key switch **MUST** be installed as a replacement item. An automotive-type switch installation may cause damage to the system.

Spark Plugs

1- Check the plug wires to be sure they are properly connected. Check the entire length of the wire/s from the plug/s to the coil. If the wire is to be removed from the spark plug, **ALWAYS** use a pulling and twisting motion as a precaution against damaging the connection.

2- Attempt to remove the spark plug/s by hand. This is a rough test to determine if the plug is tightened properly. You should not be able to remove the plug without using the proper socket size tool. Remove the spark plug/s and remember from which cylinder they were removed. Examine each plug and evaluate its condition as described in Section 5-2.

3- Use a spark tester and check for spark at each cylinder. If a spark tester is not available, hold the plug wire about 1/4" (6.3mm) from the powerhead. Turn the flywheel with a pull starter or electrical starter and check for spark. A strong spark over a wide gap must be observed when testing in this manner, because under compression a strong spark is necessary in order to ignite the air-fuel mixture in the cylinder. This means it is possible to think you have a strong spark, when in reality the spark will be too weak when the plug is installed. If there is no spark, or if the spark is weak, the trouble is most likely under the flywheel in the Phase Maker system.

ONE MORE WORD

Each cylinder has its own ignition system in a flywheel-type ignition system. This means if a strong spark is observed for one cylinder and not the other, only the weak

system is at fault. However, it is always a good idea to check and service both systems while the flywheel is removed.

Compression

Before spending too much time and money attempting to trace a problem to the ignition system, a compression check of each cylinder should be made. If the cylinder does not have adequate compression, troubleshooting and attempted service of the ignition or fuel system will fail to give the desired results of satisfactory engine performance.

SPARK PLUG LEAD

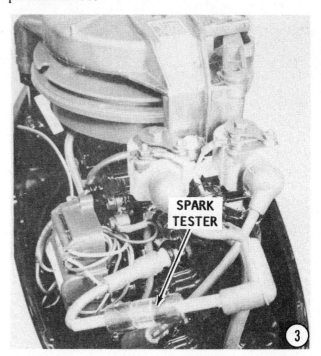

SPARK TESTER

Making a compression check should become standard procedure before any major work or tune-up is performed. Without adequate compression in each cylinder, any other work will have little affect on performance.

Remove the spark plug wire/s by pulling and twisting **ONLY** on the molded cap. **NEVER** pull on the wire because the connection inside the cap may be separated or the boot be damaged. Remove the spark plug/s. Insert a compression gauge into the cylinder spark plug hole. Crank the engine for several revolutions and note the compression reading. Repeat the procedure for each cylinder.

A variation in reading between the cylinders is far more important than the actual individual readings. If one cylinder varies more than 20 psi (138 kPa) from the other, the cylinder may be scored, the rings frozen, or the piston burned. The powerheads covered in this manual (except units with Type "C" powerheads), do not use a cylinder head. Therefore, low compression in one cylinder **CANNOT** be attributed to a blown head gasket. Compression for a one-cylinder unit should be approximately 110 psi (760 kPa).

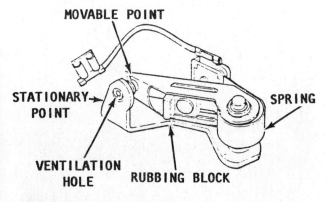

Line drawing of a typical point set with principle parts identified.

PHASE MAKER POINTS

Voltages as high as a 1000 volts can pass through "maker points" which cause rapid pitting. This pitting is not detrimental until accurate adjustment can no longer be made. 500 to 750 hrs of satisfactory operation is not uncommon! **REMEMBER**, the less an outboard engine is operated, the more care it needs. Allowing an outboard engine to remain idle will do more harm than if it is used regularly.

A breaker point set consists of two points. One is attached to a stationary bracket and does not move. The other point is attached to a moveable mount. A spring is used to keep the points in contact with each other, except when they are separated by the action of a cam built into the flywheel or machined on the crankshaft. Both points are constructed with a steel base and a tungsten cap fused to the base.

To properly diagnose spark problems, the theory of electricity flow must be understood. The flow of electricity through a wire may be compared with the flow of water through a pipe. Consider the voltage in the wire as the water pressure in the pipe and the amperes as the volume of water. Now, if the water pipe is broken, the water does not reach the end of the pipe. In a similar manner if the wire is broken the flow of electricity is broken. If the pipe springs a leak, the amount of water reaching the end of the pipe is reduced. Same with the wire. If the installation is defective or

the wire becomes grounded, the amount of electricity (amperes) reaching the end of the wire is reduced.

Check the wiring carefully, inspect the points closely, and adjust them accurately according to the Specifications in the Appendix.

Testing Preparation

FIRST, THESE WORDS

It is not necessary to remove the flywheel in order to test the breaker points in this system.

1- Remove the engine cowl or wrap-around cover. Remove the white wire from the secondary ignition coil.

2- Connect one lead of an ohmmeter to a good ground on the powerhead and connect the other lead to the white stator lead. Now, **SLOWLY** rotate the flywheel and at the same time observe movement of the ohmmeter indicator. The meter indicator **MUST** open and close as the flywheel rotates. If the points fail this test the point set must be replaced, see Disassembling and Assembling steps in this section.

Detailed Phase Maker Testing

Detailed testing of the Phase Maker system can only be properly performed after the flywheel has been removed, see Servicing, this section.

A VOA (Volt/Ohm/Ampere) meter or an ignition analyzer is required for proper testing. Before starting the testing process, the test meter **MUST** be adjusted to the set position with the leads connected.

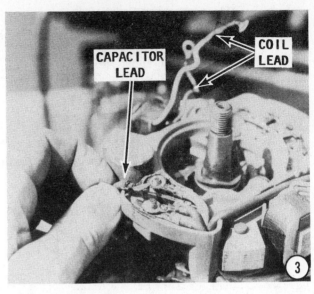

Hard Starting or
Poor Low-speed Performance

3- **TAKE TIME** to make a color code diagram of the electrical leads before disconnecting them. The color code terminals will be used during the tests and the diagram will ensure each lead will be reconnected to the proper terminal. Remove the four generating coil leads, the capacitor lead, and the heavy green leads from the electrical module.

4- Use the VOA range at Rx100 and check the low-speed resistance between the red and yellow coil leads. The reading **MUST** be 32-40.

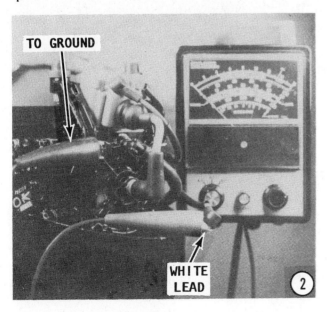

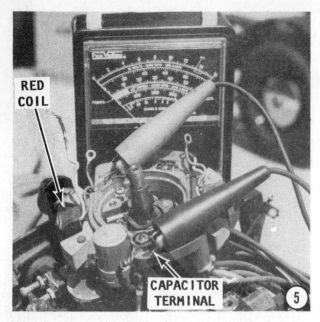

RED COIL

CAPACITOR TERMINAL

5

BLUE LEAD

CAPACITOR TERMINAL

7

5- Check the forward resistance of the rectifier: Connect the red lead of the meter to the electrical module terminal from which the red coil was disconnected. Connect the black meter lead to the terminal from which the capacitor lead was removed. Use the VOA range at Rx100 and check the resistance. The reading **MUST** be 10-15. Check the reverse resistance of the rectifier: Reverse the meter test leads, and use the VOA range at Rx100. There must be **NO** reading.

Poor High-Speed Performance

6- Check resistance: Connect the test leads between the blue and yellow coil

leads. Use the VOA range at Rx100. The reading **MUST** be 1.6-2.0.

7- Check rectifier resistance: Connect the red meter lead to the electrical module terminal from which the blue coil lead was disconnected. Connect the black meter lead to the terminal from which the capacitor lead was disconnected. Use the VOA range at Rx100. The reading **MUST** be 4.0-5.0. Check the reverse resistance of the rectifier: Reverse the meter leads and there must be **NO** reading.

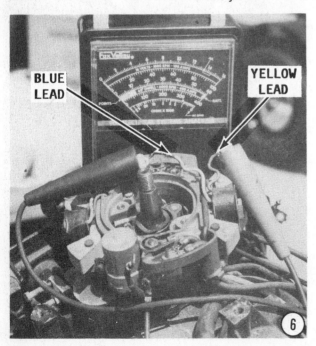

BLUE LEAD

YELLOW LEAD

6

GROUND

CAPACITOR LEAD

8

No Ignition Output

8- Connect one lead of the meter to a good ground on the powerhead. Connect the other lead to the capacitor (condenser) lead. The reading should be .45 to .55. If the capacitor (condenser) fails the test, the faulty unit must be replaced. A faulty condenser may not be detected without the use of special test equipment. The modest cost of a new condenser justifies its purchase and installation to eliminate this item as a source of trouble. To test the coil, see the Appendix for connections and specifications.

SERVICING
TYPE II IGNITION SYSTEM

General Information

The Type II Phase Maker ignition system installed on the powerheads covered in this manual will usually operate over extremely long periods of time without requiring adjustment or repair. However, if ignition system problems are encountered, and the usual corrective actions such as replacement of spark plugs and breaker points does not correct the problem, the output should be checked to determine if the unit is functioning properly.

Overhaul procedures may differ slightly on various powerheads, but the following general basic instructions will apply to all Type II high speed Phase Maker ignition systems. Some model Phase Maker units use

an electrical module, while other models do not use the module. Service procedures will cover both units and begin immediately following procedures for removal of the flywheel.

FLYWHEEL REMOVAL

1- Remove the cowling or enough of the engine cover to expose the flywheel. Remove the hand starter.

FLYWHEEL PULLER

2- Remove the nut securing the flywheel to the crankshaft. It may be necessary to use some type of flywheel strap to prevent the flywheel from turning as the nut is loosened.

3- Install the proper flywheel puller. **NEVER** attempt to use a puller which pulls on the outside edge of the flywheel or the flywheel may be damaged. After the puller is installed, tighten the center screw onto the end of the crankshaft. Continue tightening the screw until the flywheel is released from the crankshaft. Remove the flywheel.

4- **STOP,** and carefully observe the wiring layout, color coding, and terminal connections. **TAKE TIME** to make notes or a diagram of the wire routing. You may elect to follow the practice of many professional mechanics by taking a series of photographs of the engine with the flywheel removed, one from the top, and a couple from the sides showing the wiring and arrangement of parts.

PHASE MAKER WITHOUT AN ELECTRICAL MODULE

DISASSEMBLING

1- Remove the cap screw from the stator advance arm, and then disconnect the electrical leads. Remove the two Phillips head screws securing the point cover in place. Lift the cover off the point set.

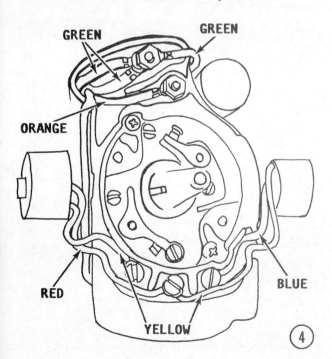

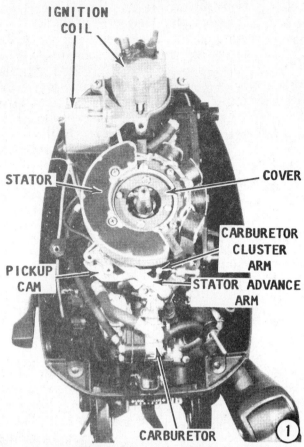

2- Remove the four screws and D-washer from the wire clamp. Lift the stator assembly from the engine.

SOME GOOD WORDS

High primary voltage in Thunderbolt ignition systems will darken and roughen the breaker points within a short period. This is not cause for alarm. Normally points in this

condition would not operate satisfactorily in the conventional magneto, but they will give good service in the Thunderbolt systems. Therefore, **DO NOT** replace the points in a Thunderbolt system unless an obvious malfunction exists, or the contacts are loose or burned. Rough or discolored contact surfaces are **NOT** sufficient reason for replacement. The cam follower will usually have worn away by the time the points have become unsatisfactory for efficient service.

3- To remove a point set: Remove the two hold-down screws and the 1/4" nut from the point set. Remove the electrical lead. Lift the point set from the armature plate. If servicing a two-cylinder unit, remove the other point set in the same manner.

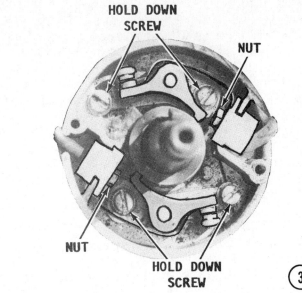

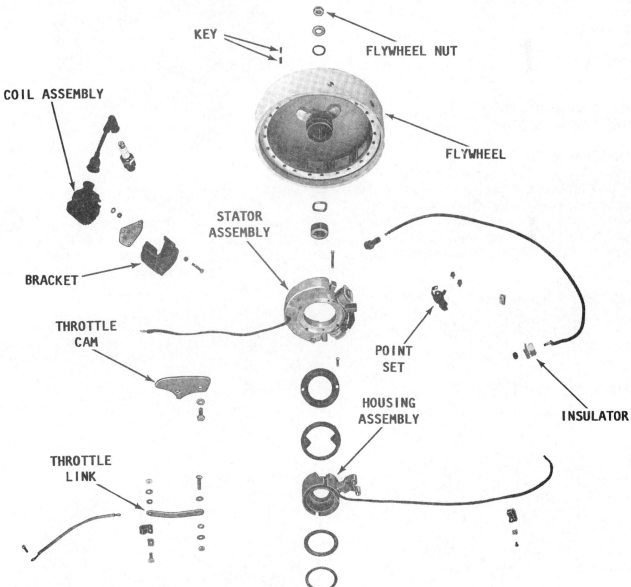

*Arrangement of principle parts for the Type II ignition system with points and **WITHOUT** an electrical module.*

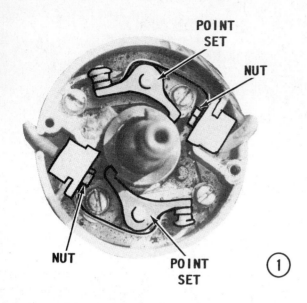

POINT SET

NUT

NUT

POINT SET

①

PHASE MAKER WITHOUT AN ELECTRICAL MODULE

ASSEMBLING

1- Place each new point set in position on the armature plate and secure it with the two hold down screws. Connect the electrical leads to the point set. Install and tighten the 1/4" nut.

2- Slowly rotate the flywheel until the rubbing block on the point set is at the high point of the cam. At this position, adjust the points to 0.020" (.51mm). Repeat the procedure for the other point set.

3- Install the stator housing onto the armature plate and secure it in place with the four Phillips screws. Place the stator cover over the stator and secure it with the two Phillips screws. Connect the ignition advance arm.

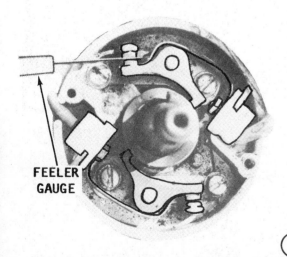

FEELER GAUGE

②

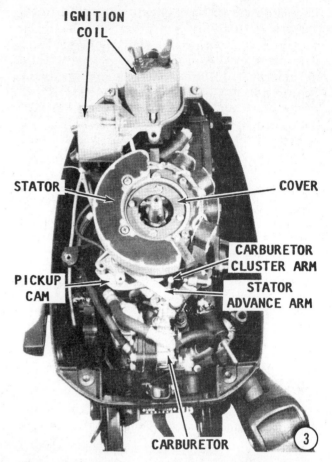

IGNITION COIL

STATOR

COVER

CARBURETOR CLUSTER ARM

STATOR ADVANCE ARM

PICKUP CAM

CARBURETOR

③

4- Place the key in the crankshaft keyway. Check the flywheel magnets to be sure they are free of any metal parts. Double check the taper in the flywheel hub and the taper on the crankshaft to verify they are clean and contain no oil, as a precaution against the flywheel "walking" on the crankshaft during operation. Slide the

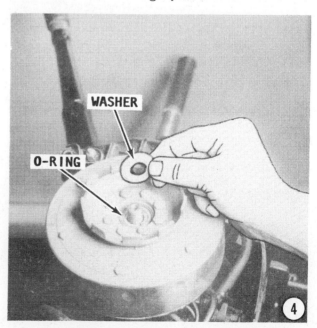

WASHER

O-RING

④

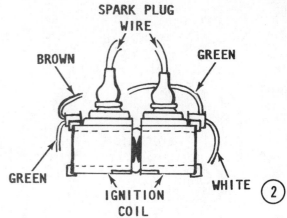

flywheel onto the crankshaft with the keyway in the flywheel aligned with the key on the crankshaft, then the **O-Ring** and washer.

5- Thread the nut onto the crankshaft, and then tighten it to the torque value given in the Specifications in the Appendix.

PHASE MAKER WITH AN ELECTRICAL MODULE

DISASSEMBLING

1- Remove the flywheel according to the procedures outlined in the first portion of this section. Remove the two Phillips attaching screws, and then remove the cover from the module.

2- Remove the stator ground strap. Disconnect the green, brown, and white electrical leads from the ignition coil/s; and the capacitor (condenser) lead from the terminal.

3- To remove the point set: Remove the two hold down screws. Remove the 1/4" nut and the electrical leads. Lift the point set from the armature plate. If servicing a

two-cylinder powerhead, remove the other point set in the same manner.

4- To remove the module: Remove the two nuts securing the electrical leads from the coil. Remove the three screws securing the stator leads. Remove the attaching screws and lift the module clear.

5- Remove the coil by simply removing the coil clips and pulling the coil from the plate. Rotate and remove the stator plate.

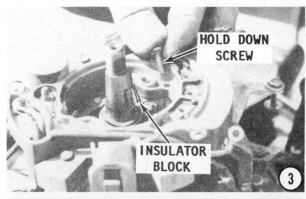

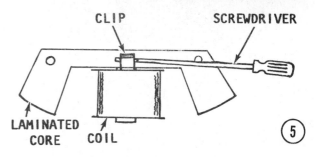

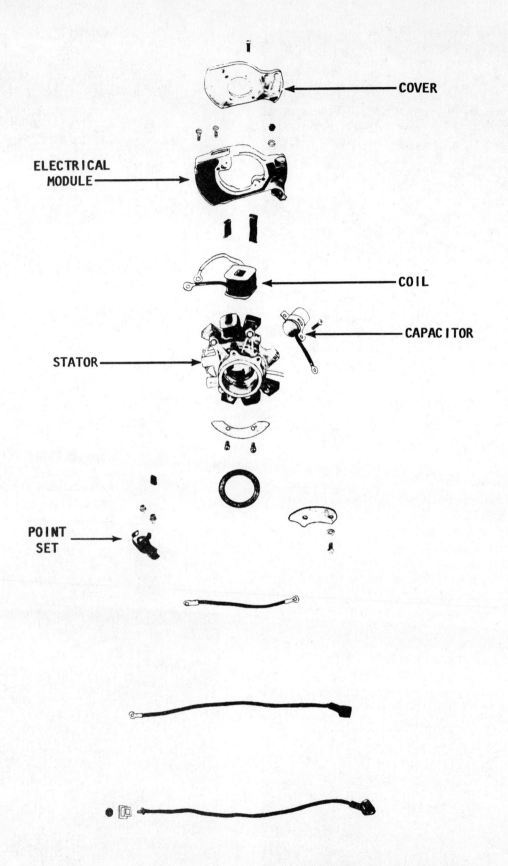

COVER

ELECTRICAL
MODULE

COIL

CAPACITOR

STATOR

POINT
SET

*Arrangement of principle parts for the Type II ignition system with points and **WITH** an electrical module.*

CLEANING AND INSPECTING

Clean the armature plate with solvent, and then blow it dry with compressed air.

Lubricate the cam with a very thin film of cam lubricant. An excessive amount of lubricant will splatter during distributor operation and will also attract dirt particles. If no lubrication is used, the cam follower will wear very quickly.

SOME GOOD WORDS: High primary voltage in Thunderbolt ignition systems will darken and roughen the breaker points within a short period. This is not cause for alarm. Normally points in this condition would not operate satisfactorily in the conventional magneto, but they will give good service in the Thunderbolt systems. Therefore, **DO NOT** replace the points in a Thunderbolt Phase Maker system unless an obvious malfunction exists, or the contacts are loose or burned. Rough or discolored contact surfaces are **NOT** sufficient reason for replacement. The cam follower will usually have worn away by the time the points have become unsatisfactory for efficient service.

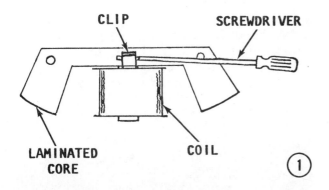

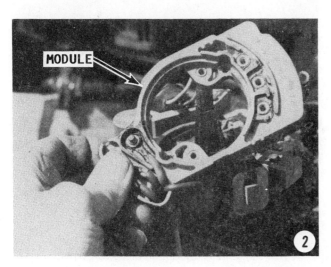

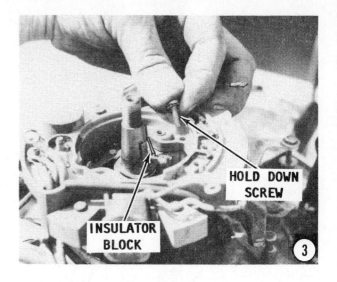

PHASE MAKER WITH AN ELECTRICAL MODULE

ASSEMBLING

1- Place the stator plate over the crankshaft and be sure the plate seats against the upper driveshaft bearing. Place the coils over the center piece of the laminated core and hook each clip over the core to secure the core.

2- Place the electrical module in position over the stator. Secure it in place with the attaching screws. Connect the coil leads and the stator leads.

3- Place each point set in position on the armature plate and secure them with the hold-down screws. Connect the electrical leads from each point set to their corresponding insulator blocks and secure them snugly with the 1/4" nuts.

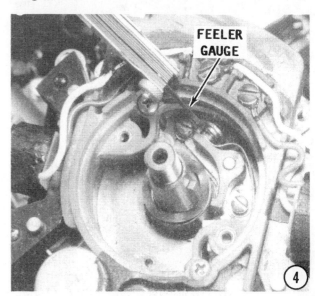

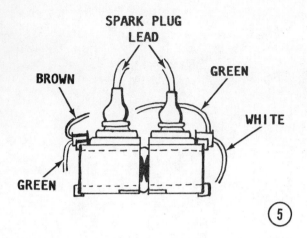

SPARK PLUG LEAD

BROWN

GREEN

WHITE

GREEN

5

4- Slowly rotate the flywheel until the rubbing block on the point set is at the high point of the cam. At this position, adjust the points to 0.020" (.51mm). Repeat the procedure for the other point set.

5- Connect the green, brown, and white wire leads to the coils. Connect the stator ground wire and the capacitor lead to the terminal.

6- Place the cover in position over the module and secure it with the two attaching Phillips screws.

7- Place the key in the crankshaft keyway. Check the flywheel magnets to be sure they are free of any metal parts. Double check the taper in the flywheel hub and the taper on the crankshaft to verify they are clean and contain no oil, as a precaution

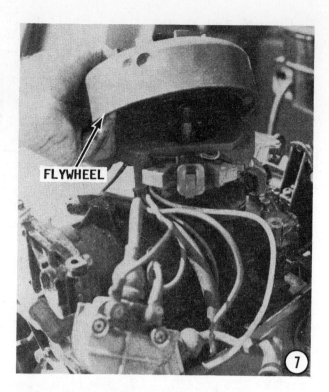

FLYWHEEL

7

against the flywheel "walking" on the crankshaft during operation. Slide the flywheel onto the crankshaft with the keyway in the flywheel aligned with the key on the crankshaft.

8- Thread the nut onto the crankshaft, and then tighten it to the torque value given in the Specifications in the Appendix. Install the hand starter. For detailed timing and synchronizing procedures, see Chapter 6. Install the cowling or engine cover.

COVER

6

TORQUE WRENCH

8

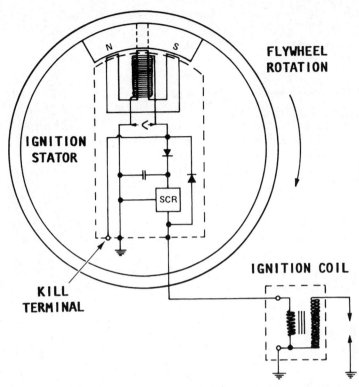

Simplified diagram of the Type III ignition system covered in this section.

5-7 TYPE III IGNITION SYSTEM
THUNDERBOLT — FLYWHEEL — CD
POINTLESS

DESCRIPTION

This ignition system is identified as Type III in the Specifications in the Appendix.

The Type III CD (capacitor discharge), ignition system consists of generating coils fastened to a stator. The stator is mounted around the crankshaft under the flywheel. The flywheel is fitted with permanent magnets inside the outer rim.

As the crankshaft and flywheel rotate, the magnets pass the generating coils. As they pass the coils, an AC voltage is generated at the coil terminals. This AC voltage is conducted to the switch box, where it is rectified and stored in a capacitor.

The trigger coil is also mounted under the flywheel, and a second set of magnets is installed in the flywheel hub. This second set of magnets causes the trigger coil to produce AC voltage as the flywheel rotates. This voltage is conducted to the switch box, where it is connected to a switch (SCR).

Polarity of the AC trigger signal determines into which ignition coil the capacitor will be discharged. An ignition coil is installed for each cylinder. The capacitor voltage is conducted to the primary winding of the ignition coil where a high strength magnetic field is built around the coil core. As the switch (SCR) turns off, there is no voltage to sustain this magnetic field, and it collapses rapidly. This rapid collapse induces a very high voltage in the secondary winding of the coil. This voltage is applied to the spark plug by a high-tension lead.

The preceding sequence occurs once per engine revolution for each cylinder.

Change in spark timing, advance or retard, is accomplished by rotating the trigger coil in relationship to the magnets on the flywheel hub.

The engine is shut down by shorting the orange wire switch box terminal or the charging coil terminal block under the edge of the flywheel to ground with the key switch, the stop button, or the ignition safety stop switch.

CRITICAL WORDS

These next two paragraphs may well be the most important words in this chapter. Misuse of the wiring harness is the most single cause of electrical problems with outboard power plants.

A wiring harness is used between the key switch and the powerhead. This harness seldom contains wire of sufficient size to allow connecting accessories. Therefore, anytime a new accessory is installed, **NEW** wiring should be used between the battery and the accessory. A separate fuse panel **MUST** be installed on the control panel. To connect the fuse panel, use two red and black No. 10 gauge wires from the battery. C.D. ignition systems require a full 12 volts for proper operation. Therefore, again let it be said, **NEVER** connect accessories through the key switch.

Key Switch

A marine type key switch **MUST** be installed as a replacement item. An automotive type switch installation may cause damage to the system.

TROUBLESHOOTING
TYPE III IGNITION SYSTEM

Always attempt to proceed with the troubleshooting in an orderly manner. The shotgun approach will only result in wasted time, incorrect diagnosis, replacement of unnecessary parts, and frustration.

Begin the ignition system troubleshooting with the spark plug/s and continue through the system until the source of trouble is located.

Spark Plugs

1- Check the plug wires to be sure they are properly connected. Check the entire length of the wire/s from the plug/s to the coils. If the wire is to be removed from the spark plug, **ALWAYS** use a pulling and twisting motion as a precaution against damaging the connection.

2- Attempt to remove the spark plug/s by hand. This is a rough test to determine if the plug is tightened properly. You should not be able to remove the plug without using the proper socket size tool. Remove the spark plug/s and keep them in order. Examine each plug and evaluate its condition as described in Section 5-2.

3- Use a spark tester and check for spark at each cylinder. If a spark tester is not available, hold the plug wire about 1/4" (6.35mm) from the engine. Turn the flywheel with a pull starter or electrical starter and check for spark. A strong spark over a wide gap must be observed when testing in this manner, because under compression a strong spark is necessary in order to ignite the air/fuel mixture in the cylinder. This means it is possible to think you have a strong spark, when in reality the spark will be too weak when the plug is installed. If there is no spark, or if the spark is weak, the trouble is most likely under the flywheel in the CD system.

ONE MORE WORD: Each cylinder has its own ignition system in a flywheel-type

SPARK PLUG LEAD

①

②

ignition system. This means if a strong spark is observed at one cylinder and not at the other, only the weak system is at fault. However, it is always a good idea to check and service both systems while the flywheel is removed.

A trigger coil failure on a two-cylinder engine will affect spark to both cylinders.

Compression

Before spending too much time and money attempting to trace a problem to the ignition system, a compression check of each cylinder should be made. If the cylinder does not have adequate compression, troubleshooting and attempted service of the ignition or fuel system will fail to give the desired results of satisfactory engine performance.

Remove the spark plug wire/s by pulling and twisting **ONLY** on the molded cap. **NEVER** pull on the wire because the connection inside the cap may be separated or the boot be damaged. Remove the spark plug/s. Insert a compression gauge into the cylinder spark plug hole. Crank the engine for several revolutions and note the compression reading. Repeat the procedure for each cylinder.

A variation in reading between the cylinders is far more important than the actual individual readings. If one cylinder varies more than 20 psi (138 kPa) from the other one, the cylinder may be scored, the rings frozen, or the piston burned. The powerheads covered in this manual (except the Model 2.2) do not have a head. Therefore, low compression in one cylinder **CANNOT** be attributed to a blown head gasket.

The compression reading for a one-cylinder powerhead should be approximately 110 psi (760 kPa).

Testing

The following tests may be made without removing the flywheel. The test will indicate if the switch box, or other flywheel components are defective. The switch box has been mounted under the flywheel since about 1980. On models previous to 1980, the flywheel does not have to be removed in order to replace the switch box.

All of the tests may be performed with a VOA (Volt/Ohm/Ampere) meter or with an ignition analyzer. The general procedure is to disconnect the leads from the switch box, or the switch box terminals, and then to make the test with the meter. Compare the results with the Specifications given in the Appendix. Replace any part that fails to meet the specifications. These tests are difficult to perform without the proper test

Taking a compression check of each cylinder should become standard procedure before any major work or tune-up is performed. Without adequate compression in each cylinder, any other work is useless.

equipment. Your local marine shop is equipped to make an accurate determination of serviceable and faulty components and thus save you money.

SERVICING
TYPE III IGNITION SYSTEM

The Type III CD ignition system installed on the powerheads covered in this manual will usually operate over extremely long periods of time without requiring adjustment or repair. However, if ignition system problems are encountered, and the usual corrective actions such as replacement of spark plugs and timing check does not correct the problem, the CD output should be checked to determine if the unit is functioning properly.

Ignition overhaul procedures may differ slightly on various powerhead models, but the following general basic instructions will apply to all Type III high-speed flywheel CD ignition systems.

REMOVAL

1- Remove enough of the engine cowling or cover to expose the flywheel.

Model 3.6 Only
Pull the carburetor front cover free of the cowling. Unscrew the fuel cap; com-

press the plastic fuel cap retainer; and then remove the fuel cap assembly. remove the cowling.

Model 3.5 Only
Remove the rubber seal from around the fuel tank filler neck. Lift up on the cowling to disengagte the two locating pins. Tilt the cowling forward to clear the starter handle and pull the cowl free.

Disconnect the leads from the battery terminals.

2- Remove the crankshaft nut in the center of the flywheel. A flywheel strap may be required to hold the flywheel securely while the nut is loosened.

3- Obtain the proper type flywheel puller. **NEVER** attempt to use a puller which pulls on the outside edge of the flywheel or the flywheel may be damaged. After the puller is installed, tighten the center screw onto the end of the crankshaft. Continue tightening the screw until the flywheel is released from the crankshaft. Remove the flywheel.

4- Determine the faulty part, if this has not been done, according to the good words in Troubleshooting, this Section. Remove the faulty part as determined from the testing in the previous section, paying particular attention to wiring connections. Take time to either make a drawing or take a polaroid picture of the area to **ENSURE**, without a doubt, the wiring will be connected in proper sequence.

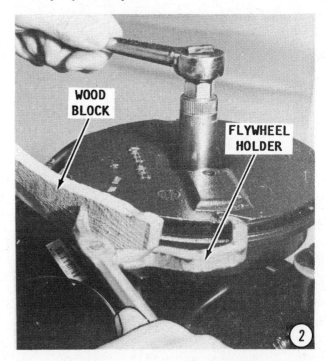

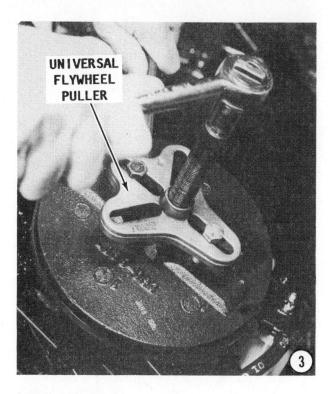

UNIVERSAL
FLYWHEEL
PULLER

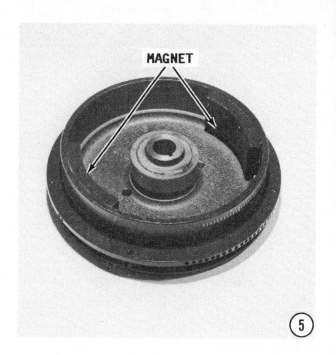

MAGNET

INSTALLATION

5- Check the inside of the flywheel for any indication of metal particles adhering to the magnets. Double check the taper in the flywheel hub and the taper on the crankshaft to verify they are clean and contain no oil, as a precaution against the flywheel "walking" on the crankshaft during operation. Insert the key into the crankshaft keyway. Slide the flywheel onto the crankshaft, with the slot in the flywheel aligned with the key on the crankshaft. Install and tighten the flywheel nut to the torque value given in the Specifications in the Appendix.

6- Install the hand starter, if one is used. Install the cowling or engine cover. Connect the battery leads to the battery terminals.

For timing and synchronizing, see Chapter 6.

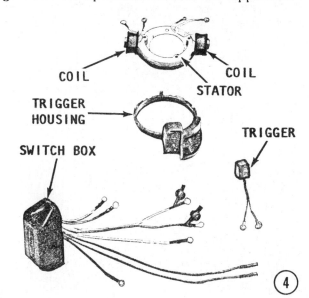

COIL

COIL

STATOR

TRIGGER
HOUSING

TRIGGER

SWITCH BOX

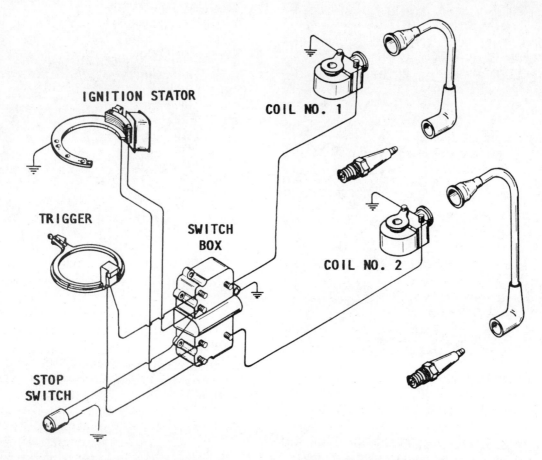

Functional diagram of the Type IV ignition system covered in this section for 2-cylinder powerheads.

5-8 TYPE IV IGNITION SYSTEM
THUNDERBOLT — FLYWHEEL — CD
COIL PER CYLINDER

DESCRIPTION

This system is identified in the Specifications in the Appendix as Type IV. The system has been used on many of the two-cylinder powerheads covered in this manual since 1973, and on some newer single cylinder powerheads since mid 1987. The Specification listings in the Appendix, indicates the models with the Type IV system installed.

The Type IV system is alternator-driven with distributorless CD (capacitor discharge).

The system for a single-cylinder powerhead consists of a capacitor charging coil, a trigger coil, a CDI unit, an ignition coil and a spark plug. The capacitor charging coil is mounted under the flywheel. The trigger coil is mounted to one side of the flywheel in an easily accessible location. The CDI unit and ignition coil are mounted on the powerhead.

Permanent magnets are installed on the inside outer rim of the flywheel. The ignition coil produces AC voltage as the permanent magnets in the flywheel pass the stationary capacitor charging coil. The AC voltage is conducted to the CDI unit where it is rectified and stored in a capacitor (condenser).

A second set of magnets is mounted around the flywheel hub. When the second set of magnets passes the trigger coil, AC voltage is produced and conducted to the CDI unit. This voltage produces a signal which discharges the capacitor voltage into the ignition coil at a precise time.

The capacitor voltage is conducted to the ignition coil primary. The ignition coil increases this voltage enough to jump the gap at the spark plug.

This sequence occurs once for each cylinder per crankshaft revolution.

No advance or retard timing is possible on these single-cylinder units.

The system for a two-cylinder powerhead consists of a stator assembly, trigger assembly, switchbox, plus one ignition coil, and one spark plug per cylinder.

The stator assembly is mounted beneath the flywheel. Permanent magnets are installed on the inside outer rim of the flywheel. The ignition coils produce AC voltage as the permanent magnets in the flywheel pass the stationary stator ignition coils. The AC voltage is conducted to the switchbox where it is rectified and stored in a capacitor (condenser).

The trigger assembly has two coils and is also mounted under the flywheel. A second set of magnets is mounted around the flywheel hub. When the second set of magnets passes the trigger coils, AC voltage is produced and conducted to an electronic switch (SCR) in the switchbox.

The switch discharges the capacitor voltage into the ignition coil at a precise time and in the proper firing order sequence.

The capacitor voltage is conducted to the ignition coil primary. The ignition coil increases this voltage enough to jump the gap at the spark plug.

This sequence occurs once for each cylinder per crankshaft revolution.

Advance or retard is accomplished by rotating the trigger coil position in relation to the permanent magnets on the flywheel hub.

CRITICAL WORDS

These next two paragraphs may well be the most important words in this chapter. Misuse of the wiring harness is the most single cause of electrical problems with outboard power plants.

A wiring harness is used between the key switch and the powerhead. This harness seldom contains wire of sufficient size to allow connecting accessories. Therefore, anytime a new accessory is installed, **NEW** wiring should be used between the battery and the accessory. A separate fuse panel **MUST** be installed on the control panel. To connect the fuse panel, use two red and black No. 10 gauge wires from the battery. CD ignition systems require a full 12 volts for proper operation. Therefore, again let it be said, **NEVER** connect accessories through the key switch.

Key Switch

A marine type key switch **MUST** be installed as a replacement item. An automotive type switch installation may cause damage to the system.

TROUBLESHOOTING
TYPE IV IGNITION SYSTEM

READ, BELIEVE, & OBEY

Because high voltage is present, **NEVER** ever touch or disconnect any ignition part on a powerhead equipped with a Type IV ignition system while:

a- The powerhead is running
b- The key switch is **ON**.
c- The battery cables are connected.

The following safety precautions are listed for your personal safety and to prevent damage to expensive parts.

NEVER reverse battery cable connections. The battery negative (-) is ground. The black cable must always be connected to this terminal. The red cable must always be connected to the positive (+) terminal.
NEVER check polarity by "sparking" the battery terminals with the battery cable connections.
NEVER disconnect the battery cables while the engine is running.
NEVER crank the engine if the switch boxes are not properly grounded to the engine.

Components of the Type IV ignition system installed on the powerheads covered in this manual may be tested with a VOA (Volt/Ohm/Ampere) meter.

ALWAYS check the following areas for sources of trouble **BEFORE** opening up the ignition system.

Check to be sure the electrical harness, ignition switch, and mercury switch are not the source of the problem.

Verify the plug-in connectors are fully engaged and the terminals are free of corrosion.

Check to be sure all electrical components and ground wires are properly grounded to the engine. Inspect visible wire connections to be sure they are tight and free of corrosion.

Observe the entire electrical system for disconnected wires, and for short or open circuits.

Ready For Troubleshooting

Always attempt to proceed with the troubleshooting in an orderly manner. The

SPARK PLUG
LEAD

"shot in the dark" approach will only result in wasted time, incorrect diagnosis, replacement of unnecessary parts, and frustration.

Begin the ignition system troubleshooting with the spark plug/s and continue through the system until the source of trouble is located.

Spark Plugs

1- Check the plug wires to be sure they are properly connected. Check the entire length of the wire/s from the plug/s to the coil. If the wire is to be removed from the

spark plug, **ALWAYS** use a pulling and twisting motion as a precaution against damaging the connection.

2- Attempt to remove the spark plug/s by hand. This is a rough test to determine if the plug is tightened properly. You should not be able to remove the plug without using the proper socket size tool. Remove the spark plug/s and keep them in order. Examine each plug and evaluate its condition as described in Section 5-2.

3- Use a spark tester and check for spark at each cylinder. If a spark tester is not available, hold the plug wire about 1/4" (6.4mm) from the powerhead. Turn the flywheel with a pull starter or electrical starter and check for spark. A strong spark over a wide gap must be observed when testing in this manner, because under compression a strong spark is necessary in order to ignite the air/fuel mixture in the cylinder. This means it is possible to think you have a strong spark, when in reality the spark will be too weak when the plug is installed. If there is no spark, or if the spark is weak, the trouble is most likely in the CD system.

ONE MORE WORD

Each cylinder has its own ignition system in a flywheel-type ignition system. This means if a strong spark is observed for one cylinder and not at the other, only the weak

SPARK PLUG
TESTER

Making a compression check should become standard procedure before any major work or tune-up is performed. Without adequate compression in each cylinder, any other work will have little affect on performance.

system is at fault. However, it is always a good idea to check and service both systems while the flywheel is removed.

Compression

Before spending too much time and money attempting to trace a problem to the ignition system, a compression check of each cylinder should be made. If the cylinder does not have adequate compression, troubleshooting and attempted service of the ignition or fuel system will fail to give satisfactory engine performance.

Remove the spark plug wire/s by pulling and twisting **ONLY** on the molded cap. **NEVER** pull on the wire because the connection inside the cap may be separated or the boot be damaged. Remove the spark plug/s. Insert a compression gauge into one cylinder spark plug hole. Crank the engine for several revolutions and note the compression reading. Repeat the procedure for the other cylinder.

A variation in reading between the two cylinders is far more important than the actual individual readings. If one cylinder varies more than 20 psi (138 kPa) from the other, the cylinder may be scored, the rings frozen, or the piston burned. With the exception of Type "C" powerheads which do not use a cylinder head. Therefore, low

compression in one cylinder **CANNOT** be attributed to a blown head gasket. A one-cylinder powerhead should have a compression reading of approximately 110 psi (760 kPa).

RESISTANCE TESTS
SINGLE-CYLINDER UNITS

Capacitor Charging Coil Resistance Tests

1- Disconnect the White and Black/Red leads leading from the capacitor charging coil under the flywheel at their quick disconnect fittings. Obtain and ohmmeter. Set the meter to the Rx10 scale. Make contact with the Red meter lead to the Black/Red lead from the coil. Make contact with the Black meter lead to the White coil lead. The meter should register between 9.3 and 14.2 ohms.

If the resistance is not within the limits given and the powerhead does not operate, the capacitor charging coil must be replaced.

To replace the coil, the hand rewind starter must be removed first, see Chapter 12, and then the flywheel "pulled". Use a strap wrench to secure the flywheel, while the flywheel nut is being loosened, and then use a flywheel puller, to "pull" the flywheel from the taper of the crankshaft.

Trigger Coil Resistance Test

2- Disconnect the Red/White lead leading from the trigger coil, mounted next to the flywheel, at its quick disconnect fitting. Remove the forward screw retaining the coil to the housing, and remove the Black lead from this retaining screw. Select the Rx10 scale on the ohmmeter. Make contact with the Red meter lead to the Black lead from the trigger coil. Make contact with

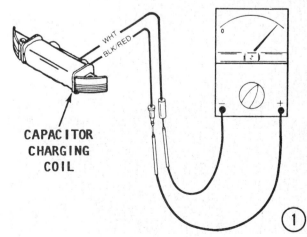

CAPACITOR
CHARGING
COIL

①

the Black meter lead to the Red/White lead from the trigger. The meter should register between 8.0 and 11.5 ohms.

If the resistance is not within the limits given and the powerhead does not operate, the trigger coil must be replaced.

To replace the trigger coil, remove the two securing screws and pull the coil up and out of the housing. Pass the two leads through the grommet in the housing.

Ignition Coil Resistance Tests

FIRST, THESE WORDS

The following two tests are best performed with the ignition coil removed from the powerhead. Therefore, the next step begins with instructions for removal.

3- Slide the Yellow/Black lead from the ignition coil at the spade fitting on the coil. Disconnect the high tension lead at the spark plug. Remove the two securing bolts and lift the coil free of the powerhead. Three small black leads will remain, which were secured by the two bolts to the coil. Make a note of the original locations of these three leads, to ensure installation back into the same location from which they were removed.

To test the primary windings: select the Rx1 scale on the ohmmeter. Make contact with the Black meter lead to the spade terminal on the coil. Make contact with the Red meter lead to the laminated core of the coil on the **SAME SIDE** as the spade terminal. The meter should register between 0.02 and 0.38 ohms.

To test the secondary windings: select the Rx1000 scale on the ohmmeter. Make

contact with the Red meter lead to the spade terminal on the coil. Make contact with the Black meter lead to the high tension lead terminal. The meter should reister between 3.0 and 4.4 ohms.

If the reading is questionable but the powerhead operates, install a spark plug tester and check the strength of the spark.

If the spark is acceptable, it is not necessary to replace the ignition coil. However, keep this area in mind for possible future trouble.

If the spark is weak, it will get weaker. Therefore, replace the coil.

Position the ignition coil against the powerhead. Install the forward securing bolt through the laminated core, slide the eyelet connectors of the three small Black leads onto the bolt and thread the bolt into the powerhead. Install the other securing bolt. Tighten both bolts to a torque valve of 5.9ft lbs (8Nm). Connect the Yellow/Black lead from the CDI unit to the spade terminal on the coil. Install the high tension lead to the spark plug.

Other Electrical Components

The CDI unit is located under the ignition coil. At press time, resistance specifications to test this unit were not available

TRIGGER COIL

COIL LEADS

2

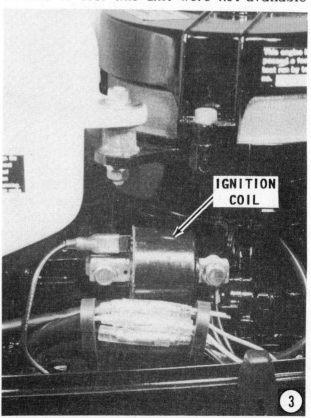

IGNITION COIL

3

from the manufacturer. Systematic resistance tests of the capacitor charging coil, trigger coil, and ignition coil, as well as inspection of quick disconnect fittings, "kill" switch action and spark plug condition may serve to eliminate ignition problems. If all areas are tested and inspected and the powerhead still does not operate in a satisfactory manner, then the CDI unit must be removed and replaced.

Some models may be equipped with a lighting coil under the flywheel. This coil may be identified by its clean laminated copper windings. The lighting coil, so named because it may be used to power the lights on the boat when used together with a voltage regulator, allows the powerhead to generate additional current to charge the battery. Manual start models may be equipped with a lighting coil, but the coil leads will be taped off and not used. Again, at press time, no resistance specifications were available for testing this coil.

RESISTANCE TESTS
TWO-CYLINDER UNITS

Stator Types

Two types of stator assemblies are installed on powerheads with the Type IV ignition system. One type has a black ground wire. The stator is grounded to the powerhead through this black ground wire.

The other type does not have the black ground wire. The stator is grounded directly to the powerhead through the stator mounting plate.

Stator Tests

FIRST, THESE WORDS

The stator **MUST** be grounded to the powerhead before making the following tests. If satisfactory meter readings are not obtained, the stator is defective and must be replaced.

1- Ground the stator to the powerhead. Refer to the appropriate table in the Appendix. Identify the leads listed under the column "Tester Leads To:". Select the appropriate scale on the VOA meter. Disconnect the lead/s at their quick disconnect fittings. Make contact with the VOA leads to the disconnected lead/s. The meter reading should be as indicated in the Appendix for the model being serviced.

If the resistance is not within the limits listed in the table and the powerhead does not operate, the stator must be replaced. Service or adjustment is not possible.

To replace the stator, the hand starter or electric cranking motor must be removed and the flywheel must be "pulled", refer to the procedures begining on Page 5-43.

Trigger Assembly Test

First, These Words

The switch box **MUST** be properly grounded to the powerhead before making this series of tests.

2- Check the small black lead from the switchbox is grounded securely to the powerhead. Refer to the appropriate table in the Appendix. Identify the leads listed under the column "Tester Leads To:" If the colors of the leads are not specified for the model being serviced, refer to the wiring diagram in the Appendix. Select the appropriate scale on the VOA meter. Disconnect the lead/s at their quick disconnect fittings. Make contact with the VOA leads to the disconnected lead/s. The meter read-

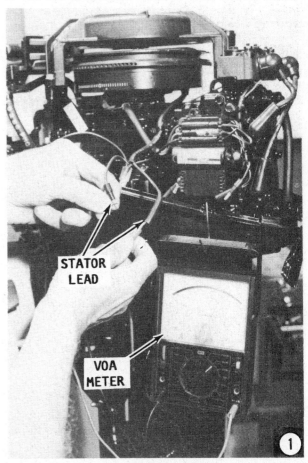

STATOR LEAD

VOA METER

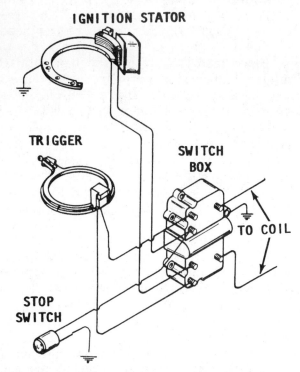

IGNITION STATOR

TRIGGER

SWITCH BOX

TO COIL

STOP SWITCH

Functional diagram of the Type IV ignition system used on 2-cylinder powerheads to show the physical relationship of the stator, trigger, switch box, and stop switch. The color code differs for various models. Therefore, check the listings in the Appendix.

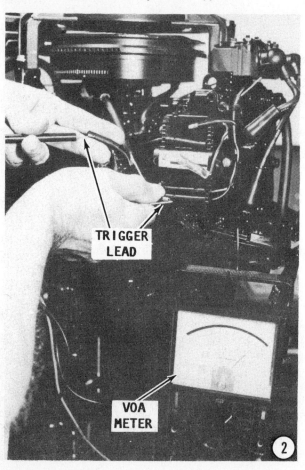

TRIGGER LEAD

VOA METER

②

ing should be as indicated in the Appendix for the model being serviced.

If the resistance is not within the limits listed in the table and the powerhead does not operate, the trigger assembly must be replaced. Service or adjustment is not possible.

To replace the assembly the hand starter or electric cranking motor must be removed and the flywheel must be "pulled", refer to the procedures begining on Page 5-43.

Ignition Coil Test

First, These Words

A VOA meter is only capable of detecting certain faults in an ignition coil. Replace the coil, if the meter readings are not as specified in the chart in the Appendix. If the coil tests are satisfactory, but the coil is still suspected, an Ignition Analyzer must be used to check the coil further. If an analyzer is not available the coil should be replaced or checked at your favorite marine dealer.

Coil terminals are often painted. Therefore, **BE SURE** to obtain good metal to metal contact when making tests.

Disconnect both small leads from the coil terminals. Remove the high-tension lead from the coil tower.

COIL

VOA METER

③

Primary Resistance Test

3- Refer to appropriate table in the Appendix. Identify the leads listed under the column "Tester Leads To:". Select the appropriate scale on the VOA meter. Make contact with the VOA leads to the disconnected lead/s. The meter reading should be as indicated in the Appendix for the model being serviced.

Secondary Resistance Test

4- Refer to appropriate table in the Appendix. Identify the leads listed under the column "Tester Leads To:". Select the appropriate scale on the VOA meter. Make contact with the VOA leads to coil tower and the appropriate testing point given in the table. The meter reading should be as indicated in the Appendix for the model being serviced.

If the reading is questionable but the powerhead still operates, install a spark plug tester and check the strength of the spark.

If the spark is acceptable, it is not necessary to replace the coil. However, keep this area in mind for possible future trouble.

If the spark is weak it will get weaker, therefore, replace the coil.

FLYWHEEL COMPONENT SERVICE TYPE IV IGNITION SYSTEM

The Type IV CD ignition system installed on the powerheads covered in this manual will usually operate over extremely long periods of time without requiring adjustment or repair. However, if ignition system problems are encountered, and the usual corrective actions such as replacement of spark plugs and timing does not correct the problem, the CD output should be checked to determine if the unit is functioning properly.

Overhaul procedures may differ slightly on various outboard models, but the following general basic instructions will apply to all Type IV high-speed flywheel CD ignition systems.

Flywheel Removal

1- Remove enough of the cowling to expose the hand starter. Remove the hand starter attaching hardware, and then lift the hand starter free of the crankshaft. Disconnect the leads from the battery terminals.

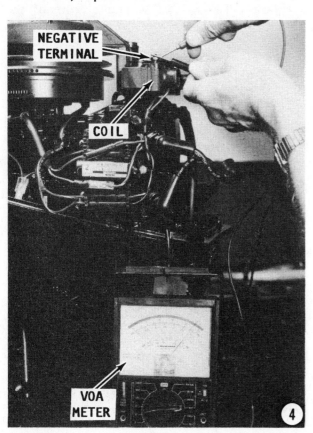

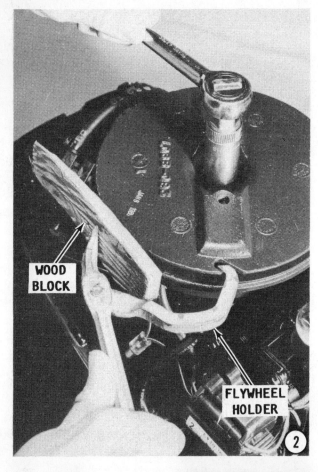

WOOD BLOCK

FLYWHEEL HOLDER

2

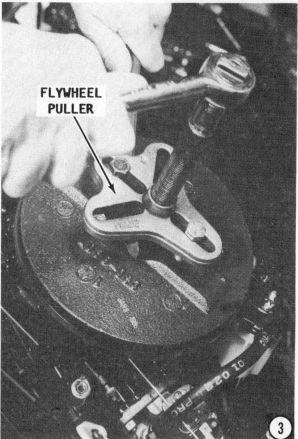

FLYWHEEL PULLER

3

2- Remove the crankshaft nut in the center of the flywheel. A flywheel strap or other device will be required to hold the flywheel securely while the nut is loosened.

3- Obtain the proper type flywheel puller. **NEVER** attempt to use a puller which pulls on the outside edge of the flywheel or the flywheel may be damaged. After the puller is installed, tighten the center screw onto the end of the crankshaft. Continue tightening the screw until the flywheel is released from the crankshaft. Remove the flywheel. **NEVER** hammer on or heat the end of the puller center bolt in an effort to remove the flywheel. Such action will damage the crankshaft or the bearings.

SPECIAL WORDS

A slight "pull" will be felt as the flywheel is lifted, due to the magnets installed on the inside rim of the flywheel.

4- STOP, and carefully observe the layout of the electrical system and associated wiring. Study how the CD system is assembled. Because there are so many different CD installation arrangements, it is not possible to illustrate all of them, and even if they were shown, you would not be able to identify the circuitry for the engine you are servicing. **TAKE TIME** to make notes of the wire routing. You may elect to follow the practice of many professional mechanics by taking a series of photographs of the engine with the flywheel removed, one from the top, and a couple from the sides showing the wiring and arrangement of parts.

4

5- Remove the stator attaching hardware, and then lift the stator and set it aside. It is not necessary to disconnect the wiring to the stator. See earlier instructions in this section to test the stator. On models equipped with an electric cranking motor, remove the alternator.

6- Lift the trigger assembly, and then disconnect the spark advance rod.

7- Unsnap the link rod from the trigger extension.

8- Disconnect all trigger assembly leads from the switch box, including the clamp on the wire harness. Lift the trigger assembly from the powerhead. See instructions in the Appendix to test the trigger assembly.

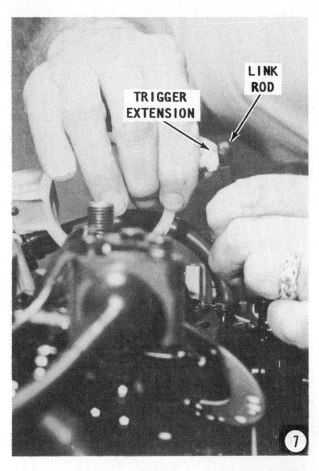

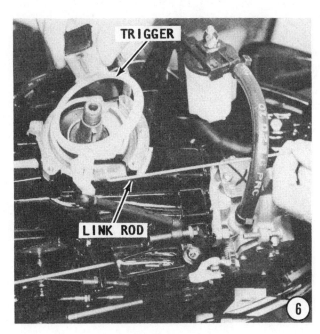

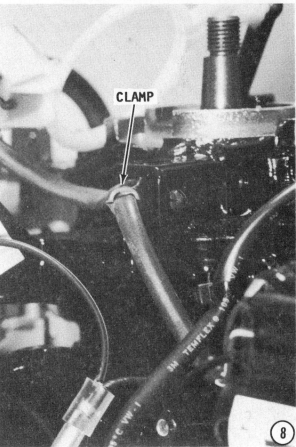

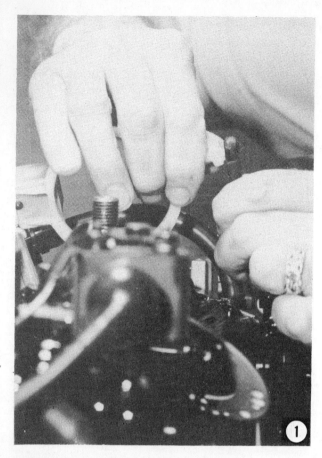

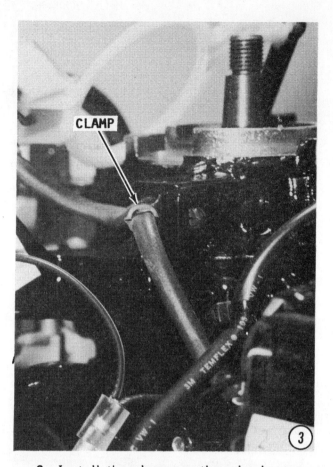

ASSEMBLING FLYWHEEL PARTS

1- Connect the link rod to the trigger assembly extension. Hook the spark advance rod onto the trigger extension. Snap the rod into place. On earlier model units the rod is held in place with a locknut.

2- Slide the trigger assembly over the end cap into place.

3- Install the clamp on the wire harness. Route the trigger wiring harness along the side of the powerhead, and then connect the wires to the proper terminals in the switch box. Refer to the diagram or photographs made before removal. If this was not done, check the wiring diagrams in the Appendix.

4- Place the stator in position over the trigger assembly. Apply a drop of Loctite

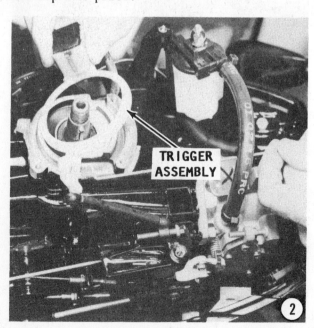

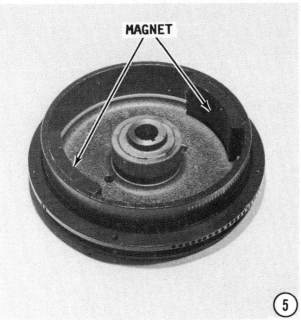

MAGNET

(5)

and then secure it in place with the attaching hardware. Connect the wires to the alternator. Connect the stator leads to the proper terminals of the rectifier and switch box. Refer to the wiring diagrams in the Appendix.

Flywheel Installation

5- Place the key in the crankshaft keyway. Check the flywheel magnets to be sure they are free of any metal parts. Double check the taper in the flywheel hub and the taper on the crankshaft to verify they are clean and contain no oil, as a precaution against the flywheel "walking" on the crankshaft during operation.

6- Slide the flywheel down over the crankshaft with the keyway in the flywheel aligned with the key on the crankshaft. Rotate the flywheel clockwise and check to be sure the flywheel does not contact any part of the magneto or the wiring.

7- Thread the flywheel nut onto the crankshaft and tighten it to the torque value given in the Appendix.

A, or equivalent, to the threads of the stator attaching screws. Install and tighten the attaching screws to a torque value of 30 in lbs (3.4 Nm). On models equipped with an electric cranking motor, position the alternator on the opposite side from the stator,

(6)

TORQUE WRENCH

HOLDING BAR

(7)

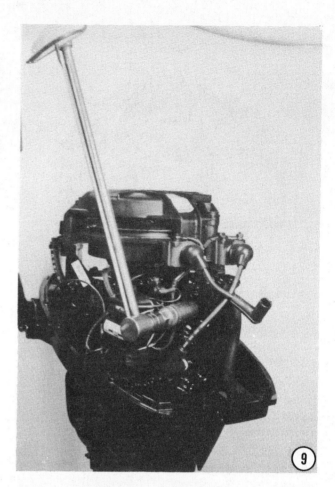

8- Install the hand starter and secure it in place with the attaching hardware.

9- Install the spark plugs and tighten them to a torque value of 20-1/2 ft lbs (27.8 Nm).   Install the electrical leads to the battery terminals.

For detailed timing and synchronizing procedures, see Chapter 6.

6
TIMING AND
SYNCHRONIZING

6-1 INTRODUCTION AND PREPARATION

Timing and the synchronization on an outboard engine is extremely important to obtain maximum efficiency. The engine cannot perform properly and produce its designed horsepower output if the fuel carburetion and ignition systems have not been precisely adjusted.

Synchronization

In simple terms, synchronization is timing the carburetion to the ignition. This means, as the throttle is advanced to increase engine rpm, the carburetor and the ignition systems are both advanced equally and at the same rate.

Therefore, anytime the fuel system or the ignition system on an engine is serviced to replace a faulty part, or any adjustments are made for any reason, engine timing and synchronization **MUST** be carefully checked and verified.

For this reason the timing and synchronizing procedures have been separated from all others and presented alone in this chapter.

Before making adjustments with the timing or synchronizing, the ignition system should be thoroughly checked according to the procedures outlined in Chapter 5, and the fuel system in good working order per Chapter 4.

Timing

The timing on many models is set at the factory and there is no way in heaven, or on earth, to make an adjustment. However, on some powerheads, the timing may be adjusted. All outboard engines have some type of synchronization between the carburetion and ignition systems.

Many models do not have timing marks on the flywheel and a dial indicator must be used to properly time the engine. On later models the flywheel does have a timing mark and the engine may be properly timed using a timing light. This method has its disadvantages, because the engine must be run at full throttle in forward gear in a test tank, or at full throttle on the boat in a waterway. The problems of roaring across a lake or down the river at full throttle and attempting to check the timing with a light can easily be imagined.

Therefore, the timing light is not too practical a method of timing and synchronizing a large horsepower engine if a test tank is not available.

On some late model powerheads, the timing is checked using a timing light while cranking the engine. Using this procedure it is not necessary to mount the engine in a test tank or to move the boat into a body of water. **HOWEVER,** a flush device **MUST** be used and the engine **NEVER** operated above 1000 rpm.

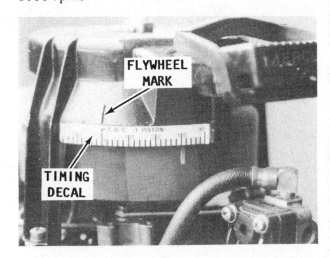

Typical flywheel mark and timing decal used on smaller powerheads covered in this manual.

PREPARATION

Timing and synchronizing the ignition and fuel systems on an outboard motor are critical adjustments. Therefore, the following equipment is essential and is called out repeatedly in this section. This equipment must be used as described, unless otherwise instructed. Naturally, they are removed following completion of the adjustments.

Dial Indicator

Top dead center (TDC) of the No. 1 (top) piston must be precisely known before the timing adjustment can be made. TDC can only be determined through installation of a dial indicator into the No. 1 spark plug opening.

Timing Light

During many procedures in this section, the timing mark on the flywheel must be aligned with a stationary timing mark on the engine while the engine is being cranked, or is running. Only through use of a timing light connected to the No. 1 spark plug, can the timing mark on the flywheel be observed while the engine is operating.

Tachometer

A tachometer connected to the engine must be used to accurately determine engine speed during idle and high-speed adjustment.

The meter readings range from 0 to 6,000 rpm in increments of 100 rpm. Tachometers have solid state electronic circuits which eliminates the need for relays or batteries and contributes to their accuracy.

Most marine outboard units have a female plug at the forward end of the shift box as a convenience for installation of a tachometer. Therefore, when purchasing a tachometer for the boat, check to be sure the adaptor plug will mate with the fitting on the shift box.

If the boat is not equipped with a tachometer, connect one lead to the primary (negative (-)) terminal of any one coil. Connect the other lead of the tachometer to a good ground on the powerhead.

Test Tank

The engine must be operated at various times during the procedures. Therefore, a test tank, flush device, or moving the boat into a body of water, is necessary.

CAUTION: Water must circulate through the lower unit to the engine any time the engine is run to prevent damage to the water pump in the lower unit. Just five seconds without water will damage the water pump.

NEVER, AGAIN, NEVER operate the engine at high speed with a flush device attached. The engine, operating at high speed with such a device attached, would **RUNAWAY** from lack of a load on the propeller, causing extensive damage.

6-2 MODEL 39, 40, 60, 75, AND 110 -- 1965-1969

Primary Pickup Synchronization

1- Mount the engine in a test tank. Connect a tachometer to the engine. Start the engine and allow it to warm to operating temperature.

CAUTION: Water must circulate through the lower unit to the engine any time the engine is run to prevent damage to the water pump in the lower unit. Just five seconds without water will damage the water pump.

Shift the engine into **FORWARD** gear. Advance the throttle by turning the twist grip until the engine is operating between 1000 and 1100 rpm. At this speed, the magneto cam should just touch the throttle lever on the carburetor. If 1000 to 1100 rpm cannot be obtained, loosen the screw on the top of the carburetor securing the throttle lever to the carburetor. Slowly move the arm inward or outward until the engine is operating at the recommended rpm.

Model 110, Serial No. 1492282 to 1580202

2- Shift the engine into **NEUTRAL**, and then adjust the neutral speed limiter stop to obtain a maximum speed of 2400 to 2700 rpm.

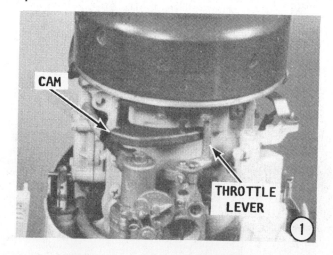

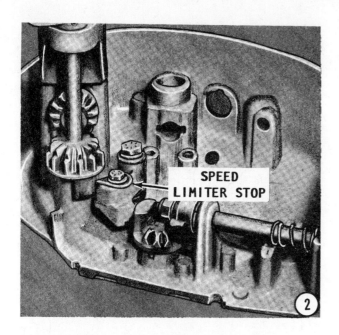

previous step, loosen the Allen screw at the bottom of the twist grip and align the twist grip. Tighten the Allen screw, and then check the adjustment by backing off the throttle to normal idle, and then advancing the throttle to the **START** position. The engine should be operating at 2400 to 2700 rpm. This procedure will eliminate any error caused by "play" in the throttle. An accurate adjustment at this time will ensure easy start of a cold engine. Disconnect the tachometer and remove the engine from the test tank.

6-3 MODEL 200 1965 - 1966

Maximum Advance Adjustment

1- Remove the engine cowling. Remove the No. 1 (top) spark plug and install a dial indicator in the opening. Rotate the flywheel **CLOCKWISE** until TDC of the piston stroke is reached. At that point, adjust the dial indicator to $0°$.

Model 60, Serial No. 1610265 and Above

3- Adjust the throttle stop screw until the threaded end of the screw extends 5/16" (8mm) thru the throttle lever control.

Model 110, Serial No. 1580203 and above

Adjust the throttle stop screw until the threaded end of the screw extends 1/4" (6mm) thru the throttle lever control.

Tiller Handle Adjustment

4- With the engine running and the shift lever in the **NEUTRAL** position, advance the throttle until the engine is operating at 2400 to 2700 rpm. When the engine is operating at this speed, the **START** position on the twist grip should align with the indicator arrow on the tiller handle. If the twist grip is not properly aligned as stated in the

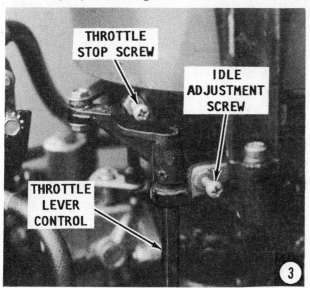

Connect one lead of an ohmmeter to the No. 1 (top) cylinder of the breaker points. Connect the other meter lead to a good ground on the engine. Rotate the flywheel **COUNTERCLOCKWISE** until the correct BTDC reading on the dial indicator is reached, according to the Specifications in the Appendix.

2- Shift the engine into **FORWARD** gear. Advance the throttle until the pin on the intermediate magneto lever is positioned as shown in the accompanying illustration. The breaker points should open at this time. If they fail to open, loosen the jam nut on the magneto lever and adjust the lever until the points just open. Tighten the jam nuts. Advance the magneto slowly until the points just open. Hold the magneto at this position and adjust the magneto advance stop screw until it just makes contact with the magneto stop. Tighten the jam nuts. Remove the dial indicator. Install the spark plugs.

6-4 MODEL 350 1965 - 1969

1- Remove the engine cowling or enough of the engine cover to expose the flywheel. The flywheel must be removed in order to adjust the maximum advance. Remove the nut securing the flywheel to the crankshaft. It may be necessary to use some type of flywheel strap to prevent the flywheel from turning as the nut is loosened.

Install the proper flywheel puller. **NEVER** attempt to use a puller which pulls on the outside edge of the flywheel or the flywheel may be damaged. After the puller is installed, tighten the center screw onto the end of the crankshaft. Continue tightening the screw until the flywheel is released from the crankshaft. Remove the flywheel.

2- Shift the engine into **FORWARD** gear. Install the flywheel nut onto the bare crankshaft. Tighten the nut only until the crankshaft begins to rotate against the compression in the cylinders.

Remove all the spark plugs to relieve compression in the cylinders. Install a dial indicator into the No. 1 (top) spark plug opening. Connect one lead of an ohmmeter to the No. 1 (top) cylinder ignition point terminal. Connect the other meter lead to a good ground on the engine.

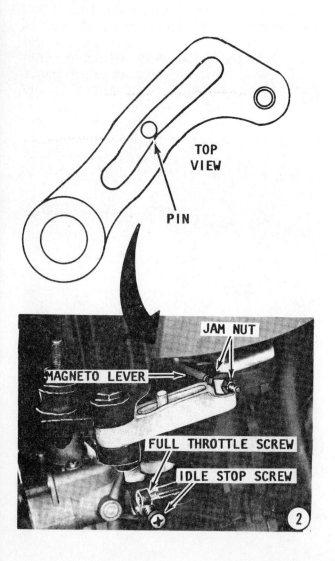

TOP VIEW

PIN

JAM NUT

MAGNETO LEVER

FULL THROTTLE SCREW

IDLE STOP SCREW

②

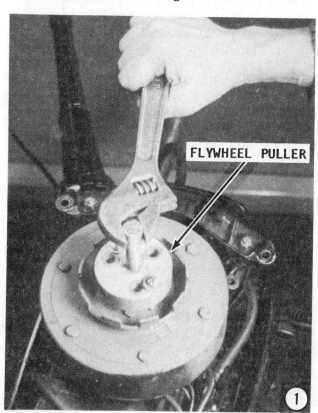

FLYWHEEL PULLER

①

With a wrench on the flywheel nut, slowly rotate the crankshaft **CLOCKWISE** until the No. 1 piston is at TDC of the stroke. At this point adjust the dial indicator to 0°.

With a wrench on the flywheel nut, slowly rotate the crankshaft **COUNTERCLOCKWISE** until the required setting BTDC, as determined from the Specifications in the Appendix, is obtained.

3- Now, slowly advance the magneto against the magneto stop. The breaker points should **OPEN**, as indicated by the ohmmeter. If the breaker points did not open or if they opened before the magneto contacted the stop, loosen the locknut and adjust the linkage until the points just open as the magneto contacts the stop. Tighten the locknut.

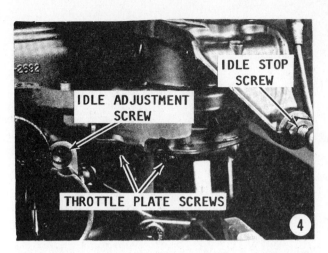

4- With the throttle still at full advance (the magneto just contacting the stop), loosen the two screws on the carburetor pickup plate and adjust the plate until the carburetor is at full throttle (the butterfly is wide open). Check for 0.010 - 0.015" (.25-.38mm) "play" in the carburetor shaft. This amount of "play" will prevent a strain on the linkage. **NEVER** bend the pickup with the nylon sleeve to make this adjustment. Linkage adjustment on the port side of the engine should not be necessary. However, if the setting has been disturbed, adjust the linkage until 1/2" (13mm) of threads remain exposed. Remove the dial indicator and install the spark plug.

5- Remove the flywheel nut from the crankshaft. This can be accomplished by placing the wrench on the nut and then striking the end of the wrench handle a

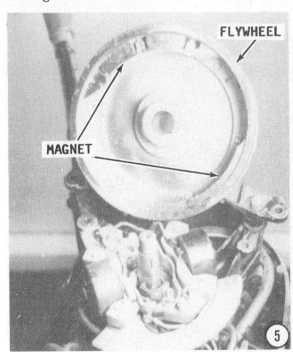

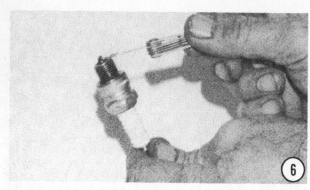

quick blow with the heel of your hand to jar it loose. Check to be sure there are no metal parts stuck to the magnets inside the flywheel. Place the key in the crankshaft keyway. Check to be sure the inside taper of the flywheel and the taper on the crankshaft is clean of dirt or oil, to prevent the flywheel from "walking" on the crankshaft during operation. Slide the flywheel on the crankshaft with the keyway in the flywheel aligned with the key. Thread the nut onto the flywheel and tighten it to the torque value given in the Specifications in the Appendix.

6- Check the gap on each spark plug according to the Specificatons in the Appendix. Install the spark plugs. Install the engine cowling.

6-5 MODEL 200 1967 - 1969

Maximum Advance Adjustment

1- Remove the engine cowling. Remove the No. 1 (top) spark plug and install a dial indicator in the opening. Rotate the flywheel **CLOCKWISE** until TDC of the piston stroke is reached. At that point, adjust the dial indicator to 0°. Connect one lead

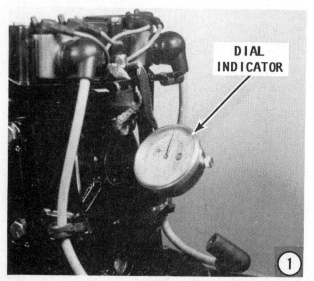

DIAL INDICATOR

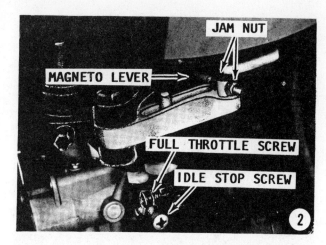

MAGNETO LEVER   JAM NUT

FULL THROTTLE SCREW

IDLE STOP SCREW

of an ohmmeter to the No. 1 (top) cylinder of the breaker points. Connect the other meter lead to a good ground on the engine.

Rotate the flywheel **COUNTERCLOCK-WISE** until the correct BTDC reading on the dial indicator is reached, according to the Specifications in the Appendix.

2- Back off the full throttle screw. Advance the throttle to the full advance position. Hold a slight pressure on the twist grip handle and at the same time, turn the full throttle screw inward until the points just begin to open, as evidenced by the ohmmeter.

3- With the throttle at the full advance position, check the carburetor shutter. The shutter **MUST** be in the full open position. Check the clearance in the carburetor cluster. The clearance must be 0.005 -0.015" (.25-.38mm). Correct the clearance, as required, by turning the adjusting screw inward or outward. The proper clearance will prevent a strain on the linkage. Remove the dial indicator. Install the spark plugs.

Tiller Handle Adjustment

4- Mount the engine in a test tank. **NEVER** use a flush device for this test.

Remove the engine cowling and connect the fuel line to a fuel source. Install a tachometer to the engine.

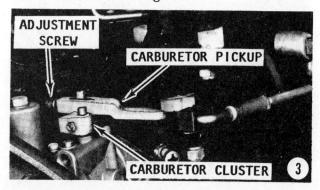

ADJUSTMENT SCREW

CARBURETOR PICKUP

CARBURETOR CLUSTER

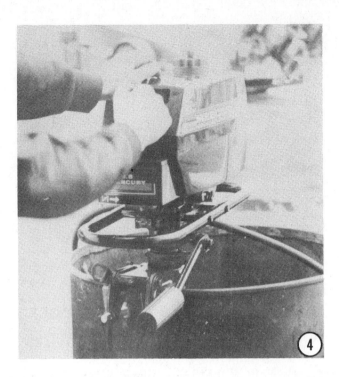

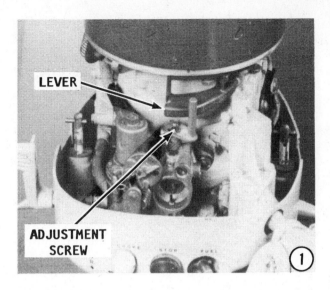

CAUTION: Water must circulate through the lower unit to the engine any time the engine is run to prevent damage to the water pump in the lower unit. Just five seconds without water will damage the water pump.

Start the engine and allow it to warm to operating temperature.

5- With the engine in **NEUTRAL**, rotate the twist grip throttle until the engine is operating at 2400 - 2700 rpm. At this engine speed, the word **START** on the twist grip should be aligned with the indicator arrow on the handle. If **START** and the arrow are not aligned, loosen the Allen screw at the bottom of the twist grip and align the word **START** with the arrow.

Tighten the Allen screw to hold the adjustment. Check the adjustment by backing off the throttle to idle speed and then advance it to **START**. Proper adjustment will assist in starting a cold engine and will eliminate any possible error caused by "play" in the linkage. Remove the tachometer. Remove the engine from the test tank.

6-6 MODEL 40 1970

Maximum Advance Adjustment
1- Shift the engine into **FORWARD** gear. With the engine **NOT** running, advance the twist grip to obtain full throttle. No adjustment is possible at full throttle. However, check for 0.050" (1.3mm) clearance between the cam and the throttle lever. If an adjustment is required, loosen the screw on top of the carburetor and adjust the carburetor lever to obtain the 0.050" (1.3mm) clearance. Tighten the screw securely to hold the adjustment. Back-off the throttle to idle speed position.

Primary Pickup Synchronization
2- Remove all spark plugs to relieve compression in the cylinders. Install a dial indicator in the No. 1 (top) spark plug opening. Now, slowly rotate the flywheel clock-

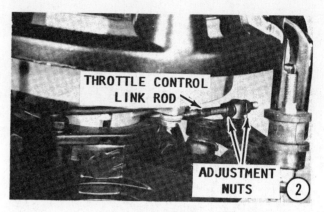

wise and determine TDC for the No. 1 piston stroke. Once TDC is reached, reset the dial indicator to 0°.

Connect one lead of an ohmmeter to the white contact lead of the point set for the No. 1 cylinder. Connect the other meter lead to a good ground on the engine. Slowly rotate the flywheel clockwise until 0.005" (.13mm) is reached on the dial indicator.

3- Rotate the twist grip to close the throttle until the ohmmeter indicates the ignition points just open. At this point, adjust the throttle lever, by loosening the screw on top of the carburetor and shifting the lever inward or outward, until the cam follower on the carburetor just makes contact with the armature plate cam. Tighten the screw securely to hold the adjustment. Remove the dial indicator. Install the spark plugs. Install the engine cowling and test the engine by operating it in a test tank.

6-7 MODEL 200 WITH MAGNETO IGNITION SYSTEM 1970 -- 1977

Maximum Advance Adjustment

1- Remove the engine cowling. Remove the No. 1 (top) spark plug and install a dial indicator in the opening. Rotate the flywheel **CLOCKWISE** until TDC of the piston

stroke is reached. At that point, adjust the dial indicator to 0°. Connect one lead of an ohmmeter to the No. 1 (top) cylinder of the maker points. Connect the other meter lead to a good ground on the engine.

Rotate the flywheel **COUNTERCLOCKWISE** until the correct BTDC reading on the dial indicator is reached, according to the Specifications in the Appendix.

2- Adjust the two elastic stop nuts on the throttle control link until the ohmmeter indicates the points just **CLOSE** at BTDC. Refer to the Specifications in the Appendix for the exact setting.

Primary Pickup Synchronization

3- Rotate the flywheel **COUNTERCLOCKWISE** until the dial indicator reads 0.002 BTDC. Close the throttle with the twist grip until the ohmmeter indicates the points just **OPEN**. At this point adjust the carburetor pickup screw until it just makes contact with the cluster actuating rod on the carburetor.

4- Rotate the twist grip to the full throttle position. Now, adjust the upper screw in the vertical shaft while vibrating the carburetor cluster until there is 0.035-0.048" (.9-1.2mm) play between the pickup screw and the cluster actuating rod on the carburetor. Remove the ohmmeter and dial indicator. Install the No. 1 spark plug.

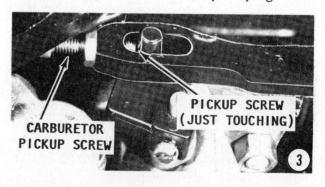

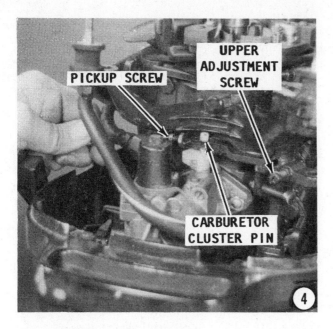

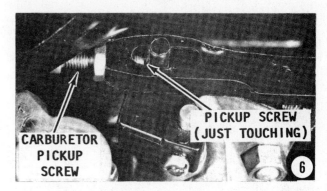

frame. Rotate the twist grip throttle to the wide open position. Check the BTDC degree indication with the Specifications in the Appendix. If adjustment is required: Loosen the two elastic stop nuts on the throttle control link rod.

Maximum Advance Timing

5- Mount the engine in a test tank. **NEVER** use a flush device for this test.

Connect the fuel line to a fuel source. Install a timing light to the No. 1 (top) spark plug.

CAUTION: Water must circulate through the lower unit to the engine any time the engine is run to prevent damage to the water pump in the lower unit. Just five seconds without water will damage the water pump.

Start the engine and allow it to warm to operating temperature.

Aim the timing light at the degree markings on the top of the cowling support

Primary Pickup Synchronizing Timing

6- Return the throttle to the idle position and check the timing against the Specifications in the Appendix. Adjust the carburetor pickup screws until the specifications are met. Shut down the engine, remove the timing light.

Tiller Handle Adjustment

7- Install a tachometer to the engine. Start the engine and allow it to warm to operating temperature. With the engine in **NEUTRAL**, rotate the twist grip throttle until the engine is operating at 2400 - 2700 rpm. At this engine speed, the word **START** on the twist grip should be aligned with the indicator arrow on the handle. If **START** and the arrow are not aligned, loosen the Allen screw at the bottom of the twist grip and align the word **START** with the arrow. Tighten the Allen screw to hold the adjustment. Check the adjustment by backing off the throttle to idle speed and then advance it to **START**. Proper adjustment will assist in starting a cold engine and will eliminate any possible error caused by "play" in the linkage. Remove the tachometer. Disconnect the fuel line. Remove the engine from the test tank.

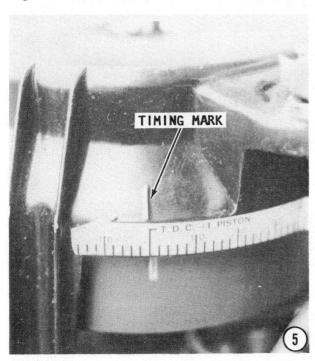

6-8 MODEL 75 AND 110 1970
MODEL 40, 75 AND 110 1971—1974
MODEL 45 1975—1979
MODEL 4.5 1979—1985
MODEL 4.0 1986—MID 1987
MODEL 90cc 1986—87

SPECIAL WORDS

For Model 4.0 Mid 1987 and on, also Model 5.0 1988 and on, see Section 6-18 toward the end of this chapter.

Maximum Advance Adjustment

1- Remove the engine cowling. Move the shift lever to the **FORWARD** position. Remove all spark plugs to relieve compression in the cylinders. Install a dial indicator in the No. 1 (top) spark plug opening.

Slowly rotate the flywheel clockwise and determine TDC for the No. 1 cylinder. Once TDC is reached, reset the dial indicator to 0°.

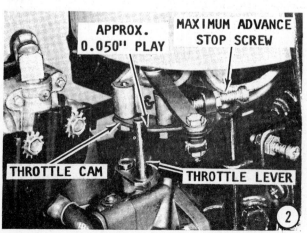

Connect one lead of an Ohmmeter to the white contact lead of the point set for the No. 1 cylinder. Connect the other meter lead to a good ground on the engine.

Slowly rotate the flywheel counterclockwise until the required BTDC position is reached on the dial indicator, as listed in the Specifications in the Appendix.

2- Advance the twist grip throttle to the wide open position. At this position adjust the maximum advance stop screw until the ohmmeter indicates the points have just opened at the required BTDC position as listed in the Specifications.

ONE MORE WORD

On the Model 45, 1976 - 1979, and Model 4.5 1979-1985, the points must **CLOSE** at the required BTDC position, as listed in the Specifications. Retain this position of the stop screw by tightening the locknut securely.

Check for 0.050" (1.3mm) clearance between the cam and the throttle lever. If an adjustment is required, loosen the screw on top of the carburetor and adjust the carburetor lever to obtain the 0.050" (1.3mm) clearance. Tighten the screw securely to hold the adjustment.

Primary Pickup Synchronization

3- Now, slowly rotate the flywheel clockwise until the dial meter indicates the piston is ATDC per the Specifications in the Appendix. Back off the twist grip throttle until the ohmmeter indicates the points just open. At this point the throttle lever should barely make contact with the ignition cam.

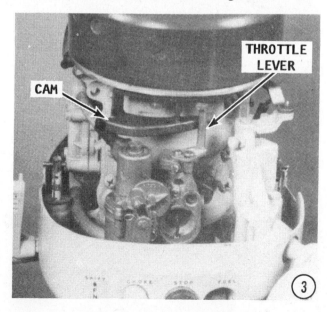

SPECIAL WORDS

On the Model 45, 1976 -1979, and Model 4.5 1979--1985, the points should just close ATDC per the Specifications.

4- Remove the dial indicator. Install the spark plugs. Install the engine cowling. Test the engine by operating it in a test tank.

Tiller Handle Adjustment

5- Mount the engine in a test tank or move the boat to a body of water. **NEVER** use a flush device for this test.

CAUTION: Water must circulate through the lower unit to the engine any time the engine is run to prevent damage to the water pump in the lower unit. Just five seconds without water will damage the water pump.

Start the engine and allow it to warm to operating temperature.

With the engine running and the shift lever in the **NEUTRAL** position, advance the throttle until the engine is operating at 2400 to 2700 rpm. When the engine is operating at this speed, the **START** position on the twist grip should align with the indicator arrow on the tiller handle.

If the twist grip is not properly aligned as stated in the previous step, loosen the Allen screw at the bottom of the twist grip

ALLEN SCREW UNDER GRIP ⑤

and align the twist grip. Tighten the Allen screw, and then check the adjustment by backing off the throttle to normal idle, and then advancing the throttle to the **START** position. The engine should be operating at 2400 to 2700 rpm. This procedure will eliminate any error caused by "play" in the throttle. An accurate adjustment at this time will ensure easy start of a cold engine. Remove the engine from the test tank.

6-9 MODEL 400 1970 - 1971

Primary Pickup Synchronizing Timing

1- Mount the engine in a test tank or move the boat into a body of water. **NEVER** use a flush device when making the primary pickup, maximum advance timing, or the secondary pickup adjustments. Remove the engine cowling. Connect the fuel line to a fuel source. Connect a tachometer to the engine. Connect a timing light to the No. 1 spark plug.

2- Check the position of the trigger plate. The plate should be approximately in the center of the elongated slots. With a feeler gauge, measure the clearance between the trigger pickup head and the trigger housing. This clearance should be 0.050

④

FUEL LINE CONNECTOR ①

PLATE IN CENTER OF SLOTS

2

- 0.060" (1.3-1.5mm). If adjustment is necessary, use shims between the trigger plate and the starter housing.

Check the position of the throttle lockout cam. Adjust the cam, if necessary by shimming the trigger plate to the starter housing.

3- Start the engine and allow it to warm to operating temperature.

CAUTION: Water must circulate through the lower unit to the engine any time the engine is run to prevent damage to the water pump in the lower unit. Just five seconds without water will damage the water pump.

Shift the engine into **FORWARD** gear. Aim the timing light at the window in the flywheel guard. Advance the idle until the white timing dots on the flywheel show

through the window. See the Specifications in the Appendix for correct timing.

With the engine running at this speed, loosen the two screws on the carburetor pickup plate and slide the plate until the cam just makes contact with the primary pickup arm on the carburetor cluster. Tighten the screws to hold the adjustment.

Maximum Advance Timing

4- Advance the twist grip throttle until the engine is operating at 5000 - 5200 rpm. Adjust the spark stop screw to align the flywheel timing white line through the window in the flywheel guard. See the Specifications in the Appendix for correct setting. Back off the throttle to normal idle speed. Shut down the engine. Disconnect the timing light, the tachometer, and the fuel line.

Secondary Pickup Adjustment

With the spark arm stop making contact with the stop, but not actuating the throttle arm, adjust the throttle arm until it just makes contact with the secondary pickup on the carburetor cluster. To make the adjustment, **CAREFULLY** bend the arm as required.

Rotate the throttle arm to the wide open position. Now, adjust the throttle stop screw until the carburetor shutter is fully open. Check for 0.010 - 0.015" (.3-.4mm) "play" in the carburetor shaft. This amount of "play" will prevent a strain on the linkage. After the proper clearance is obtained, tighten the locknut to hold the adjustment. Install the engine cowling. Remove the engine from the test tank.

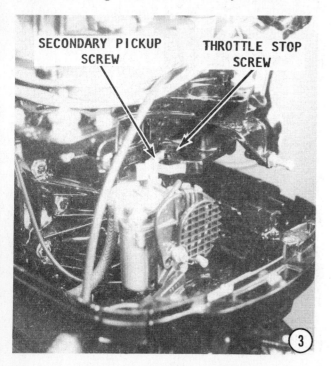

SECONDARY PICKUP SCREW   THROTTLE STOP SCREW

3

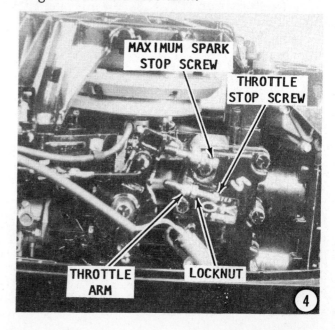

MAXIMUM SPARK STOP SCREW

THROTTLE STOP SCREW

THROTTLE ARM   LOCKNUT

4

6-10 MODEL 402 1972 - 1979
MODEL 40 1979 — 1983
MODEL 35 1984 — 1989

Primary Pickup Synchronizing Timing

1- Mount the engine in a test tank, or move the boat into a body of water. **NEVER** use a flush device when making the primary pickup, maximum advance timing, or the secondary pickup adjustments. Remove the engine cowling. Connect the fuel line to a fuel source. Connect a tachometer to the engine. Connect a timing light to the No. 1 (top) spark plug.

CAUTION: Water must circulate through the lower unit to the engine any time the engine is run to prevent damage to the water pump in the lower unit. Just five seconds without water will damage the water pump.

Start the engine and allow it to warm to operating temperature.

Shift the engine into **FORWARD** gear. Aim the timing light at the window in the flywheel guard. Advance the throttle until the timing dots on the flywheel are aligned with the notch in the timing window. See the Specifications in the Appendix for the correct setting.

2- Adjust the turnbuckle, on early models, or the adjustment screw to the left of the throttle lever on late models, until the throttle cam just makes contact with the primary pickup arm on the carburetor cluster. Tighten the turnbuckle locknuts to hold the adjustment. Back off the throttle to idle speed.

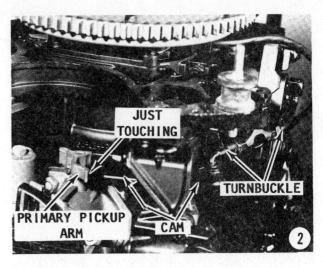

Maximum Advance Timing

3- With the engine running in **FORWARD** gear, advance the throttle to wide open. Engine rpm should be 5000 - 5500 rpm. Aim the timing light at the timing window. Adjust the maximum spark advance screw until the straight line on the flywheel is aligned with the notch in the timing window. See the Specifications in the Appendix for the proper setting. Back-off the throttle to idle speed. Shut down the engine. Disconnect the timing light and the fuel line. Remove the tachometer from the engine.

Secondary Pickup adjustment

4- With the engine shut down, position the spark advance lever until it just makes contact with the maximum spark stop. **DO NOT** actuate the throttle lever. Loosen the

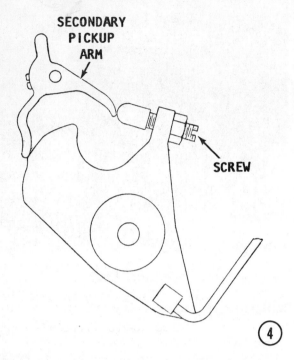

SECONDARY
PICKUP
ARM

SCREW

④

locknut and adjust the secondary throttle pickup screw until the end of the screw just makes contact with the secondary pickup arm on the carburetor cluster. Tighten the locknut to hold the adjustment.

Full Throttle Stop Screw Adjustment

5- With the engine shut down, move the throttle lever to the wide open throttle (WOT) position. Now, adjust the throttle stop screw until the carburetor shutter is fully open at the WOT. Check to be sure the carburetor shutter does not act as a throttle stop. Use a feeler gauge and check the clearance between the secondary pickup screw and the secondary pickup arm on the carburetor cluster. This clearance should be 0.010 - 0.015" (.25 - .38mm). Remove the engine from the test tank.

THROTTLE LEVER
FULL THROTTLE STOP SCREW

⑤

6-11 MODEL 40 1976 - 1981

Idle Retard Adjustment

1- Remove the engine cowling. Move the throttle lever to the full retard position and the carburetor shutter plate to the fully closed position. Loosen the throttle cam adjustment screw. Adjust the throttle cam until 0.005-0.015" (.1-.3mm) clearance is obtained between the throttle cam and the cluster pin. Tighten the adjustment screw securely to hold this adjustment.

Loosen the retard cable jam nuts, and then adjust the cable until approximately 3-4 threads (about 1/8") (3.2mm) are exposed. Tighten the jam nuts securely to hold this adjustment.

Hold the twist grip against the idle stop and at the same time adjust the advance cable until there is no slack in the cable. Tighten the jam nuts securely to hold this adjustment.

2- Remove the idle lever from the idle screw on the carburetor. Preset the carburetor idle screw at 1-1/4 turns out from a lightly seated position. In other words, slowly and carefully turn the screw in until it barely seats, then back it out 1-1/4 turns.

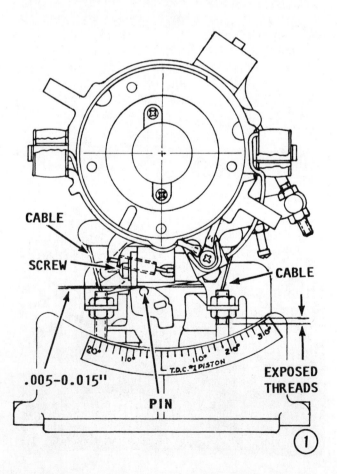

CABLE

SCREW

CABLE

.005-0.015"

PIN

EXPOSED
THREADS

①

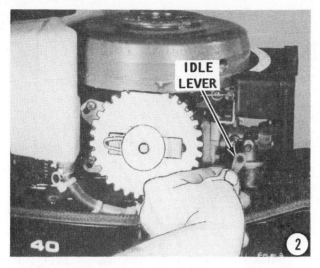

Primary Pickup Synchronizing Timing

FIRST, this procedure **MUST** be performed with the engine in a test tank.

CAUTION: Water must circulate through the lower unit to the engine any time the engine is run to prevent damage to the water pump in the lower unit. Just five seconds without water will damage the water pump.

3- Connect a timing light to the No. 1 (top) spark plug. Connect a tachometer to the engine.

Start the engine and allow it to warm to operating temperature.

Shift the engine into **FORWARD** gear, and operate at 1/2 throttle for approximately five minutes to warm the engine to operating temperature.

With the engine in gear; thoroughly warm; and operating at 1000 to 1500 rpm; observe the ignition timing while quickly closing the throttle against the idle stop. The timing should fully retard to approximately 18° to 20° ATDC, then advance slightly to 15° to 18° ATDC as the idle speed approaches 600 rpm.

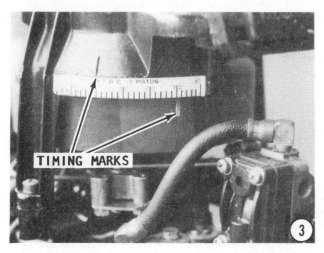

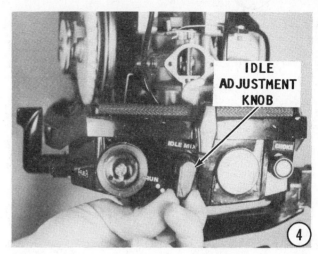

4- Adjust the trigger link rod to set the idle timing at 2° below the advance reading which is obtained as the idle speed approaches 600 rpm. An example: If the timing fully retards to 18° ATDC as the throttle is closed, and advances to 15° ATDC at 600 rpm, then set the idle timing at 17° ATDC. Turn the twist throttle grip against the idle stop. Now, advance the throttle grip until the throttle cam contacts the cluster pin, then check the timing. The timing should be $14^{o} + 2^{o}$ ATDC. If the throttle pickup is incorrect, adjust the throttle cam, as required, to obtain throttle pickup at $14^{o} + 2^{o}$ ATDC.

Turn the idle adjustment knob to the **RUN** position. Install the idle lever onto the carburetor idle screw in the 9 o'clock position.

Maximum Advance Timing

5- Advance the throttle grip until 22° BTDC is reached. At this point adjust the maximum spark advance screw until it just makes contact with the stop. Tighten the jam nut on the spark advance screw securely to hold the adjustment. Disconnect the timing light and the tachometer. Install the engine cowling. Remove the engine from the test tank.

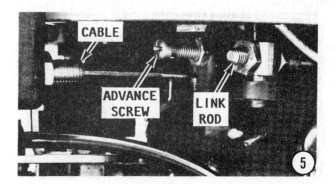

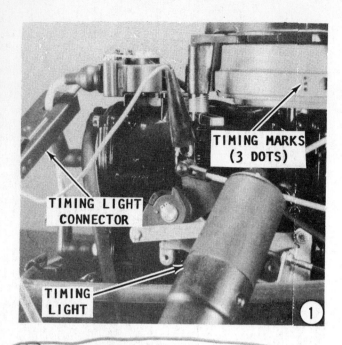

TIMING MARKS
(3 DOTS)

TIMING LIGHT
CONNECTOR

TIMING
LIGHT

1

6-12 MODEL 75 AND 110 1975 - 1979
MODEL 7.5 AND 9.8 1979 — 1985

Maximum Advance Timing

1- Mount the engine in a test tank. Remove the engine cowling. Connect the fuel line to a fuel source. Install a tachometer to the engine. Connect a timing light to the No. 1 (top) spark plug. Start the engine and allow it to warm to operating temperature.

CAUTION: Water must circulate through the lower unit to the engine any time the engine is run to prevent damage to the water pump in the lower unit. Just five seconds without water will damage the water pump.

MAXIMUM
SPARK
ADVANCE
SCREW

2

2- Move the shift lever into **FORWARD** gear. Advance the throttle until the maximum timing mark on the flywheel is aligned with the timing mark on the starter housing. Refer to the Specifications in the Appendix for recommended rpm. Adjust the maximum spark advance screw until the end of the screw just makes contact with the throttle lever. Back off the throttle to normal idle, and then shut the engine down. Disconnect the timing light.

Maximum Neutral RPM
Limiter Adjustment

The purpose of the maximum neutral rpm speed limiter is to stop the rotation of the tiller handle-twist grip at a specified rpm when the shift lever is in the **NEUTRAL** position. This adjustment must be accurate to ensure easy starting when the engine is cold and also to prevent the engine from starting at maximum operating rpm.

3- With the engine operating at idle speed, loosen the neutral speed stop bolt just enough to enable the maximum neutral speed stop to slide back and forth. Now, turn the twist grip to advance engine rpm to 2400 - 2700 rpm. **WITHOUT** turning the twist grip, position and secure the maximum neutral speed stop against the neutral speed limiter follower.

This adjustment will now stop the throttle at the recommended maximum neutral speed and assist starting a cold engine. Check the adjustment to be sure the maximum neutral speed stop does not allow the engine rpm to exceed 2400-2700 rpm. Shift the engine into **FORWARD** gear and check to be sure the neutral speed stop does not interfere with throttle operation in forward gear. If the stop does interfere with throttle operation, move the back edge of the stop **TOWARD** the cylinder block.

NEUTRAL RPM
LIMITER

BOLT

3

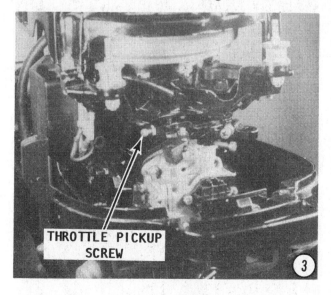

Tiller Handle and Throttle Decal Alignment

FIRST, these words: The adjustment outlined in the previous step must be completed **BEFORE** the tiller handle and throttle decal alignment can be properly made.

4- With the engine operating at idle speed and the shift lever in the **NEUTRAL** position, turn the twist grip until the throttle is against the maximum neutral stop. At this point, the arrow on the tiller handle should be aligned with the word **START** on the throttle decal. If the arrow is not properly aligned, proceed as follows:

    a- Remove and retain the two screws securing the tiller handle to the engine.

    b- Re-align the gears, as necessary to align the word **START** with the arrow.

    c- Install the tiller handle to the engine and secure it with the two screws.

Disconnect the tachometer and fuel line. Install the engine cowling and remove the engine from the test tank.

6-13 MODEL 200 WITH CD IGNITION SYSTEM 1975 -- 1981

Maximum Advance Timing

1- Mount the engine in a test tank. **NEVER** use a flush device for this test. Remove the engine cowling. Connect the fuel line to a fuel source. Connect a tachometer to the engine. Connect a timing light to the No. 1 spark plug.

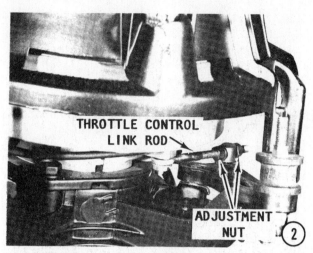

CAUTION: Water must circulate through the lower unit to the engine any time the engine is run to prevent damage to the water pump in the lower unit. Just five seconds without water will damage the water pump.

2- Start the engine and allow it to warm to operating temperature.

Shift the engine into **FORWARD** gear. Advance the throttle to the wide open position. Aim the timing light at the timing mark on the rewind housing. Adjust the two elastic stop nuts, or adjust the link rod, until the timing mark on the flywheel is aligned with the mark on the rewind housing. Refer to the Specifications in the Appendix for the exact BTDC setting. Tighten the elastic stop nuts, or the jam nuts. Back off the throttle to the normal idle position.

Primary Pickup Synchronizing Timing

3- Shift the engine into **FORWARD** gear. Advance the throttle until the timing mark on the flywheel is aligned with the specified throttle pick-up, as given in the Appendix. Now, adjust the throttle pick-up screw until the end of the screw just makes contact with the carburetor cluster pin. Tighten the nut on the adjustment screw. Back off the throttle to the normal idle position. Shut the engine down and disconnect the timing light.

Maximum Neutral RPM
Limiter Adjustment

The purpose of the maximum neutral rpm speed limiter is to stop the rotation of the tiller handle twist grip at a specified rpm when the shift lever is in the **NEUTRAL** position. This adjustment must be accurate to ensure easy starting when the engine is cold and also to prevent the engine from starting at maximum operating rpm.

4- With the engine operating at idle speed, loosen the neutral speed stop bolt just enough to enable the maximum neutral speed stop to slide back-and-forth. Now, turn the twist grip to advance engine rpm to 2400 - 2700 rpm. **WITHOUT** turning the twist grip, position and secure the maximum neutral speed stop against the neutral speed limiter follower. This adjustment will now stop the throttle at the recommended maximum neutral speed and assist starting a cold engine. Check the adjustment to be sure the maximum neutral speed stop does not allow the engine rpm to exceed 2400-2700 rpm. Shift the engine into **FORWARD** gear and check to be sure the neutral speed stop does not interfere with throttle operation in forward gear. If the stop does interfere with throttle operation, move the back edge of the stop **TOWARD** the cylinder block.

ALLEN SCREW UNDER GRIP

5

Tiller Handle and Throttle Decal Alignment

FIRST, this word: The adjustment outlined in the previous step must be completed **BEFORE** the tiller handle and throttle decal alignment can be properly made.

5- With the engine operating at idle speed, and the shift lever in the **NEUTRAL** position, turn the twist grip until the throttle is against the maximum neutral stop. At this point, the arrow on the tiller handle should be aligned with the word **START** on the throttle decal. If the arrow is not properly aligned, proceed as follows:

a- Remove and retain the two screws securing the tiller handle to the engine.

b- Align the gears, as necessary to align the word **START** with the arrow.

c- Install the tiller handle to the engine and secure it with the two screws.

Disconnect the tachometer and fuel line. Install the engine cowling and remove the engine from the test tank.

6-14 MODEL 3.5 and 3.6 HP 1980 -- 1985

SPECIAL WORDS

The timing on the 3.5 and 3.6 hp engines is not adjustable. The following steps give detailed procedures to adjust the idle speed and the idle mixture.

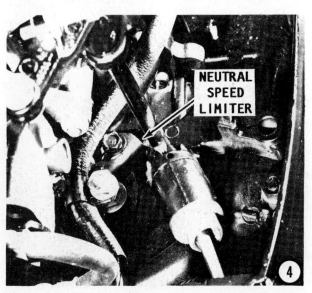

NEUTRAL
SPEED
LIMITER

4

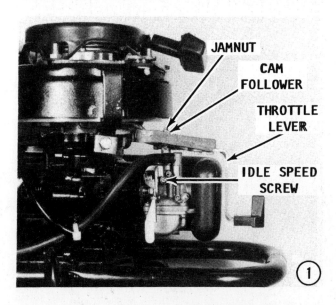

JAMNUT

CAM
FOLLOWER

THROTTLE
LEVER

IDLE SPEED
SCREW

1

Primary Pickup Synchronizing

1- Remove the carburetor silencer, if one is installed. Move the throttle lever to the starboard side (to the left as you face the the front of the engine). Loosen the jam nut and cam follower on top of the throttle valve. **DO NOT** allow the shaft to rotate. Observe through the carburetor throat and at the same time, rotate the idle speed adjusting screw **COUNTERCLOCKWISE** until the throttle valve stops its downward movement. Now, rotate the idle speed adjustment screw inward **(CLOCKWISE)**, until the screw **BARELY** touches the throttle valve. From this position, rotate the screw inward two (2) complete turns, as a rough initial adjustment.

Idle Mixture Adjustment

2- The idle mixture adjustment screw controls the air mixing with the fuel.

Rotate the idle mixture screw **CLOCKWISE** (inward) until it **BARELY** seats, and then rotate it **COUNTERCLOCKWISE** (outward) two (2) complete turns as an intial rough adjustment.

Connect a tachometer to the engine. Start the engine.

CAUTION: Water must circulate through the lower unit to the engine any time the engine is run to prevent damage to the water pump in the lower unit. Just five seconds without water will damage the water pump.

Make the fine adjustment with the idle mixture screw by rotating the screw outward **VERY** slowly until the engine rpm

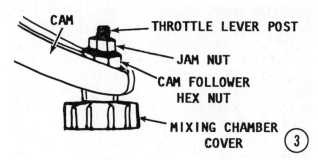

begins to slow down, then turn the screw inward until the engine rpm begins to increase. At this point **STOP**. The idle mixture is properly adjusted. If the screw is turned in too far, the engine rpm will begin to slow down again.

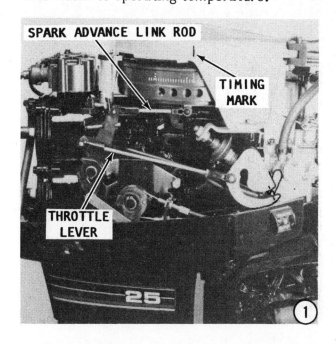

Idle Speed Adjustment

3- Rotate the idle speed adjusting screw inward or outward until the engine is operating at 750 to 850 rpm. Stop the engine. With the engine throttle lever at the lowest idle position, adjust the cam follower until it **BARELY** makes contact with the throttle lever. At this position, tighten the jam nut on the throttle valve shaft.

6-15 MODEL 18 AND 25 1980 -- 1983
MODEL 18XD and 25XD 1984 -- 1985
MODEL 20 1979 -- 1980 AND 1986 AND ON
MODEL 25 1986 AND ON

Primary Pickup Synchronizing Timing

1- Connect a timing light to the No. 1 spark plug wire. Start the engine and allow it to warm to operating temperature.

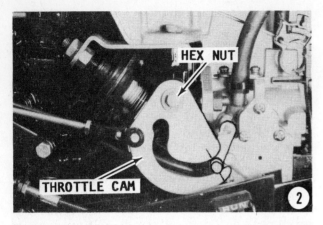

CAUTION: Water must circulate through the lower unit to the engine any time the engine is run to prevent damage to the water pump in the lower unit. Just five seconds without water will damage the water pump.

Advance the throttle to the wide open position. With the timing light aimed at the flywheel, the timing mark on the flywheel (three (3) dots), should align with the timing mark on the manual starter housing. If the marks do **NOT** align while the engine is running at wide open throttle, stop the engine.

SAFETY WORDS

Personal injury could result if the spark advance link rod should accidently make contact with the flywheel while the power-head is running. Therefore, keep the rod clear of the flywheel.

Adjust the spark advance link rod on the starboard side of the engine. To move the timing mark to the left, retard, (when the engine is running), shorten the rod. To move the timing mark to the right, advance, (when the engine is running), lengthen the rod. Start the engine and check the adjustment.

2- Stop the engine. Loosen the hex screw on the throttle cam. Adjust the cam until the center of the roller on the throttle shaft aligns with the straight line at the end of the throttle cam. (The throttle cam is elongated behind the hex attaching screw). Tighten the hex attaching screw.

Throttle Pickup Timing Adjustment

3- Start the engine and advance the throttle until the throttle pickup marks on the flywheel (two (2) dots) are aligned with the timing mark on the starter housing. The straight line at the end of the throttle cam should be at the roller or within 1/16" (1.6mm). If an adjustment is necessary, stop the engine and the adjust the throttle lever on the starboard side of the engine. The dashpot, installed on late model engines, should be fully retracted when the adjustment is completed. Again, start the engine and check the adjustments with the timing light. If the adjustments are satisfactory, stop the engine and disconnect the timing light. Adjust the idle speed screw to obtain 700 to 800 rpm. At this rpm, the stem of the dashpot should be fully depressed. If the stem is not depressed, make the necessary adjustment.

Neutral RPM Adjustment

4- Start the engine and retard the throttle to the **SLOW** position. Shift the unit into **NEUTRAL**, and move the **Primer/Fast/Idle** knob into the center detent. Manually adjust the ratchet until the rpm ranges between 1400 - 1700 rpm.

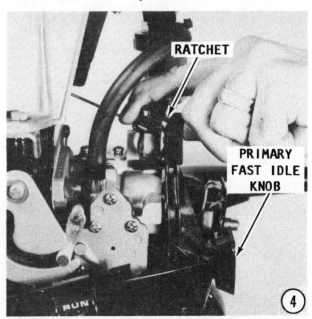

6-16 MODEL 2.2 1984—1989
MODEL 3.0 1990 AND ON

SPECIAL WORDS

The timing on the Model 2.2 and the Model 3.0 is not adjustable. The following procedures provide detailed instructions to adjust the throttle jet needle setting and the idle speed adjustment.

Throttle Jet Needle Setting

1- Remove the knobs from the throttle and choke levers, then lift away the front cover. Remove the throttle lever assembly from the top of throttle valve bracket. With a pair of pliers, unscrew the mixing chamber cover and lift out the throttle valve assembly. Compress the spring in the throttle valve assembly to allow the throttle cable end to clear the recess in the base of the throttle valve and to slide down the slot.

Disassemble the throttle valve, consisting of the throttle valve, spring, jet needle, (with E-clip normally on the second groove), jet retainer and throttle cable end. The tapered jet needle varies the fuel flow as the throttle is opened. Relocating the E-clip closer to the end of the jet needle will lean the air/fuel mixture; lowering the E-clip will enrich the mixture.

It is preferable to operate the powerhead with a too rich a mixture rather than too lean. A too lean an air/fuel mixture will cause the powerhead to overheat. A powerhead operating too hot will cause pre-ignition. If the condition is not corrected the powerhead could be severely damaged.

If the air/fuel mixture is too rich, the rpm will decrease as the throttle is opened beyond the 2/3 to full throttle range.

If the correct mixture is not obtainable by changing the position of the E-clip on the throttle needle, the main jet located inside the fuel bowl in the center of the float **MUST** be changed from the standard jet size No. 94. See table in the Appendix.

Idle Speed Adjustment

2- Connect a tachometer to the powerhead.

CAUTION: Water must circulate through the lower unit to the engine any time the engine is run to prevent damage to the water pump in the lower unit. Just five seconds without water will damage the water pump.

Start the the engine and allow it warm to operating temperature.

Adjust the throttle lever to the lowest speed, and then adjust the idle speed screw until the powerhead idles at 900-1000 rpm.

6-17 MODEL 6.0, 8.0, 9.9, 15 AND 210cc 1986 AND ON

Maximum Advance Timing

1- Mount the engine in a test tank. Remove the engine cowling. Connect the fuel line to a fuel source. Install a tachometer to the engine. Connect a timing light to the No. 1 (top) spark plug. Start the engine and allow it to warm to operating temperature.

← 1st Groove   **LEANER MIXTURE**
← 2nd
← 3rd
← 4th   **RICHER MIXTURE**

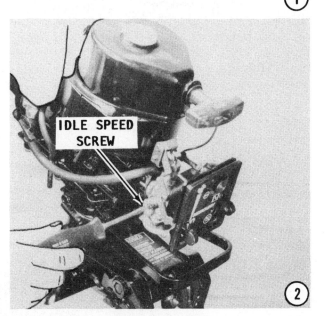

①

②

①

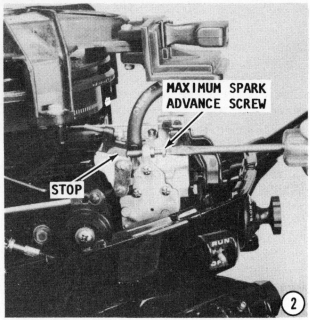

CAUTION: Water must circulate through the lower unit to the engine any time the engine is run to prevent damage to the water pump in the lower unit. Just five seconds without water will damage the water pump.

2- Move the shift lever into **FORWARD** gear. Advance the throttle until the maximum timing mark on the flywheel is aligned with the timing mark on the starter housing. Refer to the Specifications in the Appendix for recommended rpm. Adjust the maximum spark advance screw until the end of the screw just makes contact with the stop. Back off the throttle to normal idle, and then shut the engine down. Disconnect the timing light.

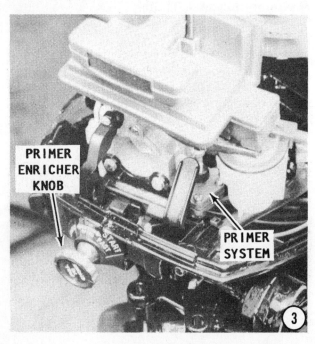

Idle Timing Adjustment

3- Push the **Primer/Enricher** knob inward and at the same time rotate the knob **COUNTERCLOCKWISE** as far as possible.

4- Reduce the powerhead rpm to idle speed, 550-650 rpm. Shift the unit into **FORWARD** gear and adjust the idle timing screw until the mark on the hand starter housing aligns with the 6° BTDC mark on the decal (2 dots).

Idle Wire Adjustment

5- Push the **Primer/Enricher** knob in, and then turn the knob **COUNTERCLOCKWISE** as far as possible. Shift the unit into **NEUTRAL** and adjust the upper screw on the ratchet to remove any clearance between the idle wire and the trigger. The wire should **BARELY** make contact with the trigger.

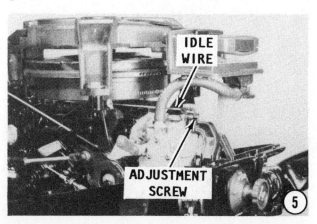

6-18 MODEL 4.0 MID 1987 AND ON
MODEL 5.0 1988 AND ON

The timing on the models listed is pre-set at the factory. This setting may be checked by performing the following procedures. However, adjustment is not possible.

If checking indicates the timing to be incorrect, the fault may be either mechanical or electrical.

Mechanical
A mechanical fault can only be directly related to the flywheel -- either the flywheel has been installed without the Woodruff key or the Woodruff key has been sheared.

Electrical
An electrical problem may be caused by a fault in the ignition system. The ignition system uses an electronic spark advancement. Detailed testing procedures for the capacitor charging coil, the trigger, and the ignition coil are listed in Chapter 5. Check the Table of Contents.

Checking Timing
1- Begin by removing the cowling. Next, remove the spark plug and install a dial indicator in the spark plug opening. Now, slowly rotate the flywheel CLOCK-WISE and determine TDC for the cylinder using the dial indicator.

After TDC has been determined, observe the two vertical lines embossed on the port side of the flywheel. The aft line should align with the split line of the cylinder block and the crankcase cover.

If the aft line is centered as just explained, the flywheel has been installed correctly and the Woodruff key is in place.

If the aft line is not centered as explained, the hand rewind starter and the flywheel must be removed and the Woodruff key checked for damage and correct installation.

Remove the dial indicator, install the spark plug and connect a timing light to the powerhead.

Mount the engine in a test tank or move the boat to a body of water. NEVER use a flush attachment while conducting the following tests.

CAUTION
Water must circulate through the lower unit to the powerhead anytime the powerhead is operating to prevent damage to the water pump in the lower unit. Just five seconds without water will damage the water pump impeller.

Idle Timing
2- Start the engine and allow it to warm to normal operating temperature.

Allow the engine to run at idle speed -- 900-1000 rpm in FORWARD gear.

Aim the timing light at the port side of the powerhead. The split line between the cylinder block and the crankcase cover should be misaligned about 1/4" (6.4mm) toward the forward side of the aft line embossed on the flywheel. This position

corresponds to 5° BTDC advance at the flywheel.

If the timing is not as indicated, and the powerhead operates roughly or misfires, the problem is most likely electrical. Detailed procedures to test the capacitor charging coil, the trigger, and the ignition coil.

WOT Timing

3- Twist the throttle grip to the **FAST** position, with the engine still in **FORWARD** gear. Aim the timing light at the port side of the powerhead.

The forward line embossed on the flywheel should align with the split line between the cylinder block and the crankcase cover.

This position corresponds to 30° BTDC advance at the flywheel.

If the timing is not as indicated, and the powerhead operates roughly or misfires, the problem is most likely electrical. Detailed procedures to test the capacitor charging coil, the trigger, and the ignition coil, are given in Chapter 5, beginning on Page 5-39.

7
ELECTRICAL

7-1 INTRODUCTION

The battery, charging system, and the cranking system are considered subsystems of the electrical system. Each of these units will be covered in detail in this chapter beginning with the battery.

7-2 BATTERIES

The battery is one of the most important parts of the electrical system. In addition to providing electrical power to start the engine, it also provides power for operation of the running lights, radio, electrical accessories, and possibly the pump for a bait tank.

Because of its job and the consequences, (failure to perform in an emergency) the best advice is to purchase a well-known brand, with an extended warranty period, from a reputable dealer.

The usual warranty covers a prorated replacement policy, which means you would be entitled to a consideration for the time left on the warranty period if the battery should prove defective before its time.

Do not consider a battery of less than **70-ampere hour** or **100-minute** reserve capacity. If in doubt as to how large your boat requires, make a liberal estimate and then purchase the one with the next higher ampere rating.

MARINE BATTERIES

Because marine batteries are required to perform under much more rigorous conditions than automotive batteries, they are constructed much differently than those used in automobiles or trucks. Therefore, a marine battery should always be the No. 1 unit for the boat and other types of batteries used only in an emergency.

Marine batteries have a much heavier exterior case to withstand the violent pounding and shocks imposed on it as the boat moves through rough water and in extremely tight turns.

The plates in marine batteries are thicker than in automotive batteries and each plate is securely anchored within the battery case to ensure extended life.

The caps of marine batteries are "spill proof" to prevent acid from spilling into the bilges when the boat heels to one side in a tight turn, or is moving through rough water.

A fully charged battery, filled to the proper level with electrolyte, is the heart of the ignition and electrical systems. Engine cranking and efficient performance of electrical items depend on a full-rated battery.

Because of these features, the marine battery will recover from a low charge condition and give satisfactory service over a much longer period of time than any type intended for automotive use.

NEVER use a "Maintenance Free" type battery with an outboard unit. The charging system is not regulated as with automotive installations and the battery may be quickly damaged.

BATTERY CONSTRUCTION

A battery consists of a number of positive and negative plates immersed in a solution of diluted sulfuric acid. The plates contain dissimilar active materials and are kept apart by separators. The plates are grouped into what are termed elements. Plate straps on top of each element connect all of the positive plates and all of the negative plates into groups.

The battery is divided into cells which hold a number of the elements apart from the others. The entire arrangement is contained within a hard-rubber case. The top is a one-piece cover and contains the filler caps for each cell. The terminal posts protrude through the top where the battery connections for the boat are made. Each of

OVERFILLING
CORROSION   FRAYED OR BROKEN CABLES
DIRT
LOOSE HOLD-DOWN
SEALING COMPOUND DEFECT
CELL CONNECTOR CORROSION
CRACKED CASE
CRACKED CELL COVER   LOW ELECTROLYTE

A visual inspection of the battery should be made each time the boat is used. Such a quick check may reveal a potential problem in its early stages. A dead battery in a busy waterway or far from assistance could have serious consequences.

the cells is connected to its neighbor in a positive-to-negative manner with a heavy strap called the cell connector.

BATTERY RATINGS

Four different methods are used to measure and indicate battery electrical capacity:

1- Ampere-hour rating
2- Cold cranking performance
3- Reserve capacity
4- Watt hour rating

The ampere-hour rating of a battery refers to the battery's ability to provide a set amount of amperes for a given amount of time under test conditions at a constant temperature of $80°$ ($27°C$). Amperes x hours equals ampere-hour rating. Therefore, if the battery is capable of suppling 4 amperes of current for 20 consecutive hours, the battery is rated as an 80 ampere-hour battery.

The ampere-hour rating is useful for some service operations, such as slow charging or battery testing.

Cold cranking performance is measured by cooling a fully charged battery to $0°F$ ($-17°C$) and then testing it for 30 seconds to determine the maximum current flow. In this manner the cold cranking amperes rating is the number of amperes available to be drawn from the battery before the voltage drops below 7.2 volts.

Reserve capacity of a battery is considered the length of time -- in minutes -- at $80°F$ ($27°C$) a 25 ampere current can be maintained before the voltage drops below 10.5 volts. This test is intended to provide an approximation of how long the engine, including electrical accessories such as bilge pump, radio, running light, could continue to operate satisfactorily if the alternator or magneto did not produce sufficient current. A typical rating is 100 minutes.

Watt-hour is a very useful rating of battery power. It is determined by multiplying the number of ampere hours times the voltage. Therefore, a 12-volt battery rated at 80 ampere-hours would be rated at 960 watt-hours (80 x 12 = 960).

If possible, the new battery should have a power rating equal to or higher than the unit it is replacing.

BATTERY LOCATION

Every battery installed in a boat must be secured in a well-protected ventilated area. If the battery area lacks adequate ventilation, hydrogen gas which is given off during charging could become very explosive. This is especially true if the gas is concentrated and confined.

BATTERY SERVICE

The battery requires periodic servicing and a definite maintenance program will ensure extended life. If the battery should test satisfactorily, but still fails to perform properly, one of five problems could be the cause.

1- An accessory might have accidently been left on overnight or for a long period during the day. Such an oversight would result in a discharged battery.

2- Slow speed engine operation for long periods of time resulting in an undercharged condition.

3- Using more electrical power than the alternator can replace would result in an undercharged condition.

4- A defect in the charging system. A faulty alternator, defective rectifier, or high resistance somewhere in the system could cause the battery to become undercharged.

5- Failure to maintain the battery in good order. This might include a low level of electrolyte in the cells; loose or dirty cable connections at the battery terminals; or possibly an excessively dirty battery top.

Electrolyte Level

The most common practice of checking the electrolyte level in a battery is to remove the cell cap and visually observe the level in the vent well. The bottom of each vent well has a split vent which will cause the surface of the electrolyte to appear distorted when it makes contact. When the distortion first appears at the bottom of the split vent, the electrolyte level is correct.

Some late-model batteries have an electrolyte-level indicator installed which operates in the following manner:

A transparent rod extends through the center of one of the cell caps. The lower tip of the rod is immersed in the electrolyte when the level is correct. If the level should drop below normal, the lower tip of the rod is exposed and the upper end glows as a warning to add water. Such a device is only necessary on one cell cap because if the electrolyte is low in one cell it is also low in the other cells. **BE SURE** to replace the cap with the indicator onto the second cell from the positive terminal.

During hot weather and periods of heavy use, the electrolyte level should be checked more often than during normal operation. Add potable (drinking) water to bring the level of electrolyte in each cell to the proper level. **TAKE CARE** not to overfill, because adding an excessive amount of water will cause loss of electrolyte and any loss will result in poor performance, short battery life, and will contribute quickly to corrosion. **NEVER** add electrolyte from another battery. Use only clean pure water.

Battery Testing

A hydrometer is a device to measure the percentage of sulfuric acid in the battery electrolyte in terms of specific gravity. When the condition of the battery drops from fully charged to discharged, the acid leaves the solution and enters the plates, causing the specific gravity of the electrolyte to drop.

It may not be common knowledge, but hydrometer floats are calibrated for use at 80°F (27°C). If the hydrometer is used at any other temperature, hotter or colder, a correction factor must be applied. (Remember, a liquid will expand if it is heated and will contract if cooled. Such expansion and contraction will cause a definite change

An explosive hydrogen gas is normally released from the cells under a wide range of circumstances. This battery exploded when the gas ignited from someone smoking in the area when the caps were removed. Such an explosion could also be caused by a spark from the battery terminals igniting the gas.

in the specific gravity of the liquid, in this case the electrolyte.)

A quality hydrometer will have a thermometer/temperature correction table in the lower portion, as shown in the accompanying illustration. By knowing the air temperature around the battery and from the table, a correction factor may be applied to the specific gravity reading of the hydrometer float. In this manner, an accurate determination may be made as to the condition of the battery.

The following six points should be observed when using a hydrometer.

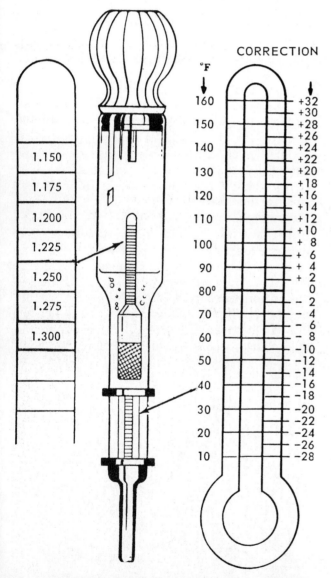

A check of the electrolyte in the battery should be on the maintenance schedule for any boat. A hydrometer reading of 1.300, or in the green band, indicates the battery is in satisfactory condition. If the reading is 1.150 or in the red band, the battery must be charged. Observe the six safety points listed in the text when using a hydrometer.

1- **NEVER** attempt to take a reading immediately after adding water to the battery. Allow at least 1/4 hour of charging at a high rate to thoroughly mix the electrolyte with the new water. This time will also allow for the necessary gasses to be created.

2- **ALWAYS** be sure the hydrometer is clean inside and out as a precaution against contaminating the electrolyte.

3- If a thermometer is an integral part of the hydrometer, draw liquid into it several times to ensure the correct temperature before taking a reading.

4- **BE SURE** to hold the hydrometer vertically and suck up liquid only until the float is free and floating.

5- **ALWAYS** hold the hydrometer at eye level and take the reading at the surface of the liquid with the float free and floating.

Disregard the light curvature appearing where the liquid rises against the float stem. This phenomenon is due to surface tension.

6- **DO NOT** drop any of the battery fluid on the boat or on your clothing, because it is extremely caustic. Use water and baking soda to neutralize any battery liquid that does accidently drop.

After withdrawing electrolyte from the battery cell until the float is barely free, note the level of the liquid inside the hydrometer. If the level is within the green band range for all cells, the condition of the battery is satisfactory. If the level is within the white band for all cells, the battery is in fair condition.

If the level is within the green or white band for all cells except one, which registers in the red, the cell is shorted internally. No amount of charging will bring the battery back to satisfactory condition.

If the level in all cells is about the same, even if it falls in the red band, the battery may be recharged and returned to service. If the level fails to rise above the red band after charging, the only solution is to replace the battery.

Cleaning

Dirt and corrosion should be cleaned from the battery just as soon as it is discovered. Any accumulation of acid film or dirt will permit current to flow between the terminals. Such a current flow will drain the battery over a period of time.

Clean the exterior of the battery with a solution of diluted ammonia or a soda solution to neutralize any acid which may be present. Flush the cleaning solution off with clean water. **TAKE CARE** to prevent any of the neutralizing solution from entering the cells, by keeping the caps tight.

A poor contact at the terminals will add resistance to the charging circuit. This resistance will cause the voltage regulator to register a fully charged battery, and thus cut down on the alternator output adding to the low battery charge problem.

Scrape the battery posts clean with a suitable tool or with a stiff wire brush. Clean the inside of the cable clamps to be sure they do not cause any resistance in the circuit.

JUMPER CABLES

If booster batteries are used for starting an engine the jumper cables must be connected correctly and in the proper sequence to prevent damage to either battery, or to the alternator diodes.

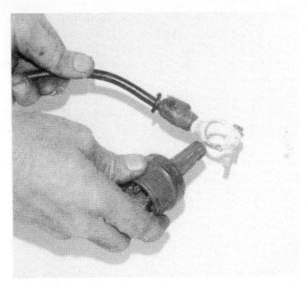

The second part of the brush shown in the previous illustration may be used to clean the cable connectors.

ALWAYS connect a cable from the positive terminal of the dead battery to the positive terminal of the good battery **FIRST**. **NEXT**, connect one end of the other cable to the negative terminal of the good battery and the other end to a good ground on the powerhead. **DO NOT** connect the negative jumper from the good battery to the negative terminal of the low battery. Such action will almost always cause a spark

One of the most effective means of cleaning the battery terminals is to use a two-part inexpensive brush designed for this specific purpose.

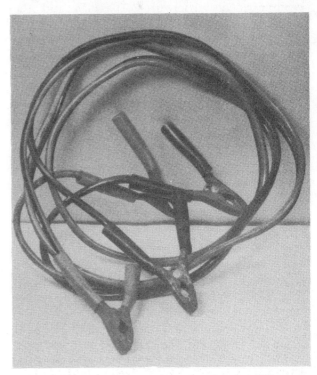

A common set of heavy-duty jumper cables. Observe the safety precautions given in the text when using jumper cables.

which could ignite gases escaping through the vent holes in the battery filler caps. Igniting the gases may result in an explosion destroying the battery and causing severe personal **INJURY.**

By making the negative (ground) connection on the powerhead, if an arc is created, it will not be near the battery.

DISCONNECT the battery ground cable before replacing an alternator or before connecting any type of meter to the alternator.

If it is necessary to use a fast-charger on a dead battery, **ALWAYS** disconnect one of the boat cables from the battery **FIRST,** to prevent burning out the diodes in the rectifier.

NEVER use a fast-charger as a booster to start the engine because the diodes in the alternator will be **DAMAGED.**

STORAGE

If the boat is to be laid up for the winter or for more than a few weeks, special attention must be given to the battery to prevent complete discharge or possible damage to the terminals and wiring. Before putting the boat in storage, disconnect and remove the batteries. Clean them thoroughly of any dirt or corrosion, and then charge them to full specific gravity reading. After they are fully charged, store them in a clean cool dry place where they will not be damaged or knocked over, preferably on a couple blocks of wood. Storing the battery up off the deck, will permit air to circulate freely around and under the battery and will help to prevent condensation.

NEVER store the battery with anything on top of it or cover the battery in such a manner as to prevent air from circulating around the fillercaps. All batteries, both new and old, will discharge during periods of storage, more so if they are hot than if they remain cool. Therefore, the electrolyte level and the specific gravity should be checked at regular intervals. A drop in the specific gravity reading is cause to charge them back to a full reading.

In cold climates, care should be exercised in selecting the battery storage area. A fully-charged battery will freeze at about 60 degrees below zero. A discharged battery, almost dead, will have ice forming at about 19 degrees above zero.

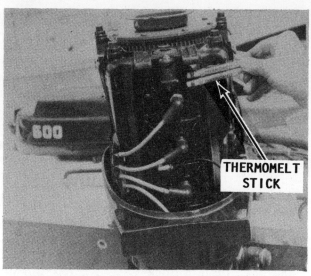

A thermomelt stick is a quick, simple, inexpensive, and fairly accurate method to determine powerhead operating temperature, as shown here on a 4-cylinder unit.

7-3 THERMOMELT STICKS

Thermomelt sticks are an easy method of determining if the powerhead is running at the proper temperature. Thermomelt sticks are not expensive and are available at your local marine dealer.

Start the engine with the propeller in the water and run it for about 5 minutes at roughly 3000 rpm.

CAUTION: Water must circulate through the lower unit to the engine any time the engine is run to prevent damage to the water pump in the lower unit. Just five seconds without water will damage the water pump.

The 140 degree stick should melt when you touch it to the lower thermostat housing or on the top cylinder. If it does not melt, the thermostat is stuck in the open position and the engine temperature is too low.

Touch the 170 degree stick to the same spot on the lower thermostat housing or on the top cylinder. The stick should not melt. If it does, the thermostat is stuck in the closed position or the water pump is not operating properly because the engine is running too hot.

If the powerhead is not equipped with a thermostat, the problem may be solved by reverse flushing to clean out the cooling system and/or servicing the water pump. For service procedures for the thermostat, see Chapter 8. For service procedures for the water pump, see Chapter 9.

7-4 TACHOMETER

An accurate tachometer can be installed on any engine. Such an instrument provides an indication of engine speed in revolutions per minute (rpm). This is accomplished by measuring the number of electrical pulses per minute generated in the primary circuit of the ignition system.

The meter readings range from 0 to 6,000 rpm, in increments of 100. Tachometers have solid-state electronic circuits which eliminates the need for relays or batteries and contributes to their accuracy. The electronic parts of the tachometer susceptible to moisture are coated to prolong their life.

Some of the outboard units covered in this manual have a female plug at the forward end of the shift box as a convenience for installation of a tachometer. Therefore, when purchasing a tachometer, check to be sure the adaptor plug will mate with the fitting on the shift box.

7-5 ELECTRICAL SYSTEM GENERAL INFORMATION

In the early days, all outboard engines were started by simply pulling on a rope wound around the flywheel. As time passed and owners were reluntant to use muscle power, it was necessary to replace the rope starter with some form of power cranking system. Today, many small engines are still started by pulling on a rope, but others have a cranking motor installed.

The system utilized to replace the rope method was an electric cranking motor coupled with a mechanical gear mesh be-

tween the cranking motor and the powerhead flywheel, similar to the method used to crank an automobile engine.

The electrical system consists of three circuits:

a- Charging circuit
b- Cranking motor circuit
c- Ignition circuit

Charging Circuit

The charging circuit consists of permanent magnets and a stator located within the flywheel; a rectifier located elsewhere on the powerhead; an external battery; and the necessary wiring to connect the units. The negative side of the rectifier is grounded. The positive side of the rectifier passes through the internal harness plug to the battery. The negative side of the battery is connected, through the connector, to a good ground on the engine.

The alternating current generated in the stator windings passes to the rectifier. The rectifier changes the alternating current (AC) to direct current (DC) to charge the 12-volt battery.

Cranking Motor Circuit

The cranking motor circuit consists of a cranking motor and a starter-engaging mechanism. A solenoid is used as a heavy-duty switch to carry the heavy current from the battery to the cranking motor. On most

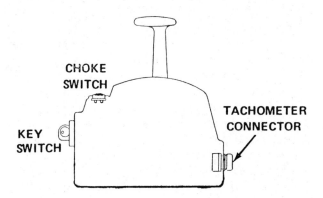

Many models of shift boxes now have a plug connector on the forward face for installation of a tachometer cable.

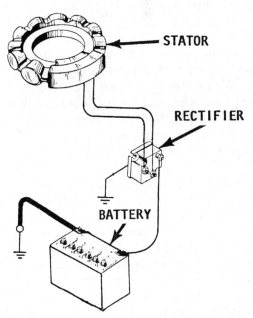

Functional diagram of a typical charging circuit showing relationship of the stator, solid state rectifier, and the battery.

models, the solenoid is actuated by turning the ignition key to the **START** position.

Ignition Circuit

The ignition circuit is covered extensively in Chapter 5.

7-6 CHARGING CIRCUIT SERVICE

The stator is located under, and protected by, the flywheel. Therefore, the stator seldom causes problems in the charging circuit. Most problems in the charging circuit can be traced to the rectifier or to the battery. If either the stator or the rectifier fails the troubleshooting tests, the defective unit cannot be repaired, it **MUST** be replaced.

STATOR SERVICE

This section provides detailed procedures for testing, removing, and installing the stator.

Stator Testing

The stator may be tested without removing the flywheel, by merely disconnecting the two yellow leads from the rectifier and using an ohmmeter. Check the resistance of the stator windings against those given in the Specifications in the Appendix. If the stator resistance does not meet the Specifications, it **MUST** be replaced.

REMOVAL

BAD NEWS: The flywheel must be removed to gain access to the stator.

Disconnect the battery leads from the battery terminals. Remove the front cowl cover and the wrap-around cowl, if one is used. Remove the top cowl. Remove the nut on the crankshaft in the center of the flywheel. A flywheel holder may be required to prevent the flywheel from turning in order to loosen the nut.

Obtain the proper flywheel puller to pull the flywheel. **NEVER** use a puller which pulls on the outside edge of the flywheel, or the flywheel may be damaged. After the puller is installed, tighten the center screw onto the end of the crankshaft. Continue tightening the screw until the flywheel is released from the crankshaft. Remove the flywheel puller. Lift the flywheel from the crankshaft.

Remove the yellow wires leading from the stator to the black terminal on the rectifier. Remove the retaining screw from the stator and lift the stator from the engine.

INSTALLATION

Place the stator in position on the powerhead. Coat the threads of the stator attaching screws with blue Loctite, or equivalent. Secure the stator with the at-

Once the flywheel is removed, major ignition components are exposed for service.

The trigger must be removed to gain access to the stator.

taching screws through the stator into the powerhead block.

Insert the flywheel key into the crankshaft keyway. Check to be sure the inside taper of the flywheel and the taper on the crankshaft are clean of dirt or oil, to prevent the flywheel from "walking" on the crankshaft during operation. Slide the flywheel down over the crankshaft with the keyway in the flywheel aligned with the key on the crankshaft. Rotate the flywheel clockwise and check to be sure the flywheel does not contact any powerhead part or the wiring. Thread the flywheel nut onto the crankshaft, and then tighten it to the torque value given in the Specifications in the Appendix. For detailed timing procedures, see Chapter 6.

7-7 CRANKING MOTOR CIRCUIT SERVICE

DESCRIPTION

As the name implies, the sole purpose of the cranking motor circuit is to control operation of the cranking motor to crank the powerhead until the engine is operating. The circuit includes a solenoid or magnetic switch to connect or disconnect the motor from the battery. The operator controls the switch with a push button or key switch.

A neutral start switch is installed into the circuit to permit operation of the cranking motor **ONLY** if the shift control lever is in **NEUTRAL**. This switch is a safety device to prevent accidental engine start when the engine is in gear.

The cranking motor is a series wound electric motor which draws a heavy current from the battery. It is designed to be used only for short periods of time to crank the engine for starting. To prevent overheating the motor, cranking should not be continued for more than 30-seconds without allowing the motor to cool for at least three minutes.

Actually, this time can be spent in making preliminary checks to determine why the engine fails to start.

Theory of Operation

Power is transmitted from the cranking motor to the powerhead flywheel through a Bendix drive. This drive has a pinion gear mounted on screw threads. When the motor is operated, the pinion gear moves upward and meshes with the teeth on the flywheel ring gear.

When the engine starts, the pinion gear is driven faster than the shaft, and as a result, it screws out of mesh with the flywheel. A rubber cushion is built into the Bendix drive to absorb the shock when the pinion meshes with the flywheel ring gear. The parts of the drive **MUST** be properly assembled for efficient operation. If the drive is removed for cleaning, **TAKE CARE** to assemble the parts as shown in the accompanying illustration. If the screw shaft assembly is reversed, it will strike the splines and the rubber cushion will not absorb the shock.

The sound of the motor during cranking is a good indication of whether the cranking motor is operating properly or not. Natural-

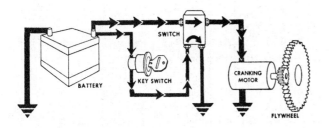

Simple functional diagram of a cranking circuit.

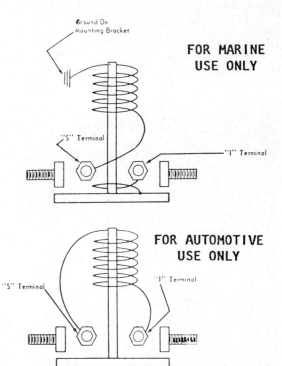

Schematic diagram of a marine solenoid (top) and an automotive solenoid (bottom). The marine solenoid has an internal ground and is therefore suitable for outboard installations.

ly, temperature conditions will affect the speed at which the cranking motor is able to crank the engine. The speed of cranking a cold engine will be much slower than when cranking a warm engine. An experienced operator will learn to recognize the favorable sounds of the powerhead cranking under various conditions.

Faulty Symptoms

If the cranking motor spins, but fails to crank the engine, the cause is usually a corroded or gummy Bendix drive. The drive should be removed, cleaned, and given an inspection.

If the cranking motor cranks the engine to slowly, the following are possible causes and the corrective actions that may be taken:

a- Battery charge is low. Charge the battery to full capacity.

b- High resistance connections at the battery, solenoid, or motor. Clean and tighten all connections.

c- Undersize battery cables. Replace cables with sufficient size.

d- Battery cables too long. Relocate the battery to shorten the run to the solenoid.

Maintenance

The cranking motor does not require periodic maintenance or lubrication. If the motor fails to perform properly, the checks outlined in the previous paragraph should be performed.

The frequency of starts governs how often the motor should be removed and reconditioned. The manufacturer recommends removal and reconditioning every 1000 hours.

Naturally, the motor will have to be removed if the corrective actions outlined under **Faulty Symptoms** above, does not restore the motor to satisfactory operation.

CRANKING MOTOR TROUBLESHOOTING

Before wasting too much time troubleshooting the cranking motor circuit, the following checks should be made. Many times, the problem will be corrected.

a- Battery fully charged.

b- Shift control lever in **NEUTRAL**.

c- All electrical connections clean and tight.

d- Wiring in good condition, insulation not worn or frayed.

Two more areas may cause the powerhead to crank slowly even though the cranking motor circuit is in excellent condition: a tight or "frozen" powerhead and water in the lower unit. The following troubleshooting procedures are presented in a logical sequence, with the most common and easily corrected areas listed first in each problem area. The connection number refers to the numbered positions in the accompanying illustrations.

Perform the following quick checks and corrective actions for following problems:

1- Cranking Motor Rotates Slowly

a- Battery charge is low. Charge the battery to full capacity.

b- Electrical connections corroded or loose. Clean and tighten.

c- Defective cranking motor. Perform an amp draw test. Lay an amp draw-gauge on the cable leading to the cranking motor. Turn the key on and attempt to crank the engine. If the gauge indicates an excessive amperage draw, the cranking motor **MUST** be replaced or rebuilt.

2- Cranking Motor Fails to Crank Powerhead Test Motor

a- Disconnect the cranking motor lead from the solenoid to prevent the powerhead from starting during the testing process.

NOTE: This lead is to remain disconnected from the solenoid during tests No. 2 thru No. 7.

b- Disconnect the black ground wire from the No. 2.

c- Connect a voltmeter between the No. 2 and a common engine ground.

d- Turn the key switch to the **START** position.

e- Observe the voltmeter reading. If there is the slightest amount of reading, check the black ground wire connection or check for an open circuit. Connect the ground wire back to the No. 2 and move to Step 7. If there is no voltmeter reading, proceed with Step 3.

3- Test Cranking Motor Solenoid

a- Connect a voltmeter between the engine common ground and the No. 3.

b- Turn the ignition key switch to the **START** position.

c- Observe the voltmeter reading. If there is the slightest indication of a reading, the solenoid is defective and must be replaced. If there is no reading, proceed with Step 4.

4- Test Neutral Start Swtich

a- Connect a voltmeter between the common engine ground and the No. 4. Turn the ignition key switch to the **START** position.

b- Observe the voltmeter. If there is any indication of a reading, the neutral start switch is open in the shift box or the yellow wire lead is open between the No. 3 and No. 4. If there is no voltmeter reading, proceed to Step 5.

5- Test Ignition Switch

a- Connect a voltmeter between a common engine ground and No. 5.

b- Observe the voltmeter. If there is the slightest indication of a reading, the ignition switch is defective and must be replaced. If there is no reading, proceed with Step 6.

6- Test for Open Wire

a- Connect a voltmeter between the common engine ground and No. 6.

b- The voltmeter should indicate 12-volts. If the meter needle flickers (fails to hold steady), check the circuit between No. 6 and common engine ground. If meter fails to indicate voltage, replace the positive battery cable.

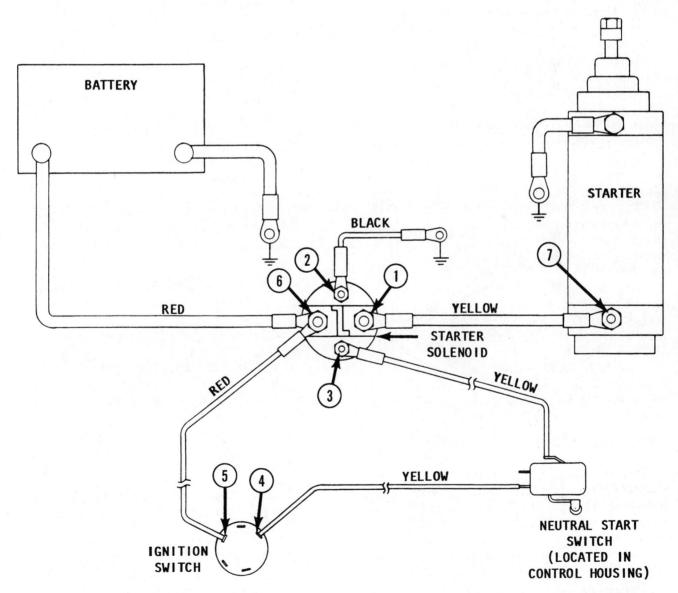

7- Further Tests for Solenoid

a- Connect the voltmeter between the common engine ground and No. 1.

b- Turn the ignition key switch to the **START** position.

c- Observe the voltmeter. If there is no reading, the cranking motor solenoid is defective and must be replaced. If a reading is indicated and a click sound is heard, proceed to Step 8.

8- Test Yellow Cable

a- Connect the yellow cable to the cranking motor solenoid.

b- Connect the voltmeter between the engine common ground and No. 7.

c- Turn the ignition key switch to the **START** position.

d- Observe the voltmeter. If there is no reading, check the yellow cable for a poor connection or an open circuit. If there is any indication of a reading, and the starter does not turn, the cranking motor must be replaced.

CRANKING MOTOR SOLENOID TROUBLESHOOTING

Description

The cranking motor solenoid is a switch between the battery and the motor. Several types of solenoids are used and many look very much alike. For marine applications, a solenoid with an internal ground is used. Grounding this type solenoid is accomplished by a wire internally connected to one of the small terminals. Connecting an external wire for a ground will serve no purpose. **NEVER** attempt to use an automotive-type solenoid, because such a unit will cause more trouble and damage than can be imagined.

When purchasing a replacement solenoid, look for a statement on the package indicating the unit is for marine use.

Illustrations of several different type solenoids are included in this section as an assist to understanding the operating functions of each type.

Solenoid Testing

The following test must be conducted with the solenoid removed from the engine.

a- Connect one test lead of an ohmmeter to each of the large solenoid terminals.

b- Connect the positive (+) lead from a fully charged 12-volt battery to the small solenoid terminal marked **S**.

c- Momentarily make contact with the ground lead from the battery to the small solenoid terminal marked **I**. If a loud "click"

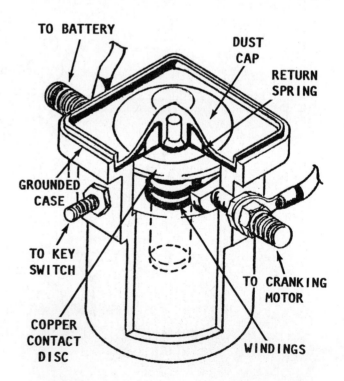

*This solenoid acts as a relay in the cranking motor circuit. If the unit is found to be defective it **MUST** be replaced.*

A cutaway drawing of a marine cranking motor solenoid, with major parts identified.

sound is heard, and the ohmmeter indicates continuity, the solenoid is in serviceable condition. If, however, a "click" sound is not heard, and/or the ohmmeter does not indicate continuity, the solenoid is defective and must be replaced **ONLY** with a **MARINE**-type solenoid.

7-8 CRANKING MOTOR SERVICE

Description

Two different cranking motors are used on the powerheads covered in this manual. The motors are almost identical in construction except for size and the manner of removing the pinion gear. The larger motor is used on the 35hp and 40hp early model units. The smaller motor is used on all other units when a cranking motor is installed.

Marine cranking motors are very similar in construction and operation to the units used in the automotive industry.

All marine cranking motors use the inertia-type drive assembly. This type assembly is mounted on an armature shaft with external spiral splines which mate with the internal splines of the drive assembly.

NEVER operate a cranking motor for more than 30-seconds without allowing it to cool for at least three minutes. Continuous operation without the cooling period can cause serious damage to the cranking motor.

Both cranking motors operate in much the same manner and the service work in-

Comparison of a large cranking motor (top) with a small motor (bottom), as referenced in this chapter.

volved in restoring a defective unit to service is almost identical. Therefore, the information in this chapter is presented for the major components under separate headings. Differences, where they occur, between the large and small motor are clearly indicated.

CRANKING MOTOR REMOVAL

Before beginning any work on the cranking motor, disconnect the positive (+) lead from the battery terminal. Remove the cowling from the powerhead. Disconnect the yellow cable at the cranking motor terminal or at the solenoid.

Remove the mounting bolts from the cranking motor housing and remove the starter from the powerhead.

Test leads connected to a solenoid in preparation to testing, as explained in the text.

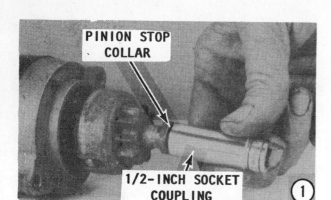

CRANKING MOTOR DISASSEMBLING

LARGE CRANKING MOTOR
USED ONLY ON 35HP AND 40HP UNITS

PINION GEAR SERVICE

This type cranking motor has the pinion gear secured to the shaft with a snap ring.

1- Remove the pinion gear from the armature. This is accomplished by sliding a deep half-inch socket onto the shaft until the end of the socket butts against the edge of the pinion stop collar. Next, tap the end of the socket to drive the stop collar away from the snap ring.

2- Pry the snap ring from the groove in the shaft, with a narrow blade screwdriver.

3- If a nut is used instead of the snap ring, slide the pinion gear upward, and then

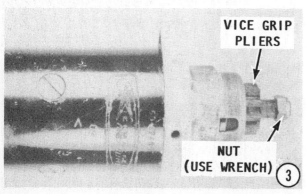

place a wrench on the retaining nut. Hold the pinion gear with a pair of vise grip pliers. Hold firm and at the same time back off the nut with the wrench. Remove and **DISCARD** the nut. **NEVER** attempt to use the same nut a second time.

4- To remove the pinion gear and spring, move the gear upward on the shaft, and then snap the spring out of the hole on the bottom side of the gear.

5- Use a screwdriver and lift the other end of the spring from the armature shaft. If other work is to be performed on the motor, move directly to the starter repair section in this chapter.

DISASSEMBLING THE FRAME
LARGE CRANKING MOTOR

GOOD NEWS

If the only motor repair necessary is replacement of the brushes, the pinion gear does not have to be removed. All cranking motors have thru-bolts securing the upper and lower cap to the field frame assembly. In all cases both caps have some type of mark or boss. These marks are used to properly align the caps with the field frame assembly.

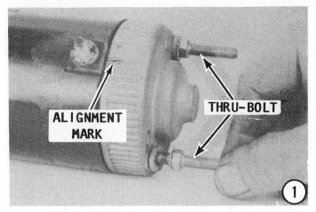

ALIGNMENT MARK

THRU-BOLT

①

1- Observe the caps and find the identifying mark or boss on each. If the marks are not visible, make an identifying mark prior to removing the thru-bolts as an essential aid during assembling. Remove the thru-bolts. On some models, the thru-bolts thread into the opposite cap, and on other models, a nut is used.

2- Remove the lower cap. Remove the

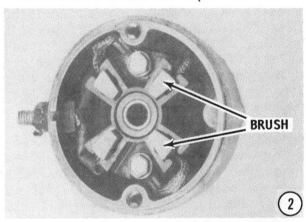

BRUSH

②

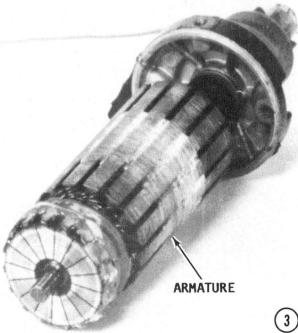

ARMATURE

③

brushes from their holders, and then remove the brush springs.

3- Pull on the armature shaft from the drive gear end and remove it from the field frame assembly.

Cleaning and Inspecting is presented in Section 7-9.

Testing of the various electrical parts is presented in Section 7-10.

Assembling is covered in Section 7-11.

DISASSEMBLING PINION GEAR AND FRAME ASSEMBLY SMALL CRANKING MOTOR

FIRST, THESE WORDS

Disassembly of the small cranking motor is very different from the large motor presented in the previous section. The pinion gear assembly is not disassembled separately from the frame. Instead, disassembly of the small motor is done almost at once with removal of the thru-bolts.

The following procedures pickup the work from the beginning, and include removal of the unit from the powerhead.

1- Disconnect the electrical leads from the battery terminal posts. Disconnect the lead from the solenoid or from the cranking motor. Remove the two mounting bolts securing the motor to the powerhead.

CRITICAL WORDS

Alignment marks **MUST** be scribed on the end cap and frame assembly to ensure the armature and end cap are properly installed back into their original position. The marks will not guarantee the thru-bolts will slide through the frame assembly on the first attempt, but without them aligned it is not possible.

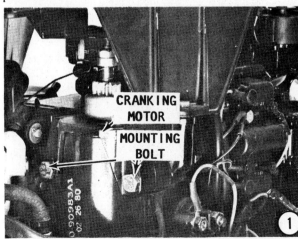

CRANKING MOTOR

MOUNTING BOLT

①

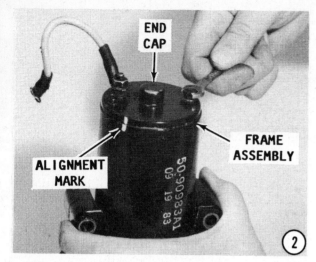

2- Scribe a mark on the motor frame and a matching mark on the end cap. These marks will **ENSURE** the end cap is installed back to the frame assembly in the exact position from which it was removed.

Remove the thru-bolts. Lift the end cap free of the frame assembly. It may be necessary to tap the end cap with a soft head mallet to jar the cap loose.

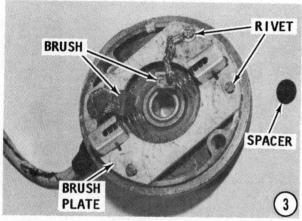

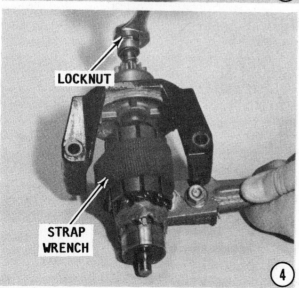

3- Carefully lift out the spacer from inside the bronze bushing in the end cap. Set the spacer in a safe place. It is a very small item, easily forgotten or lost.

The brushes of the starter shown are riveted to the brush plate. The plate in turn is riveted to the end cap. Therefore, if the brushes require replacement, the rivets securing both the end plate and the brush leads must be drilled out. The plate must be removed to ensure all fillings are cleaned from the end cap. If only the brush lead rivets are drilled out, it would be impossible to remove all the fillings from the drilling out of the end cap.

GOOD WORDS

On some small cranking motors, the brush plate and the brush leads are secured with screws. Brush replacement on these units is a simple matter.

4- Pull the armature from the frame assembly. A slight "pull" may be felt as the armature is removed, due to the magnets installed in the frame assembly. Obtain some type of strap wrench. Secure the strap wrench around the armature. Hold the armature with the strap wrench and remove the lock nut from the pinion gear end of the armature shaft. Use a quick jerk motion to "break" the nut loose.

5- After the lock nut has been removed, disassemble and keep in order the parts from the shaft, as shown: the spacer, spring, drive assembly (pinion gear), drive end cap, and washer.

Cleaning and Inspecting is presented in the next section 7-9.

Testing of the various electrical parts is presented in Section 7-10

Assembling is covered in Section 7-11.

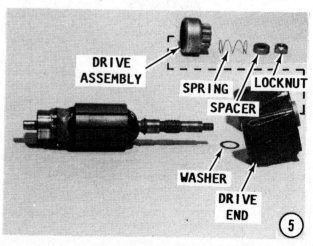

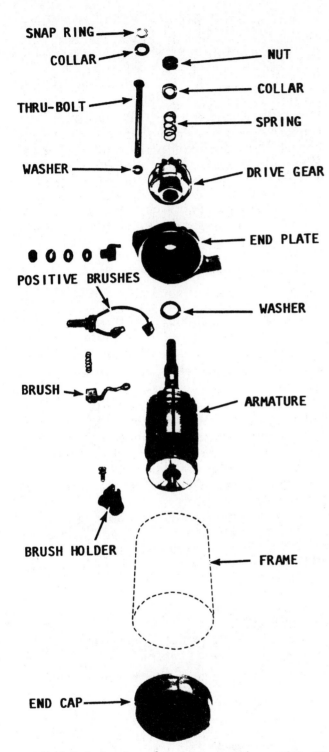

SNAP RING
COLLAR
NUT
THRU-BOLT
COLLAR
SPRING
WASHER
DRIVE GEAR
END PLATE
POSITIVE BRUSHES
WASHER
BRUSH
ARMATURE
BRUSH HOLDER
FRAME
END CAP

Exploded drawing of a large cranking motor, as identified in this chapter. Major parts are identified.

7-9 CLEANING AND INSPECTING

Pinion Gear Assembly

Inspect the pinion gear teeth for chips, cracks, or a broken tooth. Check the splines inside the pinion gear for burrs and to be sure the gear moves freely on the armature shaft.

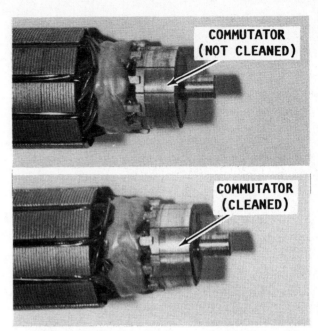

COMMUTATOR (NOT CLEANED)

COMMUTATOR (CLEANED)

Comparison of the commutator of a small cranking motor before cleaning (top), and after cleaning with crocus cloth (bottom).

BAD NEWS

This type of cranking motor pinion gear assembly cannot be repaired if the unit is defective. Replacement of the pinion gear assembly as a unit is the only answer.

Check to be sure the return spring is flexible and has not become distorted. Check both ends of the spring for signs of damage.

Clean the armature shaft and check to be sure the shaft is free of any burrs. If burrs are discovered, they may be removed with crocus cloth.

Frame Assembly

Clean the field coils, armature, commutator, armature shaft, brush-end plate and drive-end housing with a brush or compressed air. Wash all other parts in solvent and blow them dry with compressed air.

Inspect the insulation and the unsoldered connections of the armature windings for breaks or burns.

Perform electrical tests on any suspected defective part, according to the procedures outlined in Section 7-10.

Check the commutator for run out. Inspect the armature shaft and both bearings for scoring.

Turn the commutator in a lathe if it is out-of-round by more than 0.005" (0.13mm).

Check the springs in the brush holder to be sure none are broken. Check the spring tension and replace if the tension is not 32-40 ounces (900-1135gm). Check the insulated brush holders for shorts to ground. If the brushes are worn down to 1/4" (6.35mm) or less, they must be replaced.

The armature, fields, and brush holders must be checked before assembling the starter motor. See Section 7-10 for detailed procedures to test cranking motor parts.

Both the positive and negative brushes are mounted in the lower cap. The positive brushes are attached to the positive terminal and are sold as an assembled set. The negative brushes are attached to the lower cap with a bolt.

To remove the positive brushes, slip the terminal out of the slot in the cap. The negative brushes are removed by simply removing the two bolts attaching the brush lead to the lower cap. The brush leads of the small motor are riveted to the brush plate and the brush plate is riveted to the end cap.

Installation of the new positive brushes is accomplished by sliding the new positive terminal into the slot of the end cap. Install the negative brushes by positioning them in place in the lower cap, and then securing the leads with the attaching bolts. If servicing a small cranking motor, the brush leads can be soldered in place.

7-10 TESTING CRANKING MOTOR PARTS

SPECIAL WORDS

Most marine shops and all electrical motor rebuild shops will test an armature for a modest charge. If the armature has a short, it MUST be replaced.

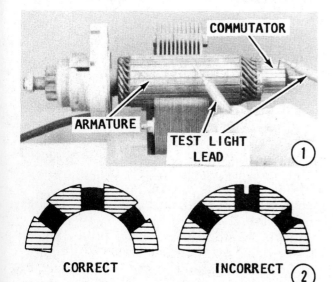

CORRECT   INCORRECT ②

Check the armature for a short circuit by placing it on a growler and holding a hack saw blade over the armature core while the armature is rotated. If the saw blade vibrates, the armature is shorted. Clean between the armature bars, and then check again on the growler. If the saw blade still vibrates, the armature must be replaced. Occasionally carbon dust from the brushes will short the armature. Therefore, blow the slots in the armature clean with compressed air.

1- Make contact with one probe of the test light on the armature core or shaft. Make contact with the other probe on the commutator. If the light comes on, the armature is grounded and must be replaced.

Turning the Commutator

2- True the commutator, if necessary, in a lathe. NEVER undercut the mica because the brushes are harder than the insulation. Undercut the insulation between the commutator bars 1/32" (0.80mm) to the full width of the insulation and flat at the bottom. A triangular groove is not satisfactory. After the undercutting work is completed, clean out the slots carefully to remove dirt and copper dust. Sand the commutator lightly with No. 00 sandpaper to remove any burrs left from the undercutting.

3- Test light probes, placed on any two commutator bars, should light and indicate continuity.

4- Check the armature a second time on the growler for possible short circuits.

GOOD WORDS

The following tests are outlined for testing a small cranking motor with one set of brushes. If the unit being serviced is a large motor with two sets of brushes, repeat each test for the second set of brushes.

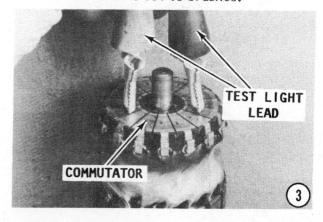

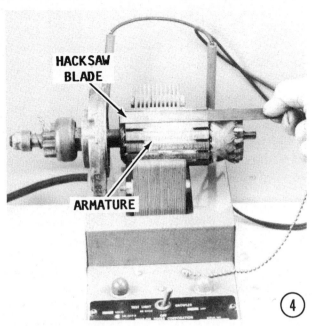

HACKSAW BLADE

ARMATURE

(4)

Positive Brushes

5- Obtain an ohmmeter. Make contact with one lead of the ohmmeter to the positive lead from the end cap. Make contact with the other test lead to the case. The ohmmeter should indicate **NO** continuity. If the meter indicates continuity, the positive lead is shorted to the case and must be replaced.

6- Again, make contact with one test lead to the positive lead from the end cap. Make contact with the other test lead to the positive brush. The ohmmeter should indi-

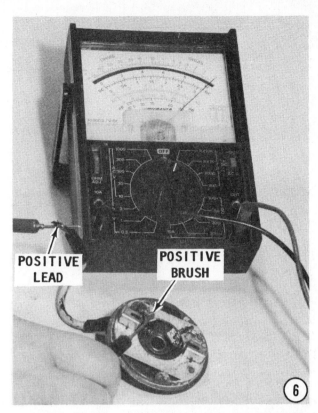

POSITIVE LEAD

POSITIVE BRUSH

(6)

cate continuity. If continuity is not indicated, there is an open in the wire. Replace the positive lead.

7- Make contact with one test to the positive brush and the other test lead to the negative brush. The meter should indicate **NO** continuity. If continuity is indicated, inspect the brush lead to the brush plate.

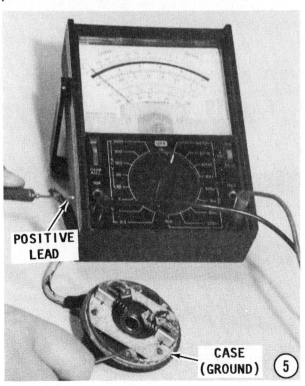

POSITIVE LEAD

CASE (GROUND)   (5)

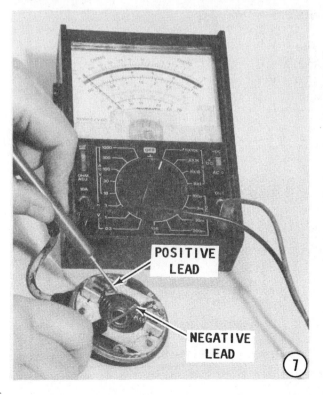

POSITIVE LEAD

NEGATIVE LEAD   (7)

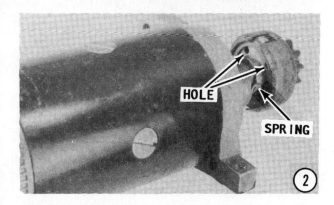

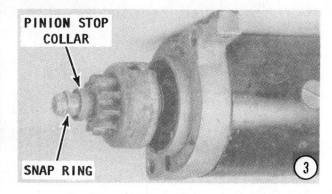

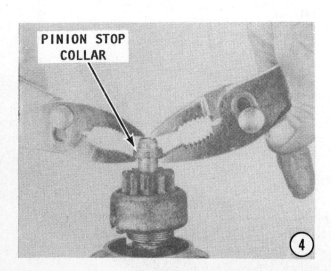

7-11 ASSEMBLING

LARGE CRANKING MOTOR

PINION GEAR

1- Lubricate the splined portion of the armature shaft with 30-weight oil. Next, place the pinion gear return spring onto the armature, with the small diameter of the spring **TOWARD** the end frame. Insert the first turn of the small end of the spring into the groove of the shaft next to the end frame. Hook the tip end of the spring into the hole at the bottom of the groove.

2- Hold the spring out of the way in its free position while the pinion gear is threaded onto the shaft in its fully disengaged position. **TAKE CARE** not to distort the spring. Wind the free end of the spring just 3/4 turn, and then hook it into the nearest of the four holes provided in the backing plate. Check to be sure the spring is securely hooked into the hole.

3- Slide the pinion stop collar onto the shaft, with the cupped surface facing **AWAY** from the pinion gear. Install the snap ring into the groove at the end of the shaft.

4- Use a pair of pliers and squeeze the snap ring to fit it into the groove properly. Position the pinion stop collar next to the snap ring, and then install a washer next to the other side of the snap ring.

GOOD WORDS: Use two pair of pliers, one on each side of the shaft, to grip the stop collar and the washer. The stop collar **MUST** rotate freely when the assembling work is complete.

5- If a nut is used instead of the snap ring, start a **NEW** nut onto the armature shaft. **DO NOT** attempt to use a nut that has been removed. Move the drive gear upward, and then clamp a pair of vise grip pliers on the drive gear. Hold the nut on top

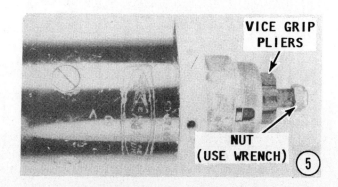

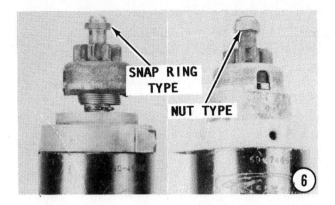

of the armature with a socket wrench. Now, hold the gear with the vise grip pliers and at the same time tighten the nut with the wrench.

6- Rotate the pinion gear against the pinion stop and relieve any turns of the spring which may be overlapping other turns. If the spring is assembled properly, the pinion gear should snap back from the engaged position.

Install the cranking motor onto the powerhead. Connect the yellow cable to the cranking motor terminal. Connect the battery cable to the battery. Install the cowling.

ASSEMBLING THE FRAME
LARGE CRANKING MOTOR
USING SPECIAL TOOL

The following procedures pickup the work after cleaning, Section 7-9, and testing, Section 7-10 have been completed.

Make a tool as shown in the accompanying illustration to prevent the brushes from

being damaged during installation of the commutator end cap. If a special tool is not possible, see the next section, Assembling a Large Cranking Motor Without a Special Tool.

1- Slide the brush springs into the brush holders, and then install the positive and negative leads. Position the special tool over the cap and brushes to hold the brushes in place.

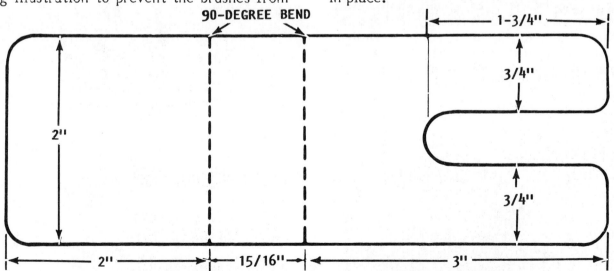

Working diagram for making a special tool to hold the brushes in place during installation of the end cap on a large cranking motor, as described in the text. The tool, in place, is shown on this page, illustration for Step 1.

2- Insert the armature into the frame assembly. Align the marks on the frame assembly with the marks on the upper end cap.

3- Position the lower end cap onto the frame assembly. Lower the cap as far as it will go, and then remove the special tool. Now, align the mark on the cap with the mark on the frame, and then install the thru-bolts and tighten them securely.

4- Place the cranking motor on the floor. To test operation of the motor, first connect one lead from a set of jumper cables to the positive terminal of a battery. Connect the other end of the same lead to the positive terminal of the motor. Connect one end of the second lead of the jumper cables to the negative terminal of the battery.

Now, hold the motor firmly on the floor with one foot, as shown, and at the same time, momentarily make contact with the other end of the second jumper lead to the cranking motor case. The pinion gear should spin rapidly.

Slide the rubber collars and spacer onto the starter, if they are used. Install the cranking motor onto the engine and secure it in place with the clamps and mounting bolts. Check to be sure the black ground cable is attached with the lower mounting bolt. Connect the black ground cable to the negative (-) cranking motor terminal and the yellow cable to the positive (+) terminal. Install the front cowling cover and the wrap around cowling, if one is used. Connect the positive (+) lead to the battery terminal.

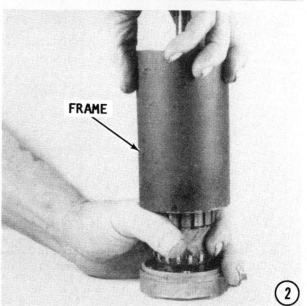

ASSEMBLING THE FRAME LARGE CRANKING MOTOR WITHOUT SPECIAL TOOL

The following procedures pickup the work after cleaning, Section 7-9, and testing, Section 7-10 have been completed.

1- Install the brush springs into the brush holder, and then place each brush on top of the springs. Lay the end cap on the bench with the brushes facing up. Pick up the armature and place the commutator on top of the brushes. Lower the armature and at the same time, work each brush into its holder. Continue to lower the armature until the full weight of the armature is on the brushes.

2- Now, very **CAREFULLY** lower the frame assembly down over the armature. **TAKE CARE** because the magnets in the frame assembly will tend to pull against the armature. When the frame makes contact with the lower cap, align the marks on the cap and the frame. Slide the upper cap washer onto the shaft.

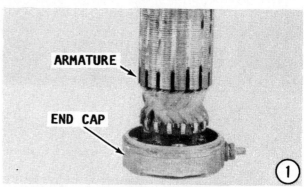

ARMATURE

END CAP

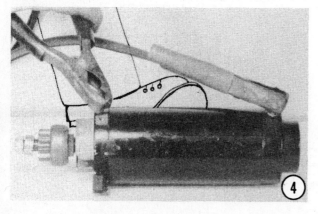

FRAME

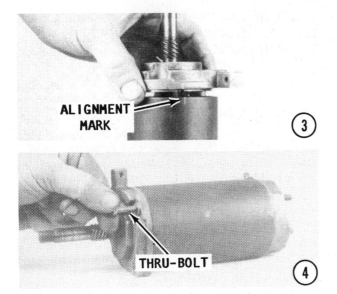

ALIGNMENT MARK

THRU-BOLT

3- Install the upper cap with the mark on the cap aligned with the mark on the frame.

4- Install the thru-bolts and tighten them securely. Install the pinion gear and secure it in place with a new locknut.

5- Place the cranking motor on the floor. To test operation of the motor, first connect one lead from a set of jumper cables to the positive terminal of a battery. Connect the other end of the same lead to the positive terminal of the motor. Connect one end of the second lead of the jumper cables to the negative terminal of the battery.

Now, hold the motor firmly on the floor with one foot, as shown, and at the same time, momentarily make contact with the other end of the second jumper lead to the case. The pinion gear should spin rapidly.

Slide the rubber collars and spacer onto the starter, if they are used. Install the cranking motor onto the powerhead and secure it in place with the clamps and mounting bolts. Check to be sure the black ground cable is attached with the lower mounting bolt. Connect the black ground cable to the negative (-) cranking motor terminal and the yellow cable to the positive (+) terminal.

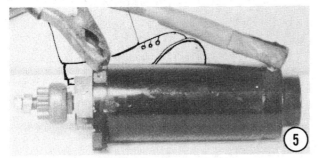

Install the front cowling cover and the wrap around cowling, if one is used. Connect the positive (+) lead to the battery terminal.

ASSEMBLING SMALL CRANKING MOTOR

The following procedures pickup the work after cleaning, Section 7-9, and testing, Section 7-10 have been completed.

1- Apply a thin coating of Multi-purpose lubricant to the splines of the armature shaft. Slide the washer onto the armature shaft, followed by the drive end cap. Next, install the drive assembly onto the shaft, then the spring and spacer. Start the locknut onto the end of the shaft.

2- Obtain a strap wrench. Hold the armature with the strap wrench and tighten the locknut on the armature shaft.

3- Insert the assembled armature into the frame assembly. A slight "pull" will be felt as the armature is installed due to the magnets in the frame assembly.

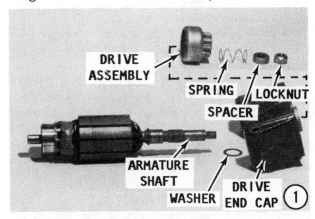

DRIVE ASSEMBLY

SPRING / LOCKNUT

SPACER

ARMATURE SHAFT

WASHER

DRIVE END CAP

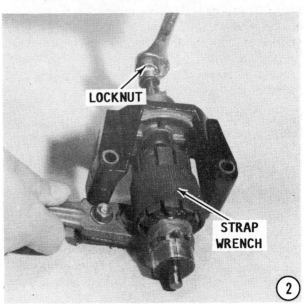

LOCKNUT

STRAP WRENCH

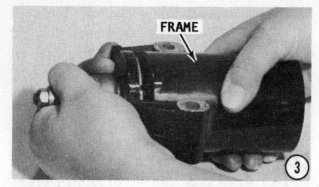

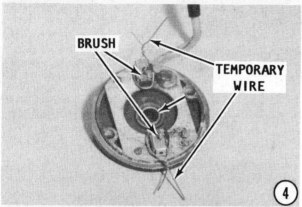

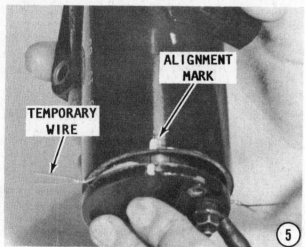

4- Push in on the brush spring, and then feed the lead of the new brush through the slot on top of the brush retainer. Rivet the end of each new brush lead to the brush plate. On some small cranking motors, the leads are held in place with a screw. After the leads have been secured to the plate, push in on each brush and secure the brush temporarily in the retracted position with a small piece of wire. A paper clip, bent to the proper shape, will do the job. Insert the small spacer inside the bronze bushing.

5- Bring the end cap together with the frame assembly. The end of the armature shaft will slide into the bronze bushing and the retracted brushes will slide over the commutator. Rotate the cap slightly to align the marks made during disassembly, Step 2. The marks **MUST** be aligned to ensure the unit is being assembled back into its original position. The marks will not guarantee the thru-bolts will slide through the frame assembly on the first attempt, but without them aligned it is not possible.

Hold the cap tightly against the frame assembly with one hand and remove the wires holding the brushes in the retracted position with the other hand. Assistance from another person would be most helpful at this time.

6- Slide the two thru-bolts through the bottom end cap, through the frame assembly and into the drive end cap. The permanent magnets in the frame assembly will "attract" the thru bolts as they are passed through. Therefore, have patience and look through the holes in the drive end cap to see how to move the bolts to permit it to thread into the cap. Once the bolts have been started in the threads, tighten them both securely.

7- Install the cranking motor to the powerhead and secure it in place with the two mounting bolts. Tighten the bolts securely. Connect the lead to the solenoid. Connect the leads to the battery terminals.

8
POWERHEAD

8-1 INTRODUCTION

The carburetion and ignition principles of two-cycle engine operation **MUST** be understood in order to perform a proper tuneup on an outboard motor or industrial engine.

The two-cycle engine differs in several ways from a conventional four-cycle (automobile) engine.

The exterior and interior of the powerhead must be kept clean, well-lubricated, and properly tuned and adjusted, if the owner is to receive the maximum enjoyment from the unit.

1- The method by which the fuel-air mixture is delivered to the combustion chamber.
2- The complete lubrication system.
3- In most cases, the ignition system.
4- The frequency of the power stroke.

These differences will be discussed briefly and compared with four-cycle engine operation.

Intake/Exhaust

Two-cycle engines utilize an arrangement of port openings to admit fuel to the combustion chamber and to purge the exhaust gases after burning has been completed. The ports are located in a precise pattern in order for them to be opened and closed at an exact moment by the piston as it moves up and down in the cylinder. The exhaust port is located slightly higher than the fuel intake port. This arrangement opens the exhaust port first as the piston starts downward and therefore, the exhaust phase begins a fraction of a second before the intake phase.

Actually, the intake and exhaust ports are spaced so closely together that both open almost simultaneously. For this reason, the pistons of most two-cycle engines have a deflector-type top. This design of the piston top serves two purposes very effectively.

First, it creates turbulence when the incoming charge of fuel enters the combustion chamber. This turbulence results in more complete burning of the fuel than if the piston top were flat. The second effect of the deflector-type piston crown is to force the exhaust gases from the cylinder more rapidly.

This system of intake and exhaust is in marked contrast to individual valve arrangement employed on four-cycle engines.

Lubrication

A two-cycle engine is lubricated by mixing oil with the fuel. Therefore, various parts are lubricated as the fuel mixture passes through the crankcase and the cylinder. Four-cycle engines have a crankcase containing oil. This oil is pumped through a circulating system and returned to the crankcase to begin the routing again.

Physical Laws

The two-cycle engine is able to function because of two very simple physical laws.

One: Gases will flow from an area of high pressure to an area of lower pressure. A tire blowout is an example of this principle. The high-pressure air escapes rapidly if the tube is punctured.

Two: If a gas is compressed into a smaller area, the pressure increases, and if a gas expands into a larger area, the pressure is decreased.

If these two laws are kept in mind, the operation of the two-cycle engine will be easier understood.

Actual Operation

Beginning with the piston approaching top dead center on the compression stroke: The intake and exhaust ports are closed by the piston; the reed valve is open; the spark plug fires; the compressed fuel-air mixture is ignited; and the power stroke begins. The reed valve was open because as the piston moved upward, the crankcase volume increased, which reduced the crankcase pressure to less than the outside atmosphere.

As the piston moves downward on the power stroke, the combustion chamber is filled with burning gases. As the exhaust port is uncovered, the gases, which are under great pressure, escape rapidly through the exhaust ports. The piston continues its downward movement. Pressure within the crankcase increases, closing the reed valves against their seats. The crankcase then becomes a sealed chamber. The air-fuel mixture is compressed ready for delivery to the combustion chamber. As the piston continues to move downward, the intake port is uncovered. A fresh air/fuel mixture

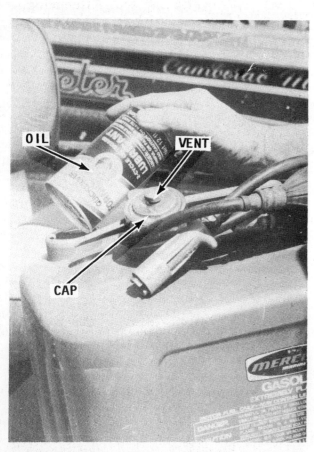

Adding approved lubricant to the fuel tank at the time the tank is being filled. Some fuel must be in the tank to prevent the oil from sticking to a dry bottom.

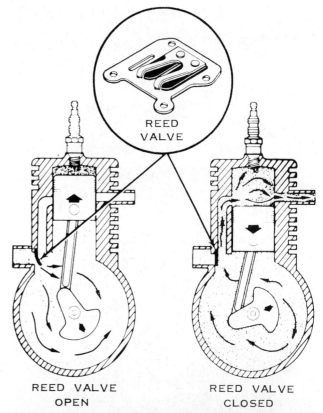

REED VALVE
OPEN

REED VALVE
CLOSED

Reed valves are used to control the flow of air/fuel into the crankcase and eventually into the cylinder. As the piston moves upward in the cylinder, the resulting suction in the crankcase overcomes the spring tension of the reed. The reed is pulled free from its seat and the air/fuel mixture is drawn into the crankcase.

rushes through the intake port into the combustion chamber striking the top of the piston where it is deflected along the cylinder wall. The reed valve remains closed until the piston moves upward again.

When the piston begins to move upward on the compression stroke, the reed valve opens because the crankcase volume has been increased, reducing crankcase pressure to less than the outside atmosphere. The intake and exhaust ports are closed and the fresh fuel charge is compressed inside the combustion chamber.

Pressure in the crankcase decreases as the piston moves upward and a fresh charge of air flows through the carburetor picking up fuel. As the piston approaches top dead center, the spark plug ignites the air-fuel mixture, the power stroke begins and one complete cycle has been completed.

Timing

The exact time of spark plug firing depends on engine speed. At low speed the spark is retarded, fires later than when the piston is at or beyond top dead center. Engine timing is built into the unit at the factory.

At high speed, the spark is advanced, fires earlier than when the piston is at top dead center. On some late models and larger engines, the timing can be changed in the field to meet advance and retard factory specifications.

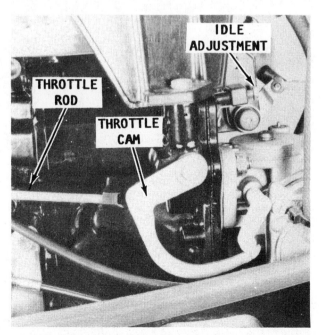

Typical location of the idle adjustment screw, throttle cam, and the throttle rod.

Summary

More than one phase of the cycle occurs simultaneously during operation of a two-cycle engine. On the downward stroke, power occurs above the piston while the ports are closed. When the ports open, exhaust begins and intake follows. Below the piston, fresh air-fuel mixture is compressed in the crankcase.

On the upward stroke, exhaust and intake continue as long as the ports are open. Compression begins when the ports are closed and continues until the spark plug ignites the air-fuel mixture. Below the piston, a fresh air-fuel mixture is drawn into the crankcase ready to be compressed during the next cycle.

CHAPTER ORGANIZATION

The remainder of this chapter is divided into four main working parts as follows:

8-2 -- Disassembling and Assembling -- Powerhead "A" -- split block without a head and with internal reed box --reeds installed around the crankshaft.

8-3 -- Disassembling and Assembling -- Powerhead "B" -- split block without a head and with external reed block.

8-4 -- Disassembling and Assembling -- Powerhead "C" -- split block with a head and with reed valves installed under the crankshaft.

8-5 -- Cleaning, Inspecting, and Service of individual powerhead parts including the block.

Specifications in the Appendix for the type powerhead on the outboard unit being serviced.

Each main section contains complete and independent instructions.

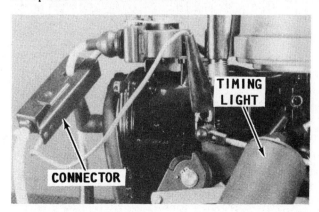

Timing light connected to the No. 1 spark plug lead in preparation for making fine timing adjustments.

Repair Procedures

Service and repair procedures will vary slightly between individual models, but the basic instructions are quite similar. Special tools may be called out in certain instances. These tools may be purchased from the local marine dealer.

Torque Values

All torque values must be met when they are specified. Detailed torque values for each powerhead covered in this manual are listed in the Appendix.

A torque wrench is essential to correctly assemble the powerhead. **NEVER** attempt to assemble a powerhead without a torque wrench. Attaching bolts **MUST** be tightened to the required torque value in three progressive stages, following the specified tightening sequence. Tighten all bolts to 1/3 the torque value, then repeat the sequence tightening to 2/3 the torque value. Finally, on the third and last sequence, tighten to the full torque value.

Powerhead Components

Service procedures for the carburetors, fuel pumps, starter, and other powerhead components are given in their respective Chapters of this manual. See the Table of Contents.

Reed Block Installation

The reed box assembly on Powerhead "A" is installed around the crankshaft. The reed valves of Powerhead "C" are installed beneath the crankshaft. This means, in order

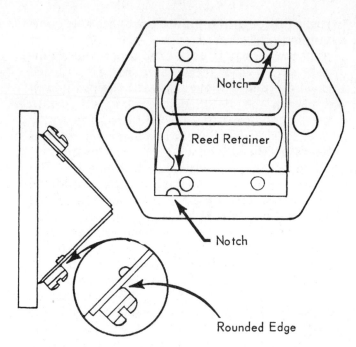

Drawing of a reed plate assembly installed externally on Type "B" powerheads.

to replace a broken reed on Powerhead "A" or "C", the powerhead must be overhauled.

The reeds on Powerhead "B" are contained in an externally mounted reed block. Therefore, the powerhead need not be disassembled in order to replace a broken reed.

Cleanliness

Make a determined effort to keep parts and the work area as clean as possible. Parts **MUST** be cleaned and thoroughly inspected before they are assembled, installed, or adjusted. Use proper lubricants, or their equivalent, whenever they are recommended.

Different type internal reed blocks installed around the crankshaft of Type "A" powerheads.

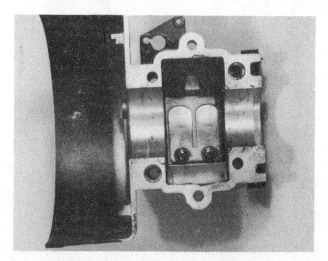

Reeds installed underneath the crankshaft on Type "C" powerheads.

8-2 POWERHEAD "A"
SPLIT BLOCK WITHOUT HEAD
INTERNAL REED BOX AROUND
THE CRANKSHAFT
(See Tune-up Specifications — Last Column)

REMOVAL AND DISASSEMBLING

ADVICE

Before commencing any work on the powerhead, an understanding of two-cycle engine operation will be most helpful. Therefore, it would be well worth the time to study the principles of two-cycle engines, as outlined briefly in Section 8-1. A Polaroid, or equivalent instant-type camera is an extremely useful item, providing the means of accurately recording the arrangement of parts and wire connections **BEFORE** the disassembly work begins. Such a record is invaluable during assembling.

POWERHEAD "A" REMOVAL

1- If a battery is used to crank the engine, disconnect the battery leads from the battery terminals. Disconnect the fuel line from the engine, or from the fuel tank.

2- Remove the front cowl cover and the wrap-around cowl, if one is used. Remove the top cowl. It may be necessary to remove the pull rope retainer, and then to feed the rope through the cowl as it is removed. After the cowl is removed, tie a figure "8" knot in the rope to prevent the rope from winding back inside the pull starter.

STOP, and carefully observe the wiring and hose connections before proceeding. Because there are so many different engines and the arrangement is slightly different on each, it is not possible to illustrate all of them. Even if they were shown, you would not be able to identify the engine being

serviced.   Therefore, **TAKE TIME** to make notes and tag the wiring and hoses.   You may elect to follow the practice of many professional mechanics by taking a series of photographs of the engine, one from the top, and a couple from the sides showing the wiring and arrangement of parts.

3- Remove the fuel line from the fuel pump. Disconnect the wiring to the coils, choke, stop button, and the cranking motor (if one is used). Remove the spark plug wires by pulling and twisting on only the molded cap. **NEVER** pull on the wire or the connection inside the cap may become separated or the boot damaged. Remove the spark plug/s. **TAKE CARE** not to tilt the socket as the plug is removed, or the insulator may be cracked. In most cases, it is best to remove the powerhead components, such as the carburetor, magneto, and cranking motor (if one is used). Removal and handling the powerhead is much easier with these items removed.

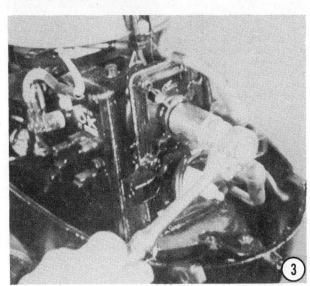

4- Disconnect the choke linkage from the carburetor choke shutter. Remove the tiller throttle handle from the engine.

5- Remove the rewind hand starter attaching hardware. Lift the assembly from the powerhead.

6- Disconnect the tattle-tale water hose from the powerhead. Remove the two nuts and bolts from the front and rear support

frame, if applicable. Remove the nut securing the flywheel to the crankshaft. It may be necessary to use some type of flywheel strap to prevent the flywheel from turning as the nut is loosened. Install the proper flywheel puller.

NEVER attempt to use a puller which pulls on the outside edge of the flywheel, or the flywheel may be damaged. After the puller is installed, tighten the center screw onto the end of the crankshaft. Continue tightening the screw until the flywheel is released from the crankshaft. Remove the flywheel. See the applicable section of Chapter 5, for the ignition system installed on the powerhead being serviced. The ignition system should be removed before the powerhead is separated from the driveshaft housing.

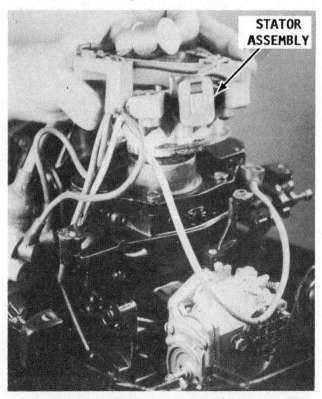

Removing the electrical components after the flywheel has been removed.

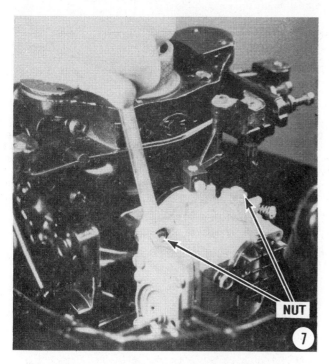

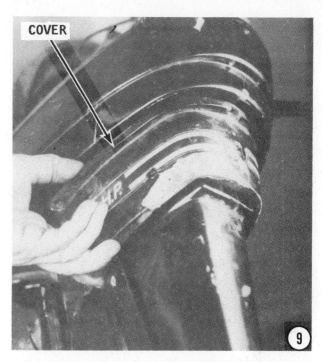

7- Remove the attaching hardware, and then remove the carburetor from the powerhead.

8- Remove the attaching hardware, and then the vertical throttle shaft, if one is installed.

9- If the unit is equipped with a driveshaft housing cover, remove the Phillips screws, and then the cover.

10- Remove the nuts from the studs along the top edge of the exhaust housing. Jar the powerhead loose from the exhaust housing. Remove and **DISCARD** the gasket.

The following procedures pickup the work for powerhead disassembling after the preliminary tasks, Step 1 thru 10 have been completed.

POWERHEAD "A" DISASSEMBLING

1- Remove the screws from the intake transfer port covers. Tap the covers to jar them loose, and then remove them.

SPECIAL WORDS

On later model units, separate the fuel pump assembly from the intake transfer plate cover. Remove the coil mounting housing and the thermostat (if so equipped).

2- Remove the cylinder block cover bolts, and then pry or tap the cover loose from the powerhead. **REMEMBER** these engines do not have a cylinder head.

3- The exhaust cover and divider plate **MUST** be removed in order to inspect the

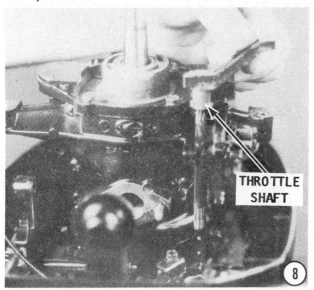

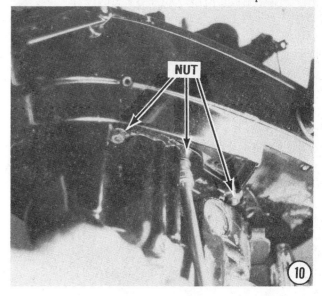

gasket and baffle plate. Remove the bolts around the cover and plate, and then pry or tap them loose from the powerhead.

4- Lay the powerhead on a bench or work surface with the flat side towards the bench, and the intake ports of the carburetor facing up. Turn the powerhead on the work surface until the flywheel end of the crankshaft is facing to the **LEFT**.

SPECIAL WORD:
The remaining disassembly and assembly procedures will assume the powerhead to be in this position.

5- Remove the crankshaft end cap bolts, but **DO NOT** remove the caps at this time. Remove the reed block attaching bolt and the crankcase cover bolts.

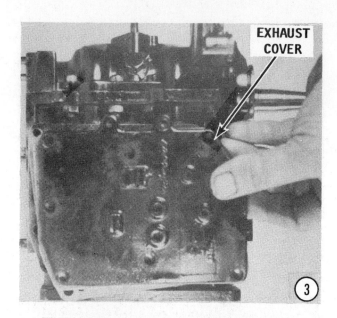

6- Remove the crankcase cover by tapping upward under the crankshaft end, but tap only enough to jar it loose. **NEVER** pry between the case and the block. Lift the case from the block and set it aside. **TAKE CARE** to protect the mating surfaces of the case and the block, from being marred or scratched. Lay the case mating surface on a soft cloth. **DO NOT** clean these surfaces at this time. See the Cleaning and Inspection section.

7- Remove the end caps by tapping them with a soft headed mallet to jar them loose.

8- Grasp each end of the crankshaft and **CAREFULLY** pull upward, to work the crankshaft, reed block and piston free of the cylinder block. If this assembly is not handled **WITH CARE**, the reed stop and piston can be damaged. Place the crankshaft in an upright position.

9- Make an identifying mark on the reed box with an awl, to ensure it will be assembled exactly as the original installation. Remove the reed block and the main bearing by first removing the two Phillips head screws. Next, tap the solid side of the reed block cage with a mallet to separate the halves from the crankshaft. **TAKE CARE** not to bend or distort the reed valve stops. Temporarily assemble the reed blocks together to ensure they remain as a matched set.

SOME GOOD WORDS:

The piston, rod, and rod caps **MUST** be kept together and installed into the same cylinder from which they were removed.

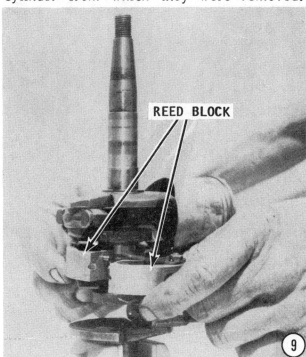

One method of identification is to use an awl and scribe a matching mark on the inside of the piston skirt, the rod, and the rod caps. The rod caps must also be installed in the same direction from which they were removed. Most rod caps do not have a rod number, but they do have an alignment mark to ensure they will be installed properly.

10- Cup both hands around the piston with your fingers at the back of the ring. Now, spread the ring open with your thumb nails inserted at the ring ends and remove the ring up over the top of the piston. **OBSERVE** the pins in the ring groove. These pins are essential to prevent the rings from rotating in the groove. If the ring were allowed to rotate, the end of the ring would become lodged in the cylinder port and break the ring.

11- Before removing the rod cap, observe the identifying matching marks on the boss of the rod and cap. These **MUST** be matched again during assembly. Remove the rod cap nuts. Place the crankshaft assembly on a clean work surface. Tap the rod bolts with a mallet to separate the rod halves. Keep the needle bearings with the same rod, or stow them in a numbered tray. Place the cap back onto the rod to ensure it will be installed with the same rod during assembly.

12- A punch and hammer may be used to remove the crankshaft seal retainer from the end of the crankshaft. **HOWEVER,** it must be done very carefully. Using the punch, tap alternately on each side to work the retainer free. Slide the oil seal and ball bearing from the crankshaft.

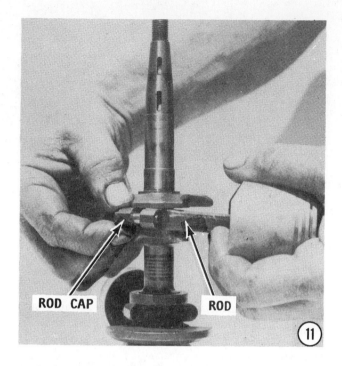

ROD CAP   ROD

⑪

13- WEAR eye protection glasses while removing the piston pin lockrings, because the lockring is made of spring steel and may slip out of the pliers or pop out of the groove with considerable force. Remove the two G-type lockrings using a pair of needle-nose pliers.

SEAL   BEARING   ⑫

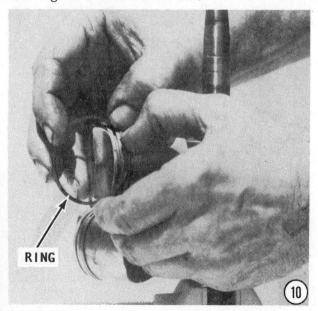

RING   ⑩

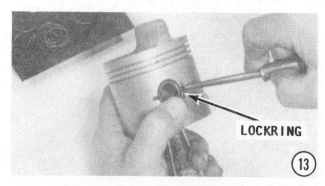

LOCKRING   ⑬

14- Remove the piston from the rod by first placing a pressing pin into the top recess of the piston. Next, support the bottom of the piston with one hand and drive the pin thru the piston with a mallet and the pressing pin. **BAD NEWS**

On some models it may be necessary to heat the top of the piston to approximately 190° with hot water or a torch lamp before the piston pin can be driven out. On some models needle bearings are used with the piston pin. These bearings will fall out after the pin has been removed.

GOOD WORDS:

New needle bearings should be installed in connecting rods, even though they appear to be in good condition. If the old needle bearings are to be re-used, be sure they are identified and installed onto the same crankpin throw from which they were removed.

15- Drive the seals from the cap with a punch or other blunt tool. Use a bearing puller and remove the bearings from the end cap.

CLEANING AND INSPECTING

Detailed instructions for cleaning and inspecting all parts of the powerhead, including the block, will be found in the last section of this chapter, Section 8-5.

ASSEMBLING POWERHEAD "A"

FIRST, THESE WORDS

Be sure all parts to be re-used have been carefully cleaned and thoroughly inspected, as outlined in Section 8-8. Parts that have not been properly cleaned, or parts not suitable for service can damage a good powerhead within a few minutes after starting the engine.

NEW gaskets **MUST** always be used during an overhaul.

A torque wrench is essential to correctly assemble the powerhead. **NEVER** attempt to assemble a powerhead without a torque wrench. Attaching bolts **MUST** be tightened to the required torque value in three progressive stages, following the specified tightening sequence. Tighten all bolts to 1/3 the torque value, then repeat the sequence tightening to 2/3 the torque value. Finally, on the third and last sequence, tighten to the full torque value.

1- Place the crankshaft in an upright position. Check to be sure the reed stop has the proper opening, per the inspecting instruction in Section 8-8 and the Reed Stop Opening Chart in the Appendix. Lubricate each half of the reed box/s and the crankshaft journal with Multipurpose Lubricant, or equivalent. Place each half of the reed box around the crankshaft journal. Install and tighten the two screws securing the box to the crankshaft. On engines using bolts and nuts to secure the reed box to the crankshaft, the bolts **MUST** be installed with the heads up and the bolts down, to ensure clearance with the crankshaft.

2- Place the piston in a container of hot water, approximately 190°F. Leave the piston in the hot water, ready for installation later in Step 4. If hot water is used, be sure to blow the piston dry with compressed air prior to installation.

3- To replace the piston pin and the needle bearings into the piston, first lay the needles on a clean piece of paper. Next, coat the sleeve portion of a piston pin installation tool with a small amount of Multipurpose Lubricant, or equivalent. If a piston pin tool is not available, a drift slightly smaller in diameter than the pin, may be used.

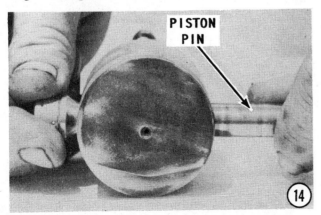

PISTON PIN

⑭

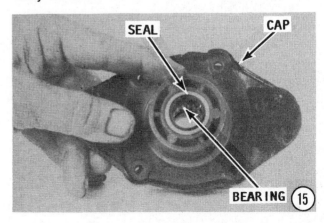

SEAL   CAP

BEARING   ⑮

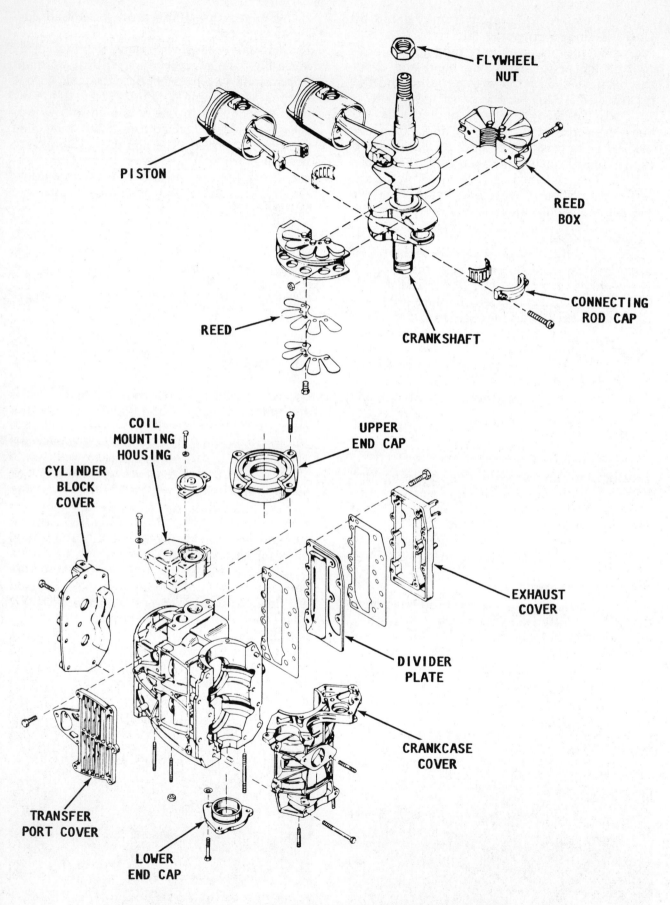

Exploded drawing of Type "A" powerhead with major parts identified.

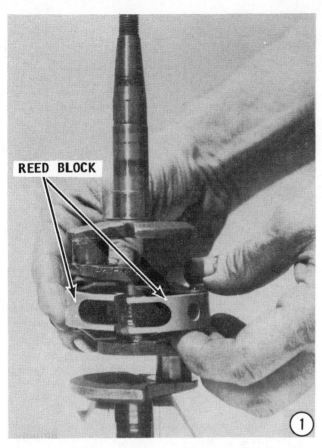

Install the retainer and the needle bearings onto the end of the tool. Leave **ONE** needle bearing **OUT.** Push the bearings and the tool into the rod piston pin bore. Now, install the last needle bearing. Place the top retainer on the side of the rod. Finally, ease the tool out of the rod, and at the same time hold onto the needle bearing retainer washers.

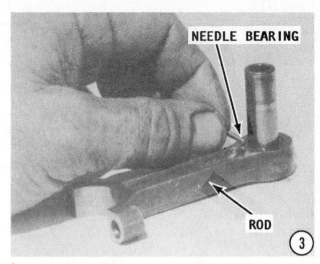

4- Carefully position the piston over the end of the rod end, paying attention that the retainer washers remain in place, and at the same time ease the needle bearing tool down through the piston pin bore and through the rod. **DO NOT** force the tool through the needle bearings.

STOP: Check to be sure the piston is facing the proper direction, the slope area towards the exhaust port of the block and the intake side of the deflector towards the intake port. Now, bring the piston pin up to the bottom side of the piston and through the piston pin bore of the piston. Hold the piston pin tool tightly down against the piston pin, and at the same time use a mallet and drive the piston pin up and through the piston and rod bearing. As soon as the pin is almost flush with the piston surface, remove the piston pin tool and use it to drive the piston pin upward until the backing ring grooves are visible.

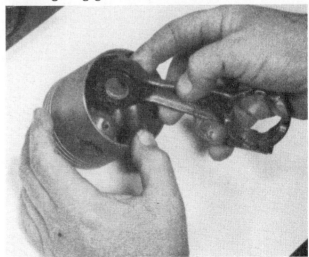

Inserting the rod into the piston with care to ensure none of the needle bearings is lost.

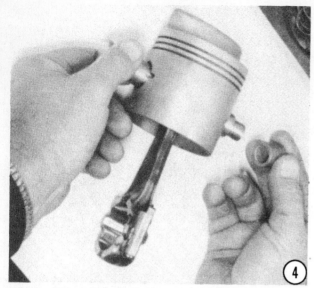

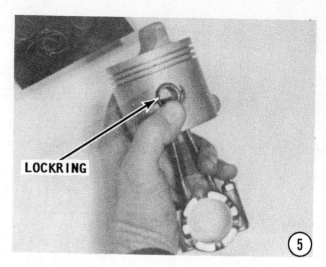

5- Snap the locking clips into the groove of the piston pin bore in the piston. The clip retains the piston pin in place after the piston becomes hot during operation.

6- Expand the ring slightly with the fingernail of each of your thumbs and at the same time support the back of the ring with your fingers. Now, slide the ring down over the piston and into its proper groove. After the rings are in place, each ring should rotate freely. Lubricate the piston, the rings, and the cylinder bore with a good grade of outboard motor oil. Rotate the rings in their grooves until the ends of each ring is over the locating pin in the groove of the piston.

Repeat Steps 2 thru 6 for the other piston (if working on a 2-cylinder unit).

7- Place the rod cap in position on the rod. Coat the inside surface of the rod with a small amount of Multi-purpose Lubricant, or equivalent. Place the needle bearings into the inside surface of the rod bearing, one at a time. On some models, the needle bearings are retained in a cage. In this case, place the cage on the inside of the rod with the needles in place.

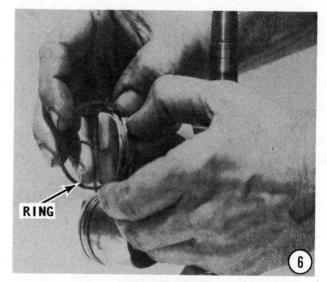

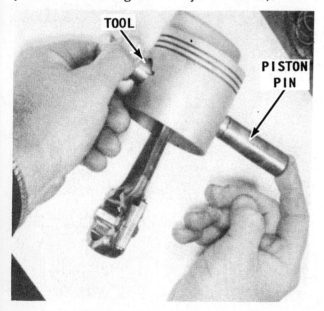

Inserting the rod into the piston with care to ensure none of the needle bearings is lost.

Needle bearing assembled around the crankshaft throw.

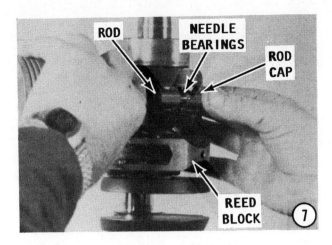

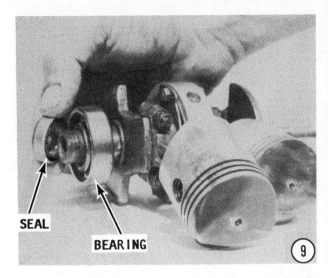

GOOD WORDS

Needle bearings are used throughout the outboard powerheads covered in this manual. Each needle must be able to move around the bearing surface and to rotate at the same time. If a needle bearing ceases to move or to rotate, even for a very short time, the needle will develop a flat spot.

NEVER intermix new and used needle bearings in the same connecting rod. If some of the bearings need to be replaced, then replace **ALL** of the bearings.

8- Remove the cap from the rod, it should have been lightly attached to the rod during disassembly to ensure installation with the proper rod. Place the rod and cap in position around the crankshaft journal. Verify the markings on the cap and the rod match. Check to be sure the rod bolts are clean.

DO NOT apply any oil to the threads. Use a new nut and tighten the nuts alternately, in three stages, to the torque value given in the Specifications in the Appendix. Use a small diameter wire and check for missing needle bearings. After the proper torque value has been reached, rotate the

connecting rod and check for binding or any "rough" spots. If the rod does not rotate smoothly, the rod and cap must be removed and the needle bearings and crankshaft journal checked. Repeat Steps 2 thru 8 for the other rod.

9- Install the ball bearing and oil seal, with the lip of the the seal towards the end of the crankshaft. Install the seal retainer onto the lower end of the crankshaft.

WORDS FROM EXPERIENCE

There is no easy way to install the assembled pistons and crankshaft into the block. Without a ring compressor tool, the job is even more difficult. A Mercury ring compressor tool is usually available at modest cost from the local Mercury dealer.

10- Place the ring compressor tool over the crankshaft and down the skirt of the piston onto the rings. Begin to tighten the

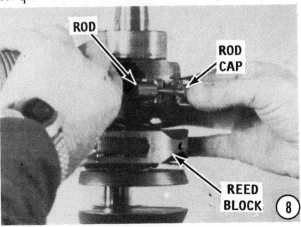

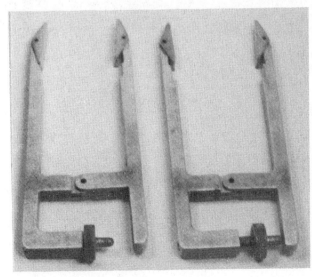

Ring compressors used to install the pistons into the cylinders.

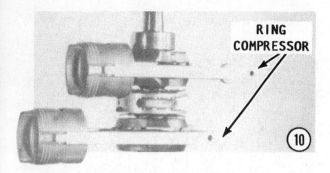

tool onto the rings, and at the same time, check to be sure the ring ends are over the piston pin. Continue to tighten the tool until the rings are almost flush with the surface of the piston. Repeat this step for the other piston.

11- Place the reed block locating pins into the block, if they were removed. Now, lower the piston and crankshaft assembly into the block with one piston down and the other up. Work the "down" piston into the cylinder bore and at the same time, lower the crankshaft, reed block, and the other piston downward.

12- Now, remove the ring compressor from the piston that has entered the bore. Repeat this procedure for the other piston. When the second piston reaches the cylinder bore in the cylinder block, work both pistons

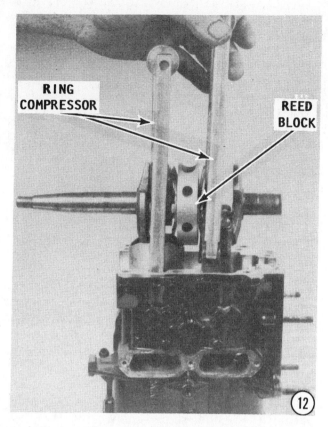

and the crankshaft assembly downward. At the same time the crankshaft is being lowered, align the reed block with the locating pin in the powerhead block. After the pistons are in the cylinder bores and the reed block is in position, remove the other ring compressor tool.

13- Check to be sure each piston ring has spring tension. This is accomplished by **CAREFULLY** pressing on each ring with a screwdriver extended through the intake ports. If spring tension cannot be felt (the

spring fails to return to its original position), the ring was probably broken during the piston and crankshaft installation process. **TAKE CARE** not to burr the piston rings while checking for spring tension.

14- Install the upper end cap roller bearing. The proper size mandrel from Bearing Removal and Installation Kit No. C-91-31229A1, must be used for seal installation. Press the bearing into place from the **NUMBERED** side of the race. Press the ball bearing into the end cap with a press block, as shown. On Mercury Model 402 (1972-78), 40 (40 hp 1979-83), and 35 hp (1984 & on) only: Install the upper ball bearing onto the crankshaft, as shown.

15- Apply a thin coating of Loctite Type "A" to the outer diameter of the oil seal. Use the proper size mandrel from Bearing Removal and Installation kit and press the seal into the end cap with the lip of the seal **TOWARD** the roller bearing. On all models except the Mercury Model 402 (1972-78), 40 (40 hp 1979-83), and 35 hp (1984 & on) only: **TEMPORARILY** install the upper end cap assembly and the original shim material to allow the crankshaft end play to be checked.

16- With one hand, hold the crankshaft from bouncing back, and with the other hand and a mallet, tap the threaded end of the crankshaft to cause it to bottom-out at the other end.

Now, hold the crankshaft in that position, and check the crankshaft end play between the end cap inner ball bearing race and the thrust face on the crankshaft. Recommended end play is 0.004-0.012" (0.10-0.30mm). If the end play is greater than 0.012" (0.30mm), remove shims between the crankcase and the end cap. If the end play is less than 0.004" (0.10mm), add shim material between the crankcase and the end cap. Shim kits include 0.002, 0.003, 0.005, and 0.010" (0.05, 0.08, 0.13, and 0.25mm) shim material.

UPPER CAP

⑮

After the end play has been checked and adjusted, remove the end cap. Coat a **NEW** O-ring with a thin layer of Multipurpose Lubricant, and then install the **O**-ring. Make a final installation of the end cap. Install the bottom cap. **DO NOT** push the caps against the block. Leave approximately 1/8" (3.18mm) between the cap and the block to allow for the crankcase cover to be installed.

17- Before mating the crankcase cover and the block, check to be sure:

a- The mating surfaces are clean. **NEVER** use any kind of tool or abrasive material to clean the surfaces. Use only solvent and elbow grease.

b- The crankshaft is properly seated.

c- The rod caps have been correctly installed.

d- The reed block is properly seated into the cylinder block.

Use Loctite Primer "T" to clean the crankcase cover and cylinder block surfaces

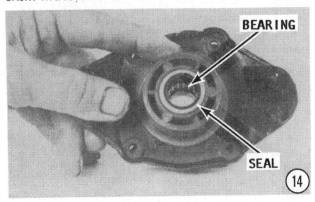

BEARING

SEAL

⑭

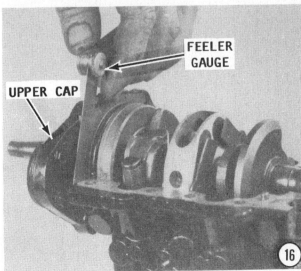

FEELER GAUGE

UPPER CAP

⑯

SEALER

which **DO NOT** have a gasket strip. Install a
gasket strip into the groove in the crankcase
cover on the Model 402 (1972-78), 40 (40 hp
1979-83), and 35 hp (1984 & on) **ONLY.** Lay
down a coating of Permatex 2C-12 evenly
onto the crankcase covers, which have a
gasket strip. An alternate method is to
apply a continuous bead of Loctite 504-31 to
the crankcase covers which **DO NOT** have a
gasket strip.

18- Apply a continuous bead along the
inside of the mounting bolt holes. If a void
should occur while the bead is being applied,
either remove the entire bead with a rag, or
apply an additional bead parallel to the void
and overlapping the first bead layed down.

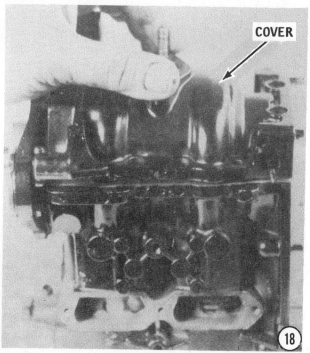

COVER

REED BOLT   LOCKWASHER

Now, assemble the crankcase cover to the
crankcase without making any lateral movement. Wipe off any excess sealer remaining
outside the assembly. Check to be sure the
crankcase cover is down and flush with the
block. **DO NOT** use the crankcase bolts to
pull the cover down.

19- Center the reed cage in the carburetor inlet, and then secure it in place
with the attaching screw.

20- Tighten the crankcase bolts. Begin
at the center and work alternately towards
both ends, as shown in the accompanying

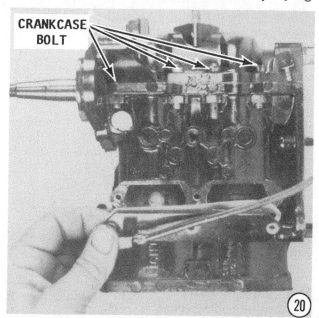

CRANKCASE BOLT

TORQUE SEQUENCES
FOR TYPE "A" POWERHEAD

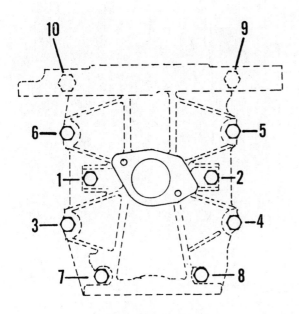

CRANKCASE COVER
200in lb (22Nm)

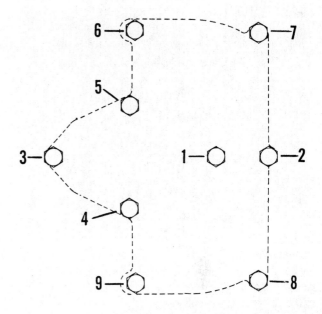

TRANSFER PORT COVER
50in lb (6Nm)

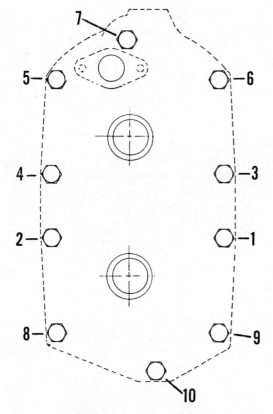

CYLINDER BLOCK COVER
100in lb (11.3Nm)

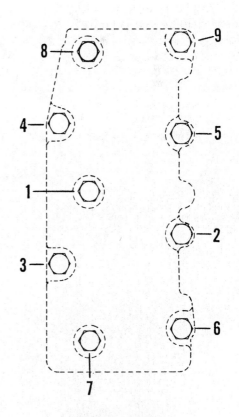

EXHAUST COVER
200in lb (22Nm)

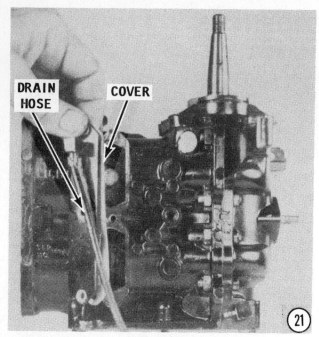

illustration on Page 8-19, to prevent distorting the cover. Tighten the cover bolts to the torque value given in the Specifications in the Appendix

A torque wrench is essential to correctly assemble the powerhead. **NEVER** attempt to assemble a powerhead without a torque wrench. Attaching bolts **MUST** be tightened to the required torque value in three progressive stages, following the specified tightening sequence. Tighten all bolts to 1/3 the torque value, then repeat the sequence tightening to 2/3 the torque value. Finally, on the third and last sequence, tighten to the full torque value.

Install and tighten the end cap bolts in three stages to the torque value given in the Appendix.

SPECIAL WORDS

Notice in the Appendix the Model 40 (40 hp) and Model 35 (Serial No. 5823918 and up), the upper and lower end caps have different torque values given.

After the bolts have been tightened, rotate the crankshaft several times to verify it does not bind or have any "rough" spots.

21- Use a **NEW** gasket and install the intake transfer port cover. Tighten the attaching hardware in three stages to the torque value given in the Appendix, in the sequence shown in the accompanying illustration on Page 8-19.

SPECIAL WORDS

On newer model units, attach the fuel pump to the intake transfer port cover. Replace the coil mounting housing and thermostat (if so equipped).

22- Install the cylinder block cover with a **NEW** gasket and tighten the attaching hardware in three stages to the torque value given in the Appendix , in the sequence shown in the accompanying illustration on Page 8-19.

23- Install the exhaust cover and plate. Tighten the bolts in the sequence pattern given in the accompanying illustration on Page 8-19 in three stages and to the torque value given in the Appendix. On the Mercury Model 402 (1972-78), 40 (40 hp 1979-83), and 35 hp (1984 & on) **ONLY**: Install the valve, gaskets and plate onto the thermostat.

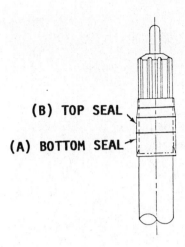

(B) TOP SEAL

(A) BOTTOM SEAL

①

POWERHEAD "A" INSTALLATION

1- Install the bottom and top seal onto the driveshaft. Apply a small amount of Multilubricant onto the driveshaft splines. Wipe the top of the driveshaft clean so excessive lubricant will not be trapped in the clearance space between the crankshaft and the driveshaft. Any trapped lubricant will cause an excessive pre-load between the shafts when the gear housing nuts are tightened. An excessive pre-load will cause damage to either the gear housing or the powerhead, or both when the powerhead is operated.

2- Position a **NEW** gasket onto the bottom of the powerhead. Place the powerhead onto the exhaust housing.

3- Thread the nuts onto the powerhead studs and tighten them to the torque value

NUT

③

GASKET

②

listed in the Appendix in three successive stages.

4- Install the exhaust housing cover, if one is used.

5- If a vertical throttle shaft is used, install it onto the block.

6- Position a **NEW** gasket in place on the intake manifold with the vacuum pressure hole aligned with the hole in the manifold. Install the carburetor. Tighten the carburetor attaching nuts alternately to the torque value given in the Appendix.

7- Replace and connect the electrical components on top of the engine. Use the diagram or photographs made prior to disassembly and see Chapter 5 and 7.

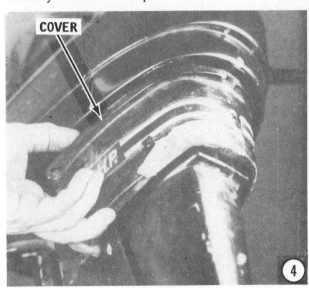

COVER

④

THROTTLE SHAFT

5

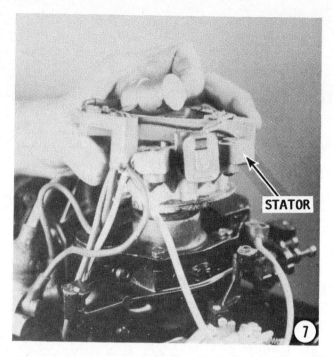

STATOR

7

8- Insert the flywheel key in the crankshaft keyway. Check the inside rim of the flywheel to be sure metal particles are not stuck to the flywheel magnets. Check to be sure the inside taper of the flywheel and the taper on the crankshaft are clean of dirt or oil, to prevent the flywheel from "walking" on the crankshaft during operation.

Slide the flywheel down over the crankshaft with the keyway in the flywheel aligned with the key on the crankshaft. Rotate the flywheel clockwise and check to be sure the flywheel does not contact any part of the magneto or the wiring. Thread the

flywheel nut onto the crankshaft. Use the proper holding tool to prevent the flywheel from rotating and tighten the flywheel nut to the torque value given in the Specifications in the Appendix. Install the front and rear support frame nuts and bolts, if they are used. Install the water tattle-tale hose to the block.

GASKET

6

TORQUE WRENCH

8

9- Install and secure the rewind starter onto the powerhead with the attaching hardware. If the rewind starter requires service, see Chapter 12.

10- If the tiller handle was removed, install it onto the engine housing. Connect the choke linkage to the choke shutter.

11- Connect the fuel lines. Install the electric choke, stop button, and the starter, if one is used. Check the gap on each spark plug according to the Specifications in the Appendix. (For many models, the gap is set.) Install the spark plugs and tighten them to a torque value of 20-1/2 ft-lbs (30.51 Nm). Connect the high-tension leads to the spark plugs. For synchronizing procedures, see Chapter 6.

12- Mount the engine in a test tank.

CAUTION: Water must circulate through the lower unit to the engine any time the engine is run to prevent damage to the water pump in the lower unit. Just five seconds without water will damage the water pump.

Attempt to start and run the engine without the cowl installed. Perhaps the flywheel will accept an emergency-type pull cord, or you can figure some way to rotate the flywheel until it starts. This will give you the opportunity to check for fuel and oil leaks, without the cowl in place. Follow the break-in procedures.

After you are satisfied the engine is operating properly, install the top engine cowl, and at the same time, feed the rewind cord through the housing. Attach the pull handle onto the rewind cord.

TILLER HANDLE →

(10)

HAND STARTER

(9)

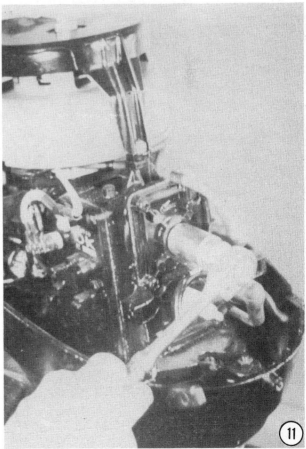

(11)

Break-in Procedures

As soon as the engine starts, **CHECK** to be sure the water pump is operating. If the water pump is operating, a water mist will be discharged from the exhaust relief holes at the rear of the drive shaft housing.

DO NOT operate the engine at full throttle except for **VERY** short periods, until after 10 hours of operation as follows:

a– Operate at 1/2 throttle, approximately 2500 to 3500 rpm, for 2 hours.

b– Operate at any speed after 2 hours **BUT NOT** at sustained full throttle until another 8 hours of operation.

c– Mix gasoline and oil during the break-in period, total of 10 hours, at a ratio of 25:1. If the unit being serviced is equipped with the Auto Blend oil injection system, the manufacturer recommends oil be added to the fuel in a ratio of 50:1 in **ADDITION** to the oil injection system. The oil in the fuel will give a ratio of 25:1 during the break-in period.

d– While the engine is operating during the initial period, check the fuel, exhaust, and water systems for leaks.

e– See Chapter 6 for synchronizing procedures.

After the test period, disconnect the fuel line. Remove the engine from the test tank. Install the engine cowl.

8-3 POWERHEAD "B" SPLIT BLOCK WITHOUT HEAD EXTERNAL REED BLOCK
(See Tune-up Specifications -- Last Column)

REMOVAL AND DISASSEMBLING

ADVICE

Before commencing any work on the powerhead, an understanding of two-cycle engine operation will be most helpful. Therefore, it would be well worth the time to study the principles of two-cycle engines, as outlined briefly in Section 8-1. A Polaroid, or equivalent instant-type camera is an extremely useful item, providing the means of accurately recording the arrangement of parts and wire connections **BEFORE** the disassembly work begins. Such a record is invaluable during the assembly work.

POWERHEAD "B" REMOVAL

1– Remove the gas cap, and then the top cowl. Disconnect the high tension leads from the spark plugs. **ALWAYS** use a pulling and twisting motion as a precaution against damaging the connection. Close the fuel shutoff valve, and then siphon the fuel from the engine mounted fuel tank. Remove the four screws securing the sound box cover to the sound box, and then remove the cover and gasket. Remove the two screws securing the sound box to the carburetor, and then remove the sound box.

2– Remove the hand starter retaining screws, and then remove the starter. Re-

Sound box prior to removal.

move the idle adjusting lever from the carburetor idle adjustment screw.

3- STOP, and carefully observe the wiring and hose connections before proceeding. Because there are so many different engines and the arrangement is slightly different on each, it is not possible to illustrate all of them. Even if they were shown, you would

not be able to identify the engine being serviced. Therefore, **TAKE TIME** to make notes and tag the wiring and hoses. You may elect to follow the practice of many professional mechanics by taking a series of photographs of the engine, one from the top, and several from the sides showing the wiring and arrangement of parts.

4- Loosen the locknuts on the advance and retard throttle cables. Remove the two cables from the mounting bracket and the

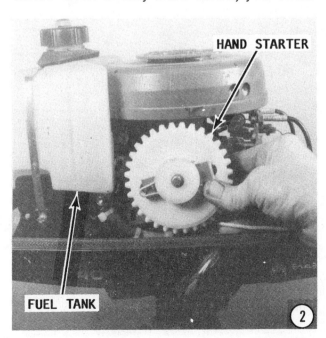

Fuel shutoff valve located at the front of the engine. Other models have this valve installed on the starboard side.

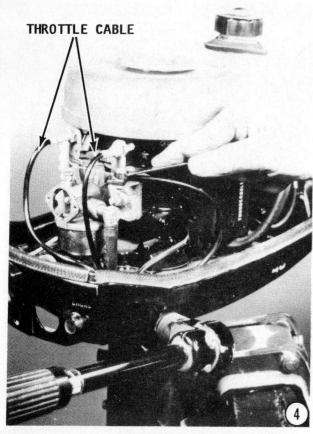

THROTTLE CABLE

FUEL TANK

5- Open the fuel tank mounting strap and lift the fuel tank from the support bracket. Allow the tank to hang over the side of the bottom cowl. Disconnect the four primary wires from the coils. Remove the sta-strap securing the primary wire harness to the fuel tank support bracket.

6- Remove the nut and flat washer securing the flywheel to the crankshaft. It may be necessary to use some type of flywheel strap to prevent the flywheel from turning as the nut is loosened. Install the proper flywheel puller. **NEVER** attempt to use a puller which pulls from the outside edge of the flywheel or the flywheel may be damaged. After the puller is installed, tighten the center screw onto the end of the crankshaft. Continue tightening the screw until the flywheel is released from the crankshaft. Remove the flywheel.

7- Remove the flywheel key from the keyway in the crankshaft. Remove the two

cable ends from the ball sockets. Disconnect the choke cable from the mounting bracket. Disconnect the cable end from the carburetor screw. Remove the black steering handle wire and the black switch box, wire from the ground on the side of the exhaust cover. Disconnect the orange, steering handle, wire from the side of the switch box. Cut the sta-strap and disconnect the fuel line from the carburetor.

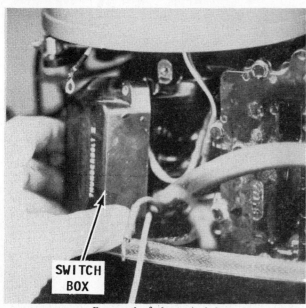

SWITCH BOX

Removal of the switch box.

FLYWHEEL PULLER

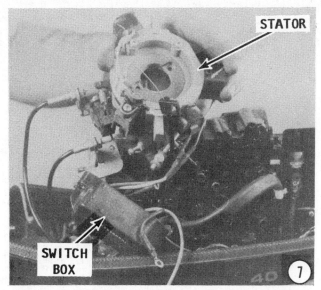

STATOR

SWITCH BOX

7

NUT

9

bolts securing the switch box to the power-head. Remove the two screws attaching the stator assembly to the powerhead. Lift off the stator assembly with the attached trig-ger, the throttle linkage, and the switch box and allow it to hang over the side of the bottom cowl.

8- Remove the carburetor attaching locknuts and flat washers, and then remove the carburetor. Disconnect the bleed hose from the fitting on the carburetor adaptor. Remove the carburetor adaptor, reed block, and gaskets from the powerhead.

GOOD WORDS

The reed block can be adequately in-spected without disassembling. If inspection indicates a part requires replacement, see Section 8-8, Cleaning and Inspecting, at the end of this chapter. Remove the filler block from the crankcase, if one is installed.

9- Remove the six nuts securing the powerhead and the bottom cowl to the driveshaft housing. Hold the bottom cowl down, and at the same time pry up on the powerhead to break the gasket between the powerhead and the bottom cowl.

10- Lift the powerhead straight up from the bottom cowl to prevent tearing the gasket. If the crankshaft and drive shaft splines bind together and do not separate smoothly when the powerhead is being lift-ed, it is possible the driveshaft pulled up/out of the gear housing and disengaged the wat-er pump impeller key.

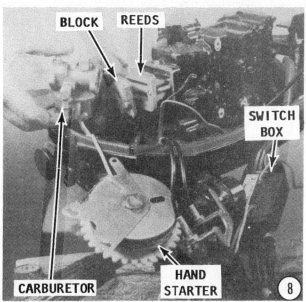

BLOCK   REEDS

SWITCH BOX

CARBURETOR   HAND STARTER

8

POWERHEAD

SWITCH BOX

10

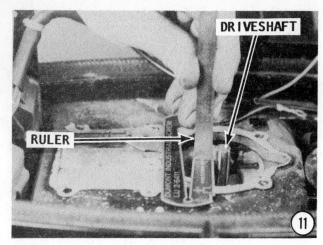

11- Measure the height of the driveshaft to determine if the driveshaft has pulled up/out of the gear housing as follows: Place a straight edge on the powerhead mounting surface of the bottom cowl, and then measure the distance between the straight edge and the top of the driveshaft, as shown. Next, push down on the driveshaft until the driveshaft bottoms-out in the gear housing. If the driveshaft moves downward more than 1/4" (6.35mm), it will be necessary to remove the gear housing and reinstall the water pump impeller key. To install the water pump impeller key, see Chapter 9.

POWERHEAD "B" DISASSEMBLING

1- Remove the bolts securing the exhaust cover and exhaust manifold to the cylinder block. Remove the exhaust cover, manifold and gasket from the cylinder block. **NEVER** attempt to pry the cover off the block. Tap sideways with a soft mallet to loosen the cover. Disconnect the tattletale hose (if equipped) from the fitting at the bottom of the cowl.

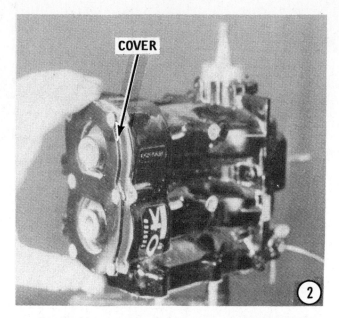

2- Remove the six screws securing the cylinder block cover to the cylinder block. Remove the cylinder block cover and gasket from the cylinder block. **NEVER** attempt to pry the cover off the block. Tap sideways with a soft mallet to loosen the cover. Later model powerheads will have an intake cover on the starboard side secured with two bolts. If the block is to be serviced, this cover and the gasket should be removed.

3- Remove the six bolts securing the crankcase cover to the cylinder block. **NEVER** pry between the two block halves with any kind of tool. Tap lightly on the bottom side of the crankshaft to jar the block halves apart. After they have separated, lift the crankcase cover from the cylinder. Catch and save the needle bearings which may fall out of the center main bearing.

4- Remove the center main bearing liner from the crankcase cover.

5- Lift the crankshaft assembly from the cylinder block. Save the needle bearings from the center main bearing. By count, there **MUST** be 25 needle bearings. If the count is short one or more may still be inside the block. Remove the bearing liner from the cylinder block.

ADVICE

New needle bearings should be installed in the connecting rods, even though they may appear to be in serviceable condition. New bearings will ensure lasting service after the overhaul work is completed. If it is necessary to install the used bearings, keep them separate and identified to ensure they will be installed onto the same crankpin throw and with the same connecting rod from which they were removed.

6- Slide the upper oil seal and the crankshaft roller bearing off the crankshaft. **DISCARD** the oil seal.

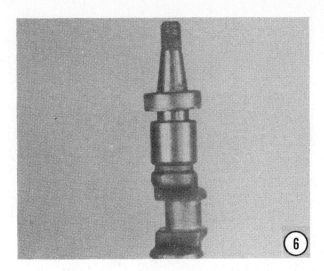

7- Cup both hands around the piston with your fingers at the back of the ring. Now, spread the ring open with your thumbnails inserted at the ring ends and remove the ring up over the top of the piston. **OBSERVE** the pins in the ring groove. These pins are essential to prevent the rings from rotating in the groove. If the ring was allowed to rotate, the end of the ring would become lodged in the cylinder port and break the ring.

8- Before removing the rod cap, observe the identifying matching marks on the boss of the rod and cap. These **MUST** be matched again during assembly.

9- Bend the tabs away from the connecting rod screws, and then remove the screws. Place the crankshaft assembly on a clean work surface. Separate the connecting rod cap from the connecting rod. Remove the connecting rod, cap, and needle

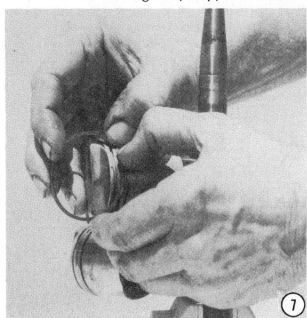

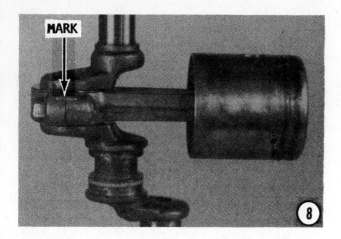

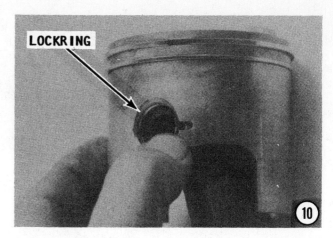

bearings from the crankpin throw. By count, there should be 25 needle bearings. Keep the bearings with the same rod, or stow them in an identified container to ensure they will be installed with the same rod during assembly. Attach the cap back onto the rod to ensure it will be installed with the same rod during installation.

10- WEAR eye protective glasses while removing the piston pin lockrings, because the lockring is made of spring steel and may slip out of the pliers or pop out of the groove with considerable force. Remove the two lockrings. A lockring tool, Mercury part No. 91-5252A1, may be used to remove the lockring. An alternate method is to use a punch to pop the ring out. If a punch is used **TAKE CARE** not to damage the piston. **DISCARD** the lockrings, because they should not be used a second time.

11- Remove the piston from the rod by first inserting a punch into the hollow end of the piston pin. Next, support the bottom of the piston with one hand and drive the pin through the piston with a mallet and the punch.

BAD NEWS

On some models it may be necessary to heat the top of the piston to approximately 190° with hot water or a heat lamp before the piston pin can be driven out. On some models, needle bearings are used with the piston pin. These bearings will fall out after the pin has been removed. **ALWAYS** handle the piston with care, because the piston can be distorted with careless treatment.

12- DO NOT remove the roller bearing from the piston pin end of the connecting rod unless it is necessary to replace the bearing. If inspection determines the bearing must be replaced, support the connecting rod on a suitable mandrel and press the bearing out with Bearing Tool C-91-77584.

Rod with the wrist pin bearing still in place. This bearing should not be removed unless it is unfit for further service.

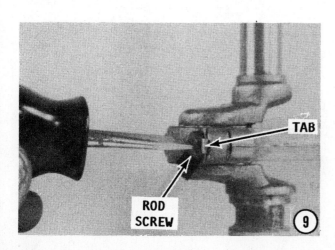

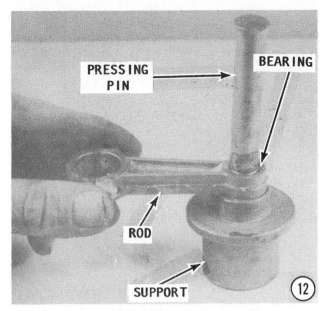

13- Remove the coupler from the end of the crankshaft by installing a Universal Puller Plate C-91-37241 between the coupler and the oil seal. Support the puller plate in a press and press the crankshaft out of the coupler. TAKE CARE to support the crankshaft while it is being pressed out of the coupler to prevent it from falling to the floor and being damaged.

14- Secure the coupler in a vise equipped with soft jaws. Remove the retainer seal from the end of the coupler by tapping on the retainer seal with a flat end punch and hammer. Discard the seal.

15- Remove and DISCARD the O-ring from the crankshaft. Remove and DIS- CARD the oil seal from the crankshaft. Remove the retaining ring from the crankshaft.

16- DO NOT remove the crankshaft ball bearing, unless it is necessary to replace the bearing. If inspection determines the bearing must be replaced, first, remove the retaining ring securing the bearing onto the crankshaft. The retainer can be removed with a pair of expanding type snap ring pliers. Install Universal Puller Plate, C-91-3724 below the crankshaft bearing. Position the crankshaft assembly in an arbor press and support the crankshaft with the puller plate. TAKE CARE to support the crankshaft while it is being pressed out of the bearing to prevent it from falling to the floor and being damaged. Remove the plastic sealing ring from the crankshaft.

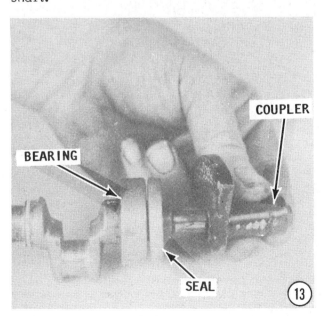

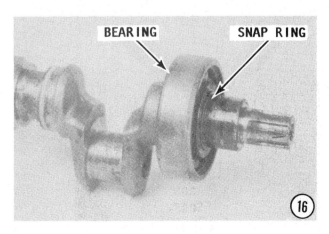

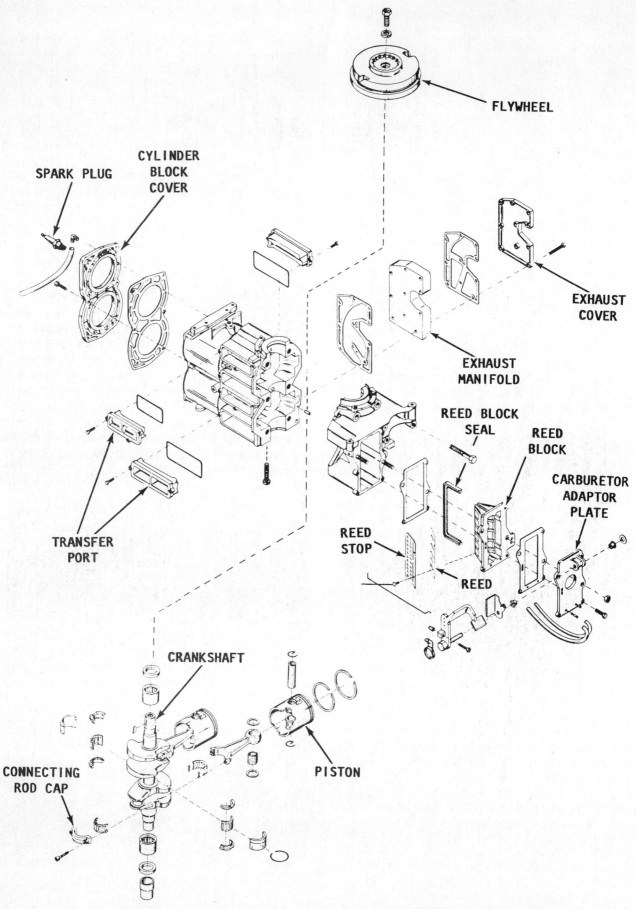

FLYWHEEL

SPARK PLUG

CYLINDER
BLOCK
COVER

EXHAUST
COVER

EXHAUST
MANIFOLD

REED BLOCK
SEAL

REED
BLOCK

CARBURETOR
ADAPTOR
PLATE

TRANSFER
PORT

REED
STOP

REED

CRANKSHAFT

PISTON

CONNECTING
ROD CAP

Exploded drawing of early model Type "B" powerhead with major parts identified.

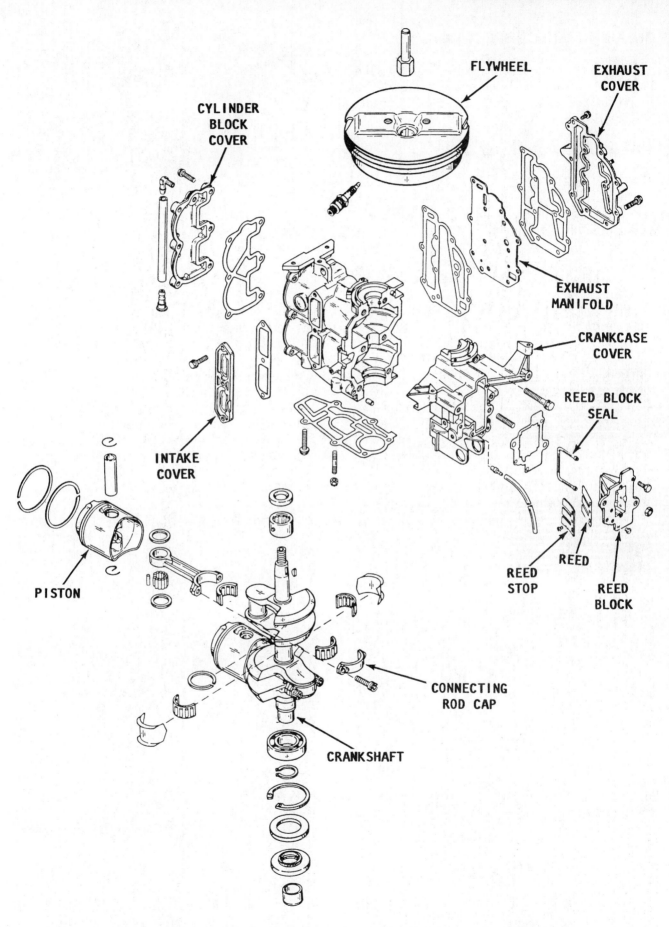

Exploded drawing of late model Type "B" powerhead with major parts identified.

CLEANING AND INSPECTING

Detailed instructions for cleaning and inspecting all parts of the powerhead will be found in the last section of this Chapter, Section 8-5.

ASSEMBLING POWERHEAD "B"

FIRST, THESE WORDS

Be sure all parts to be re-used have been carefully cleaned and thoroughly inspected, as outlined in Section 8-8. Parts that have not been properly cleaned, or parts not suitable for service can damage a good powerhead within a few minutes after starting the engine.

NEW gaskets **MUST** always be used during an overhaul.

A torque wrench is essential for correct assembly of the powerhead. **NEVER** attempt to assemble a powerhead without a torque wrench. Attaching bolts for covers **MUST** be tightened to the required torque value in three progressive steps, following the specified tightening sequence. On the first round, tighten to 1/3 the torque value. On the second round, tighten to 2/3 the total torque value. Finally, on the third and last round, tighten to the full torque value.

1- Place each piston in a container of hot water, approximately 190°F, or carefully heat it with a bottle torch. Leave the piston in the hot water, ready for installation of the piston pins later in Step 8. Install a new crankshaft ball bearing assembly onto the bottom end of the crank-

shaft, if it was removed. Support the crankshaft in a press between the No. 2 throw and directly under the lower crankshaft end. Now, using a piece of tubing as a mandrel, press the lower ball bearing assembly onto the crankshaft until the bearing is seated firmly against the shoulder. Press **ONLY** on the inner race of the bearing. Remove the crankshaft from the press. Using a suitable pair of expanding-type Snap Ring Pliers, install the retaining ring to secure the bearing onto the crankshaft, if one is used.

2- Place the large retaining ring in position on the crankshaft. Lubricate the lip of the lower crankshaft oil seal that contacts the crankshaft, with light-weight oil. Now, slide the lubricated oil seal onto the drive end of the crankshaft with the lip **TOWARD** the driveshaft end. Lubricate the O-ring with light-weight oil, and then slide it onto the splined end of the crankshaft until it makes contact with the shoulder on the crankshaft.

3- The coupler **MUST** be installed onto the splined end of the crankshaft in a precise manner as follows: When installed, the

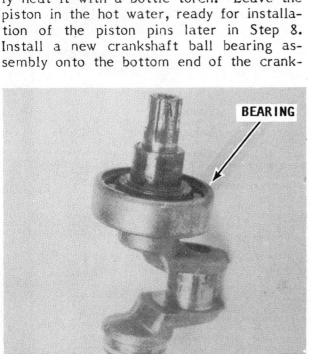

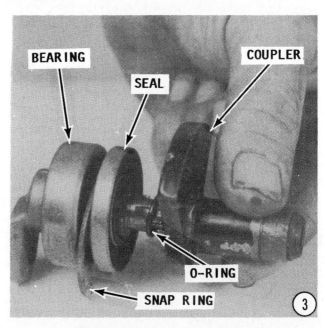

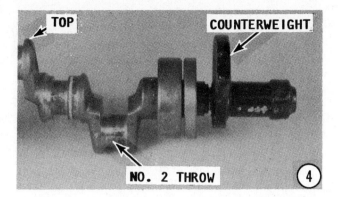

TOP   COUNTERWEIGHT

NO. 2 THROW   **4**

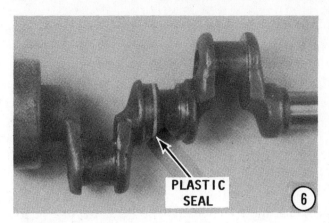

PLASTIC SEAL   **6**

centerline of the counterweight must be positioned 180° from the centerline of the No. 2 throw on the crankshaft, as shown. The No. 2 throw is the bottom throw on the crankshaft.

4- Support the crankshaft in a press between the No. 2 throw and directly under the lower crankshaft end. Press the coupler onto the splined end of the crankshaft. **CHECK** to be sure:

The coupler is over the O-ring.

The coupler is seated against the crankshaft shoulder.

The centerline of the counterweight is 180° from the centerline of the No. 2 throw.

5- Install the retainer seal onto the end of the coupler, as shown. Support the coupler in a press with a universal puller plate, and then press the retainer seal onto the coupler with a 1/2" (13mm) socket.

6- Install the plastic sealing ring into the groove of the crankshaft, as shown.

7- Install the roller bearing into the connecting rod, if it was removed. Use Tool C91-74582 to press the bearing into the connecting rod until the bearing is centered in the rod.

8- Remove the piston from the hot water it was placed into at the beginning of Step 1. Position the piston on the connect-

ing rod and push the piston pin through with the hollow end of the pin toward the **BOTTOM** of the crankshaft. In this position, the hollow will not collect fuel entering the chamber.

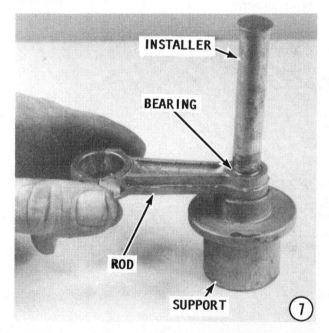

INSTALLER

BEARING

ROD

SUPPORT   **7**

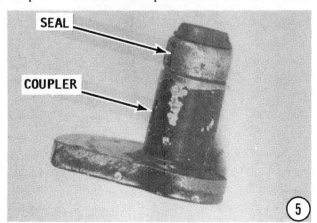

SEAL

COUPLER   **5**

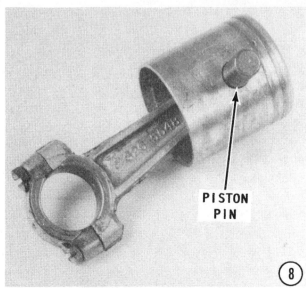

PISTON PIN   **8**

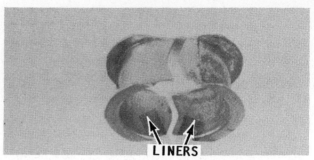

*Two bearing retainer liners with the matching **V**s clearly visible. These **V**s must be matched during installation.*

9- **WEAR** eye protection glasses while installing the piston pin lockrings, because the lockring is made of spring steel and may slip out of the pliers or pop out of the groove during installation with considerable force. Install **NEW** lockrings into the groove in each end of the piston pin bore. Check to be sure the lockrings are fully seated in the grooves.

10- Expand the ring slightly with the fingernail of each of your thumbs and at the same time support the back of the ring with your fingers. Now, slide the ring down over the piston and into the proper grooves on the piston. After the rings are in place, each ring should rotate freely. Lubricate the piston, the rings, and the cylinder bore with a good grade of outboard motor oil. Rotate the rings in their grooves until the ends of each ring is over the locating pin in the groove of the piston.

SPECIAL WORDS

Since 1986, the pistons for the 6 hp, 8 hp, 9.9 hp, 15 hp, and the 210 cc do not have a locating pin in the ring groove.

Repeat Steps 8 thru 10 for the other piston.

WORDS FROM EXPERIENCE

There is no easy way to install the assembled pistons and crankshaft into the block. Without a ring compressor tool, the job is even more difficult. A Mercury ring compressor tool is usually available at modest cost from the local Mercury dealer.

11- Position the connecting rod liner **WITHOUT** the hole in it, on the rod. Position the liner, **WITH** the hole in it, on the cap, with the hole aligned with the hole in the cap. Check to be sure the dovetail ends of the liners will match when the rod and cap are matched together with the boss marks aligned.

NEVER intermix new needle bearings with used bearings in the same connecting rod. If a bearing is not fit for service, then the entire set **MUST** be replaced.

12- With the crankshaft in a horizontal position, carefully place the needle bearings around the crankshaft throw. A total of 25 needle bearings are required for each rod.

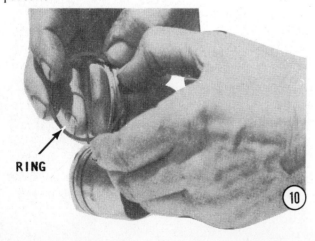

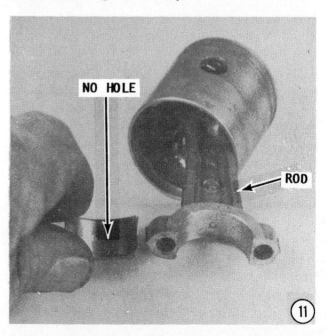

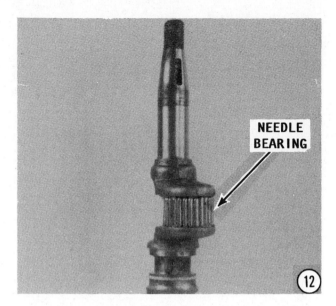

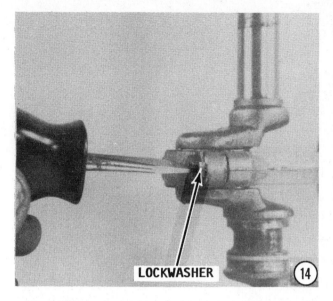

Lower the connecting rod and cap over the needle bearings on the throw, with the boss marks aligned, **AND** without disturbing the bearings.

13- Secure the connecting rod and cap together with the two rod screws using **NEW** locking tab washers. Tighten the screws alternately to the torque value given in the Specifications in the Appendix. **DO NOT** reposition the connecting rod screws, if the locking tab on the washer does not fit against the flat on the screw.

14- The locking tab **MUST** be positioned to conform to the screw, not the screw to the tab. **CHECK** the work thus far, by attempting to spread the needle bearings apart with a small pointed rod or a piece of

wire. If the needle bearings do not spread the width of a needle bearing, the correct number of bearings has been used. Lock the connecting rod screws in place by bending the tab washers up against the screws.

Repeat Steps 6 thru 10 for the other rod.

ADVICE

Always handle pistons with great care. Do not force the piston into the cylinder bore until the cylinder has been honed to the correct size. A piston can be distorted through rough treatment.

15- Slide the roller bearing onto the upper end of the crankshaft with the bearing end containing the letters facing **TOWARD** the **TOP** of the crankshaft. Place the ring

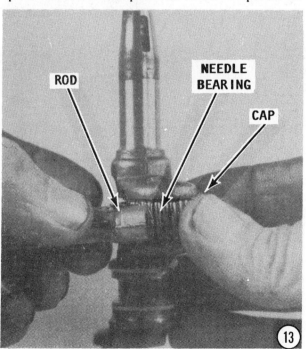

Bending the locking tab to secure the rod cap screw in place.

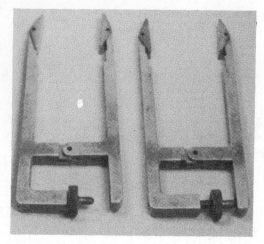

Ring compressors used to install the pistons into the cylinders.

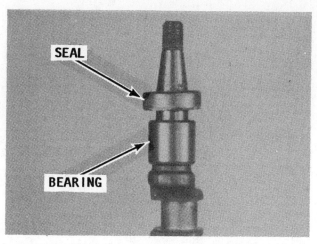

*The top oil seal which should **NOT** be installed at this time. This seal is installed after the crankcase is assembled.*

compressor tool over the crankshaft and down the skirt of the piston onto the rings. Begin to tighten the tool onto the rings, and at the same time, check to be sure the ring ends are over the piston pin. Continue to tighten the tool until the rings are almost flush with the surface of the piston. Repeat the procedure for the other piston. Leave the ring compressor tool on each piston.

GOOD WORDS

Do not install the upper oil seal at this time. The oil seal is installed **AFTER** the crankshaft cover is installed onto the cylinder block. The center main bearing liner is installed into the cylinder block **AFTER** the crankshaft assembly is installed.

16- With the top of the crankshaft positioned toward the top of the cylinder block, install the crankshaft assembly into the cylinder block in this manner: Lower the piston and crankshaft assembly into the block with one piston down and the other up. Work the "down" piston into the cylinder bore and at the same time, lower the crankshaft and the other piston downward. When

the second piston reaches the cylinder bore in the block, work both pistons and the crankshaft assembly downward. After the pistons are in the cylinder bores, remove the ring compressor tool.

17- Check to be sure each piston ring has spring tension. This is accomplished by **CAREFULLY** pressing on each ring with a screwdriver extended through the intake ports. If spring tension cannot be felt (the

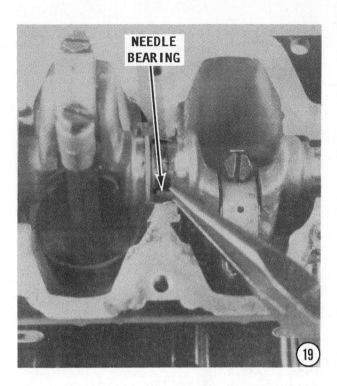

spring fails to return to its original position) the ring was probably broken during the piston and crankshaft installation process. **TAKE CARE** not to burr the piston rings while checking for spring tension.

18- Install the center main bearing liner, with the hole, into the cylinder block between the cylinders. Install the liner with the **"V"** end of the liner **TOWARD** the exhaust cover side of the cylinder block.

19- Lubricate the 25 center main needle bearings with Multipurpose Lubricant, or equivalent. Now, install the needle bearings by pushing them one at a time around the bottom of the center main bearing surface, and then placing them around the top of the bearing surface.

BAD NEWS

The retaining ring **MUST** be installed as just described in Step 18 with the open end of the ring toward the bleed groove, or the

bleed system will be blocked. The lower oil seal must be installed tight against the retaining ring.

20- Apply a thin coating of Loctite Type "A" to the outer diameter of the lower oil seal, which contacts the cylinder block. Position the crankshaft in the cylinder block. Rotate the retaining ring until the open part of the ring is **TOWARD** the bleed groove. This position will prevent the retaining ring from blocking the bleed groove. Push the lower oil seal tight against the retaining ring.

21- Install the center main bearing liner onto the crankcase cover. When the liner is in position, the dovetail ends should match when the crankcase cover and the cylinder

Main needle bearings installed onto the crankshaft throw.

SNAP RING
GROOVE
20

block are mated. Before mating the crank-case cover and the cylinder block, check to be sure:

   a- The mating surfaces are clean. **NEVER** use any kind of tool or abrasive material to clean the surfaces. Use only solvent and elbow grease.

   b- The crankshaft is properly seated.

   c- The rod caps have been correctly installed.

   22- Coat the mating surface of the crankcase cover with a thin layer of Loctite No. 514, or equivalent. **TAKE CARE** not to get any Loctite in the bleed hole. Wipe off any excess Loctite. Check again, to be sure the lower oil seal on the crankcase is pushed tight against the retaining ring.

   23- Position the crankcase cover in place on the cylinder block. Install and tighten the six cover bolts to the torque value given in the Specifications in the Appendix. Follow the tightening sequence

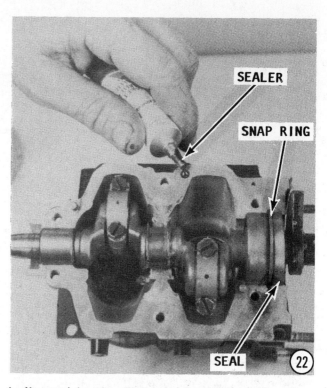

SEALER
SNAP RING
SEAL
22

indicated in the diagram on Page 8-41. Attaching bolts **MUST** be tightened to the required torque value in three progressive stages, following the specified tightening sequence. Tighten all bolts to 1/3 the torque value, then repeat the sequence tightening to 2/3 the torque value. Finally, on the third and last sequence, tighten to the full torque value.

   Clean off any excess Loctite from the crankcase cover and cylinder block. Insert the flywheel key into the crankshaft keyway, and then slide the flywheel into place.

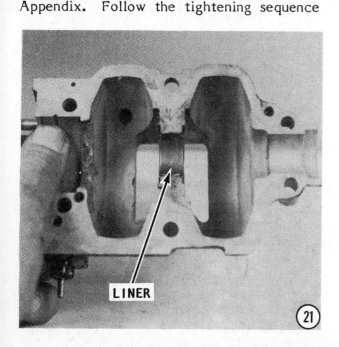

LINER
21

COVER
23

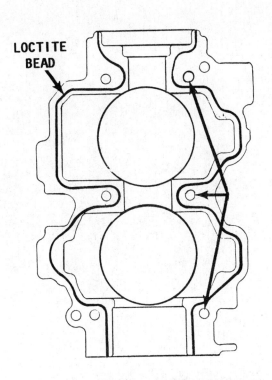

LOCTITE MASTER GASKET SEALANT
BEAD ON CRANKCASE COVER

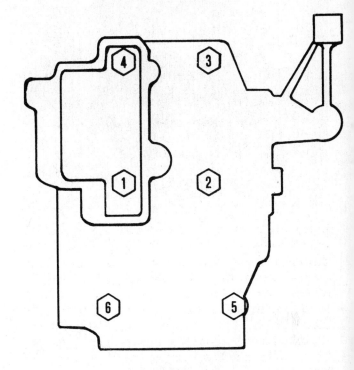

CRANKCASE COVER
16.6 ft lb
(22.6 Nm)

TIGHTENING SEQUENCE
TYPE "B" POWERHEAD

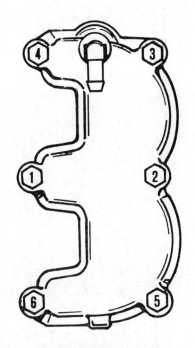

CYLINDER BLOCK COVER
60 in lb
(6.8 Nm)

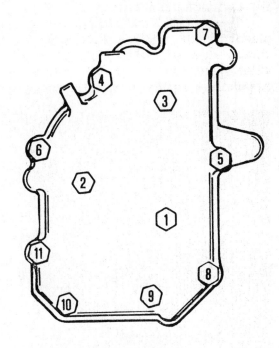

EXHAUST COVER
60 in lb
(6.8 Nm)

TOP
SEAL

24

EXHAUST
COVER

26

Now, rotate the crankshaft several turns and check to be sure it turns freely with no binding or "rough" spots. Blow compressed air through the bleed system to prevent any Loctite that may have entered the bleed system, from blocking the system. Remove the flywheel and key.

24- Lay down a thin bead of Loctite 514, or equivalent, to the outer diameter of the upper oil seal. This is the surface that contacts the powerhead. Place the seal into the cylinder block. Position a 9/16" socket onto the metal end of the seal and lightly tap the seal into the cylinder block. Clean away any excess Loctite.

25- Position a **NEW** gasket in place on the cylinder block. Install the cylinder block cover and secure it with the six attaching bolts. Tighten the bolts evenly, alternately, and in three stages to the

torque value given in the Specifications in the Appendix. Follow the tightening sequence indicated in the diagram on Page 8-41. Connect the tattle-tale hose (if so equipped) to the fitting at the bottom of the cowl.

26- Use a **NEW** gasket and install the exhaust cover and exhaust manifold to the cylinder block. Tighten the bolts evenly to the torque value given in the Specifications in the Appendix. Follow the tightening sequence indicated in the diagram on Page 8-41. Install the filler block, if one was removed during disassembly, into the crankcase cover. If the intake cover was removed in Step 13, Disassembling, install the cover using a **NEW** gasket.

COVER

25

POWERHEAD

1

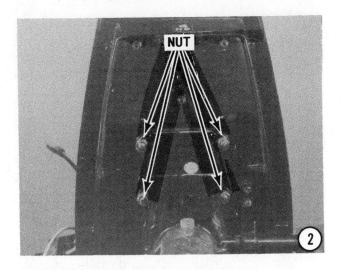

INSTALLATION POWERHEAD "B"

1- Observe if the bottom cowl is separated from the driveshaft housing. If it is separated, install a **NEW** gasket between the cowl and the housing. Position a **NEW** gasket over the powerhead studs and into place on the powerhead base. Apply a thin coating of Multipurpose Lubricant, or equivalent, to the driveshaft splines. Install the powerhead to the bottom cowl and driveshaft housing. If necessary, rotate the flywheel slightly to allow the coupler splines to index with the driveshaft splines and allow the powerhead to become fully seated.

2- Secure the powerhead to the bottom cowl and driveshaft housing with the six hex nuts. Tighten the nuts evenly, alternately, and in three stages to the torque value given in the Specifications in the Appendix. Tighten all bolts to 1/3 the torque value, then repeat the sequence tightening to 2/3

the torque value. Finally, on the third and last sequence, tighten to the full torque value.

3- Use a **NEW** gasket and install the reed block assembly onto the carburetor mounting studs. Use a **NEW** gasket and install the carburetor adaptor onto the carburetor mounting studs. Connect the bleed hose between the crankcase and the carburetor adaptor. Use a **NEW** gasket and install the carburetor onto the mounting studs. Secure the carburetor with the two flat washers and locknuts. Tighten the nuts to the torque value given in the Specifications in the Appendix.

4- Install the stator and trigger assembly to the top of the powerhead. Secure the assembly in place with the two attaching screws. Route both trigger leads and the yellow stator wire down behind the switch box to prevent them from rubbing against the flywheel. Install the switch box

WASHER

GREEN WIRE

BLUE WIRE

to the side of the power and secure it in place with the two flat washers and bolts. Check to be sure the wires are not pinched behind the switch box. Insert the flywheel key in the crankshaft keyway.

5- Check the inside rim of the flywheel to be sure metal particles are not stuck to the flywheel magnets. Check to be sure the inside taper of the flywheel and the taper on the crankshaft are clean of dirt or oil, to prevent the flywheel from "walking" on the crankshaft during operation. Slide the flywheel down the crankshaft with the keyway in the flywheel aligned with the key on the

crankshaft. Rotate the flywheel clockwise and check to be sure the flywheel does not contact any part of the magneto or the wiring.

6- Slide a flat washer onto the crankshaft, and then thread the flywheel nut onto the crankshaft.

7- Tighten the flywheel nut to the torque value given in the Specifications in the Appendix.

8- Connect the four primary lead wires from the switch box to the coils, using the lockwashers and hex nuts. If necessary, refer to the notes or photograph taken prior to disassembly, or to the Wiring Diagram in the Appendix. No. 1 cylinder is the top cylinder. Cover all connections with a

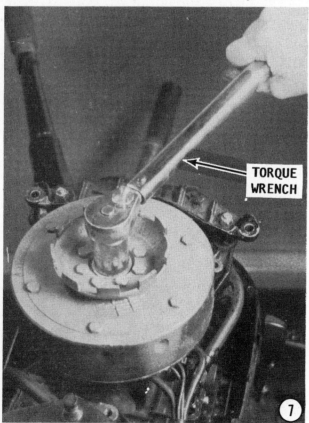

TORQUE WRENCH

FUEL TANK

coating of Liquid Neoprene. Install the rubber boots over the terminals.

9- Install the fuel tank to the mounting bracket and secure it in place with the mounting strap.

10- Install the rewind starter assembly to the powerhead. (If the rewind starter requires service, see Chapter 12.) Secure it in place with the three attaching screws.

11- Connect the fuel line to the carburetor and secure it in place with the sta-strap. Connect the orange steering handle wire, the yellow stator wire, and the orange switch box wire together onto the switch box terminal. Remove the lower bolt from the exhaust cover and connect the black steering handle ground wire and the black switch box ground wire to the bolt. Install the bolt and tighten it to the specified torque value.

12- Install the choke cable end into the choke lever on the carburetor. Use the hole closest to the choke lever pivot. Secure the choke cable to the cable mounting bracket. Now, pull out the choke knob and check to

be sure the carburetor plate is fully closed. Push the choke knob in and check to be sure the plate is fully open. Adjust the choke cable, if necessary to obtain the desired results. Route the advance throttle cable under the fuel line, and then install the advance cable and the retard cable into the mounting bracket and the cable ends into the ball sockets. The retard cable is the shorter of the two cables. Adjust the advance and retard throttle cables. See Chapter 4 to make carburetor adjustments and

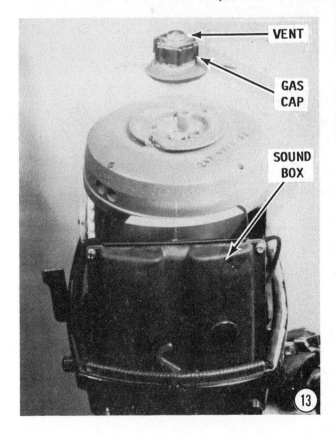

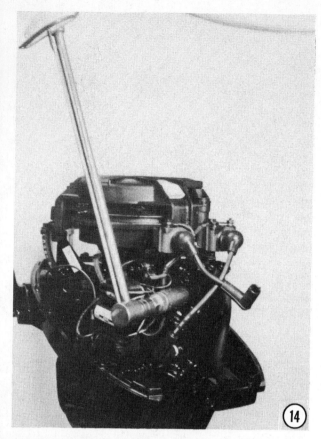

Chapter 6 for timing and synchronizing. Apply Loctite Type "A" to the threads of the two sound box attaching bolts.

13- Install the sound box to the carburetor and secure it in place with the two bolts. Install the sound box cover and gasket. Secure the cover with the four screws.

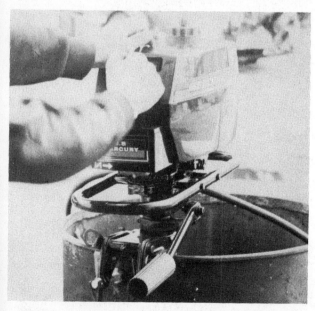

Testing engine operation in a test tank following service work. The break-in instructions given in this section should always be followed after any completed work.

Only Mercury Quicksilver engine oil should be added to the fuel for Mercury Outboards.

14- Install the spark plugs and tighten them to 20-1/2 ft-lbs (27.8 Nm). Connect the high-tension leads to the proper spark plugs. Mount the engine in a test tank. Turn the fuel shut-off valve to the **ON** position. Start the engine and follow the break-in procedures given after the Caution.

CAUTION: Water must circulate through the lower unit to the engine any time the engine is run to prevent damage to the water pump in the lower unit. Just five seconds without water will damage the water pump.

Break-in Procedures

As soon as the engine starts, **CHECK** to be sure the water pump is operating. If the water pump is operating, a water mist will be discharged from the exhaust relief holes at the rear of the drive shaft housing.

During the first 10 hours of operation, **DO NOT** operate the engine at full throttle (except for **VERY** short periods). Perform the break-in as follows:

a- Operate at 1/2 throttle, approximately 2500 to 3500 rpm, for 2 hours.

b- Operate at any speed after 2 hours **BUT NOT** at sustained full throttle until another 8 hours of operation.

c- Mix gasoline and oil during the break-in period, total of 10 hours, at a ratio of 25:1.

d- While the engine is operating during the initial period, check the fuel, exhaust, and water systems for leaks.

e- Refer to Chapter 6 for synchronizing procedures.

After the test period, disconnect the fuel line. Remove the engine from the test tank. Install the engine cowl.

8-4 POWERHEAD "C"
SPLIT BLOCK WITH HEAD
REEDS INSTALLED UNDER
THE CRANKSHAFT
(See Tune-up Specifications Last Column)

ADVICE

Before commencing any work on the powerhead, an understanding of two-cycle engine operation will be most helpful. Therefore, it would be well worth the time to study the principles of two-cycle engines, as outlined briefly in Section 8-1. A Polaroid, or equivalent instant-type camera is an extremely useful item, providing the means of accurately recording the arrangement of parts and wire connections **BEFORE** the disassembly work begins. Such a record is invaluable during assembling.

REMOVAL POWERHEAD "C"

1- Unsnap the spark plug access cover and set it aside.

2- Remove the four screws securing the two sides of the powerhead cowling. There are five more pairs of screws. Each pair is attached to a clip across the seam where the two halves come together. Remove only **ONE** of each pair of screws to permit the clips to remain in place while the cowling is being removed. Separate the two halves, and then pull them free of the powerhead.

SPARK PLUG ACCESS COVER

3- Pull the spark plug lead free of the spark plug. Use a pulling and twisting motion on the molded cap portion. **NEVER** pull on the wire or the connection inside the cap may become separated or the boot damaged.

4- Remove the spark plug. **TAKE CARE** not to tilt the socket as the plug is removed, or the insulator may be cracked.

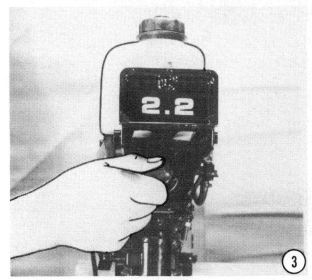

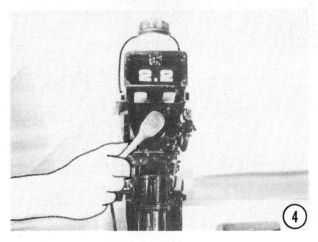

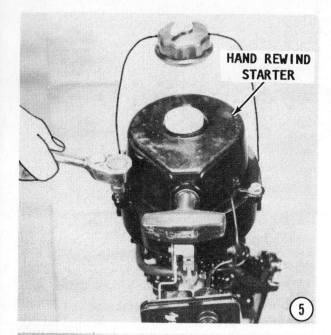

5

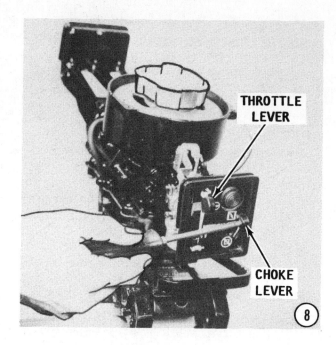

8

6

5- Remove the three bolts securing the hand rewind starter to the powerhead. Lift the hand starter free of the powerhead.

6- Rotate the fuel shut-off valve to the **OFF** position.

7- Remove the clamp securing the fuel line to the carburetor. Lift the fuel tank free of the powerhead.

8- Remove the screws securing the knob to the throttle and choke levers. Set the front plate to one side of the carburetor.

9- Separate the quick-disconnect fitting from the carburetor to the coil. The front plate of the carburetor is now free. Loosen the screw on the clamp securing the carburetor. Pull the carburetor away from the powerhead.

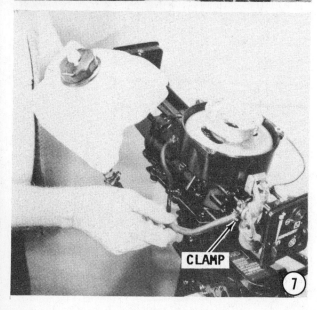

7

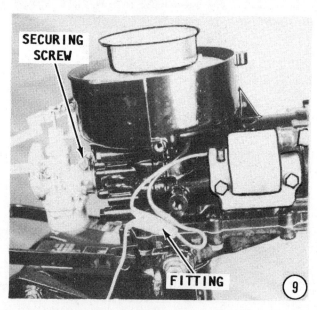

9

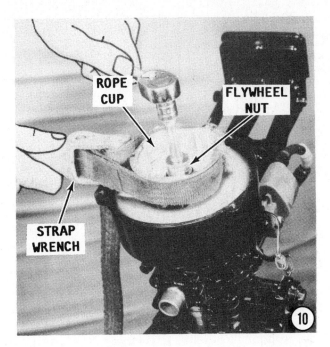

10

12

10- Obtain a strap wrench or equivalent tool. Hold the rope cup steady with the strap wrench and "break" the flywheel nut loose, then remove the nut with the proper size socket. The nut has standard right-hand threads.

11- Continue holding the rope cup with the strap wrench, and remove the three bolts securing the rope cup.

12- Obtain and set-up the proper type wheel puller. Use the holes from which the rope cup bolts were removed to secure the puller. **NEVER** attempt to use a puller which pulls on the outside edge of the flywheel, or the flywheel may be damaged. After the puller is installed and ready, tighten the center screw onto the end of the crankshaft. Continue tightening the screw until the flywheel is released from the crankshaft.

13- After the flywheel is released, remove the puller, and then lift the flywheel free of the crankshaft. A "pull" may be felt as the flywheel is lifted due to the permanent magnets installed on the inside rim of the flywheel. Handle the flywheel carefully because any sudden shock (just as dropping the flywheel) will lessen the strength of the magnets. A weak magnet will seriously affect the ignition circuit.

14- Remove the secondary coil from the side of the powerhead. Disconnect the wire leading to the stator assembly.

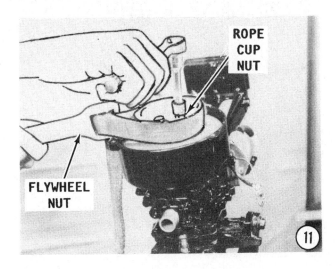

11

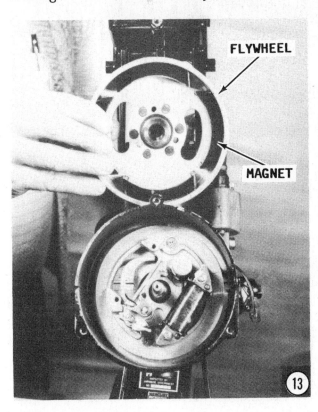

13

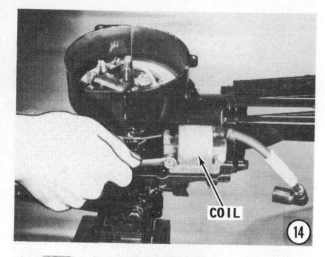

COIL

14

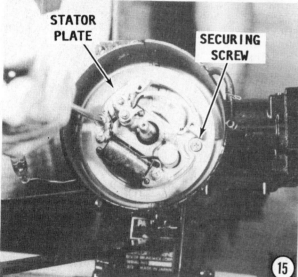

STATOR PLATE

SECURING SCREW

15

15- Remove the two screws securing the stator assembly. Lift the stator plate, and at the same time feed the wire connecting the stator plate to the coil through the opening in the cylinder block cover.

16- Remove the six bolts securing the powerhead to the driveshaft housing.

17- **CAREFULLY** pry the powerhead free of the driveshaft housing. It may be necessary to tap on the joint with a soft head mallet to break the powerhead loose. Lift the powerhead straight up and clear of the driveshaft.

BAD NEWS

If the unit is several years old, or if it has been operated in salt water, or has not had proper maintenance, or shelter, or any number of other factors, then separating the powerhead from the driveshaft housing may not be a simple task. An air hammer may be required on the studs to shake the corrosion loose; heat may have to be applied to the casting to expand it slightly; or other devices employed in order to remove the powerhead.

One very serious condition would be the driveshaft "frozen" with the crankshaft. In this case, a circular plug-type hole must be drilled and a torch used to cut the driveshaft. Let's assume the powerhead will come free on the first attempt.

The following procedures pickup the work after the powerhead is on the work bench.

POWERHEAD

DRIVESHAFT HOUSING

16

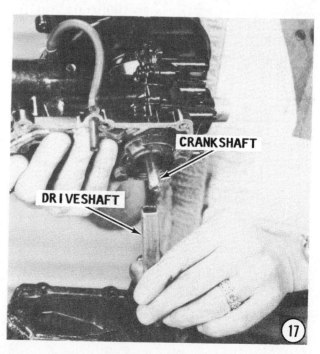

CRANKSHAFT

DRIVESHAFT

17

DISASSEMBLING POWERHEAD "C"

1- Thoroughly clean any gasket material adhering to the mating surfaces of the powerhead and the driveshaft housing.

2- Remove the two bolts securing the lower end cap to the powerhead. It is possible the end cap may not be removed until after the crankshaft has been removed from the powerhead. These two bolts **MUST** be removed at this time in order to separate the two halves of the crankcase.

3- Remove the six bolts holding the crankcase cover to the cylinder block. One of these bolts secures a wire harness retainer in place. Remember this particular bolt.

4- Separate the two halves of the crankcase. It may be necessary to carefully tap the joint with a soft-head mallet to "break" them free of each other.

5- Remove the two screws securing the reed retainers and reeds to the crankcase.

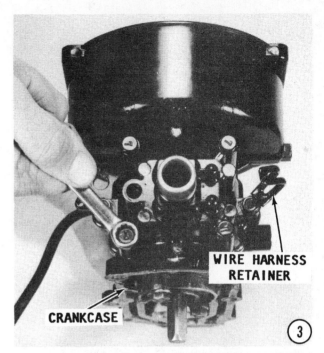

WIRE HARNESS RETAINER

CRANKCASE

POWERHEAD

CYLINDER BLOCK

CRANKCASE COVER HALF

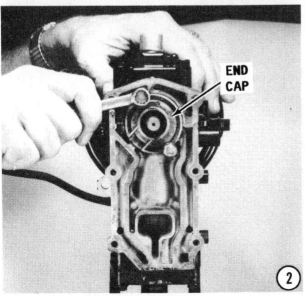

END CAP

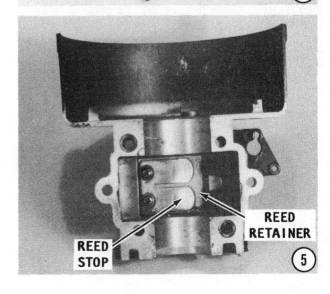

REED STOP

REED RETAINER

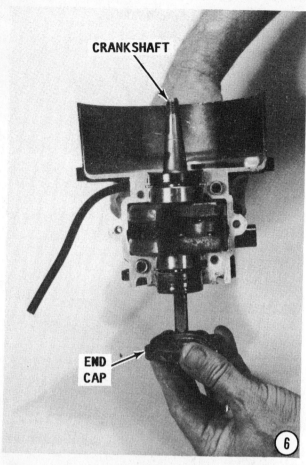

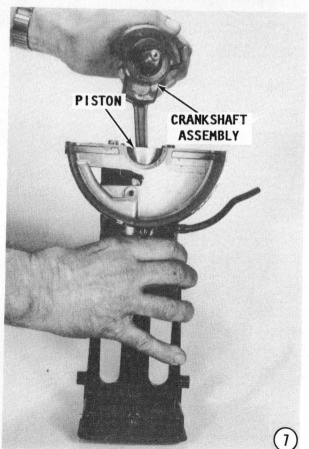

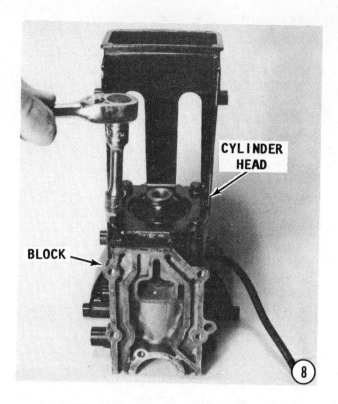

6- Slide the end cap off the crankshaft.

7- Grasp the crankshaft firmly and pull the assembly, together with the piston, out of the cylinder.

CRITICAL WORDS

The rod is an integral part of the crankshaft. The two are manufactured together and **CANNOT** be separated.

8- Remove the four bolts securing the cylinder head to the block. **DISCARD** the gasket.

9- Slide the upper and lower crankcase seal free of the crankshaft.

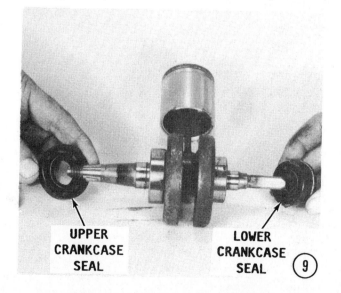

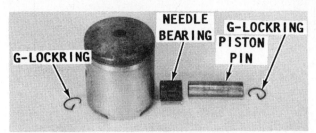

Arrangement of associated piston parts. The G-lockrings should be replaced each time the piston is disassembled.

free using a drift pin or blunt punch. Use sharp, quick, hard blows with a hammer. Your legs will absorb the shock without damaging the piston.

SAFETY WORDS

The piston pin lockrings are made of spring steel and may slip out of the pliers or pop out of the groove with considerable force. Therefore, **WEAR** eye protection glasses while removing the piston pin lockrings.

10- Remove the G-lockring from both ends of the piston pin using a pair of needle nose pliers.

11- Assume a sitting position in a chair, on a box, whatever. Lay a couple towels over your legs. Hold your legs tightly together to form a cradle for the piston above your knees. Set the piston and crankshaft assembly between your legs, as shown in the illustration. Now, drive the piston pin

GOOD WORDS

If the piston pin fails to move after a few hammer blows, as just described, it may be necessary to heat the piston in order to remove the piston pin. Heating the piston in a container of boiling water for about 10-minutes or with a bottle torch for about a minute, will cause the metal in the piston to expand ever so slightly and ease the task of removing the piston pin. If a bottle torch is used, keep the torch moving around the piston to prevent excessive heat in any one area.

12- Use a piston ring expander, tool No. 91-24697 to remove the piston ring. If the tool is not available, expand the ring with your fingers or a pair of reverse pliers enough to slip the ring up and free of the piston.

CLEANING AND INSPECTING

Detailed instructions for cleaning and inspecting all parts of the powerhead, including the block, will be found in the last section of this chapter, Section 8-8.

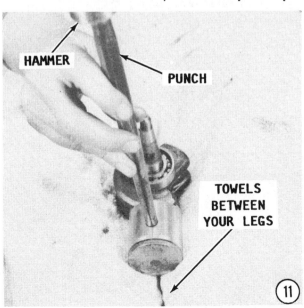

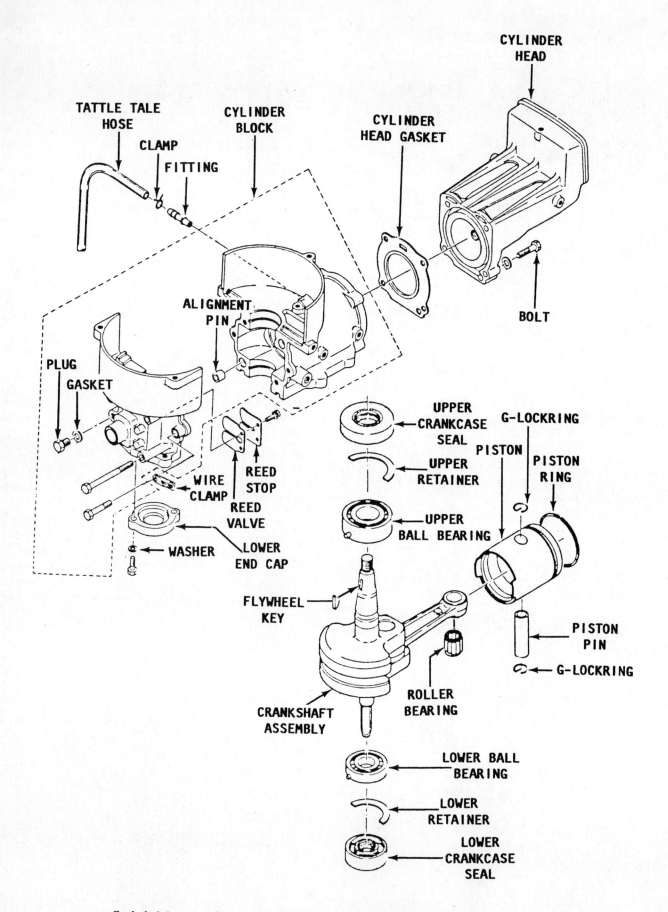

Exploded drawing of Type "C" powerhead with major parts identified.

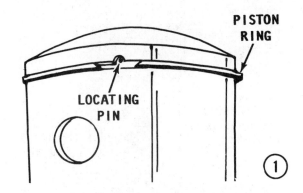

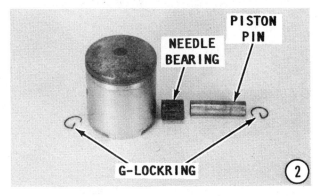

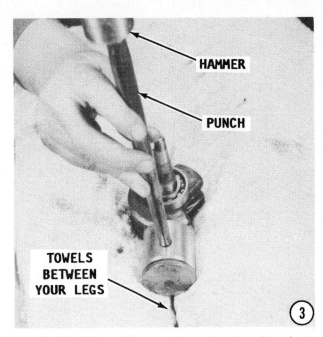

ASSEMBLING POWERHEAD "C"

1- Install a new piston ring using piston ring expander tool No. 91-24697. The ring must be installed with the groove end of the ring facing **UPWARD**. If the ring expander tool is not available, expand the ring slightly with the fingernail of each of your thumbs and at the same time support the back of the ring with your fingers. Now, slide the ring down over the piston and into the groove.

2- Position the new needle bearing into the upper end of the connecting rod. Position the piston on top of the rod so the arrow on the piston crown will face **TOWARD** the exhaust port when the crankshaft assembly is installed in the crankcase.

3- Assume a sitting position in a chair, on a box, whatever. Lay a couple towels over your legs. Hold your legs tightly together to form a cradle for the piston above your knees. Set the piston and crankshaft assembly between your legs, as shown in the illustration. Now, drive the piston pin into place through the piston using a drift pin or blunt punch. Use sharp hard blows with a hammer. Your legs will absorb the shock without damaging the piston.

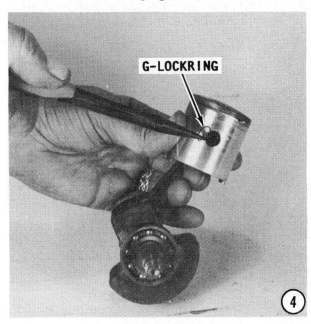

The arrow on the piston crown **MUST** *face toward the exhaust port when installed.*

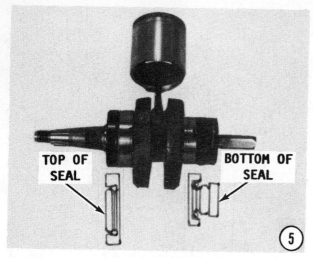

TOP OF SEAL   BOTTOM OF SEAL

(5)

To make this task a little easier, heat the piston, either in a container of boiling water for about 10-minutes, or about a minute with a bottle torch. Heating the piston will expand the metal in the piston slightly and the piston pin will go through more smoothly. If a bottle torch is used, keep the torch moving to prevent excessive heating of one area.

SAFETY WORDS

WEAR eye protection glasses while installing the piston pin lockrings, because the lockring is made of spring steel and may slip out of the pliers with considerable force.

4- Insert the two G-lockrings into the groove on both sides of the piston with a pair of needle nose pliers. The lockrings will secure the piston pin in place.

5- Lubricate the rims of **NEW** oil seals with engine oil. Install the narrow seal onto the upper bearing and the wider seal onto

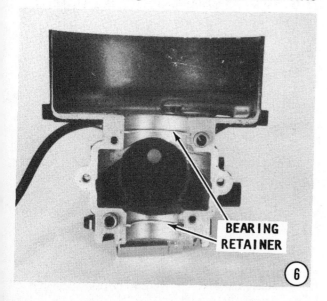

BEARING RETAINER

(6)

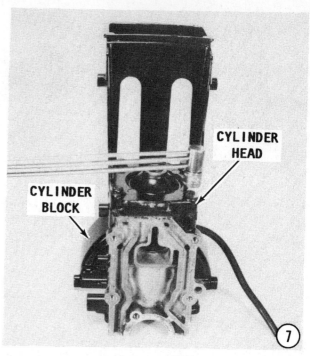

CYLINDER HEAD

CYLINDER BLOCK

(7)

the lower bearing with the flat side of the seal facing toward the crankshaft throw, as shown in the accompanying illustration. The upper end of the crankshaft can be easily identified as the end with threads for the flywheel nut.

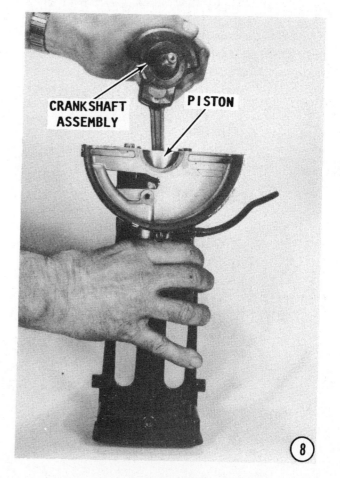

CRANKSHAFT ASSEMBLY   PISTON

(8)

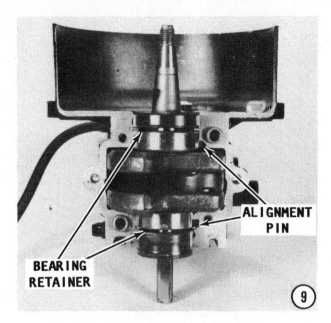

BEARING RETAINER

ALIGNMENT PIN

9

6- Install **NEW** bearing retainers into place in both halves of the cylinder block.

7- Place a **NEW** head gasket on the cylinder block. Mate the head to the block and start the retaining bolts. Tighten the bolts alternately in three stages to the torque value given in the Appendix. Tighten to 1/3 the torque value on the first stage and to 2/3 the torque value on the second stage. On the third and final stage, tighten the bolts to the full torque value.

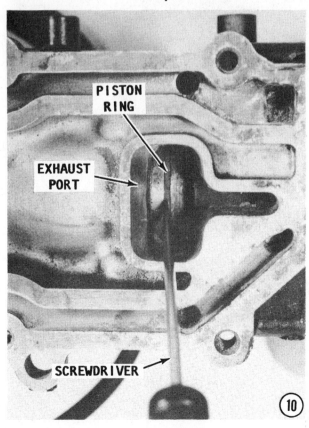

PISTON RING

EXHAUST PORT

SCREWDRIVER

10

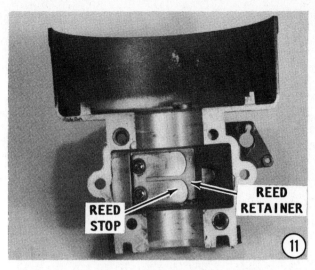

REED STOP

REED RETAINER

11

8- Lower the assembled crankshaft into the cylinder block. It is not necessary to use a piston ring compressor for this task. Lower the crankshaft assembly with one hand and at the same time squeeze the piston ring around the piston, as the piston moves into the cylinder. Keep the crankshaft **HORIZONTAL** with the block while it is being lowered.

9- Seat the crankshaft onto the bearing retainers. Rotate the bearings to allow the two alignment pins to index into their respective notches.

10- Insert a long narrow screwdriver into the exhaust port. Apply a small amount of pressure onto the piston ring with the screwdriver. You should be able to feel the ring compress and expand as pressure is applied and released.

BAD NEWS

If the piston ring fails to expand and return to its original position, the ring was probably broken during the installation process. The crankshaft assembly and piston must then be removed and a new ring installed.

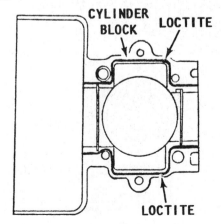

CYLINDER BLOCK   LOCTITE

LOCTITE

12

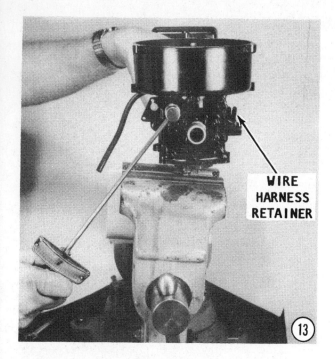

WIRE HARNESS RETAINER

⑬

11- Apply a thin coating of Loctite Type "A" onto the screws securing the reed valve and reed stops in place. Install the reed valves and reed stops in place and secure them with the two screws.

12- Apply a continuous bead of Loctite No. 514, or equivalent, onto the mating surface of the cylinder block.

13- Position the crankcase cover in place onto the cylinder block. Install and tighten the six securing bolts hand-tight. Tighten the bolts alternately and evenly to

the torque value given in the Appendix. Tighten the bolts to the required torque value in the usual manner of three stages, 1/3, 2/3, and final value. Remember, one of the bolts secures the wire harness retainer. Clean away any excess Loctite from the seam of the cover and cylinder block.

14- Slide the lower end cap over the end of the crankshaft and secure it in place with the two attaching bolts. Tighten the bolts alternately and evenly to 50 in lbs (6 Nm) in three stages. Rotate the crankshaft through several revolutions. As the crankshaft is rotated, be sensitive for any binding or "rough" spots.

POWERHEAD "C" INSTALLATION

1- Slide a **NEW** gasket down the driveshaft into position on the driveshaft housing. Now, lower the powerhead onto the driveshaft housing with the lower end of the crankshaft indexed into the hollow core of the driveshaft.

2- Install and tighten the six bolts securing the powerhead to the driveshaft housing.

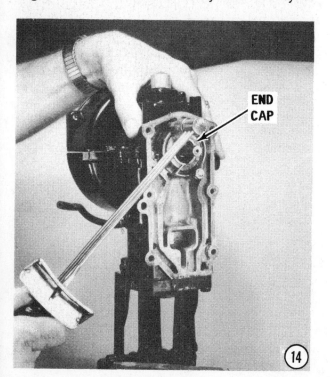

END CAP

⑭

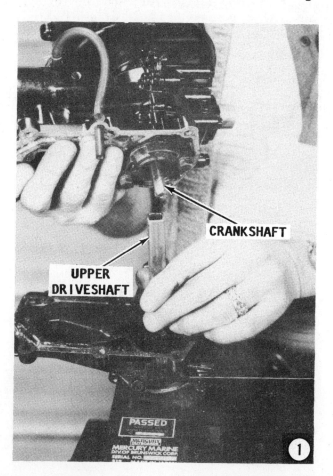

CRANKSHAFT

UPPER DRIVESHAFT

①

POWERHEAD

DRIVESHAFT HOUSING

2

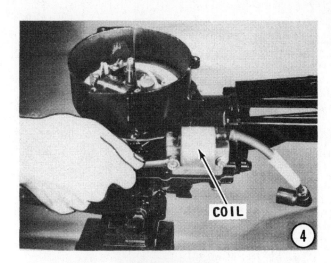

COIL

4

FLYWHEEL

5

3- Hold the stator plate assembly over the powerhead. Feed the wire from the stator plate through the opening in the base of the cylinder block cover. Now lower the stator plate down over the crankshaft and into place. Secure the stator plate with the two Phillips head screws.

4- Install the secondary coil to the side of the powerhead with the two attaching bolts. Connect the wire leading to the stator assembly.

5- Slide the flywheel down the crankshaft with the keyway in the flywheel aligned with the key on the crankshaft. Rotate the flywheel **CLOCKWISE** and check to be sure the flywheel does not make contact with any part of the magneto or the wiring.

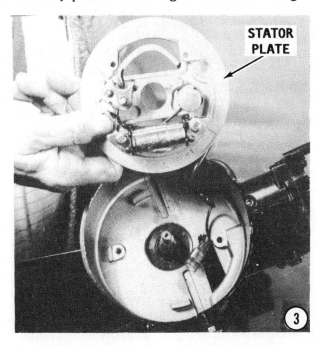

STATOR PLATE

3

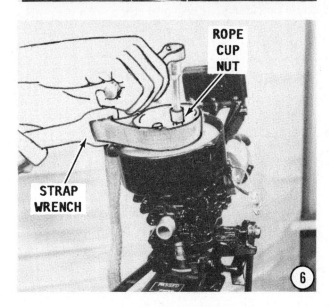

ROPE CUP NUT

STRAP WRENCH

6

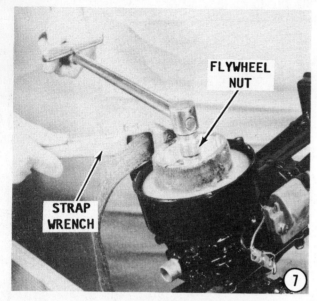

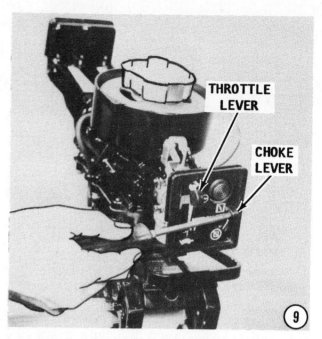

6- Install the rope cup and secure it in place with the three attaching bolts. Use a strap wrench or similar tool to prevent the rope cup from turning while the bolts are being tightened. Thread the flywheel nut onto the crankshaft.

7- Continue to hold the rope cup with the strap wrench and tighten the flywheel nut to the torque value listed in the Appendix. Remove the strap wrench.

8- Slide the carburetor onto the intake manifold and secure it in place with the clamp. Connect the wires from the front plate of the carburetor to the coil with the quick-disconnect fitting.

9- Position the front plate of the carburetor over the throttle and choke levers. Attach the throttle knob to the throttle lever and the choke knob to the choke lever.

10- Place the fuel tank in position behind the powerhead, and then connect the

fuel line to the carburetor fitting. Turn the fuel shut-off valve to the ON position.

11- Install the hand rewind starter onto the powerhead and secure it in place with the three attaching bolts.

12- Thread a new spark plug into the powerhead and tighten to the torque value listed in the Appendix.

13- Install the high-tension lead onto the spark plug.

Mount the engine in a test tank.

CAUTION: Water must circulate through the lower unit to the engine any time the engine is run to prevent damage to the water pump in the lower unit. Just five seconds without water will damage the water pump.

Attempt to start and run the engine without the cowl installed. This will provide the opportunity to check for fuel and oil leaks, without the cowl in place. Follow the break-in procedures.

14- After you are satisfied the engine is operating properly, install both halves of the cowling and secure them in place with the attaching hardware.

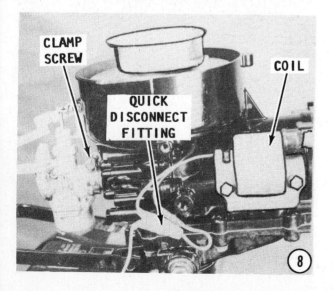

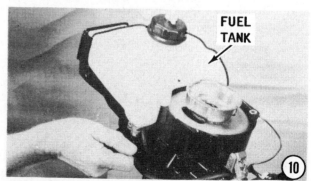

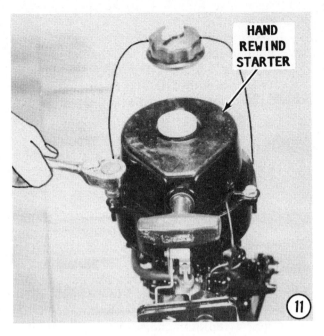

Figure 11

Figure 13

15- Snap the spark plug access cover into place.

Break-in Procedures

As soon as the engine starts, **CHECK** to be sure the water pump is operating. If the water pump is operating, a fine stream will be discharged from the exhaust relief hole at the rear of the drive shaft housing.

DO NOT operate the engine at full throttle except for **VERY** short periods, until after 10 hours of operation as follows:

a- Operate at 1/2 throttle, approximately 2500 to 3500 rpm, for 2 hours.

b- Operate at any speed after 2 hours **BUT NOT** at sustained full throttle until another 8 hours of operation.

c- Mix gasoline and oil during the break-in period, total of 10 hours, at a ratio of 25:1.

d- While the engine is operating during the initial period, check the fuel, exhaust, and water systems for leaks.

e- See Chapter 2 for tuning procedures.

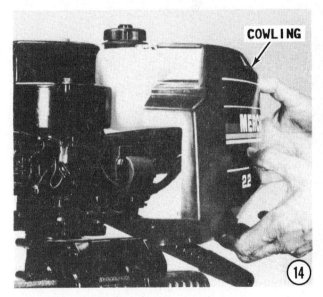

Figure 14

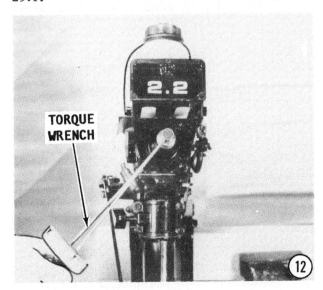

Figure 12

Figure 15

8-5 CLEANING AND INSPECTING ALL POWERHEADS

The success of the overhaul work is largely dependent on how well the cleaning and inspecting tasks are completed. If some parts are not thoroughly cleaned, or if an unsatisfactory unit is allowed to be returned to service through negligent inspection, the time and expense involved in the work will not be justified with peak engine performance and long operating life. Therefore, the procedures in the following sections should be followed in detail and the work performed with patience and attention to detail.

REED BLOCK SERVICE
REED BOX INSTALLED AROUND
THE CRANKSHAFT
POWERHEAD "A"

Secure the reed blocks together with screws and nuts tightened to the torque value given in the Appendix.

Check for chipped or broken reeds. Observe that the reeds are not preloaded or standing open. Satisfactory reeds will not adhere to the reed block surface, but still there is not more than 0.007" (0.18mm) clearance between the reed and the block surface.

DO NOT remove the reeds, unless they are to be replaced. ALWAYS replace reeds in sets. NEVER turn a used reed over to be used a second time.

Check the reed location over the reed block openings to be sure the reed is centered.

See "Reed Stop Settings" in the Specifications in the Appendix and adjust the reed stops as required.

Using a drill bit shank to measure the reed stop height, if a measurement scale is not available. Refer to the Specifications in the Appendix for the proper height.

GOOD WORDS

If the engine shows evidence of having overheated, check the condition of the plastic locating pins. If the pins are damaged (melted) the pins will affect engine performance by poor idle, hard starting, etc.

REED BLOCK SERVICE
EXTERNAL REED BLOCK
POWERHEAD "B"

Disassemble the reed block by first removing the screws securing the reed stops and reeds to the reed block, and then lifting the reed stops and reeds from the block.

Clean the gasket surfaces of the reed block. Check the surfaces for deep grooves, cracks and any distortion that could cause leakage. Replace the reed block if it is damaged.

After new reeds have been installed and the reed stop and attaching screws have been tightened to the required torque value, check the new reeds as outlined in the next step.

Check to be sure the reeds are not preloaded. They should not adhere to the block, and still the clearance between the reed and the block surface should not be more than 0.007" (0.18mm). DO NOT remove the reeds, unless they are to be replaced. ALWAYS replace reeds in sets.

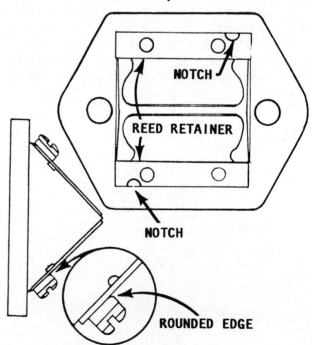

Drawing of a reed plate assembly installed externally on Type "B" powerheads.

NEVER turn used reed over to be used a second time. Install new reeds according to the procedures outlined in this Section.

Check the reed stop opening of each reed stop by measuring the distance from the top of the closed reed to the inside of the reed stop. Compare your measurement with the Specifications in the Appendix. **CAREFULLY** bend the reed stop until the required opening is obtained. **TAKE CARE** not to damage the reed.

REED BLOCK SERVICE
REED VALVES INSTALLED UNDER
THE CRANKSHAFT
POWERHEAD "C"

Inspect the reed valves for signs of wear. The reed valves should fit flush or nearly flush against the seat.

BAD NEWS

Check for signs of wear (indentation marks) on the face of the seat on the crankcase cover. If there is evidence of wear, the crankcase cover (both halves) and the cylinder block **MUST** be replaced. These three items are a matched set, line-bored, and should never be mismatched by using a different crankcase cover or cylinder block.

Measure the reed stop opening, as shown in the accompanying illustration. Replace the reed stop if the opening is not within specifications.

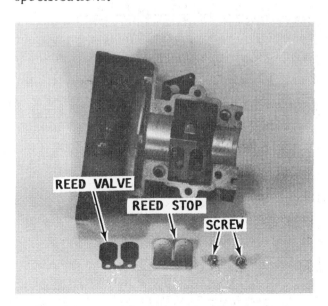

Reeds, reed stops, and attaching screws installed underneath the crankshaft on powerhead "C."

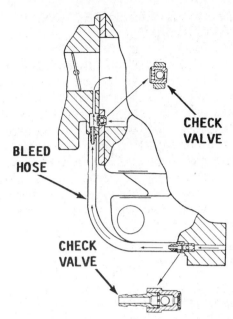

Cross-section drawing of the bleed system installed on some powerhead "C" units.

BLEED SYSTEM SERVICE
(If equipped)

Check the condition of the rubber bleed hose. Replace the hose if it shows signs of deterioration or leakage. Check the operation of the two check valves. The air/fuel mixture should be able to pass through the valve in only one direction. The check valves are located in the mating surfaces of the intake manifold-to-reed block and the crankcase-to-carburetor.

Defective check valves cannot be serviced. If defective, they **MUST** be replaced.

CRANKSHAFT SERVICE

Inspect the splines for signs of abnormal wear. Check the crankshaft for straightness. Inspect the crankshaft oil seal surfaces to be sure they are not grooved, pitted or scratched. Replace the crankshaft if it is

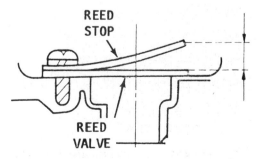

If the reed stop opening is not as specified in the Appendix, GENTLY bend the reed stop to achieve the correct dimension.

Crankshaft assembly for the powerhead "C". The rod is manufactured as an integral part of, and cannot be separated from, the crankshaft.

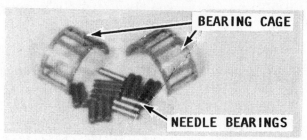

Main needle bearings and cage. The bearings should be thoroughly cleaned and closely inspected.

One piece needle bearing cage. If the cage is damaged in any manner, the unit must be replaced.

severely damaged or worn. Check all crankshaft bearing surfaces for rust, water marks, chatter marks, uneven wear or overheating. Clean the crankshaft surfaces with 320-grit carborundum cloth. **NEVER** spindry a crankshaft ball bearing with compressed air.

SPECIAL WORDS

The connecting rod of a type "C" powerhead is an integral part of the crankshaft and **CAN-NOT** be removed. Therefore, if either the crankshaft or the rod is no longer fit for service, the complete unit must be purchased and installed.

Clean the crankshaft and crankshaft ball bearing with solvent. Dry the parts, but not the ball bearing, with compressed air. Check the crankshaft surfaces a second time. Replace the crankshaft if the surfaces cannot be cleaned properly for satisfactory service. If the crankshaft is to be installed for service, lubricate the surfaces with light oil. **DO NOT** lubricate the crankshaft ball bearing at this time.

CRANKSHAFT AND END CAP BEARINGS (IF USED)

After the crankshaft has been cleaned, grasp the outer race of the crankshaft ball bearing installed on the lower end of the crankshaft, and attempt to work the race back-and-forth. There should not be excessive play. A very slight amount of side play is acceptable because there is only about 0.001" (.025mm) clearance in the bearing.

Lubricate the ball bearing with light oil. Check the action of the bearing by rotating the outer bearing race. The bearing should have a smooth action and no rust stains. If the ball bearing sounds or feels rough or catches, the bearing should be removed and discarded.

Cleaning the inside diameter of the rod and rod cap with crocus cloth.

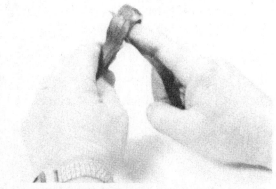

Cleaning the inside diameter of the piston pin end with crocus cloth.

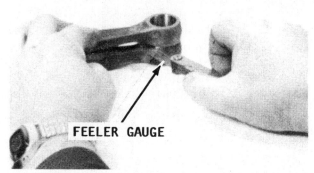

Checking for rod warpage at the piston pin end. This is accomplished by laying one rod on top of a known good rod, and then checking for clearance between the two with a feeler gauge. There should be NO clearance.

Clean the crankshaft centermain roller bearings with solvent, and then dry them thoroughly **BUT NOT** with compressed air. Lubricate the bearings with light-weight oil. **NEVER** intermix halves of upper and lower crankshaft centermain roller bearings. The bearings **MUST** be replaced only in pairs.

Inspect the centermain roller bearings. Replace the bearings in pairs if they are rusted, fractured, worn, galled, or badly discolored.

Clean the crankshaft roller bearings installed in the upper end cap with solvent, and then dry them **BUT NOT** with compressed air. Lubricate the bearings with light-weight oil.

Inspect the upper end cap roller bearing to be sure it is not rusted, fractured, worn, galled, or badly discolored. If the bearing is damaged, it should be removed and discarded.

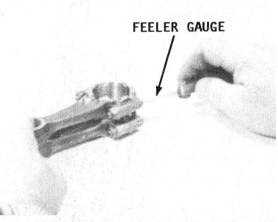

Check for rod warpage by placing one on top of a known good rod, and then attempting to insert a feeler gauge between the two surfaces. There should be NO clearance between the two rods.

CONNECTING ROD SERVICE

SPECIAL WORDS FOR POWERHEAD "C"

The connecting rod of the type "C" powerhead is an integral part of the crankshaft and **CAN-NOT** be removed. Therefore, only the upper bearing surface may be serviced, if necessary. If either the crankshaft or the rod is no longer fit for service, the complete unit must be purchased and installed.

ALL OTHER UNITS

Stand each connecting rod on a surface plate and check the alignment. The rod is bent and unfit for further service, if:

a- Light can be seen under any portion of the machined surfaces, the surfaces which mate with the rod cap.

b- The rod has a slight wobble on the plate.

c- A 0.002" (.051mm) feeler gauge can be inserted between the machined surface and the surface plate,

Inspect the connecting rod bearings for rust or signs of bearing failure. **NEVER** intermix new and used bearings. If even one bearing in a set needs to be replaced, all bearings at that location **MUST** be replaced.

Inspect the bearing surface of the rod and the rod cap for rust and pitting.

Inspect the bearing surface of the rod and the rod cap for water marks. Water marks are caused by the bearing surface being subjected to water contamination, which causes "etching". The etching resembles the size of the bearing as shown in the accompanying illustration.

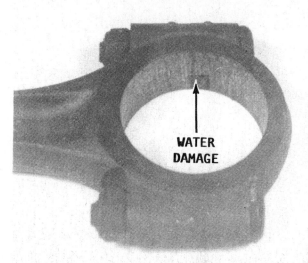

Water scoring on the inside of the rod and rod cap. Such damage is caused by water in the crankcase. This rod set MUST be replaced.

Inspect the bearing surface of the rod and rod cap for signs of spalling. Spalling is the loss of bearing surface, and resembles flaking or chipping. The spalling condition will be most evident on the thrust portion of the connecting rod in line with the I-beam. Bearing surface damage is usually caused by improper lubrication.

Check the bearing surface of the rod and rod cap for signs of chatter marks. This condition is identified by a rough bearing surface resembling a tiny washboard. The condition is caused by a combination of low-speed low-load operation in cold water, and is aggravated by inadequate lubrication and improper fuel. Under these conditions, the crankshaft journal is hammered by the connecting rod. As ignition occurs in the cylinder, the piston pushes the connecting rod with tremendous force, and this force is transferred to the connecting rod journal. Since there is little or no load on the crankshaft, it bounces away from the connecting rod. The crankshaft then remains immobile for a split second, until the piston travel causes the connecting rod to catch up to the waiting crankshaft journal, then hammers it. In some instances, the connecting rod crankpin bore becomes highly polished.

While the engine is running, a "whirr" and/or "chirp" sound may be heard when the engine is accelerated rapidly from idle speed to about 1500 rpm, then quickly returned to idle. If chatter marks are discovered, the crankshaft and the connecting rods should be replaced.

Inspect the bearing surface of the rod and rod cap for signs of uneven wear and possible overheating. Uneven wear is usually caused by a bent connecting rod or by improper shimming of the crankshaft end play, failure to maintain the same amount of shim material under each end cap. This improper shimming causes the crankshaft journal not to be centered over the cylinder bore. Overheating is identified as a bluish bearing surface color and is caused by inadequate lubrication or operating the engine at excessive high rpm.

Service the connecting rod bearing surfaces according to the following procedures and precautions.

a- Align the etched marks on the connecting rod with the etched marks on the connecting rod cap.

b- Tighten the connecting rod cap attaching bolts securely.

c- Two types of bearings are used on the crankpin end of the rod. One is a non-caged type with individual needles. The other is a caged type with separate rollers. Clean the caged type with 320 grit carborundum cloth.

d- Use **ONLY** crocus cloth to clean bearing surface at the crankshaft end of the connecting rod. **NEVER** use any other type of abrasive cloth on the caged type.

e- Clean the inside diameter of the piston pin end of the connecting rod with crocus cloth.

f- Clean the connecting rod **ONLY** enough to remove marks. **DO NOT** continue, once the marks have disappeared.

Piston and rod damaged from operating the engine at too high an rpm without sufficient load on the propeller shaft. This combination caused the powerhead to literally "blow apart". Operating the engine above an idle speed with a flush attachment connected to the lower unit could result in the same type of internal destruction.

The rings on this piston became stuck due to lack of adequate lubrication, incorrect timing, or overheating.

g- Clean the piston pin end of the connecting rod using the method described in Step "e", above, using 320 grit carborundum cloth.

h- Thoroughly wash the connecting rods to remove abrasive grit. After washing, check the bearing surfaces a second time.

i- If the connecting rod cannot be cleaned properly, it should be replaced.

j- Lubricate the bearing surfaces of the connecting rods with light-weight oil to prevent corrosion.

PISTON SERVICE

Inspect each piston for evidence of scoring, cracks, metal damage, cracked piston pin boss, or worn pin boss. Be especially critical during inspection if the engine has been submerged. If the piston pin is bent, the piston and pin **MUST** be replaced as a set for two reasons. First, a bent pin will damage the boss when it is removed. Secondly, a piston pin is not sold as a separate item.

Check the piston ring grooves for wear, burns, distortion, or loose locating pins. During an overhaul, the rings should be replaced to ensure lasting repair and proper engine performance after the work has been completed. Clean the piston dome, ring grooves and the piston skirt. Clean carbon deposits from the ring grooves using the recessed end of a broken piston ring.

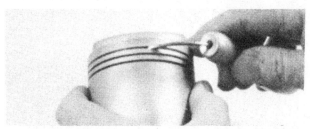

Cleaning the piston ring grooves using the end of a broken ring inserted into a wooden handle, for safety.

*The pick is pointing to the locating pin in the piston ring groove. The ring **MUST** straddle this pin to prevent it from rotating around the piston. If the open end of the ring is not properly positioned, the ring will break when the piston is installed into the cylinder.*

The pitted damage to this piston was caused by the needle bearings working loose from the piston pin.

NEVER use a rectangular ring to clean the groove for a tapered ring, or use a tapered ring to clean the groove for a rectangular ring.

NEVER use an automotive-type ring groove cleaner, because such a tool may loosen the piston ring locating pins.

Clean carbon deposits from the top of the piston using a soft wire brush, carbon removal solution or by sand blasting. If a wire brush is used, **TAKE CARE** not to burr or round machined edges. Clean the piston skirt with crocus cloth.

Inspect the piston ring locating pins to be sure they are tight. There is one locating pin in each ring groove. If the locating pins are loose, the piston must be replaced.

SPECIAL WORDS

The pistons of the Model 6.0, 8.0, 9.9, 15 and 210 cc do not have a locating pin in the ring groove.

It is believed, this crown siezed with the cylinder wall when the unit was operated at high rpm and the timing was not adjusted properly. At the same instant, the rod apparently pulled the lower part of the piston downward, severing it from the crown.

Ring End Gap Clearance

Check each ring to be sure the end gap is not excessive. The end gap may be checked by placing the ring squarely in the cleaned cylinder bore, and then measuring the end gap with a feeler gauge, as shown in the accompanying illustration on this page. Standard acceptable end gap is 0.005" (0.13mm) per inch of bore. To determine the exact amount of end gap, simply multiply the cylinder bore by 0.005".

Example: Bore = 2.00" times 0.005" equals an acceptable end gap of 0.010" (0.26mm)

Oversize Piston and Piston Rings

Scored cylinder blocks can be saved for further service by reboring and installing oversize pistons and piston rings. **HOW-EVER**, if the scoring is over 0.0075" deep, the block cannot be effectively rebored for continued use. **ONE MORE WORD:** Oversize pistons and rings are not available for all engines. Check with the parts department at your local dealer for the model engine you are servicing.

This sectioned cylinder shows an ideal cross hatch pattern on the cylinder wall. The pattern is necessary to seat the ring/s against the cylinder wall to provide an adequate seal for maximum compression.

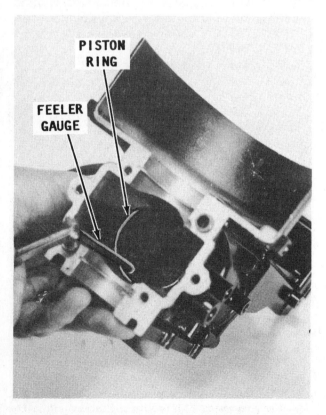

PISTON RING

FEELER GAUGE

Using a feeler gauge to measure ring gap while the ring is in the cylinder, as explained in the text.

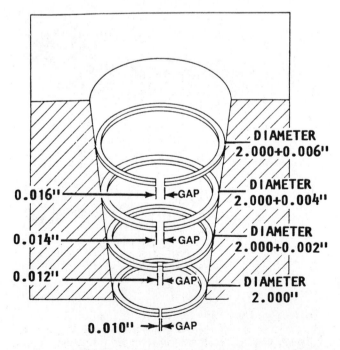

DIAMETER 2.000+0.006"

0.016"   GAP   DIAMETER 2.000+0.004"

0.014"   GAP   DIAMETER 2.000+0.002"

0.012"   GAP   DIAMETER 2.000"

0.010"   GAP

The cylinder taper drastically affects ring end gap, as shown in this cross-section line drawing.

HONING PROCEDURES

To ensure satisfactory engine perfomance and long life following the overhaul work, the honing work should be performed with patience, skill, and in the following sequence:

a- Follow the hone manufacturer's recommendations for use of the hone and for cleaning and lubricating during the honing operation.

b- Pump a continuous flow of honing oil into the work area. If pumping is not practical, use an oil can. Apply the oil generously and frequently on both the stones and work surface.

c- Begin the stroking at the smallest diameter. Maintain a firm stone pressure against the cylinder wall to assure fast stock removal and accurate results.

d- Expand the stones as necessary to compensate for stock removal and stone wear. The best cross-hatch pattern is ob-

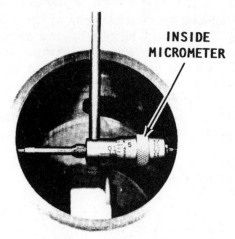

INSIDE MICROMETER

Using a hone to clean the cylinder walls. The secret of honing is to keep the hone moving in long even strokes the full length of the cylinder AND to keep the stones wet with an ample amount of lubricant.

tained using a stroke rate of 30 complete cycles per minute. Again, use the honing oil generously.

e- Hone the cylinder walls **ONLY** enough to de-glaze the walls.

f- After the honing operation has been completed, clean the cylinder bores with hot water and detergent. Scrub the walls with a stiff bristle brush and rinse thoroughly with hot water. The cylinders **MUST** be thoroughly cleaned to prevent any abrasive material from remaining in the cylinder bore. Such material will cause rapid wear of new piston rings, the cylinder bore, and the bearings.

Using a hone to clean the cylinder walls. The secret of honing is to keep the hone moving in long even strokes the full length of the cylinder AND to keep the stones wet with an ample amount of lubricant.

Cleaning the crankshaft mating surface of a 2-cylinder powerhead with solvent and a rag. A tool or abrasive material should NEVER be used to clean these surfaces.

g- After cleaning, swab the bores several times with engine oil and a clean cloth, and then wipe them dry with a clean cloth. **NEVER** use kerosene or gasoline to clean the cylinders.

h- Clean the remainder of the cylinder block to remove any excess material spread during the honing operation.

CYLINDER BLOCK SERVICE

Inspect the cylinder block and cylinder bores for cracks or other damage. Remove carbon with a fine wire brush on a shaft attached to an electric drill or use a carbon remover solution.

STOP: If the cylinder block is to be submerged in a carbon removal solution, the crankcase bleed system **MUST** be removed from the block to prevent damage to hoses and check valves.

Use an inside micrometer or telescopic gauge and micrometer to check the cylinders for wear. Check the bore for out-of-round and/or oversize bore. If the bore is tapered, out-of-round or worn more than 0.003" - 0.004" (0.08 - 0.10mm) the cylinders should be rebored to 0.015" (0.38mm) oversize and oversize pistons and rings installed.

GOOD WORDS

Oversize piston weight is approximately the same as a standard size piston. There-

fore, it is **NOT** necessary to rebore all cylinders in a block just because one cylinder requires reboring. The APBA (American Power Boat Association) accepts and permits the use of 0.015" (0.38mm) oversize pistons.

Hone the cylinder walls lightly to seat the new piston rings, as outlined in the Honing Procedures Section in this chapter. If the cylinders have been scored, but are not out-of-round or the sleeve is rough, clean the surface of the cylinder with a cylinder hone as described in Honing Procedures.

SPECIAL WORDS:

If overheating has occurred, check and resurface the spark plug end of the cylinder block, if necessary. This can be accomplished with 240-grit sand paper and a small flat block of wood.

Cylinder sleeves are an integral part of the die cast cylinder block and **CANNOT** be replaced. In other words, the cylinder cannot be resleeved. The cylinder may be rebored to 0.015" (0.38mm) oversize, unless it is scored more than 0.0075" (0.16mm) deep.

Check the bleed holes inside the transfer ports to be sure the plastic restrictors are in the upper two holes. If these restrictors are missing, it will cause a flooding-type condition in the upper cylinders and affect idle-speed operation.

Many times the manufacturer recommends the use of Loctite (a sealing compound) in place of a gasket when an airtight seal is required. TAKE CARE when using such a substance to prevent an excess of the material onto the surface. Such an excess could find it way into the crankcase and possibly cause plugged reeds or small pellets in the combustion chamber.

Checking the crankshaft "end play" with a feeler gauge. An excessive amount of "play" will cause the powerhead to "clank" during operation. Too little play will cause excessive wear and crankshaft failure. The proper amount of "play" will extend powerhead life, increase efficiency, and the unit will "purr" like a kitten.

9
LOWER UNIT

my 7.5 & 9.8
are type B
page 9.25

9-1 CHAPTER ORGANIZATION

The lower unit is considered as that part of the outboard below the exhaust housing. The unit contains the propeller shaft, the driven and pinion gears, the drive shaft from the powerhead and the water pump. The lower unit of all 2-cylinder units and the 4 hp 1-cylinder unit are equipped with shifting capabilites. The forward and reverse gears together with the clutch, shift assembly, and related linkage are all housed within the lower unit.

Shifting on units equipped with remote controls is accomplished through a cable arrangement from a shift box, installed near the helm, to the engine. The cable hookup involves two individual cables, one for shifting and the other for throttle control.

The lower unit may be removed and serviced without disturbing the remainder of the outboard motor.

SPECIAL WORDS

The water pump on the 2.2 hp unit is located on the propeller shaft. Therefore, the impeller may be removed and a new impeller installed without disturbing any other areas of the lower unit.

CHAPTER COVERAGE

Three distinctly different lower units, identified as Type A, Type B, and Type C are covered in this chapter with separate sections for each. Each section is presented with complete detailed instructions for removal, disassembly, cleaning and inspecting, assembling, adjusting, and installation of only one unit. There are no cross-references. Each section is complete from removal of the first item to final test operation.

Check the Lower Unit Type and Backlash Table in the Appendix for the lower unit used on the model being serviced.

BE SURE to check the section heading for the model unit being serviced to ensure the correct procedures are followed.

The sections and the type unit covered are as follows:

9-4 TYPE A -- No reverse capability -- operator swings the outboard 180° to move boat sternward.

9-5 Type B -- Reverse capability -- operator may shift into reverse gear.

9-6 Type C -- Reverse capability -- unique shift arrangement.

9-7 Type D -- No reverse capability -- water pump installed on propeller shaft.

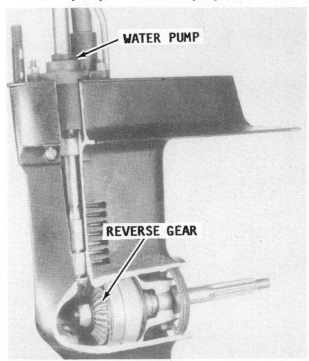

WATER PUMP

REVERSE GEAR

Cutaway view of a lower unit without a forward gear. This type illustration is used extensively throughout this chapter as an aid to seeing and understanding the working relationship of the various parts.

ILLUSTRATIONS

Because this chapter covers such a wide range of models over an extended period of time, the illustrations included with the procedural steps are those of the most popular lower units. In some cases, the unit being serviced may not appear to be identical with the unit illustrated. However, the step-by-step work sequence will be valid in all cases. If there is a special procedure for a unique lower unit, the differences will be clearly indicated in the step.

SPECIAL WORDS

All threaded parts are right-hand unless otherwise indicated.

If there is any water in the lower unit or metal particles are discovered in the gear lubricant, the lower unit should be completely disassembled, cleaned and inspected.

9-2 TROUBLESHOOTING

Troubleshooting **MUST** be done **BEFORE** the unit is removed from the powerhead to permit isolating the problem to one area. Always attempt to proceed with troubleshooting in an orderly manner. The shot-in-the-dark approach will only result in wasted time, incorrect diagnosis, frustration, and

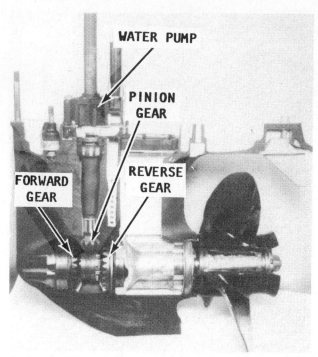

Classroom type cutaway view of a lower unit with major parts, including the propeller and water pump, installed. Notice how the forward, reverse and pinion gears all are "bevel cut".

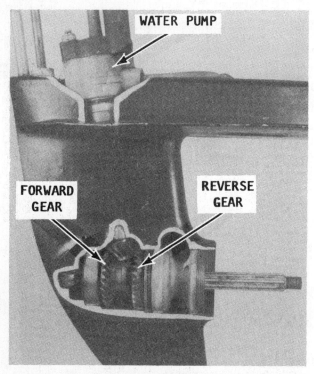

Lower unit used with the small horsepower engines. This unit has forward, neutral, and reverse gear.

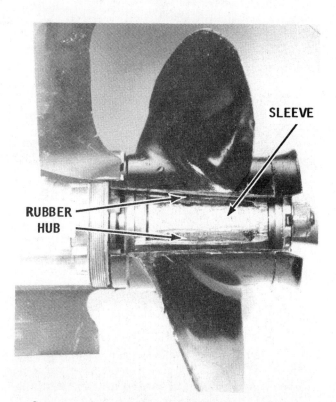

*Cutaway view showing the rubber hub and sleeve. The rubber hub protects the lower unit if the propeller should strike an underwater object. If the rubber hub loses its holding power with the inner hub of the propeller, the propeller hub **MUST** be replaced.*

RUBBER HUB

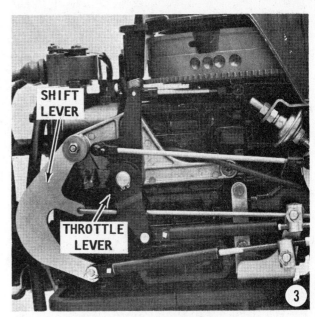

SHIFT LEVER

THROTTLE LEVER

replacement of unnecessary parts.

The following procedures are presented in a logical sequence with the most prevalent, easiest, and less costly items to be checked listed first.

1- Check the propeller and the rubber hub. See if the hub is shredded. If the propeller has been subjected to many strikes against underwater objects, it could slip on its hub. If the hub appears to be damaged, replace it with a **NEW** hub. Replacement of the hub must be done by a propeller rebuilding shop equipped with the proper tools and experience for such work.

2- **Shift mechanism check:** Verify that the ignition switch is **OFF**, to prevent possible personal injury, should the engine start.   Shift the unit into **REVERSE** gear (if so equipped), and at the same time have an assistant turn the propeller shaft to ensure the clutch is fully engaged. If the shift handle is hard to move, the trouble may be in the lower unit remote control cable, or the shift box.

3- **Isolate the problem:** Disconnect the remote-control cable at the engine and then lift off the remote-control shift cable. Operate the shift lever. If shifting is still hard, the problem is in the shift cable or control box, see Chapter 11. If the shifting

KEY SWITCH

merControl

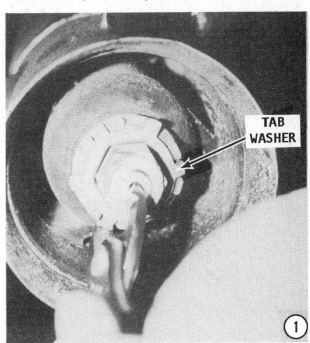

TAB WASHER

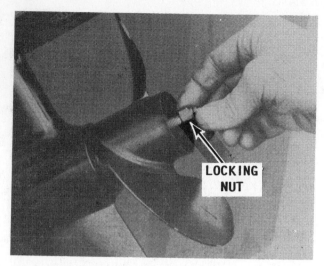

A self-locking propeller nut loses its locking ability and should not be used a second time.

feels normal with the remote-control cable disconnected, the problem must be in the lower unit. To verify the problem is in the lower unit, have an assistant turn the propeller and at the same time move the shift cable back-and-forth. Determine if the clutch engages properly.

9-3 PROPELLER REMOVAL

1- Bend the locking tabs forward out of the locking washer. Some lower units have a locknut installed instead of a lockwasher. **NEVER** pry on the edge of the propeller. Any small distortion will affect propeller performance.

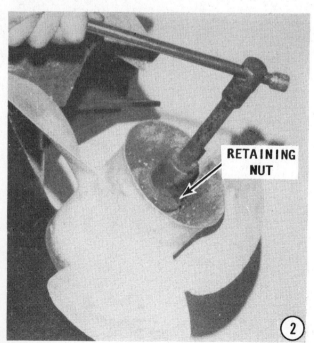

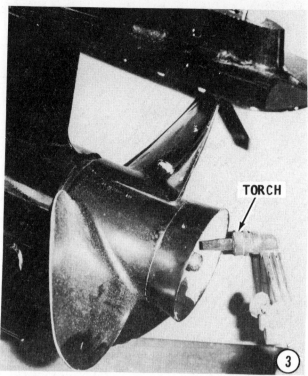

2- Place a block of wood between one blade of the propeller and the anti-cavitation plate to keep the shaft from turning. Use a socket and breaker bar to loosen the retaining nut. Remove the nut, tab washer, splined washer, and then the propeller.

3- If the propeller is frozen to the shaft, heat must be applied to the shaft to melt out the rubber inside the hub. Using heat will destroy the hub, but there is no other way. As heat is applied, the rubber will expand and the propeller will actually be blown from the shaft. Therefore, **STAND CLEAR** to avoid personal injury.

4- Use a knife and cut the hub off the inner sleeve.

5- The sleeve can be removed by cutting it with a hacksaw, or it can be removed with a puller. Again, if the sleeve is frozen, it may be necessary to apply heat. Remove

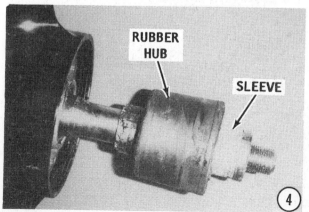

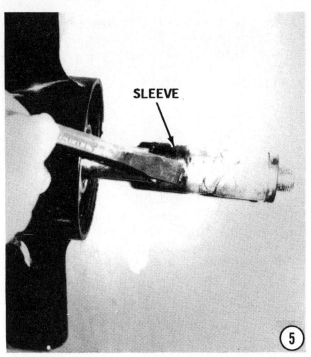

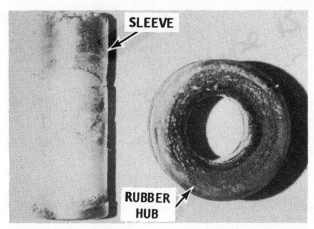

This rubber hub had to be cut, the sleeve heated and then cut loose, because of extensive corrosion.

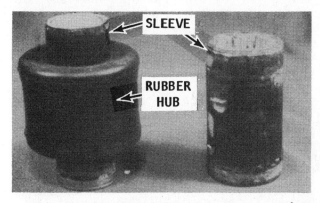

the thrust hub from the propeller shaft.

Procedures for propeller installation are given at the end of this chapter, after the lower unit has been installed.

The left sleeve and rubber hub was successfully removed with a puller. The sleeve on the right was removed with a chisel after the rubber hub was cut away.

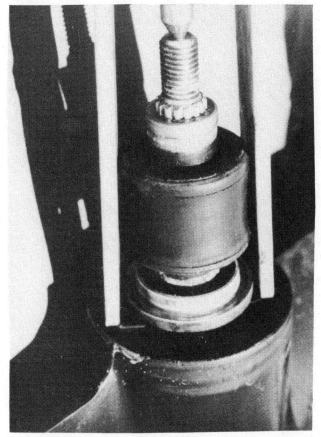

A standard puller in position on the thrust washer in preparation to removing the sleeve and rubber hub from the propeller shaft. A puller is required because sealing compound was not used on the shaft during the previous installation.

Inside view of a propeller removed for service work. The hub and sleeve remained on the propeller shaft.

9-4 LOWER UNIT TYPE "A"
NO REVERSE GEAR
(See Lower Unit Table in Appendix)

DESCRIPTION

The Type "A" lower unit is a direct drive unit. This means the unit has the capability of being shifted only between a neutral position and forward. The 3.6 model does not have shift capability even though all the parts are present. There is no provision in the lower unit for operation in reverse gear. Reverse action of the propeller is accomplished by the boat operator swinging the engine with the tiller handle a full 180° and holding it in this position while the boat moves sternward. When the operator is ready to move forward again, he simply swings the tiller handle back to the normal forward position.

WORDS OF WISDOM

Before beginning work on the lower unit, take time to **READ** and **UNDERSTAND** the information presented in Section 9-1, this chapter, and check the Lower Unit Type and Backlash Table in the Appendix to ensure the proper procedures are being followed.

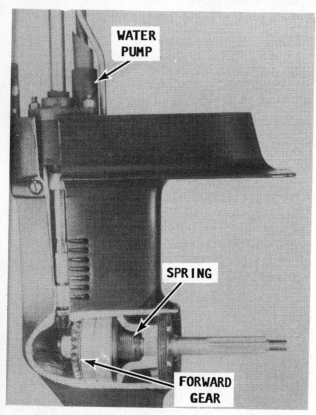

Lower unit from a small horsepower installation with major parts exposed.

Disconnect the high tension spark plug leads and remove the spark plugs before working on the lower unit.

LOWER UNIT REMOVAL

The shift shaft extends into the torpedo bore of the lower unit. Therefore, the lower unit **MUST** be removed from the exhaust housing before the propeller shaft assembly can be removed from the gear housing.

1- Position a suitable container under the lower unit, and then remove the **FILL** screw and the **VENT** screw. Allow the gear lubricant to drain into the container. As the lubricant drains, catch some with your fingers, from time-to-time, and rub it between your thumb and finger to determine if any metal particles are present. If metal is detected in the lubricant, the unit must be completely disassembled, inspected, and the damaged parts replaced.

Check the color of the lubricant as it drains. A whitish or creamy color indicates the presence of water in the lubricant. Check the drain pan for signs of water separation from the lubricant. The presence of any water in the gear lubricant is bad news. The unit must be completely disassembled, inspected, the cause of the problem determined, and then corrected.

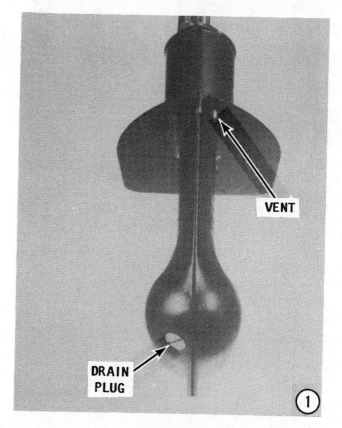

Propeller Removal

2- Check to be sure the spark plugs have been removed to prevent the possibility of engine start during propeller removal. Remove the propeller nut by first placing a block of wood between one of the propeller blades and the anti-cavitation plate to prevent the propeller from turning, and then remove the nut. Remove the splined washer.

Remove the outer thrust hub from the propeller shaft. If the thrust hub is stubborn and refuses to budge, use two **PADDED** pry bars on opposite sides of the hub and work the hub loose. **TAKE CARE** not to damage the lower unit. Remove the propeller. If the propeller is "frozen" to the shaft, see Section 9-3 for special procedures to break it loose.

Remove the inner thrust hub. If this hub is also stubborn, use padded pry bars and work the hub loose. Again, **TAKE CARE** not to damage the lower unit.

Lower Unit Removal

3- Remove the two locknuts securing the lower unit to the exhaust housing. Separate the lower unit slightly from the shaft housing in order to completely remove the upper locknut. After the upper locknut has been removed, separate the two housings. Notice how the water tube and the shift shaft remain in the exhaust housing. **NEV-**

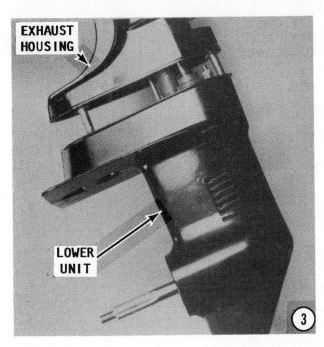

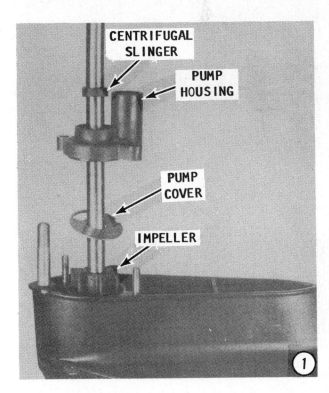

ER lift or carry the lower unit by the driveshaft because the driveshaft may pull out of the lower unit.

Water Pump Removal

1- Remove the two locknuts and flat washers securing the water pump cover to the lower unit. Slide the rubber centrifugal slinger, water pump cover and the face plate upward on the driveshaft.

Reach inside the water pump housing, and slide the impeller upward out of the

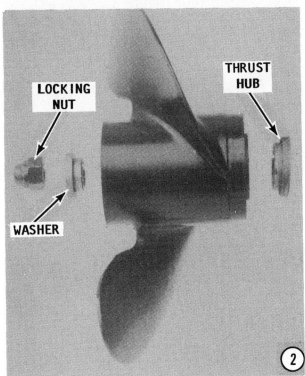

housing. Now, slide the face plate, water pump cover, centrifugal slinger, and impeller off the driveshaft.

2- Pull the driveshaft out of the lower unit. Observe that the impeller key is still attached to the driveshaft.

3- Remove the locknut and flat washer securing the water pump base to the gear housing. **CAREFULLY** pry the water pump base loose from the housing with a screwdriver, as shown. Remove and **DISCARD** the water pump base gasket from the lower unit. Clean any part of the gasket stuck to the water pump base surface.

Water Pump Disassembling

4- Work the oil seal out of the water pump cover with a "bent-end" screwdriver. Remove the water tube seal from the water pump cover by first removing the water tube guide from the water pump cover. This can be accomplished by pulling and turning the guide with a pair of pliers.

5- Remove the water tube seal by pushing in and up on the seal tab with a small tapered punch.

6- Work the oil seal out of the water pump base using a "bent-end" screwdriver.

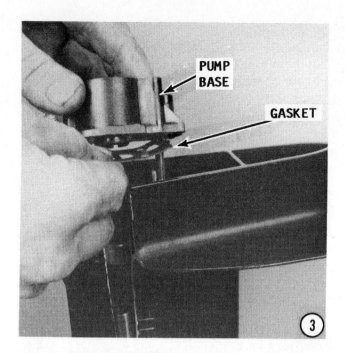

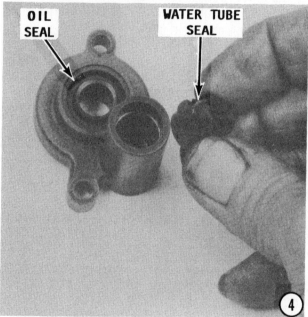

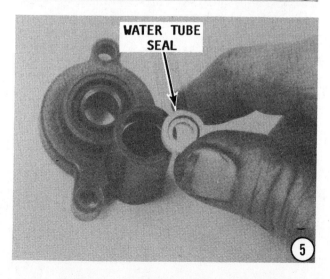

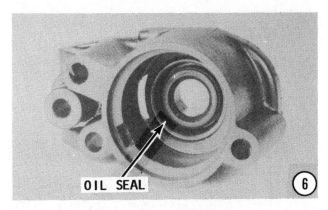

OIL SEAL   6

7- Remove the flat washer and seal from the shift shaft hole in the water pump base. **DISCARD** the seal, but retain the flat washer.

Propeller Shaft and Bearing Carrier Removal

8- Clamp the lower unit in a vertical position in a vise equipped with soft jaws. Remove the lower unit cover nut using Cover Nut Tool C-91-74588.

9- If the nut refuses to move, carefully apply heat to the housing, and try again while it is hot. If the nut still cannot be moved, drill the nut as shown in the accompanying captioned illustration.

10- Clamp the propeller shaft assembly in a vise equipped with soft jaws to protect the shaft. Now, strike the lower unit approximately midway between the anti-cavitation plate and the torpedo bore with a soft mallet to remove the propeller shaft assembly from the lower unit. If the bearing carrier is frozen, it may be necessary to carefully apply heat to the housing in order to break the carrier loose.

SPECIAL WORDS: When the propeller shaft assembly is removed from the gear housing, the bearing carrier alignment key will come out with shaft assembly. However, the drive gear bearing shim may or may not come out with the shaft assembly.

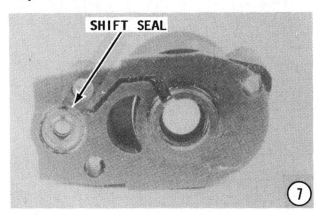

SHIFT SEAL   7

COVER NUT TOOL   8

11- Remove the pinion gear, thrust bearing, and washer from the torpedo bore. Remove any shim material left in the torpedo bore that did not come out with the propeller shaft assembly.

12- Release the propeller shaft from the vise. Separate and slide the bearing carrier,

COVER NUT   9

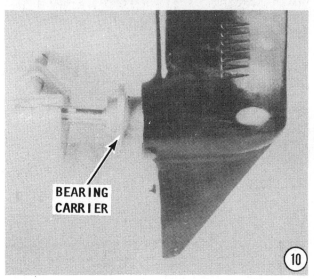

BEARING CARRIER   10

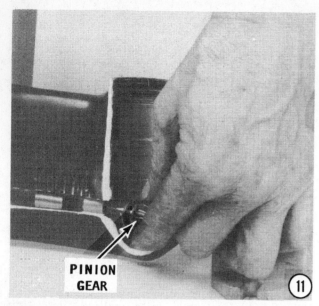

PINION GEAR ⑪

ROLLER BEARING ⑬

O-ring, and spacer from the propeller shaft. **DISCARD** the O-ring.

Bearing Carrier Disassembling

13- Clamp the bearing carrier in a vise equipped with soft jaws, as shown. Tap the oil seal out of the carrier with a punch and hammer. **DISCARD** the oil seal. Inspect the bearing carrier roller bearing. Remove the bearing carrier roller bearing **ONLY** if it is damaged and unfit for service.

14- If inspection of the bearing carrier roller bearing reveals the bearing must be replaced, remove the bearing by pressing it out of the bearing carrier with a suitable mandrel.

Propeller Shaft Disassembling

15- Clamp the propeller shaft in a vise equipped with soft jaws to protect the shaft. Remove and **DISCARD** the drive gear locknut. Remove the special flat washer from the end of the propeller shaft. Retain the washer.

16- Remove the drive gear assembly from the propeller assembly by pulling and turning the drive gear assembly **COUNTER-CLOCKWISE**. Inspect the drive gear ball bearing. Remove the ball bearing from the drive gear **ONLY** if the bearing is damaged and unfit for service.

17- Pull and turn the clutch spring **COUNTERCLOCKWISE** to remove it from the propeller shaft.

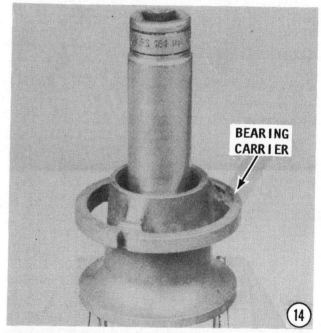

BEARING CARRIER ⑭

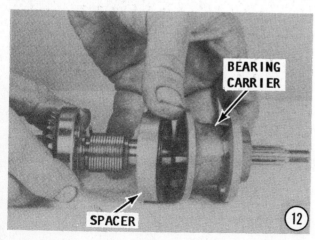

BEARING CARRIER

SPACER ⑫

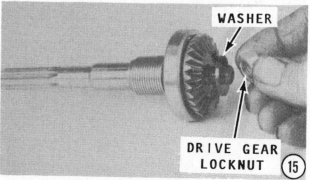

WASHER

DRIVE GEAR LOCKNUT ⑮

Drive Gear Disassembling

18- If inspection reveals the drive gear ball bearing requires replacement: Clamp the drive gear in a vise equipped with soft jaws. Use Retainer Tool C-91-74589 and remove the bearing retainer nut. If the tool is not available, use a punch, by setting it into one of the slots of the retainer nut, and then backing the nut out by striking the punch with a hammer.

19- Release the drive gear from the vise and install Universal Puller Plate C-91-22115 between the drive gear and the ball bearing. Now, position the drive gear, bearing and puller plate on an arbor press. Use a suitable mandrel and press the drive gear out of the ball bearing.

Exhaust Housing Bearing and Pinion Gear Bearing Removal

20- DO NOT remove these bearings unless they are damaged and unfit for service. Special Tools: Bearing Tool C-91-77644 and a piece of 1/4x20-8" threaded stock are required. If the tool is not available, a 3/8" round Allen head bolt may be used AFTER the center of the bolt has been drilled and

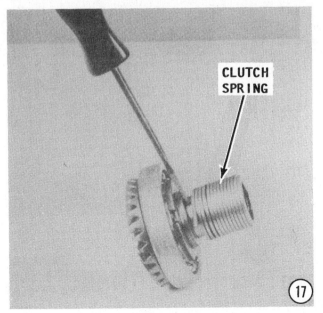

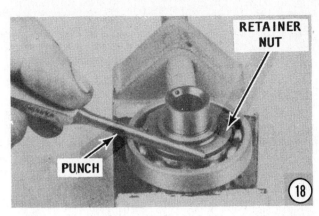

taped to 1/4x20 thread. Clamp the lower unit in a vertical position in a vise equipped with soft jaws. Thread a 1/4x20 hex nut with a flat washer onto one end of the threaded stock material. Insert the other end through Puller Plate C-91-29310, the drive shaft bearing, pinion coupler and the pinion bearing. If the special tool is not available, a piece of raw stock approximately 1/8x1x2" long with a 3/8" hole drilled in the center may be used.

After the piece of stock material is in place with the end extending into the torpedo bore, thread Bearing Tool C-91-77644, or the special tool made in the first part of this step, onto that end of the stock. Tight-

Homemade tools used to remove both driveshaft bearings and to install the lower driveshaft bearing (left). The other tool (right) is used to install the upper driveshaft bearing. The text explains how these two items can be made in most shops.

Homemade parts used to remove the driveshaft bearings and coupler. The parts are assembled on the lower unit.

en the upper hex nut against the puller plate and draw the pinion bearing, coupler, and upper drivesahft bearing up and out of the lower unit.

CLEANING AND INSPECTING

Clean all water pump parts with solvent, and then dry them with compressed air. Inspect the water pump cover and base for cracks and distortion, possibly caused from overheating. Inspect the face plate and water pump insert for grooves and/or rough surfaces. If possible, **ALWAYS** install a new water pump impeller while the lower is disassembled. A new impeller will ensure extended satisfactory service and give "peace of mind" to the owner. If the old

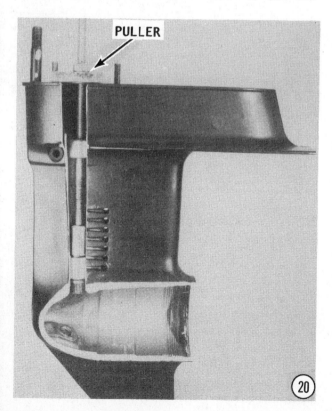

PULLER

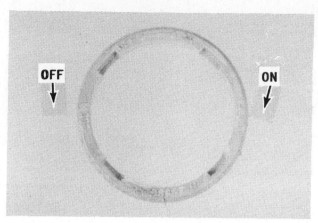

Bearing carrier retaining nut. The rotation direction for installation or removal is stamped on the face of the nut.

impeller must be returned to service, **NEVER** install it in reverse to the original direction of rotation. Installation in reverse will cause premature impeller failure.

Inspect the impeller side seal surfaces and the ends of the impeller blades for cracks, tears, and wear. Check for a glazed or melted appearance, caused from operating without sufficient water. If any question exists, and as previously stated, install a new impeller if at all possible.

Clean all bearings with solvent, dry them with compressed air, and inspect them carefully. Be sure there is no water in the air line. Direct the air stream through the bearing. **NEVER** spin a bearing with compressed air. Such action is highly dangerous and may cause the bearing to score from lack of lubrication. After the bearings are clean and dry, lubricate them with Formula 50 oil, or equivalent. Do not lubricate tapered bearing cups until after they have been inspected.

*A worn water pump impeller no longer fit for service. The importance of installing a **NEW** impeller, whenever the lower unit is disassembled, cannot be overemphasized.*

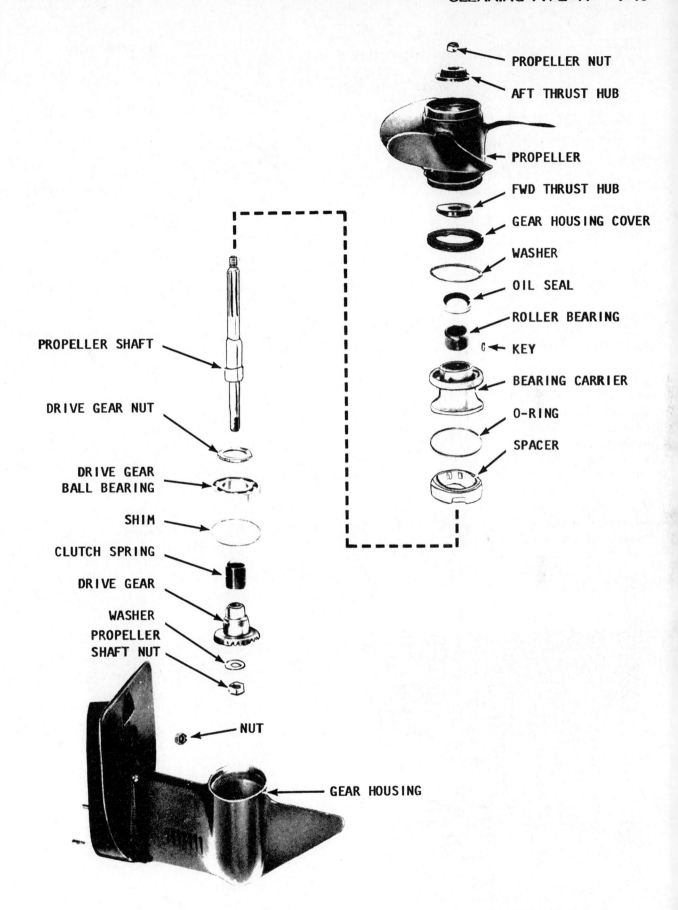

PROPELLER NUT

AFT THRUST HUB

PROPELLER

FWD THRUST HUB

GEAR HOUSING COVER

WASHER

OIL SEAL

ROLLER BEARING

KEY

BEARING CARRIER

O-RING

SPACER

PROPELLER SHAFT

DRIVE GEAR NUT

DRIVE GEAR
BALL BEARING

SHIM

CLUTCH SPRING

DRIVE GEAR

WASHER
PROPELLER
SHAFT NUT

NUT

GEAR HOUSING

Exploded view of a typical small horsepower lower unit showing arrangement of major propeller shaft parts.

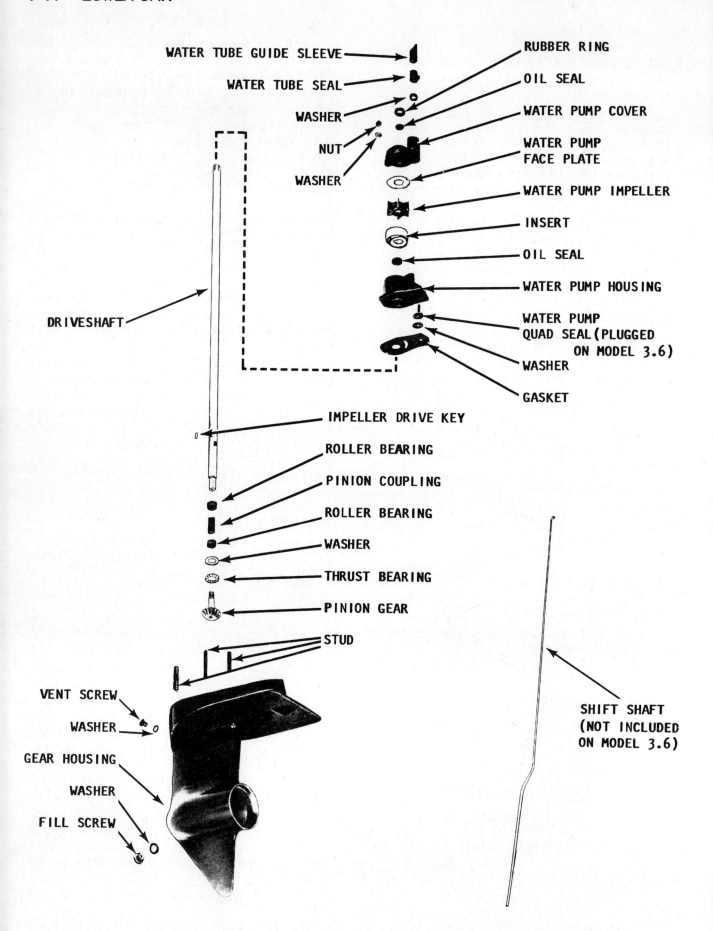

WATER TUBE GUIDE SLEEVE

WATER TUBE SEAL

WASHER

NUT

WASHER

RUBBER RING

OIL SEAL

WATER PUMP COVER

WATER PUMP
FACE PLATE

WATER PUMP IMPELLER

INSERT

OIL SEAL

WATER PUMP HOUSING

WATER PUMP
QUAD SEAL (PLUGGED
ON MODEL 3.6)

WASHER

GASKET

DRIVESHAFT

IMPELLER DRIVE KEY

ROLLER BEARING

PINION COUPLING

ROLLER BEARING

WASHER

THRUST BEARING

PINION GEAR

STUD

VENT SCREW

WASHER

GEAR HOUSING

WASHER

FILL SCREW

SHIFT SHAFT
(NOT INCLUDED
ON MODEL 3.6)

Exploded view of a typical small horsepower lower unit showing major driveshaft and water pump parts.

Inspect all ball bearings for roughness, catches, and bearing race side wear. Hold the outer race, and work the inner bearing race in-and-out, to check for side wear.

Determine the condition of tapered bearing rollers and inner bearing race, by inspecting the bearing cup for pitting, scoring, grooves, uneven wear, imbedded particles, and discoloration caused from overheating. **ALWAYS** replace tapered roller bearings as a set.

Inspect the bearing surface of the shaft roller bearings support. Check the shaft surface for pitting, scoring, grooving, imbedded particles, uneven wear and discoloration caused from overheating. The shaft and bearing must be replaced as a set if either is unfit for continued service.

Good shop practice requires installation of new O-rings and oil seals **REGARDLESS** of their appearance.

Clean the bearing carrier, pinion gear, drive gear clutch spring, and the propeller shaft with solvent. Dry the cleaned parts with compressed air.

Check the pinion gear and the drive gear for abnormal wear. Apply a coating of light-weight oil to the roller bearing. Rotate the bearing and check for cracks or catches.

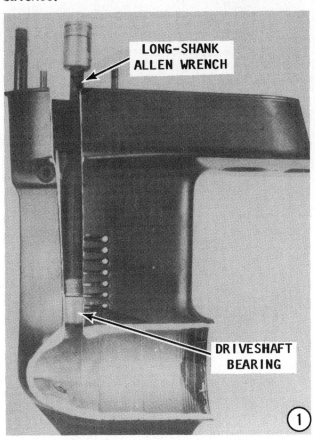

LONG-SHANK ALLEN WRENCH

DRIVESHAFT BEARING

①

Inspect the propeller shaft oil seal surface to be sure it is not pitted, grooved, or scratched. Inspect the roller bearing contact surface on the propeller shaft for pitting, grooves, scoring, uneven wear, imbedded metal particles, and discoloration caused from overheating.

ASSEMBLING TYPE "A" UNIT

Pinion Gear Bearing Installation

1- Clamp the lower unit in a vertical position in a vise equipped with soft jaws. Coat the driveshaft roller bearing bore in the lower unit with Formula 50 oil, or equivalent. Position the 9/16" pinion roller bearing onto Bearing Tool C-91-77548 with the lettered side of the bearing **AGAINST** the shoulder of the bearing tool. If the special tool is not available, the tool made during disassembly in Step 20 may be used by attaching it to the end of a long-shank Allen wrench, as shown. Press the roller bearing into the driveshaft bore by tapping **LIGHTLY** on the end of the bearing tool with a mallet. The bearing is properly seated when the bearing tool bottoms-out in the driveshaft bore.

Driveshaft Shimming

Special tools and some simple arithmetic are required to properly shim the drive gear.

a- Gauge Rod Adpator C-23-75484.
b- Gauge Rod C-91-74585.
c- Gauge Adaptor C91-74586.
d- Depth micrometer.
e- An example of the simple subtraction involved is given at the end of the step. The procedures must be followed closely in order to determine the correct amount of shim material required.

2- Insert the Gauge Rod Adaptor into the top of the driveshaft bore. Insert the Gauge Rod through the rod adaptor and the pinion roller bearing. Insert the Gauge Adaptor into the torpedo bore until it bottoms-out in the bore. Now, using the depth micrometer, measure the distance from the gauge rod to the outer end of the gauge adaptor. As a guide: The dimension from the gauge rod to the outer end of the gauge adaptor should be 3.676" + 0.005". This dimension is referenced "$\overline{A}$" in the illustration.

To determine the proper amount of shim material required, subtract 3.062" from the dimension "A" measured with the depth micrometer. Consider the answer to the subtraction as dimension "B". Now, subtract dimension "B" from 0.629" and the answer is the amount of shim material required to properly adjust the drive gear.

EXAMPLE

3.676" "A" (from micrometer)
-3.062" Factory number
.614" "B" Simple subtraction

.629" Factory number
-.614" "B"
.015" Shim material required

The shim material will be installed later in Step 13.

Exhaust Housing Bearing Installation

3- Position the pinion coupler into the driveshaft bore. The coupler can go in either way.

4- Place the 11/16" OD driveshaft roller bearing onto Bearing Tool C-91-77548 with the lettered side of the bearing **AGAINST** the bearing tool shoulder. If the special tool is not availble, a 7/16" round-head Allen bolt may be used. This is the same bolt used to secure the trim tab on the larger Mercury outboard engines. Press the roller bearing into the driveshaft bore by tapping lightly

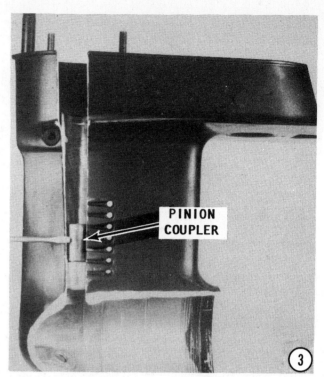

GAUGE ROD ADAPTOR
GAUGE ROD
PINION BEARING
GAUGE ADAPTOR
DEPTH MICROMETER
DIMENSION "A"
②

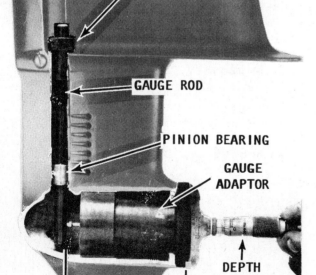

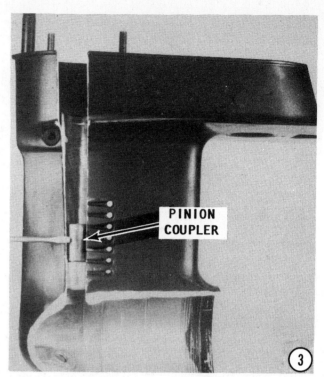

PINION COUPLER
③

Homemade tools used to remove both driveshaft bearings and to install the lower driveshaft bearing (left). The other tool (right) is used to install the upper driveshaft bearing. The text explains how these two items can be made in most shops.

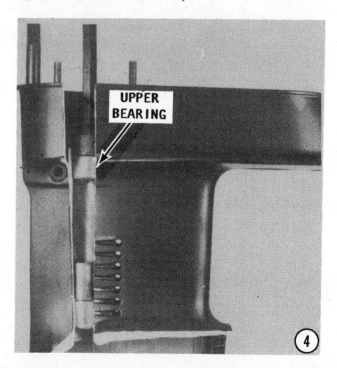

UPPER BEARING
④

on the end of the bearing tool, or 7/16" bolt, with a mallet. The driveshaft roller bearing is properly seated when the bearing tool bottoms-out in the driveshaft bore.

Propeller Shaft Assembling

5- Install the ball bearing onto the drive gear by first positioning the drive gear in a press, and then pressing the bearing onto the drive gear using a suitable mandrel, or a socket, as shown. **ALWAYS** press on the inner race of the bearing and be sure the bearing is firm against the drive gear.

6- Remove the drive gear from the press after Step 4 is completed, and clamp the drive gear in a vise equipped with soft jaws. Coat the threads of the retainer nut with Loctite Type "A", and then install the nut onto the drive gear. Use Retainer Tool C-91-74589 and a torque wrench to tighten the retainer nut to the value given in the Specifications in the Appendix. Clean any excess Loctite from the retainer nut and the drive gear.

7- Install the clutch spring onto the drive gear with the extended end of the spring **TOWARD** the bearing. This is accomplished by pushing and turning the spring **COUNTERCLOCKWISE.**

8- Install the drive gear assembly onto the propeller shaft by pushing and turning the propeller shaft through the drive gear from the clutch spring end of the gear.

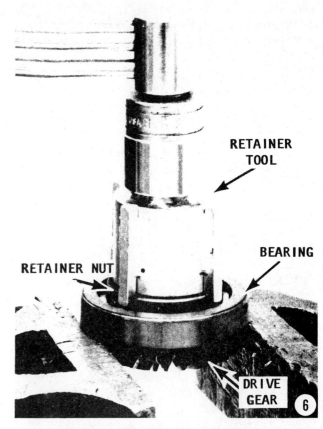

9- Clamp the propeller shaft in a vise equipped with soft jaws. Install the special flat washer and a **NEW** locknut on the propeller shaft. Tighten the locknut until it seats.

10- Back it off the locknut to allow 0.005" to 0.010" clearance between the locknut and the flat washer. This clearance is a critical dimension to ensure proper operation of the drive gear. After the required clearance has been obtained and checked, set the assembly aside for later installation.

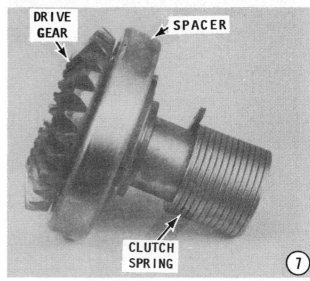

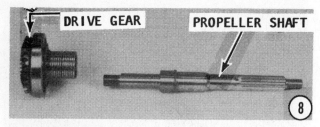

GOOD WORDS

Assembling the bearing carrier and the propeller shaft is not a difficult task, but attention to detail is critical.

The cutaway photographs, like many used in this chapter, will be a great asset during the work. Some of the parts, such as the roller bearing, oil seal, and O-ring, MUST be installed in only one way, as the procedures indicate.

Once the work is started, make an attempt to continue without interruption.

Bearing Carrier Assembling

11- Install the roller bearing into the bearing carrier by first applying a coating of light-weight oil to the carrier bore, and then pressing the roller bearing into the carrier from the LETTERED side using Bearing Tool C-91-74582, or the appropriate size socket.

12- Apply a coating of Loctite Type "A" to the outer diameter of the bearing carrier

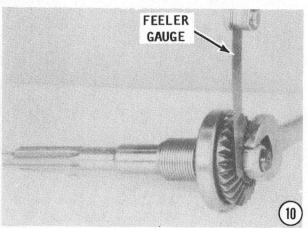

oil seal. Use Oil Seal Tool C-91-74583 and press the oil seal into the bearing carrier with the lips of the seal DOWN. Clean any excess Loctite from the bearing carrier and the oil seal. Set the assembly aside for later installation.

Propeller Shaft Installation

13- Clamp the lower unit in a horizontal position in a vise equipped with soft jaws. Install the thrust bearing and thrust washer onto the pinion gear. Install the pinion gear into the pinion bearing in the torpedo bore.

14- Install the drive gear shims of the proper thickness, as determined in Step 2, into the torpedo bore.

15- Install the propeller shaft into the torpedo bore. Rotate the propeller shaft in order to mesh the drive gear with the pinion gear.

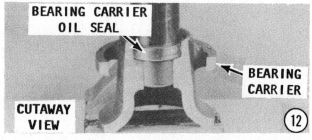

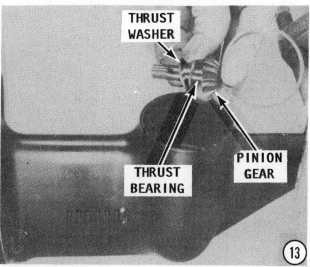

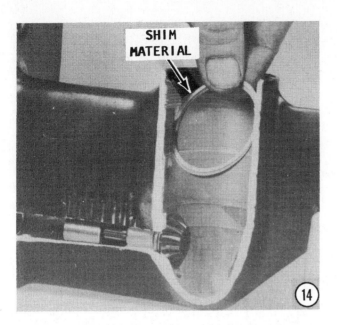

SHIM MATERIAL

14

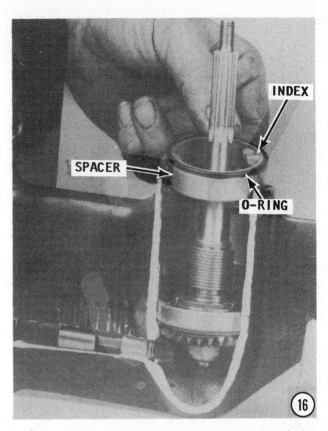

INDEX

SPACER

O-RING

16

16- Install the propeller shaft spacer into the torpedo bore and check to be sure:

a- The O-ring groove is facing **AWAY** from the drive gear bearing.

b- The shift shaft hole in the spacer is aligned with the shift shaft hole in the gear housing.

Apply a coating of Multipurpose Lubricant, or equivalent, to the O-ring, and then install the O-ring onto the ring groove of the propeller shaft spacer. Coat the outside diameter of the bearing carrier (the area

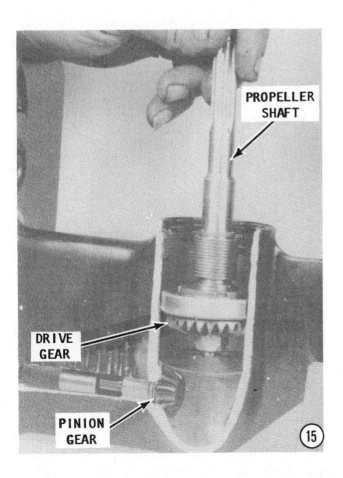

PROPELLER SHAFT

DRIVE GEAR

PINION GEAR

15

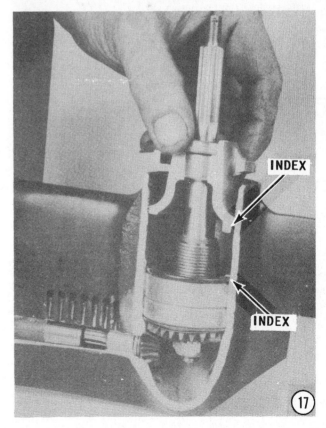

INDEX

INDEX

17

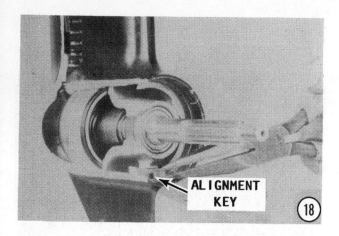

ALIGNMENT KEY

⑱

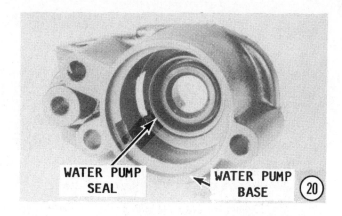

WATER PUMP SEAL   WATER PUMP BASE   ⑳

where the carrier contacts the lower unit) with Perfect Seal, or equivalent. **TAKE-CARE** to prevent any Perfect Seal from contacting the oil seal or the bearing.

17- Install the bearing carrier into the torpedo bore with the alignment tab in back of the bearing carrier indexed between the alignment fingers in the propeller shaft spacer.

18- Align the keyway in the bearing carrier with the keyway in the lower unit, and then install the alignment key.

19- Install the cover nut flat washer, if one is used, against the bearing carrier.

Coat the threads of the cover nut with Perfect Seal. Start the cover nut in the torpedo bore by hand with the words **OFF** and **ON** and the arrows visible. After the nut has been started, continue to tighten the nut to the torque value given in the Specifications in the Appendix. Use Cover Nut Tool C-91-74588 and a torque wrench to attain the required torque value.

Water Pump Base Assembling

20- Coat the outside diameter of the water pump oil seal with Loctite Type "A". Position the oil seal onto Oil Seal Tool C-91-74583 with the lips of the seal **AGAINST** the oil seal tool. Press the oil seal into the water pump base. Clean any excess Loctite from the oil seal and the water pump base.

21- Insert the seal and the flat washer into the shift shaft hole in the water pump base.

COVER NUT   ⑲

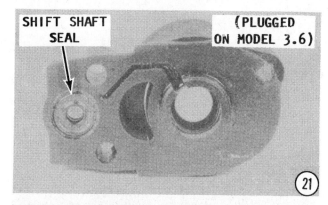

SHIFT SHAFT SEAL   (PLUGGED ON MODEL 3.6)

㉑

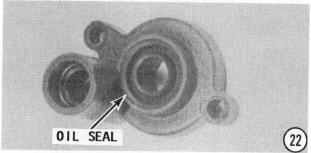

OIL SEAL   ㉒

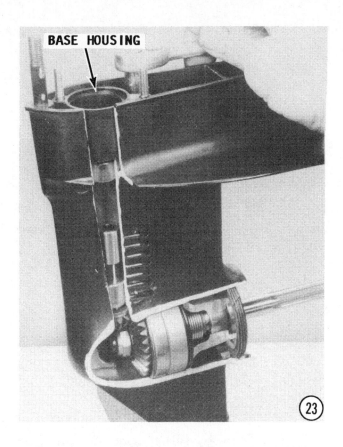

BASE HOUSING

(23)

Water Pump Cover Assembling

22- Apply a coating of Loctite Type "A" to the water pump cover oil seal. Position the oil seal in place on Oil Seal Tool C-91-74583 with the lips of the seal **AGAINST** the oil seal tool. Press the oil seal into the water pump cover. Clean any excess Loctite from the oil seal and the water pump cover.

Water Pump Installation

23- Clamp the lower unit in a vertical

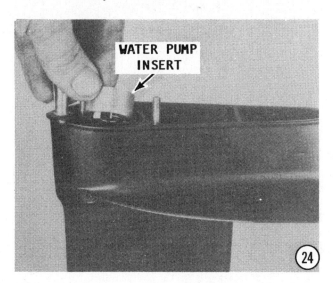

WATER PUMP INSERT

(24)

WATER PUMP IMPELLER

(25)

position in a vise equipped with soft jaws. Position a **NEW** water pump base gasket over the water pump mounting studs. Install the water pump base and secure it to the lower unit with the locknuts and flat washers. Tighten the locknuts to the torque value given in the Specifications in the Appendix.

24- Slide the water pump insert into the water pump base with the bent-tabs on the insert indexed into the slots in the base.

25- Install the water pump impeller into the water pump insert.

Advice: **ALWAYS** install a **NEW** water pump impeller, if possible, during overhaul work on the lower unit. If a new impeller is not available, **NEVER** install the impeller in reverse to its original position of rotation. Such action will cause premature impeller failure in a very short time.

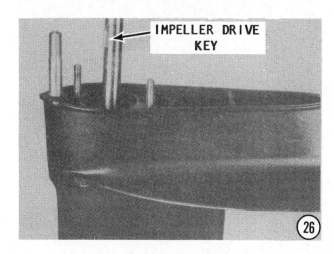

IMPELLER DRIVE KEY

(26)

26- Position the impeller drive key on the flat of the driveshaft. Use some Multi-purpose Lubricant to hold the drive key in place. Now, carefully insert the driveshaft through the impeller into the lower unit with the drive key aligned with the keyway in the impeller. Rotate the driveshaft to engage the driveshaft splines with the pinion coupler splines.

27- Slide the water pump face plate down over the impeller with the bent-tab on the face plate indexed with the slot in the water pump base.

28- Position the water pump cover over the water pump base and mounting studs. Slide the flat washers onto the mounting studs. Thread the locknuts onto the mounting studs, and then tighten them to the torque value given in the Specifications in the Appendix.

29- Install the water seal tube tab into the water pump cover. Work the water tube grommet into the cover with the tab on the grommet indexed into the guide hole in the water pump cover housing.

30- Slide the rubber centrifugal slinger down the driveshaft until it is against the water pump cover. Install the water tube guide into the water pump cover.

WATER PUMP COVER

28

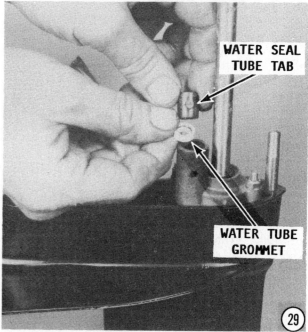

WATER SEAL TUBE TAB

WATER TUBE GROMMET

29

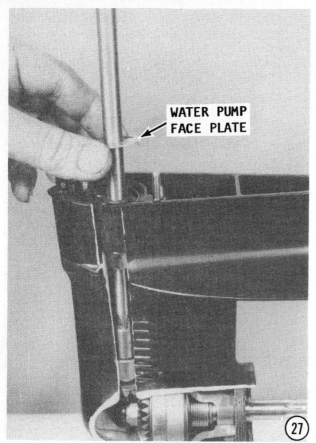

WATER PUMP FACE PLATE

27

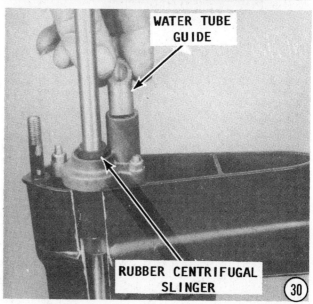

WATER TUBE GUIDE

RUBBER CENTRIFUGAL SLINGER

30

Filling Lower Unit with Lubricant

31- Remove any gasket material from the **FILL** and **VENT** screws and the lower unit surface.

CRITICAL WORDS: Never add lubricant to the lower unit without first removing the **VENT** screw and having the lower unit secured to the exhaust housing. Add lubricant **ONLY** when the lower unit is in the vertical position.

Slide **NEW** gaskets onto the **FILL** and **VENT** screws. Slowly add Super-Duty Gear Lubricant to the lower unit through the **FILL** hole until the lubricant flows out the **VENT** hole and no air bubbles are visible. Install the **VENT** screw. Remove the grease tube, or hose, from the **FILL** hose and **QUICKLY** install the **FILL** screw. Check the lower unit for oil leaks.

INSTALLATION TYPE "A" UNIT

32- Move the engine to the full up position and engage the tilt lock lever. Coat the driveshaft splines with a thin layer of Multipurpose Lubricant. **DO NOT** carry the lower unit by the driveshaft because the shaft may be pulled out of the housing.

BAD NEWS: Any excess lubricant on top of the driveshaft to crankshaft splines will be trapped in the clearance space. This trapped lubricant will not allow the driveshaft to fully engage with the crankshaft. As a result, when the lower unit nuts are tightened, a load will be placed on the driveshaft/crankshaft and will cause damage to either the powerhead or the lower unit or both. Therefore, any lubricant **MUST** be cleaned from the top of the driveshaft.

33- Move the lower unit into position with the driveshaft protruding into the exhaust housing. Raise the lower unit upward toward the exhaust housing and at the same time guide the shift shaft, water tube, and the upper and lower mounting studs into place. If necessary, rotate the flywheel slightly to allow the driveshaft splines to index with the crankshaft splines.

34- Start the two locknuts securing the lower unit to the exhaust housing, but **DO NOT** tighten them at this time. Check operation of the shift mechanism:

a- Shift the outboard into **FORWARD** gear, and then rotote the flywheel **CLOCKWISE**. The propeller shaft should rotate counterclockwise.

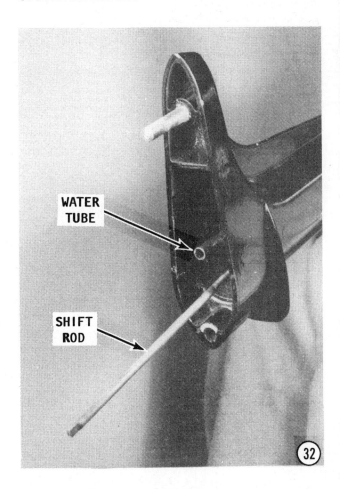

WATER TUBE

SHIFT ROD

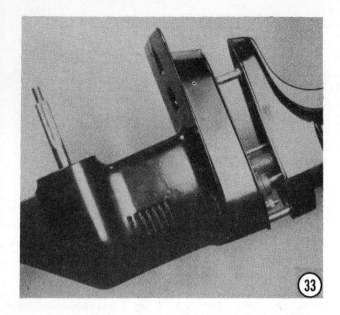

b- Shift the outboard into **NEUTRAL**, and then rotate the flywheel **CLOCKWISE**. The propeller shaft should stop rotating counterclockwise within one complete revolution of the propeller shaft.

c- **NEVER** attempt to check the shift mechanism by rotating the propeller shaft.

d- If the shifting mechanism fails to operate properly, the lower unit must be removed, and the assembly work checked.

If the shifting mechanism is operating properly, tighten the locknuts to the torque value given in the Specifications in the Appendix.

Propeller Installation

35- Disconnect the spark plug wires to prevent the engine from starting when the

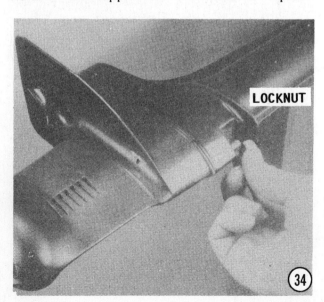

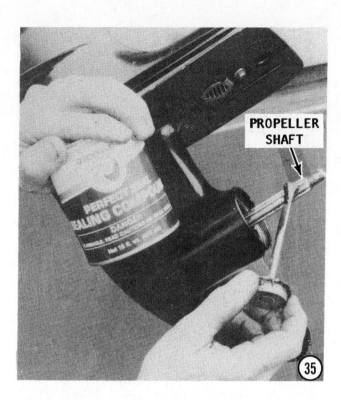

propeller and shaft are rotated. As an aid to removing the propeller the next time, apply a liberal coating of Perfect Seal to the propeller shaft splines.

36- Install the forward thrust hub, the propeller, and then the aft thrust hub onto the propeller shaft splines. Thread the propeller locknut onto the shaft and tighten it to the torque value given in the Specifications in the Appendix. Connect the spark

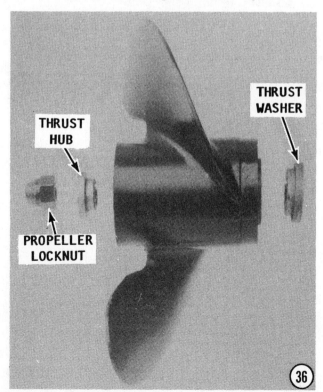

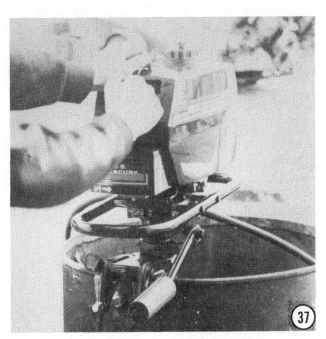

tions in the Appendix. Connect the spark plug wires to the spark plugs.

37- Mount the engine in a test tank or body of water.

CAUTION: Water must circulate through the lower unit to the engine any time the engine is run to prevent damage to the water pump in the lower unit. Just five seconds without water will damage the water pump.

Start the engine and check the completed work for satisfactory operation and no leaks.

9-5 LOWER UNIT TYPE "B" REVERSE CAPABILITY
(See Lower Unit Table in Appendix)

DESCRIPTION

In addition to the normal forward and neutral capabilities of the lower units outlined in Section 9-3, the lower unit Type "B" is equipped with a reverse gear. A clutch and the necessary shift linkage for efficient operation is contained within the lower unit.

ADVICE

Before beginning work on the lower unit, take time to **READ** and **UNDERSTAND** the information presented in Section 9-1, this chapter, and check the Lower Unit Type and Backlash Table in the Appendix to ensure the proper procedures are being followed.

Disconnect the high tension spark plug leads, remove the spark plugs, and disconnect the leads at the battery terminals, before working on the lower unit.

Shift the engine into **FORWARD** gear. Raise the lower unit upward until the tilt lever can be actuated, and then engage the tilt stop.

Propeller Removal
SAFETY WORDS: An outboard engine may start very easily. Therefore, anytime the propeller is to be removed or installed, check to be sure:
a- Shift mechanism is in **NEUTRAL**.
b- Key switch is in **OFF** position.

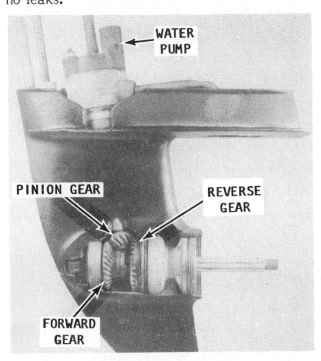

Lower unit used with medium-size Mercury outboard engines. This unit has forward, neutral, and reverse capabilities.

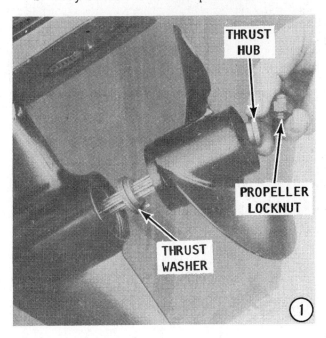

c- Spark plug wires are disconnected.

d- Electrical leads disconnected at the battery terminals.

1- Remove the propeller nut by first placing a block of wood between one of the propeller blades and the anti-cavitation plate to prevent the propeller from turning, and then remove the nut. Remove the splined washer. Remove the outer thrust hub from the propeller shaft. If the thrust hub is stubborn and refuses to budge, use two **PADDED** pry bars on opposite sides of the hub and work the hub loose. **TAKE CARE** not to damage the lower unit. Remove the propeller. If the propeller is "frozen" to the shaft, see Section 9-3 for special procedures to break it loose.

Remove the inner thrust hub. If this hub is also stubborn, use padded pry bars and work the hub loose. Again, **TAKE CARE** not to damage the lower unit.

2- Position a suitable container under the lower unit, and then remove the **FILL** screw and the **VENT** screw. Allow the gear lubricant to drain into the container. As the lubricant drains, catch some with your fingers from time-to-time, and rub it between your thumb and finger to determine if any metal particles are present. If metal is detected in the lubricant, the unit must be completely disassembled, inspected, and the damaged parts replaced. Check the color of the lubricant as it drains. A whitish or creamy color indicates the presence of water in the lubricant. Check the drain pan for signs of water separation from the lubricant. The presence of any water in the gear lubricant is bad news. The unit must be completely disassembled, inspected, the

cause of the problem determined, and then corrected.

3- **MODEL 200 ONLY:** Remove the plastic cap at the lower rear edge of the exhaust housing. Use an allen wrench to loosen the bolt securing the anodic plate to the lower unit. With a socket wrench, remove the nuts from the bottom of the anti-cavitation plate.

Loosen the nut between the lower unit and the exhaust housing, as shown. **DO NOT** attempt to remove the nut at this time. Separate the lower unit from the exhaust housing as far as the loosened nut will allow. Continue to loosen the nut and at the same time separate the two units. Continue working the nut and housing apart until the nut can be removed. After the nut is removed, the exhaust housing and the lower unit can be completely separated. Save the water tube and the water guide tube.

Water Pump Removal

4- Remove the centrifugal slinger from above the water pump cover. Remove the two 7/16" nuts and washers from the pump cover, and then lift the pump cover off the studs.

MODEL 200 ONLY: Remove the three 1/2" nuts and the flush screw securing the water pump base to the housing.

5- Remove the water pump face plate. Pry the impeller out of the pump base using two screwdrivers. Withdraw the impeller drive pin with a pair of needle-nose pliers. Check the water pump insert closely for wear or damage. A new impeller should

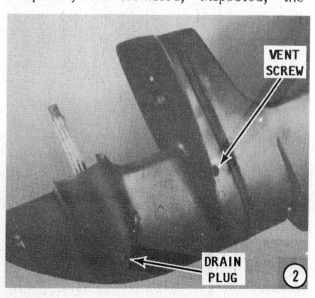

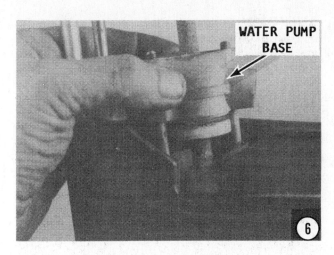

MODEL 200: Remove the pump insert by driving it out from the driveshaft opening in the pump cover.

MODEL 45 THRU 110:   Remove the water tube seal from the pump base. Set the pump aside for disassembly later.

ALWAYS be installed during overhaul work on the lower unit.

6- Pry the water pump base upward and out of the lower unit using two long-shank screwdrivers, as shown. **TAKE TIME** to pad the areas where the screwdrivers will contact the mating surface of the lower unit with the exhaust housing.

Remove the water pickup tube from the seal at the rear of the lower unit. Lift the pump base up and off the driveshaft. Remove the water tube seal and cupped washer from the lower unit. Remove and save the shims from the water pump side of the driveshaft ball bearing. Some shims may have remained with the pump base.

Remove the O-ring from the groove in the pump base. Remove the water pump base oil seal by prying it out or by driving it toward the bottom side of the base.

MODEL 45 THRU 110:   Remove the water pump insert by driving it out from the oil seal side of the pump base.

Propeller Shaft
and Bearing Carrier Disassembling

7- Clamp the lower unit in a vertical position in a vise between two blocks of wood to protect the finished surface of the unit.

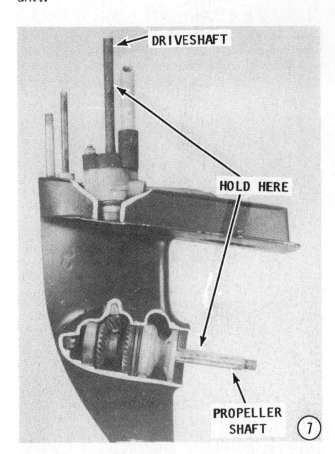

STOP: The reverse gear-to-pinion gear backlash should be checked **PRIOR** to disassembly. The reverse gear backlash can be checked only at a point of propeller shaft rotation where it is not possible to shift from neutral gear into reverse. This position is attained by slowly rotating the propeller shaft and at the same time attempting to shift into the reverse gear position. Once this position is reached:

a– Push down on the driveshaft.

b– Pull outward on the propeller shaft.

c– Hold pressure on the lower shift shaft toward reverse.

d– Lightly rotate the propeller shaft clockwise and counterclockwise.

The amount of free play felt is the reverse gear-to-pinion gear backlash. For the correct amount of backlash allowable for the unit being serviced, check the Specifications in the Appendix. Record the amount of backlash felt because it may affect the shimming of the reverse gear during assembly.

8– Remove the lower unit cover nut using the correct cover nut tool for the unit being serviced. **OBSERVE** the cover nut has cast-in arrows and embossed letters **OFF** to indicate the direction the nut is to be turned for removal.

MODEL 45 THRU 110: the nut is turned **CLOCKWISE!** This means the nut has left-hand threads.

MODEL 200: the nut is turned counterclockwise because the threads are standard right-hand.

If the nut refuses to budge, strike the handle of the tool with a mallet to jar it loose. If striking the tool handle with a mallet is not successful in breaking the nut loose, apply heat to the lower unit and continue to strike the tool handle.

9– If the nut still remains "frozen" it will be necessary to drill the cover nut, as shown. Remove the cover nut and the flat anti-galling washer installed under the nut.

10– Clamp the propeller shaft in a horizontal position in a vise between two blocks of soft wood. Strike the lower unit approximately midway between the anti-cavitation plate and the bearing carrier cavity with a soft-face mallet. **TAKE CARE** the lower unit does not fall. The propeller shaft and the bearing carrier will remain in the vise. Now, if the lower unit refused to move, indicating the bearing may be "frozen" in

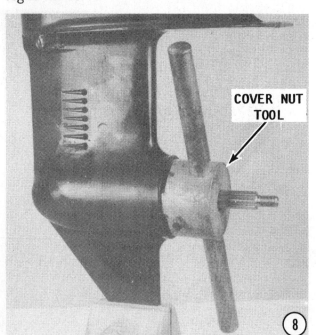

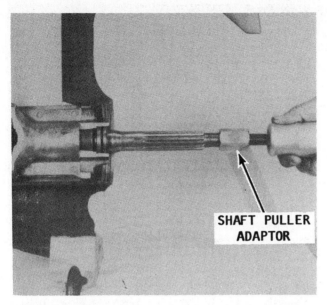

Using a shaft puller adaptor and slide hammer to remove the bearing carrier.

the unit, carefully apply heat to the lower unit and at the same time continue to strike the lower unit in an effort to separate the carrier from the unit.

11- Retain all of the shims. Some shims may remain in the lower unit and others may come out with the bearing carrier thrust washer. Release the propeller shaft and bearing carrier from the vise. Slide the propeller shaft out of the bearing carrier.

Set the propeller shaft and bearing carrier aside for disassembling later.

Driveshaft Pinion Gear and Forward Gear Removal

12- Check the forward gear-to-pinion gear backlash **PRIOR** to removal. The backlash is checked by first pushing the driveshaft down and holding it in that position

with one hand, and at the same time, rocking the forward gear to left-and-right with the other hand. The amount of free play felt is considered the forward-to-pinion gear backlash. The backlash should be from 0.003" to 0.005".

13- Bend the pinion gear locktab away from the head of the pinion gear retainer bolt. Clamp the driveshaft in a vise, equipped with soft jaws, as close as possible to the water pump studs. Loosen the pinion gear retainer bolt with a wrench, and then release the driveshaft from the vise.

Now, turn the driveshaft and at the same time hold the pinion gear retainer bolt. Remove the bolt and tab washer.

14- Clamp the driveshaft back in the vise. Hold a block of wood against the lower unit mating surface and strike the block of wood with a mallet to drive the lower unit off the driveshaft ball bearing. **DO NOT** allow the lower unit to fall. Remove, save, and tag the shims from the lower unit driveshaft bearing cavity or on the driveshaft ball bearing. **TAKE EXTRA CARE** not to intermix these shims with those from above the driveshaft ball bearing.

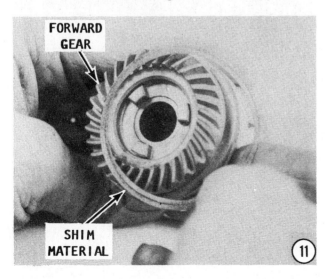

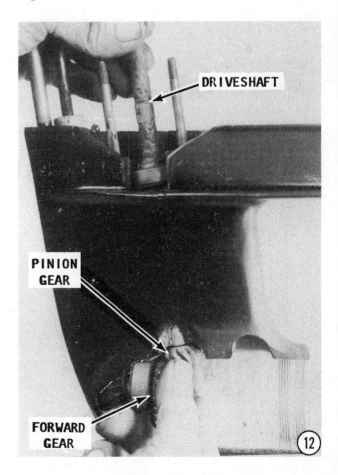

Release the driveshaft from the vise. Reach into the lower unit and lift out the pinion gear. Set the driveshaft aside for disassembly later.

15- Place a block of wood at the edge of the work bench. Hold the lower unit firmly with the anti-cavitation plate away from you and the aft edge of the plate facing down, as shown. Now, with the lower unit in this position, raise it above the block of wood, and then bring it down squarely and sharply against the wood. The shock will jar the forward gear and the bearing loose and they will drop onto the block of wood. If they failed to come out of the lower unit, repeat the shock treatment until they are released. Set the forward gear and bearing aside for later service.

Shift Shaft Removal

FIRST THESE WORDS: Take time to observe the position of the shift cam in relation to the lubricant **FILL** hole as an aid to correct installation. On 1971, '72, and '73 models, the shift cam position is opposite to earlier models.

16- Remove the shift shaft bushing retainer screw.

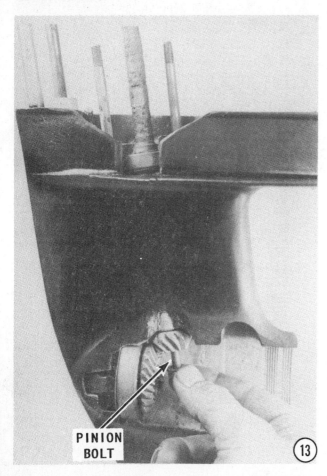

PINION BOLT   13

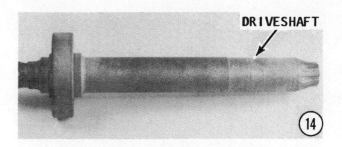

DRIVESHAFT   14

17- Clamp the shift shaft in a vise equipped with soft jaws. Drive the lower unit off the shift shaft bushing using a soft-face mallet or a block of wood and a ball-pean hammer. **TAKE CARE** to prevent the lower unit from falling. Remove the shift cam from the lower unit cavity. Remove the shift shaft from the vise, and then slide the bushing off the shaft. Remove the **E**-clip. Remove the **O**-ring from the groove in the shift shaft bushing. Pry the oil seal out of the shift shaft bushing toward the top side.

Bearing Carrier Disassembling

18- Clamp the bearing carrier in a vise equipped with soft jaws. Install Water Pump Cartridge Puller, C-91-27780 into the reverse gear. Now, pull the reverse gear out of the bearing carrier by striking the tool with a mallet. If the reverse gear ball bearing remains in the bearing carrier, in-

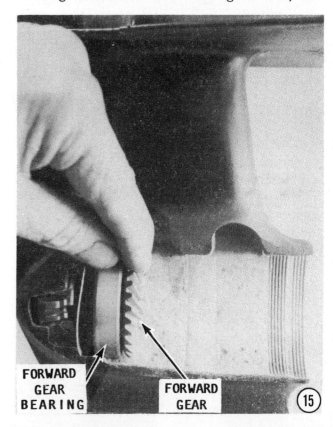

FORWARD GEAR BEARING   FORWARD GEAR   15

BUSHING RETAINER SCREW

16

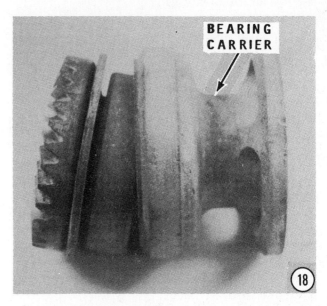

BEARING CARRIER

18

stall the puller into the bearing, and pull it out by striking the tool with a mallet.

19- If the ball bearing remains with the reverse gear, install Universal Puller Plate C-91-37241 between the reverse gear and the bearing. Place the gear, bearing, and puller plate on a press. Use a suitable mandrel and press the gear out of the bearing.

20- Remove the propeller shaft roller bearing only if it is unfit for service. On **MODEL 200:** use Bearing Mandrel C-91-24147A1. On **MODEL 45-75 and 110:** use Mandrel C-91-24273. Press the propeller

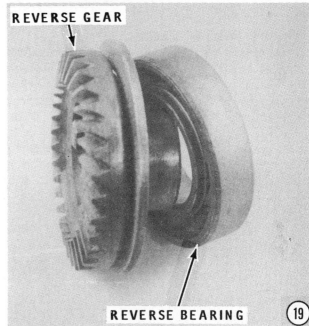

REVERSE GEAR

REVERSE BEARING   19

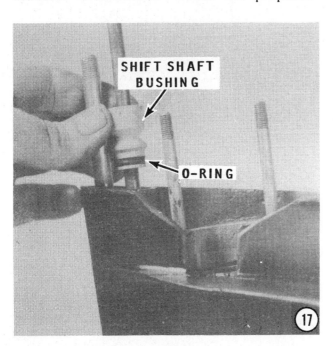

SHIFT SHAFT BUSHING

O-RING

17

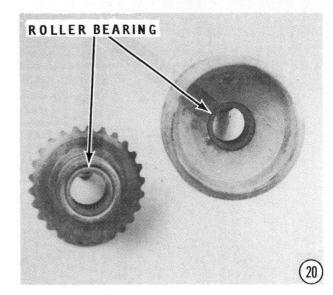

ROLLER BEARING

20

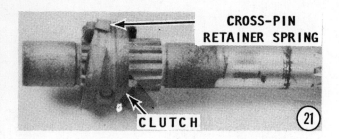

CROSS-PIN
RETAINER SPRING

CLUTCH ㉑

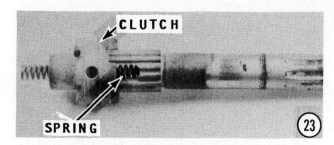

CLUTCH

SPRING ㉓

shaft roller bearing toward the aft end of the bearing carrier using the proper mandrel. The propeller shaft oil seals will be removed as the roller bearing is pressed out. If the propeller shaft oil seals are to be removed, and the roller bearing does not require removal, the seals can be pried or driven toward the aft end of the bearing carrier.

Propeller Shaft Disassembling

21- Remove the cross pin retainer spring or clip. **TAKE CARE** not to over-stretch the spring or clip.

22- Place the cam follower in position for removal in the propeller shaft. Set the cam follower against a solid object such as a vise. Now, push against the cam follower to compress the spring inside the propeller shaft, then hold this position and at the same time use a punch to remove the cross pin.

23- Release the pressure on the cam follower, then slide the clutch off the propeller shaft. Raise the propeller shaft, spline end upward. The cam follower and spring will slide out of the shaft.

Forward Gear Disassembling

24- Position Universal Puller Plate C-91-37241 between the ball bearing and the forward gear. Place the puller plate, with the gear to the bottom, on a press. Press the gear out of the bearing using a suitable mandrel. Clamp the forward gear in a vise.

25- Remove the forward gear roller bearing **ONLY** if it is unfit for further service. Once the bearing is removed, it **CANNOT** be used again. Drive the roller bearing out of the forward gear with a punch and hammer.

Driveshaft Disassembling

26- Remove the driveshaft ball bearing **ONLY** if it is unfit for further service. Remove the driveshaft ball bearing snap ring with Snap Ring Pliers C-91-24283. Place the driveshaft loosely in a vise with the ball bearing above the vise head, as shown. **DO NOT** tighten the vise on the shaft. Now, strike the end of the driveshaft with a mallet to force the driveshaft out of the bearing.

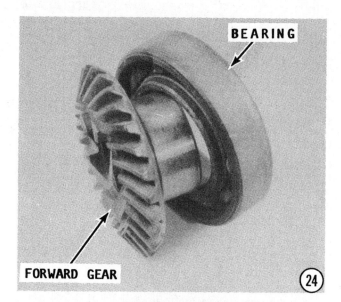

BEARING

FORWARD GEAR ㉔

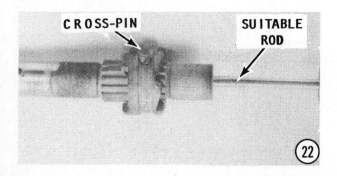

CROSS-PIN   SUITABLE
ROD

㉒

PINION
GEAR

DRIVESHAFT
BALL BEARING ㉕

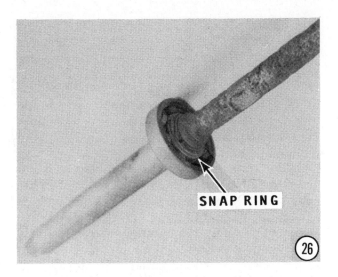

After the shift rod and cam have been cleaned, check to be sure the splines mate smoothly.

Driveshaft Housing Bearing Disassembling

27- Remove the bearing **ONLY** if it is unfit for continued service. **MODEL 45-75** and **110:** use bearing mandrel C-91-24273; **MODEL 200:** use mandrel C-91-24147A1. Insert the proper mandrel into the driveshaft roller bearing from the water pump side. Drive the roller bearing into the bearing carrier cavity with a mallet.

CLEANING AND INSPECTING

Good shop practice requires installation of new O-rings and oil seals **REGARDLESS** of their appearance.

Clean all water pump parts with solvent, and then dry them with compressed air. Inspect the water pump cover and base for cracks and distortion, possibly caused from

overheating. Inspect the face plate and water pump insert for grooves and/or rough surfaces. If possible, **ALWAYS** install a new water pump impeller while the lower unit is disassembled. A new impeller will ensure extended satisfactory service and give "peace of mind" to the owner. If the old impeller must be returned to service, **NEVER** install it in reverse to the original direction of rotation. Installation in reverse will cause premature impeller failure.

If installation of a new impeller is not possible, check the three seal surfaces. All must be in good condition to ensure proper pump operation. Check the upper, lower, and ends of the impeller vanes for grooves, cracking, and wear. Check to be sure the bonding of the impeller hub to the impeller will not allow the hub to slip inside the impeller.

Clean all bearings with solvent, dry them with compressed air, and inspect them carefully. Be sure there is no water in the air line. Direct the air stream through the bearing. **NEVER** spin a bearing with compressed air. Such action is highly dangerous and may cause the bearing to score from lack of lubrication. After the bearings are clean and dry, lubricate them with Formula 50 oil, or equivalent. Do not lubricate tapered bearing cups until after they have been inspected.

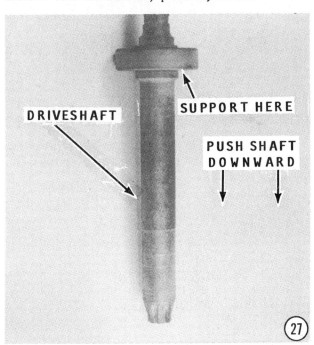

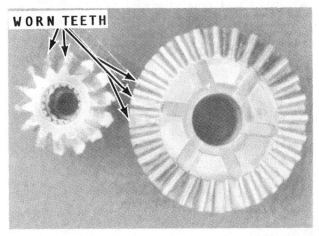

Excellent example of worn gear teeth resulting from improper shimming.

Inspect all ball bearings for roughness, scratches and bearing race side wear. Hold the outer race, and work the inner bearing race in-and-out, to check for side wear.

Determine the condition of tapered bearing rollers and inner bearing race, by inspecting the bearing cup for pitting, scoring, grooves, uneven wear, imbedded particles, and discoloration caused from overheating. **ALWAYS** replace tapered roller bearings as a set.

Clean the forward gear with solvent, and then dry it with compressed air. Inspect the gear teeth for wear. Under normal conditions the gear will show signs of wear but it will be smooth and even.

Clean the bearing carrier with solvent, and then dry it with compressed air. **NEVVER** spin bearings with compressed air. Such action is highly dangerous and may cause the bearing to score from lack of lubrication. Check the gear teeth of the reverse gear for wear. The wear should be smooth and even.

Check the clutch "dogs" to be sure they are not rounded-off, or chipped. Such damage is usually the result of poor operator habits and is caused by shifting too slowly or shifting while the engine is operating at high rpm. Such damage might also be caused by

improper shift cable adjustments.

Rotate the reverse gear and check for catches and roughness. Check the bearing for side wear of the bearing races.

Inspect the roller bearing surface of the propeller shaft. Check the shaft surface for pitting, scoring, grooving, embedded particles, uneven wear and discoloration caused from overheating.

Clean the driveshaft with solvent, and then dry it with compressed air. **NEVER** spin bearings with compressed air. Such action is dangerous and could damage the bearing. Inspect the bearing for roughness, scratches, or side wear. If the bearing shows signs of such damage, it should be replaced. If the bearing is satisfactory for further service coat it with oil.

Inspect the driveshaft splines for excessive wear. Check the oil seal surfaces above and below the water pump drive pin area for grooves. Replace the shaft if grooves are discovered.

Inspect the driveshaft roller bearing surface above the pinion gear splines for pitting, grooves, scoring, uneven wear, embedded metal particles and discoloration caused by overheating.

Inspect the propeller shaft oil seal surface to be sure it is not pitted, grooved, or scratched. Inspect the roller bearing contact surface on the propeller shaft for pitting, grooves, scoring, uneven wear, embedded metal particles, and discoloration caused from overheating.

Inspect the propeller shaft splines for wear and corrosion damage. Check the propeller shaft for straightness.

Inspect the following parts for wear, corrosion, or other signs of damage:
Shift shaft splines.
Shift shaft bushing seal surface.
Shift shaft bushing and seal.
Shift cam.

Check the **E**-clip to be sure it is not bent or stretched. If the clip is deformed, it must be replaced.

Clean all parts with solvent, and then dry them with compressed air.

Inspect:

All bearing bores for loose fitting bearings.

Gear housing for impact damage.

Cover nut threads for cross-threading and corrosion damage.

Labyrinth seal for corrosion or physical damage.

*Worn water pump impeller no longer fit for service. A **NEW** impeller, installed any time the lower unit is disassembled, is the cheapest insurance against failure of the cooling system.*

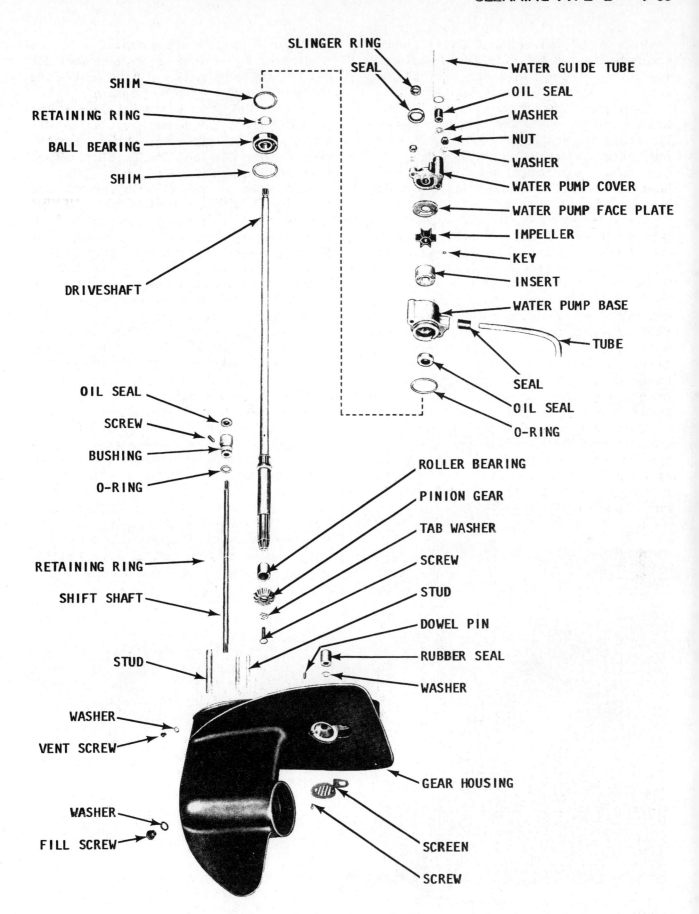

SLINGER RING

SEAL

WATER GUIDE TUBE

SHIM

RETAINING RING

BALL BEARING

SHIM

OIL SEAL

WASHER

NUT

WASHER

WATER PUMP COVER

WATER PUMP FACE PLATE

IMPELLER

KEY

INSERT

WATER PUMP BASE

TUBE

DRIVESHAFT

SEAL

OIL SEAL

O-RING

OIL SEAL

SCREW

BUSHING

O-RING

ROLLER BEARING

PINION GEAR

TAB WASHER

SCREW

STUD

DOWEL PIN

RUBBER SEAL

WASHER

RETAINING RING

SHIFT SHAFT

STUD

WASHER

VENT SCREW

GEAR HOUSING

WASHER

FILL SCREW

SCREEN

SCREW

Exploded view of a typical medium-size (up to about 20-horsepower) lower unit showing arrangement of major driveshaft and water pump parts.

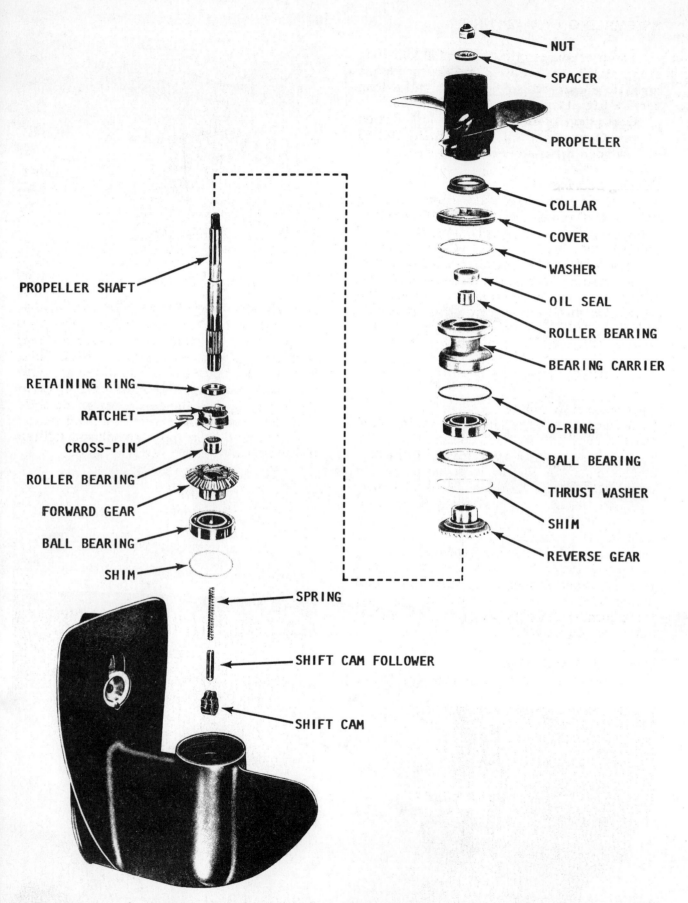

Exploded view of a typical medium-size (up to about 20-horsepower) lower unit showing arrangement of major propeller shaft parts.

ASSEMBLING TYPE "B" UNIT

The procedures outlined in the following steps **MUST** be followed in the sequence given to ensure proper operation and a long service life after the work is completed.

Good shop practice requires installation of new O-rings and oil seals **REGARDLESS** of how good the used item may appear.

Housing Bearing

1- Coat the driveshaft roller bearing bore with Formula 50 oil. **MODEL 45:** thru **110:** use Bearing Mandrel C-91-24273; for **MODEL 200:** use mandrel C-91-24147A1. Position the driveshaft roller bearing on the proper mandrel with the numbered side **TO-WARD** the mandrel shoulder. Insert the mandrel with the bearing into the drivesahft cavity. Drive the roller bearing into the driveshaft cavity with a mallet until the bearing is approximately 1/8" above the bottom of the bearing bore.

Shift Shaft Assembling

2- Coat the outer diameter of the shift shaft bushing oil seal with Loctite Type "A". Position the bushing on a press. Place the oil seal over the bushing with the lip of the seal **TOWARD** the outside. Press the seal into the bushing using a suitable mandrel. Install a **NEW** O-ring into the groove in the shift shaft bushing. Coat the seal and bushing with Multipurpose Lubricant.

Install the E-clip into the shift shaft groove. When properly installed the E-clip should be flush with the base of the shift shaft bushing after the shift shaft is installed in the lower unit.

Shift Shaft Installation

3- Insert the shift cam into the lower

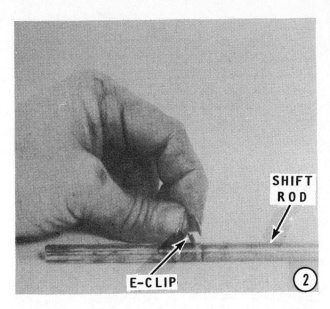

unit with the ramps of the shift cam visible from the rear of the lower unit. The high ramp of the cam, which is the reverse ramp, **MUST** be toward the lubricant **FILL** hole. Slide the shift shaft into the shift cam with the blank spline on the shaft indexed with the blank spline of the cam. Install the shift shaft bushing over the shift shaft and seat it firmly in the lower unit. Install and tighten the shift shaft bushing retainer screw.

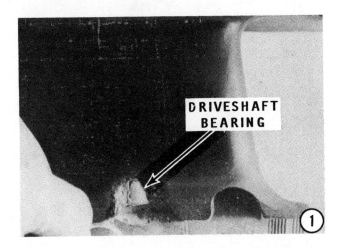

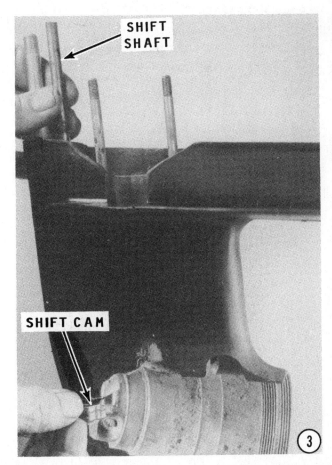

Bearing Carrier Assembling

The bearing carrier is used as a pilot to ensure proper alignment of the forward gear. Therefore, assembling of the bearing carrier is necessary at this time to permit proper installation of the forward gear.

4- Position the reverse gear on a press. Place a thrust washer over the reverse gear. If one side of the thrust washer is beveled, the bevel side must be **TOWARD** the reverse gear. Use a suitable mandrel and press the reverse gear bearing onto the reverse gear from the **NUMBERED** side. Coat the reverse gear bearing bore and the propeller shaft roller bearing bore of the bearing carrier with Formula 50 oil. Press the propeller shaft roller bearing into the bearing carrier from the **NUMBERED** side, using the correct size bearing mandrel.

5- Coat the outer diameter of the propeller shaft oil seal with Loctite Type "A". Press the oil seal into the bearing carrier using a suitable mandrel. Clean any excess Loctite from the seal and the carrier. Press the reverse gear and bearing into the bearing carrier until the bearing is seated against the shoulder. Set the assembly aside for installation of the forward gear later.

Forward Gear Assembling

6- Press the forward gear ball bearing onto the forward gear from the **NUMBERED** side, using a suitable mandrel.

Use a suitable mandrel and press the roller bearing into the forward gear from the **NUMBERED** side. The roller bearing is fully seated when the outer race is against the shoulder in the forward gear.

REVERSE GEAR

BEARING CARRIER

5

Forward Gear Installation

7- Insert the same number of shims retained during disassembly into the forward gear bearing bore.

Advice: If the shims were lost or, if a new gear housing is being installed, begin the shimming by inserting from 0.008" to 0.010" shim material. If the forward gear backlash was incorrect prior to disassembly and the same gear housing and gears are to be installed, make a shim change at this time. Add or remove shim material at an approximate ratio of 0.001" shim for each 0.0015" backlash change required.

Coat the forward gear bearing bore with Formula 50 oil. Clamp the lower unit in a horizontal position in a vise equipped with soft jaws or between a couple blocks of soft

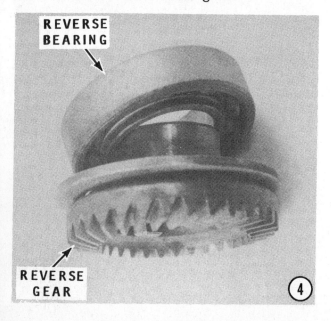

REVERSE BEARING

REVERSE GEAR

4

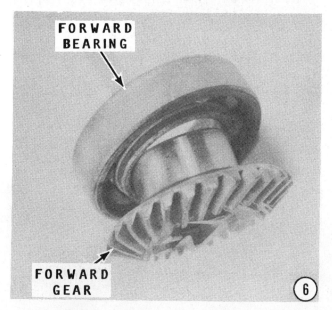

FORWARD BEARING

FORWARD GEAR

6

wood. Now, set the forward gear ball bearing squarely over the bearing bore. Slide the propeller shaft into the forward gear. Actually it is not necessary for the propeller shaft to be assembled at this time.

Slide the bearing carrier over the propeller shaft and into the lower unit to act as a pilot. **TAKE CARE** to prevent the oil seal from being damaged.

Strike the end of the propeller shaft with a mallet several times to seat the bearing against the shims. Remove the propeller shaft and bearing carrier.

Driveshaft Bearing Assembling

8- Position the driveshaft ball bearing over the driveshaft. Place the drivesahft between open jaws of a vise with the bearing above the vise jaws and with the pinion gear end **UP**. Use a mallet to force the driveshaft into the bearing until the bearing is seated against the shoulder of the driveshaft. Remove the driveshaft from the vise. Position the snap ring over the driveshaft, and then use snap ring pliers to install the ring into the snap ring groove in the driveshaft.

Driveshaft Installation

9- Clamp the lower unit in a horizontal position in a vise equipped with soft jaws or between a couple blocks of soft wood. Insert the shims retained during disassembly into

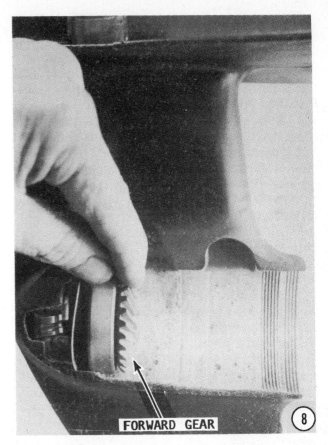

FORWARD GEAR   8

the drivesahft bearing bore. Use a small amount of Multipurpose Lubricant to hold the shims in position.

Advice: If the shims were lost, or if a new housing is being installed, start the shimming with approximately 0.015" of shim material.

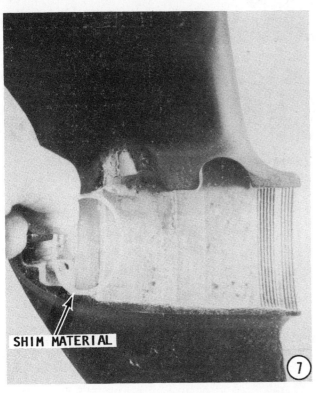

SHIM MATERIAL   7

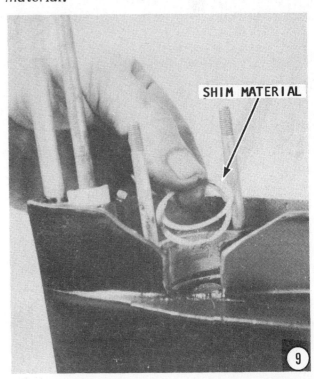

SHIM MATERIAL   9

Hold the pinion gear in the lower unit with the teeth of the pinion gear indexed with the teeth of the forward gear. Continue to hold the pinion gear with one hand and with the other hand insert the drivesahft into the lower unit. Tap the driveshaft into the lower unit with a mallet. Rotate the driveshaft after each tap in an attempt to engage the driveshaft splines with the splines of the pinion gear.

10- After the driveshaft and pinion gear splines are fully indexed (meshed), seat the driveshaft bearing against the shims. Set the tab washer on the pinion retainer bolt with the anti-rotation tabs **AWAY** from the bolt head. Thread the retainer bolt into the driveshaft. Position the anti-rotation tabs into the pinion gear splines. Tighten the pinion retainer bolt hand-tight, and at the same time check to be sure the tab washer is flush with the bottom of the pinion gear.

Clamp the driveshaft in a vise equipped with soft jaws. Tighten the pinion retainer bolt to the torque value given in the Specifications in the Appendix. **DO NOT** bend the

lock tab at this time. Use a punch and hammer and seat the drivesahft outer bearing race agaisnt the shims. **TAKE CARE** not to damage the ball bearing with the punch.

PINION GEAR AND BACKLASH ADJUSTMENTS

The following procedures are to be performed for proper pinion gear and backlash adjustments. Because a variety of possibilities may develop as the work progresses, step-by-step illustrations would only be confusing. Therefore, captioned illustrations are included with the instructions to provide an overall view of how the adjustments are to be made.

Begin by pushing down on the driveshaft and checking the pinion gear depth. The pinion gear **MUST** contact the forward gear on the full length of the gear tooth. If full length contact is not made, the pinion gear must be repositioned by adding or subtracting shim material from under the driveshaft ball bearing. This is accomplished by holding a downward pressure on the driveshaft

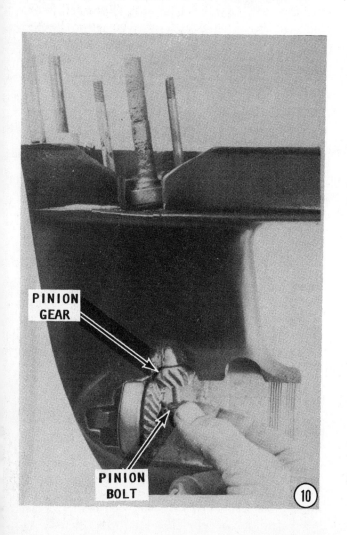

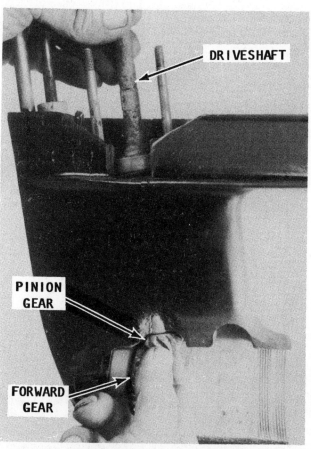

Checking the backlash of the forward gear as explained in the text.

and at the same time pushing against the forward gear and rocking the gear back-and-forth. Free play will be felt in the forward gear movement and it is this movement that is known as backlash. The proper backlash should be between 0.003" and 0.005".

If the pinion gear depth or the forward gear backlash is incorrect, follow the shimming procedures under the heading for the specific condition discovered. **DO NOT** bend the tab washer.

Pinion Gear Too Deep
but Gear Backlash Within Specifications

Remove the driveshaft, pinion gear and the forward gear. Add shim material under the driveshaft ball bearing to correct the pinion gear depth. Add an **EQUAL** amount of shim material in front of the forward gear bearing.

Assemble the parts and again check the pinion depth and forward gear backlash.

Pinion Gear NOT Deep Enough
but Gear Backlash Within Specifications

Remove the driveshaft, pinion gear, and the forward gear. Remove enough shim

Proper tooth mesh of the pinion gear with the forward gear. Detailed shimming procedures outlined in the text must be followed to ensure satisfactory service.

material from under the driveshaft bearing to correct the pinion gear depth. Remove an **EQUAL** amount of shim material from under the forward gear bearing.

Assemble the parts and again check the pinion gear depth and forward gear backlash.

ADVICE: A shim change of 0.001" will result in approximately 0.0015" change in the gear backlash.

Pinion Gear Depth Correct but
Forward Gear Backlash
LESS Than Specificiations

Remove the driveshaft, pinion gear, and the forward gear. Remove enough forward gear shim material to correct the forward gear backlash but **DO NOT** change the driveshaft ball bearing shimming.

Assemble the parts and again check the forward gear backlash.

Pinion Gear Depth Correct but Forward
Gear Backlash MORE Than Specifications

Remove the driveshaft, pinion gear, and the forward gear. Add enough forward gear shim material to decrease the gear backlash. **DO NOT** change the driveshaft ball bearing shimming.

Assemble the parts and again check the forward gear backlash.

Worn gear tooth pattern resulting from normal wear. Notice how the pattern is centered on the tooth and does not extend over the end of the gear.

FINALLY

After the correct pinion gear depth and forward gear backlash is obtained, bend the lock tab against the flat of the pinion gear retainer bolt head.

Water Pump Base Assembling
Models 39 thru 110 also Models 4, 7.5, & 9.8

Coat the outside diameter of the water pump base oil seal with Loctite Type "A". Use a suitable mandrel and press the oil seal into the water pump base with the lip of the seal **TOWARD** the impeller side of the base. Clean any excess Loctite from the seal and the pump base. Coat the insert area of the pump base with Perfect Seal. Install the insert into the pump base with the tab on the insert aligned with the hole in the pump base. Clean any excess Perfect Seal from the insert and the base.

Water Pump Base Assembling
Model 200 & Model 20 Only

Place the water pump base on a press with the impeller side **DOWN.** Coat the outside diameter of the oil seals with Loctite Type "A". Position the smaller OD oil seal on the pump base with the lip of the seal **TOWARD** the impeller side. Press the

seal into the pump base using a suitable mandrel.

Position the larger OD oil seal on the pump base with the lip of the seal **TOWARD** the driveshaft bearing side. Press the oil seal into the pump base using a suitable mandrel.

Remove any excess Loctite from the seals and the pump base.

Install a **NEW** O-ring into the groove in the driveshaft bearing side of the pump base. Coat the seals and O-ring with Multi-purpose Lubricant.

Water Pump Shimming

Proper shimming of the water pump is **CRITICAL** in order to maintain the correct pinion gear depth and forward gear backlash. Follow **ONLY** the procedures under the heading for the Mercury model being serviced.

Depth Micrometer Method

11– First Measurement

Models 39 thru 110
Also Models 4, 7.5, & 9.8

Remove the rear water pump stud. Use a depth micrometer and measure the depth of the cavity above the driveshaft ball bearing to the water pump mating surface. Record this measurement and identify it as **"A"**.

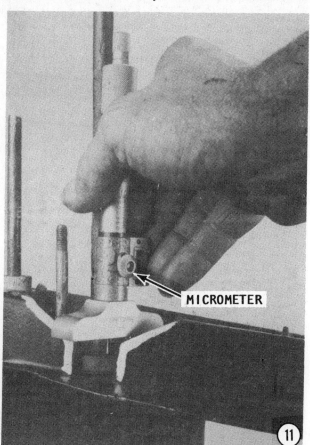

MICROMETER ⑪

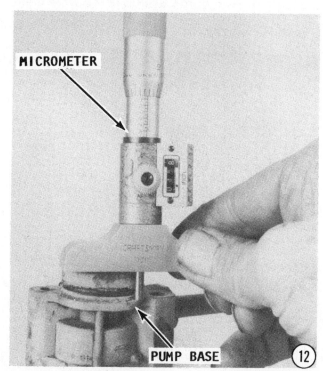

MICROMETER

PUMP BASE ⑫

Model 200 & 20 Only

Slide a **NEW** water pump base gasket down the water pump studs onto the surface of the lower unit. Now, measure the distance between the driveshaft ball bearing outer race and the water pump base gasket. Record this measurement and identify it as **"A"**.

12- Second Measurement

Models 39 thru 110,
Also Models 4, 7.5, & 9.8

Use a depth micrometer and measure the distance between the mating surface of the pump base and the top of the boss which will enter the driveshaft bearing cavity. Record this measurement and identify it as **"B"**.

Subtract measurement **"B"** from measurement **"A"** and the difference will be the amount of shim material required.

EXAMPLE

0.434" Measurement "A"
0.413" Measurement "B"
0.021" Required shim material

The proper amount of shim material will result in zero clearance between the driveshaft ball bearing and the water pump base. It will also maintain the proper pinion gear depth and forward gear backlash.

Model 200 Only

Measure the depth of the shoulder on the water pump base, as shown. This measurement will be the distance from the water

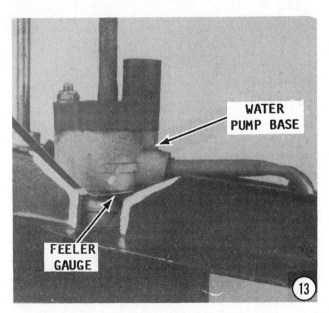

WATER PUMP BASE

FEELER GAUGE

pump base gasket surface to the mating surface where the shims will be installed. Record this measurement and identify it as **"B"**.

Subtract measurement **"B"** from measurement **"A"**. From the answer, subtract 0.002". The final figure will be the required amount of shim material. This amount of shim material will compensate for compression of the gasket and allow the gasket to seal properly. The correct amount of shim material will also maintain the proper pinion gear depth and forward gear backlash.

EXAMPLE

0.451" Measurement "A"
0.439" Measurement "B"
0.012" Simple subtraction
0.002" Factory number
0.010" Subtraction and shim required

Feeler Gauge Method

13- First and Only Measurement

Models 39 thru 110
Also Models 4, 7.5, & 9.8

Install shim material above the driveshaft ball bearing **EQUAL** to the amount removed during disassembly. Now, add approximately 0.020" shim material. Temporarily install the water pump base into the lower unit. Seat the pump base firmly on the driveshaft shims. With the pump in this position, measure the distance between the pump base and the lower housing with a feeler gauge. This measurement will be equal to the amount of shim material that must be removed to obtain zero clearance between the driveshaft ball bearing and the water pump base.

Model 200 & 20 Only

Slide a **NEW** water pump base down over the water pump studs and onto the surface of the lower unit. Install shim material **EQUAL** to the amount removed during disassembly. Add approximately 0.020" shim material. This amount will result in an over-shim condition.

Remove the O-ring from the water pump base to ease installation. This is **ONLY** for measurement purposes. Slide the water pump base down over the studs and seat it firmly.

Use a feeler gauge and measure the

distance between the water pump base and the water pump gasket. Add 0.002" to this measurement. The answer equals the correct amount of shim material that must be **REMOVED** from above the driveshaft bearing. After the shim material is removed there will be zero clearance between the driveshaft ball bearing and the water pump base with 0.002" allowed for compression of the water pump base gasket.

Water Pump Base Installation

14- Install the rear water pump stud, if it was removed. Install a **NEW** O-ring into the water pump base groove. Coat the O-ring and the oil seal with Multipurpose Lubricant, or equivalent. Position the water tube seal washer into the lower unit recess with the **CUPPED** side visible after installation.

Install **NEW** water tube seals in the pump base and lower unit water pickup. Coat the water tube seals with Multipurpose Lubricant. Insert the water tube into the water pickup seal.

Lower the water pump base over the driveshaft and water pump base studs, and at the same time start the water pickup tube into the lower unit water seal.

Model 200 Only

Install the special flush plug from the outside edge of the lower unit into the water pump base.

STOP and make the following checks before proceeding with the assembling work.

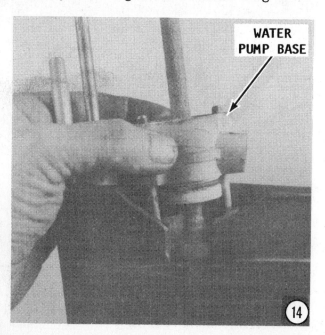

WATER PUMP BASE

14

Check to be sure the drive pin has entered the drive pin groove. Rotate the driveshaft **CLOCKWISE** and check to be sure the water pump impeller turns with the driveshaft. If the impeller fails to turn, repeat Step 14, because the water pump base was not installed properly. Check to be sure the impeller has fully entered the pump insert. The impeller should be almost flush with the top of the insert.

Ready to Proceed

After the checks outlined in the previous paragraph have been successfully completed, apply just a very small amount of Multipurpose Lubricant onto the flat area of the driveshaft and lubricate the inside of the pump insert.

15- Position the impeller drive pin on the driveshaft. Slide a **NEW** impeller down over the driveshaft. Position the impeller drive pin groove approximately 1/8-turn **COUNTERCLOCKWISE** from the impeller drive pin. Exert a downward pressure on the impeller and at the same time rotate the impeller clockwise and slide it down over the drive pin.

Water Pump Cover Assembling

Model 39 thru 110

Coat the outer diameter of the driveshaft oil seal with Loctite Type "A". Use a suitable mandrel and press the oil seal into the pump cover with the lip of the seal **AWAY** from the impeller side. Remove any excess Loctite from the oil seal and the pump cover.

WATER PUMP IMPELLER

15

KROGER
Personal
F I N A N C E®

Double Points*
on all your everyday shopping at Kroger Family of Stores

Earn Rewards every time you shop with the 1-2-3 REWARDS® Visa® Card

Let us reward you for shopping you're already doing. Your Rewards points will add up fast – there's no limit on the rewards you can earn.

With the 1-2-3 REWARDS® Visa® Card, you can earn unlimited rewards. Here's a sampling of how your points can add up quickly and easily **in one month**:

Outside Kroger Family of Stores	**Earn 1 point per $1 you spend**
You Spend	$250 for dining, gas or paying bills
You Earn	250 Rewards points
At Kroger Family of Stores†	**Earn 2 points per $1 you spend**
You Spend	$300 for groceries and other purchases at Kroger Family of Stores
You Earn	600 Rewards points
On Kroger Family of Brands	**Earn 3 points per $1 you spend**
You Spend	$75 for our Kroger Family of Brand products
You Earn	225 Rewards points
Total points for one month	**1,075 Rewards points = $5 Rewards Certificate**

the re-
ith the
water
al into

of the
all the
er with
hole in
Perfect

drive-

ss the
er with
Clean
nd the
eal and

r tube
purpose

plate
all the
e with

pump

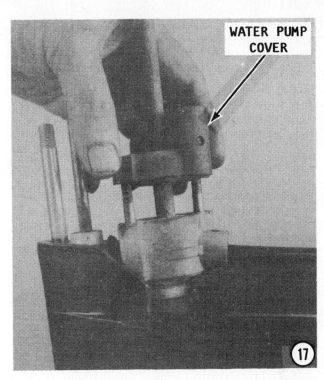

17- Slide the water pump cover down over the driveshaft and the pump studs. Install the washers and nuts onto the studs, and then tighten the nuts to the torque value given in the Specifications in the Appendix.

18- Install the centrifugal slinger over the driveshaft and against the pump cover. Install a **NEW O**-ring into the groove at the top of the driveshaft. Some models may not have the groove and therefore, do not use the O-ring. Place the water tube guide into the inlet of the water pump cover.

Check the forward gear backlash again at this time.

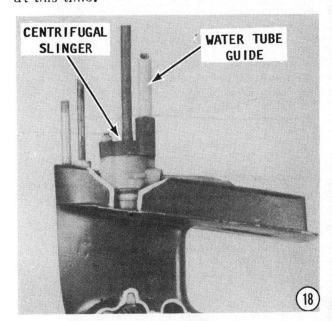

Model 200 Only

Position a **NEW** gasket on the face plate and one on the pump base. Install the face plate and the water pump cover. Apply a small amount of Multipurpose Lubricant to the drive pin area of the driveshaft.

Position the impeller drive pin on the flat area of the driveshaft. Slide the impeller over the driveshaft and down over the drive pin. With a thin blade screwdriver, check to be sure the drive pin did not slide off the flat area on the driveshaft.

Slide the water pump cover over the driveshaft and down over the impeller and at the same time rotate the driveshaft **CLOCKWISE**. Install the washers and nuts onto the water pump studs, and then tighten them to the torque value given in the Specifications in the Appendix. Install the centrifugal slinger. Install a **NEW** O-ring into the groove on the driveshaft. Some models may not have the groove and, therefore do not use the O-ring.

Propeller Shaft Assembling

19- Install the sliding clutch onto the propeller shaft with the ramp side of the clutch **TOWARD** the forward end of the shaft. Now, position the clutch until the pin hole aligns with the slot in the propeller shaft. Insert the spring into the forward end of the propeller shaft.

Insert a small punch through the hole in the sliding clutch and propeller shaft slot, and then pry the spring toward the aft end of the shaft until the cross-pin can be inserted from the opposite side of the sliding punch. Start the pin into the hole and remove the punch.

20- After the cross-pin is started, place the cam follower into the forward end of the propeller shaft. Push against the cam follower until the cross-pin is aligned with the hole on the opposite side of the sliding clutch and then push the cross-pin on in. Install the cross-pin retainer.

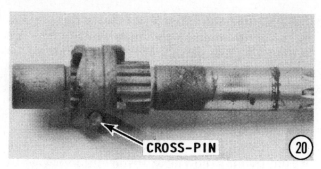

CROSS-PIN

20

21- On some models the retainer is a spring and on other models the retainer is a clip. **TAKE CARE** not to over-stretch the retainer.

CROSS-PIN
RETAINER SPRING

21

22- Place the shift cam in the **FORWARD** gear position. Insert the propeller shaft complete with cam follower into place in the center of the forward gear assembly.

Install shim material into the lower unit. If the reverse gear backlash was correct prior to disassembly and the pinion gear depth has not been changed, install shim material **EQUAL** to the amount removed during disassembly. The correct reverse gear backlash is 0.003" to 0.005".

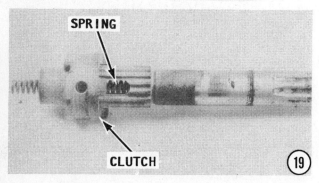

SPRING

CLUTCH

19

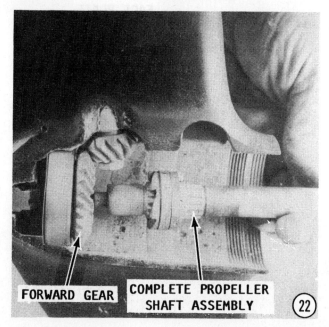

FORWARD GEAR   COMPLETE PROPELLER
SHAFT ASSEMBLY

22

Pinion Gear and Driveshaft Shimming

Shim material must be added or removed from the lower unit to obtain the proper pinion gear depth and reverse gear backlash. The following conditions outline possible circumstances that may have been encountered and the corrective action to be taken for each.

a- The pinion gear was lowered and the backlash was correct. Add shim material **EQUAL** to the amount removed from under the driveshaft ball bearing.

b- The pinion gear was raised and the gear backlash was correct. Remove shim mateial **EQUAL** to the amount added under the driveshaft ball bearing.

c- The pinion gear depth was unchanged and the gear backlash was **MORE** than specifications. Remove shims at a ratio of 0.001" shim material for each 0.0015" of gear lash change required.

d- The pinion gear depth was unchanged and the gear backlash was **LESS** than specifications. Add shim material at a ratio of 0.001" material for each 0.0015" gear lash change required.

Bearing Carrier Installation

23- Install a **NEW** O-ring between the thrust washer and the bearing carrier. Coat the O-ring and the propeller shaft oil seals with Multipurpose Lubricant. Coat the outer diameter of the bearing carrier with Perfect Seal. **TAKE CARE** to prevent any Perfect Seal from contacting the O-ring, oil seals, or the bearings.

Slide the bearing carrier over the propeller shaft and at the same time work the lips of the propeller shaft oil seal over the propeller shaft with a small screwdriver.

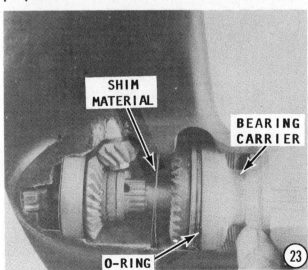

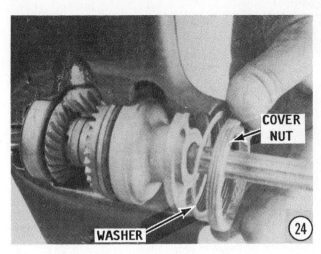

Push the bearing carrier into the lower unit and at the same time rotate the driveshaft **CLOCKWISE** to index the reverse gear with the pinion gear.

24- Coat the threads of the cover nut with Perfect Seal. Position the lower unit cover washer over the bearing carrier. Slide the cover nut over the propeller shaft with the words **OFF ON** and the arrows visible. To avoid the possiblility of cross-threading, start the cover nut by hand, **COUNTERCLOCKWISE** for **MODEL 45 thru 110** and **CLOCKWISE** for **MODEL 200**.

25- For **MODEL 45 thru 110:** use Gear Case Cover Tool C-91-48830; for **MODEL 200:** use tool C-91-33450, and tighten the cover nut to the torque value given in the Specifications in the Appendix.

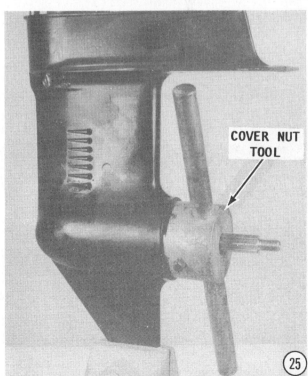

Shimming Reverse Gear Backlash

All MODEL 45 thru 200

26- Shift the lower unit into neutral gear. Rotate the propeller shaft and at the same time attempt to shift into reverse gear. The reverse gear backlash can only be checked at the point of propeller shaft rotation where it is impossible to shift into reverse gear, the clutch dogs are aligned, as shown. When this point is reached, push down on the driveshaft and pull out on the propeller shaft, and at the same time maintain pressure on the lower shift shaft toward the reverse gear position. Now, lightly rock the propeller shaft right and left. The amount of free play felt is the clearance between the reverse gear and the pinion gear teeth, commonly referred to as the backlash. The correct amount of backlash should be 0.003" to 0.005".

If the reverse gear backlash it not correct, remove the bearing carrier and add shim material to increase backlash or remove shim material to decrease backlash. For each 0.001" of shim material change, the backlash will change approximately 0.0015"

After making a shim change, recheck the backlash as described in this step.

Filling the Lower Unit

27- **ALWAYS** fill the lower unit with lubricant and check for leaks **BEFORE** installing the lower unit to the powerhead.

Take time to remove any old gasket material from the **FILL** and **VENT** recesses and from the screws. Place the lower unit in an upright vertical position. Fill the lower unit with Super-Duty Lubricant, or equivalent, through the **FILL** hole at the bottom of the unit. **NEVER** add lubricant to the lower unit without first removing the **VENT** screws **AND** having the unit in its normal operating position (vertical). Failure to remove the vent screw will result in air becoming trapped within the lower unit. Trapped air will not allow the proper amount of lubricant to be added. Continue filling **SLOWLY** until the lubricant flows out the **VENT** hole with no air bubbles visible.

Use a **NEW** gasket and install the **VENT** screw. Slide a **NEW** gasket onto the **FILL** screw. Remove the lubricant tube and **QUICKLY** install the **FILL** screw.

Check the lower unit for leaks.

WORDS OF ADVICE

The oil level (vent) screw should be removed periodically and a piece of wire used to determine oil level.

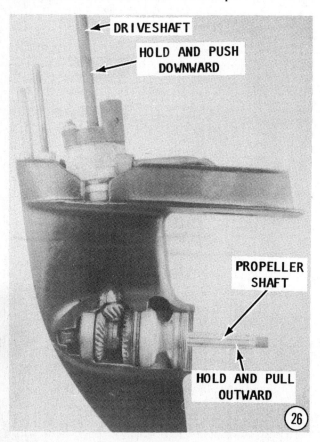

DRIVESHAFT

HOLD AND PUSH DOWNWARD

PROPELLER SHAFT

HOLD AND PULL OUTWARD

26

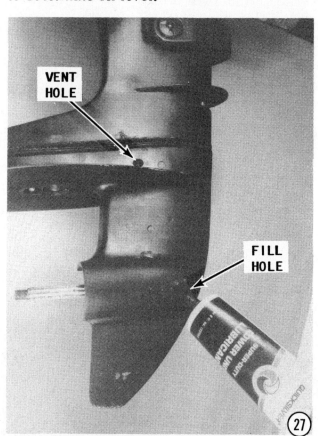

VENT HOLE

FILL HOLE

27

INSTALLATION TYPE "B" UNIT

28- Swing the exhaust housing outward until the tilt lock lever can be actuated, and then engage the tilt lock. Move the shift lever to the **FORWARD** gear position.

Cover the driveshaft and shift shaft splines with a light coating of Multipurpose Lubricant. **TAKE CARE** not to use an excessive amount of lubricant on the driveshaft splines on any Mercury Outboard.

BAD NEWS: An excess amount of lubricant on top of the driveshaft to crankshaft splines will be trapped in the clearance space. This trapped lubricant will not allow the driveshaft to fully engage with the crankshaft. As a result, when the lower unit nuts are tightened, a load will be placed on the driveshaft/crankshaft and will cause damage to either the powerhead or the lower unit or both. Therefore, any lubricant **MUST** be cleaned from the top of the driveshaft.

Install the water guide tube into the water pump cover with the tube entered into the water passage in the bottom cowl.

Position the driveshaft into the exhaust housing and at the same time align the water tube with the water guide tube and maintain the lower unit mating surface parallel with the exhaust housing mating surface. Push the flower unit toward the exhaust housing and at the same timer rotate the flywheel to permit the crankshaft splines to index (mate) with the driveshaft splines.

Hold the lower unit in position with one hand and check the shift operation with the other hand. Shift into neutral gear: In neutral the propeller shaft must be free to turn in either direction. Shift into reverse gear: The propeller shaft should **NOT** be free to rotate more then 1/2-turn. Shift into forward gear: The propeller shaft should not be free to turn counterclockwise.

If the lower unit fails any of the shift tests just described, the upper and lower shift shafts are not aligned properly. Remove the lower unit and repeat steps 25 and 26.

29- Start the 9/16" self-locking nut at the leading edge of the exhaust housing. **MODEL 45 thru 110:** use a screwdriver and push the water pickup screen over the rear driveshaft stud. Install the 1/2" rear stud. **DO NOT** tighten the nut at this time. Tighten the 9/16" front retainer nut, and then tighten the rear nut. Tighten both nuts to the torque value given in the Specifications in the Appendix.

MODEL 200: install the nut on the stud inside the anodic plate recess and install the anodic plate.

GOOD WORDS

A new anodic plate should be installed following a lower unit overhaul.

Insert the plastic cap into the hole at the lower rear edge of the exhaust housing. Release the tilt lock and lower the engine to the normal operating position.

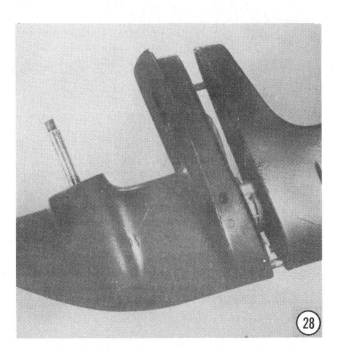

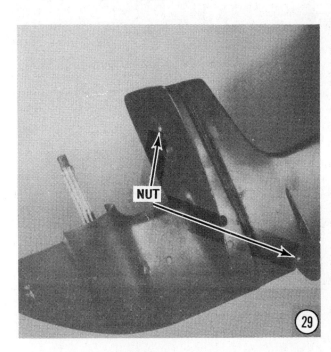

Propeller Installation

30- Disconnect the spark plug wires to prevent the engine from starting when the propeller and shaft are rotated. As an aid to removing the propeller the next time, apply a liberal coating of Perfect Seal to the propeller shaft splines.

31- Install the thrust hub into the center of the propeller, and then slide the propeller and thrust hub onto the propeller shaft. Slide the washer onto the shaft, and then thread the propeller locknut onto the shaft and tighten it to the torque value given in the Specifications in the Appendix. Use a block of wood between a propeller blade and the anti-cavitation plate to prevent the propeller shaft from rotating while the nut is being tightened.

Connect the spark plug wires to the spark plugs. Connect the electrical lead to the battery terminal. Mount the engine in a test tank or body of water.

CAUTION: Water must circulate through the lower unit to the engine any time the engine is run to prevent damage to the water pump in the lower unit. Just five seconds without water will damage the water pump.

Start the engine and check the completed work for satisfactory operation and no leaks.

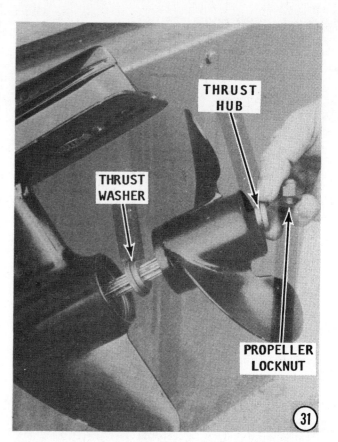

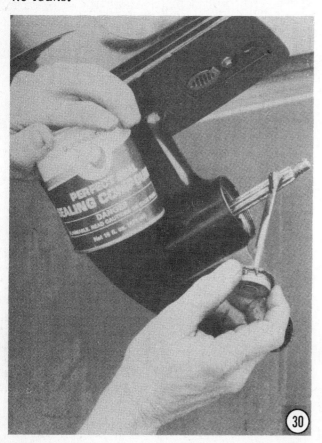

*Testing a small horsepower engine in a fifty-gallon tank. The engine should **NEVER** be run above idle or possibly moderate speed in such a test tank.*

9-6 LOWER UNIT "C"
REVERSE CAPABILITY
UNIQUE SHIFT ARRANGEMENT
(See Lower Unit Table in Appendix)

DESCRIPTION

In addition to the normal forward and neutral capabilities of the smaller engines, as outlined in Section 9-4, the lower unit of these model Mercury outboards are equipped with a reverse gear. A clutch and the necessary shift linkage for efficient operation is contained within the lower unit.

Since mid 1987, the manufacturer has introduced minor changes in the design of this type lower unit. The newer lower units may be easily identified from the exterior by the following features:

a- A rubber plug on the starboard side of the intermediate housing, just above the anti-cavitation plate. This plug conceals the shift coupler.

b- A sacrificial anode, resembling a large washer, on the starboard side of the lower unit, just above the anti-cavitation plate, next to the oil level vent plug.

c- Two bolts securing the bearing carrier to the lower unit.

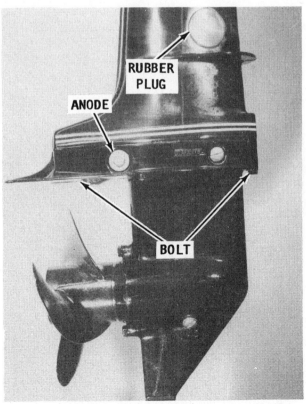

Exterior veiw of a "newer" type lower unit (manufactured since Mid 1987), with major differences identified.

If servicing a lower unit exhibiting these features, the following design differences will be encountered during disassembling of the unit:

a- Two piece shift rod, connected together with an externally accessible coupler.

b- Additional gaskets in the water pump.

c- An additional water pump base, containing a new design double lip oil seal plus additional gaskets.

d- New design sliding clutch at the forward end of the propeller shaft.

e- New design bearing carrier with a new design double lip oil seal.

When these differences affect the disassembly procedures, of the unit being serviced, separate steps will be given and identified as "newer" units. Models manufactured before mid 1987 will be identified as "older" units.

ADVICE

Before beginning work on the lower unit, take time to **READ** and **UNDERSTAND** the information presented in Section 9-1, this chapter, and check the Lower Unit Type and Backlash Table in the Appendix to ensure the proper procedures are being followed.

Disconnect the high tension spark plug leads, remove the spark plugs, and disconnect the leads at the battery terminals, before working on the lower unit.

Shift the engine into **FORWARD** gear. Raise the lower unit upward until the tilt lever can be actuated, and then engage the tilt stop.

Propeller Removal

SAFETY WORDS: An outboard engine may start very easily. Therefore, anytime the propeller is to be removed or installed, check to be sure:

a- Shift mechanism is in **NEUTRAL.**

b- Key switch is in **OFF** position.

c- Spark plug wires are disconnected.

d- Electrical leads disconnected at the battery terminals.

1- Remove the propeller nut by first placing a block of wood between one of the propeller blades and the anti-cavitation plate to prevent the propeller from turning, and then remove the nut. Remove the splined thrust hub.

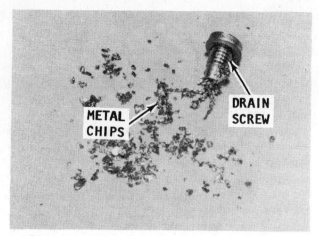

*The drain screw is magnatized to pickup metal particles in the lubricating fluid. The presence of metal particles on the screw or in the fluid is **BAD NEWS!** The lower unit must be disassembled and the damaged parts replaced.*

SPECIAL WORDS FOR OLDER UNITS

The splined thrust hub will be used later as a special tool to remove the bearing carrier from the lower unit.

Position a suitable container under the lower unit, and then remove the **FILL** screw and the **VENT** screw. Allow the gear lubricant to drain into the container. As the lubricant drains, catch some with your fingers from time-to-time, and rub it between your thumb and finger to determine if any metal particles are present. If metal is detected in the lubricant, the unit must be completely disassembled, inspected, and the damaged parts replaced. Check the color of the lubricant as it drains. A whitish or creamy color indicates the presence of water in the lubricant. Check the drain pan for signs of water separation from the lubricant. The presence of any water in the gear lubricant is bad news. The unit must be completely disassembled, inspected, the

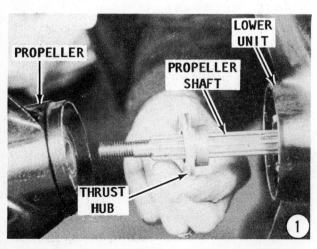

cause of the problem determined, and then corrected.

Older units: Remove the shift lever coupler located in the bottom cowling behind the carburetor.

Newer units: Pry the plastic plug from the starboard side of the intermediate housing. Shift the unit into **REVERSE GEAR.** Loosen, but **DO NOT REMOVE** the exposed bolt.

Older units: Remove the four attaching bolts securing the lower unit to the driveshaft housing.

Newer units: Remove the two attaching bolts, one located at the leading edge of the intermediate housing and the other directly aft of the water pickup on the underside of the anti-cavitation plate.

All units: Ease the lower unit from the intermediate housing.

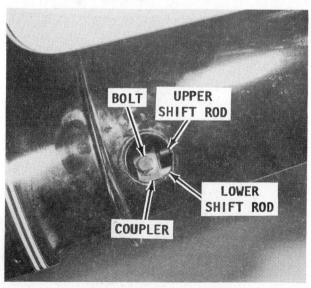

Access to the coupler for the upper and lower shift rods is gained by removing the rubber plug on the starboard side of the intermediate housing.

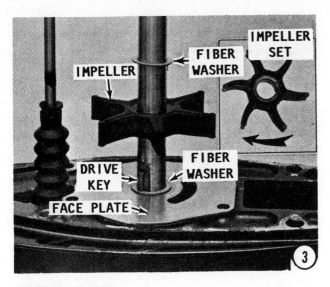

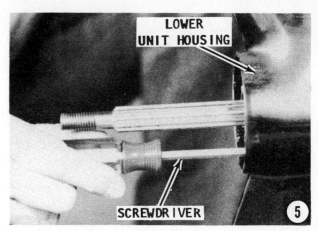

WATER PUMP

Removal

2- Slide the centrifugal rubber slinger up and free of the driveshaft (older units only). Remove the four screws securing the water pump cover. Slide the water pump cover up and free of the driveshaft.

3- Remove the impeller and the Woodruff key.

SPECIAL WORDS
FOR "OLDER" MODELS

The impeller is sandwiched between two fiber washers. **TAKE CARE** not to lose these washers.

Newer units: Remove the insert cartridge, water pickup tube and grommet from the water pump housing. Remove the gasket on top of the water pump face plate.

All units: Remove the water pump face plate and the gasket installed under the lower plate.

Propeller Shaft and Bearing Carrier Removal
Newer units:

4- Remove the two bolts securing the bearing carrier to the lower unit. Clamp the propeller shaft in a vise equipped with soft jaws. Obtain a soft headed mallet and strike the lower unit just above the bearing carrier to jar the bearing carrier free.

An alternate method of releasing the bearing carrier is as follows: Clamp the skeg in a vice equipped with soft jaws. Obtain puller P/N 91-27780 or a slide hammer with a jaw expander attachment. Hook the jaws inside the bearing carrier and use the slide hammer to jar the bearing carrier free.

A third method of releasing the bearing carrier is as follows: Rotate the bearing carrier cap 90° (1/4 turn), and then gently tap the bearing carrier cap with a soft headed mallet to jar the carrier free.

Remove the propeller shaft from the bearing carrier. Remove and discard the O-ring.

Older units:
Steps 5 thru 9

5- Remove the four slotted screws from the lower unit housing cavity.

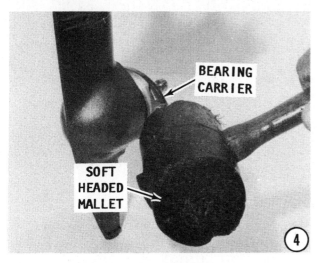

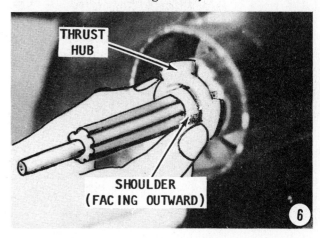

ROTATE CLOCKWISE TO REMOVE BEARING CARRIER

⑦

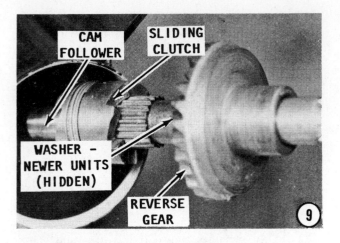

CAM FOLLOWER SLIDING CLUTCH

WASHER - NEWER UNITS (HIDDEN)

REVERSE GEAR

⑨

6- Slide the splined thrust hub onto the propeller shaft with the shoulder facing outward, as shown. Engage the ears of the thrust hub into the recesses of the bearing carrier.

7- Slide the propeller onto the propeller shaft until the splines of the propeller are fully engaged with the splines of the propeller shaft.

CRITICAL WORDS

The threads on the bearing carrier are **LEFT-HAND.** Therefore, the carrier must be rotated **CLOCKWISE** to remove.

Using the proper size socket or end wrench on the propeller, rotate the propeller **CLOCKWISE.** Continue rotating the propeller until the bearing carrier is loose.

8- Slide the bearing carrier free of the propeller shaft.

9- Pull outward on the propeller shaft. The shaft, reverse gear, sliding clutch, and cam follower, will all come out with the propeller shaft.

Bearing Carrier Disassembling
All units:

10- Remove the reverse gear from the bearing carrier. Newer models have a washer in front of the reverse gear, take care not to lose this washer. **WATCH** for and **SAVE** any shim material from the back side of the reverse gear. The shim material is critical in obtaining the correct backlash during assembling. On some models, the reverse gear and ball bearing assembly is pressed into the bearing carrier. The ball bearing is pressed onto the back of the reverse gear and must be separated using a bearing separator.

If the reverse gear and ball bearing assembly is pressed into the bearing carrier, the assembly may be "pulled" from the carrier using a slide hammer with jaw expander attachment, or special puller assembly and plate tool P/N 91-83165M. This tool has jaws similar to the slide hammer attachment, but uses a long screw and plate to extract the bearing, instead of a "hammer" action.

When using either tool check to be **SURE** the jaws are hooked onto the inner race.

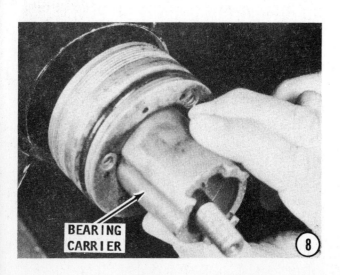

BEARING CARRIER

⑧

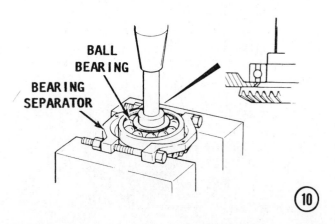

BALL BEARING

BEARING SEPARATOR

⑩

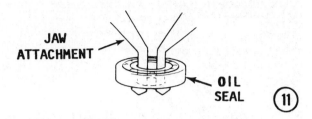

JAW ATTACHMENT

OIL SEAL

⑪

CRITICAL WORDS

Perform the following work only if the seal/s have been damaged and are no longer fit for service. Removal of the seal/s destroys sealing qualities. Therefore, the seal/s cannot be installed a second time. Be absolutely sure a new seal/s are available before removing the old seal/s.

11- Inspect the condition of the two seals (older units) or the double lip seal (newer units) in the bearing carrier. If the seal/s appear to be damaged and replacement is required, use the same slide hammer and jaw attachment or special tool as used in the previous step to remove the seal/s.

SPECIAL WORDS

Perform the following work only if the needle bearing (newer models) or bushing (older models) is damaged and is no longer fit for further service. Removal of the bearing or seal will distort it and therefore it cannot be reinstalled. Be absolutely sure a new part is available before removing the bearing or bushing. Unfortunately, the oil seals, good or bad, must be removed before the bearing or bushing can be driven out, and of course they must be replaced with new ones.

12- Note the position of the bearing or bushing in relation to the carrier. Look for any embossed numbers or letters and which shoulder of the bearing or bushing needs to be flush with the carrier. Remove the bearing or bushing using the same method and tools as described for the removal of the reverse gear bearing and oil seal/s, or obtain a suitably sized socket and driver, and drive the bearing or bushing from the carrier.

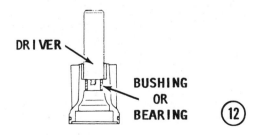

DRIVER

BUSHING OR BEARING

⑫

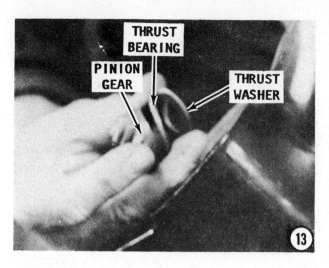

THRUST BEARING

PINION GEAR

THRUST WASHER

⑬

All Units

13- Reach into the lower unit cavity and grasp the pinion gear. Pull upward on the driveshaft with your other hand and the pinion gear (all units), bearing, and bearing race (older units only) will come free into your hand. When removing the driveshaft **WATCH** for and **SAVE** any shim material found on top of the driveshaft bearing. This shim material is critical to obtaining the correct backlash during installation. On newer units, the driveshaft has a ball bearing in addition to the needle bearing inside the lower unit. This ball bearing, sleeve and bushing are not serviceable. If any part of the driveshaft is defective, it must be replaced with a new driveshaft assembly.

SPECIAL WORDS

The shift shaft used on older units has a great many more parts on it than the shaft

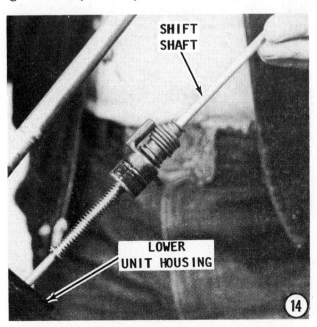

SHIFT SHAFT

LOWER UNIT HOUSING

⑭

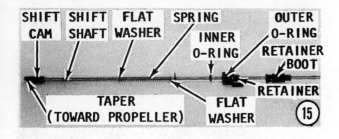

SHIFT CAM | SHIFT SHAFT | FLAT WASHER | SPRING | INNER O-RING | OUTER O-RING | RETAINER BOOT | RETAINER

TAPER (TOWARD PROPELLER)   FLAT WASHER

(15)

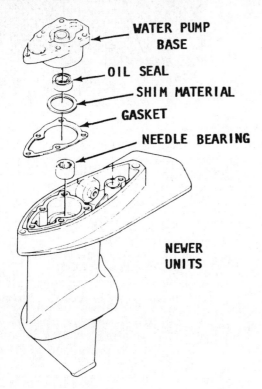

WATER PUMP BASE
OIL SEAL
SHIM MATERIAL
GASKET
NEEDLE BEARING

NEWER UNITS

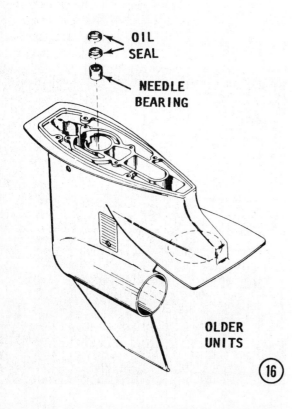

OIL SEAL
NEEDLE BEARING

OLDER UNITS

(16)

used in newer units. However, the cam at the lower end is identical for all units.

Shift Shaft Removal

14- **Newer units:** Remove the bolt, washer and retaining ring from the shift shaft.

All units: Pull upward on the shift shaft until it is clear of the lower unit.

15- Disassemble the shift shaft assembly only if a part shows excessive signs of wear.

Newer units: Slide the plastic bushing from the shaft. Remove the O-ring from the water pump base.

Older units: The assembly consists of the shift shaft and from left to right, shift cam, flat washer, spring, flat washer, inner O-ring, outer O-ring, retainer, and retainer boot, as shown. Change the inner O-ring **ONLY** if the lower unit lubricant shows signs of water indicating the O-ring has failed. The O-ring is very difficult to install without being damaged.

Water Pump Base Removal
Newer units:

16- Remove the water pump base. Remove and discard the gasket. There is a single oil seal contained within the water pump base.

CRITICAL WORDS

Perform the following work only if the seal has been damaged and is no longer fit for service. Removal of the seal destroys its sealing qualities. Therefore, the seal cannot be installed a second time. Be absolutely sure a new seal is available before removing the old seal.

Inspect the condition of the seal in the water pump base. If the seal appears to be damaged and replacement is required, use the same slide hammer and jaw attachment or special tool as used in step 10 to remove the seal.

Newer units: A single needle bearing is used at this location, just above the pinion gear.

Older units: A single needle bearing and two oil seals are stacked on top of the bearing. If the oil seals or bearings are fit for further service, they are best left alone. Removing either the oil seals or the bearings will destroy them. If only the oil seals must be removed (older units only), obtain puller P/N 91-27780 or a slide hammer with expanding jaw attachment to "pull" the seal. If the bearing must be removed, obtain needle bearing installation/removal tool P/N 91-17351 or a suitably sized 3/8" drive socket which will rest on the outer bearing cage. Use the tool and attached guide plate to remove the bearing. If using the alternate method, install the socket on a very long extension and drive the bearing down into the lower unit cavity.

CRITICAL WORDS

If a two piece bearing, such as the type installed in "older units", is to be replaced, the bearing and the race **MUST** be replaced as a matched set. Renove the bearing only if it is unsuitable for further service.

Water Pump Base Removal
Older units:

17- Tilt the lower unit. The forward gear tapered roller bearing will fall into your hand. If the bearing fails to fall free, strike the open end of the lower unit on a block of wood to jar the bearing free. **WATCH** for and **SAVE** any shim material found behind the forward gear. This shim material is critical to obtaining the correct backlash during installation. The shim material will be located between the forward gear and the bearing.

Newer units: The forward gear bearing is a one piece ball bearing and will not -- or should not -- just fall out of the lower unit cavity. The bearing must be "pulled" from the lower unit only if unfit for further service. Obtain puller P/N 91-27780 or a slide hammer with a jaw expander attachment. Hook the jaws inside the inner bearing race and use the slide hammer to jar the race free. **WATCH** for and **SAVE** any shim material found behind the forward gear. This shim material is critical to obtaining the correct backlash during installation.

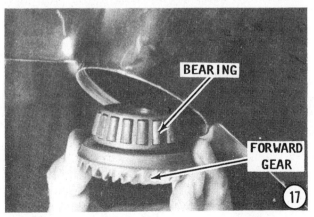

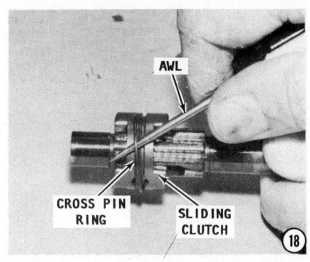

The shim material will be located between the forward gear and the bearing.

Older units: Obtain puller P/N 91-27780 or a slide hammer with a jaw expander attachment. Hook the jaws inside the bearing race and use the slide hammer to jar the race free.

Clutch Mechanism Disassembly

18- Remove the cam follower (all units) and guide (newer units only) from the propeller shaft.

Older units: Insert an awl under the end loop of the cross pin ring and unwind the ring from the sliding clutch.

19- Use a long pointed punch and press out the cross pin **LEAVING** the punch inside the propeller shaft once the pin has been driven free. Press the end of the propeller shaft down against the work bench and slow-

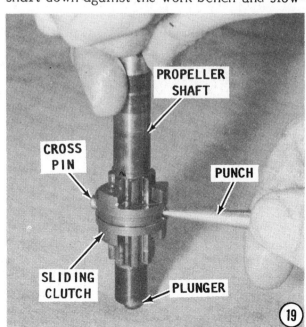

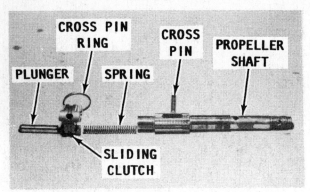

Arrangement of shift parts on the propeller shaft.

ly extract the punch. In this way, the compression spring inside the shaft will be contained and will not fly out. Observe how the clutch is installed on the shaft before sliding the clutch from the shaft.

CLEANING AND INSPECTING

Clean all water pump parts with solvent, and then dry them with compressed air. Inspect the water pump cover and base for cracks and distortion, possibly caused from overheating. Inspect the face plate and water pump insert for grooves and/or rough surfaces. If possible, **ALWAYS** install a new water pump impeller while the lower unit is disassembled. A new impeller will ensure extended satisfactory service and give "peace of mind" to the owner. If the old impeller must be returned to service, **ALWAYS** install the impeller in the original direction of rotation. Installation in reverse will cause premature impeller failure.

If installation of a new impeller is not possible, check the three seal surfaces. All must be in good condition to ensure proper pump operation. Check the upper, lower, and ends of the impeller vanes for grooves, cracking, and wear. Check to be sure the bonding of the impeller hub to the impeller will not allow the hub to slip inside the impeller.

Clean all bearings with solvent, dry them with compressed air, and inspect them carefully. Be sure there is no water in the air line. Direct the air stream through the bearing. **NEVER** spin a bearing with compressed air. Such action is highly dangerous and may cause the bearing to score from lack of lubrication. After the bearings are clean and dry, lubricate them with Formula 50 oil, or equivalent.

Inspect all ball bearings for roughness, scratches and bearing race side wear. Hold the outer race, and work the inner bearing race in and out, to check for side wear.

Determine the condition of tapered bearing rollers and inner bearing race, by inspecting the bearing cup for pitting, scoring, grooves, uneven wear, embedded particles, and discoloration caused from overheating. **ALWAYS** replace tapered roller bearings as a set.

Clean the forward gear with solvent, and then dry it with compressed air. Inspect the gear teeth for wear. Under normal conditions the gear will show signs of wear but it will be smooth and even.

Clean the bearing carrier with solvent, and then dry it with compressed air. **NEVER** spin bearings with compressed air. Such action is highly dangerous and may cause the bearing to score from lack of lubrication. Check the gear teeth of the reverse gear for wear. The wear should be smooth and even.

Check the teeth on the sliding clutch to be sure they are not rounded-off, or chipped. Such damage is usually the result of poor operator habits and is caused by shifting too slowly or shifting while the engine is operating at high rpm. Such damage might also be caused by improper shift cable adjustments.

Inspect the roller bearing surface of the propeller shaft. Check the shaft surface for pitting, scoring, grooving, embedded particles, uneven wear and discoloration caused from overheating.

Clean the driveshaft with solvent, and then dry it with compressed air. Inspect the bearing for roughness, scratches, or side wear. If the bearing shows signs of such damage, it should be replaced. If the bearing is satisfactory for further service coat it with oil.

Inspect the driveshaft splines for excessive wear. Check the oil seal surfaces above and below the water pump drive pin area for grooves. Replace the shaft if grooves are discovered.

Inspect the driveshaft roller bearing surface above the pinion gear splines for pitting, grooves, scoring, uneven wear, embedded metal particles and discoloration caused by overheating.

Inspect the propeller shaft oil seal surface to be sure it is not pitted, grooved, or scratched. Inspect the roller bearing con-

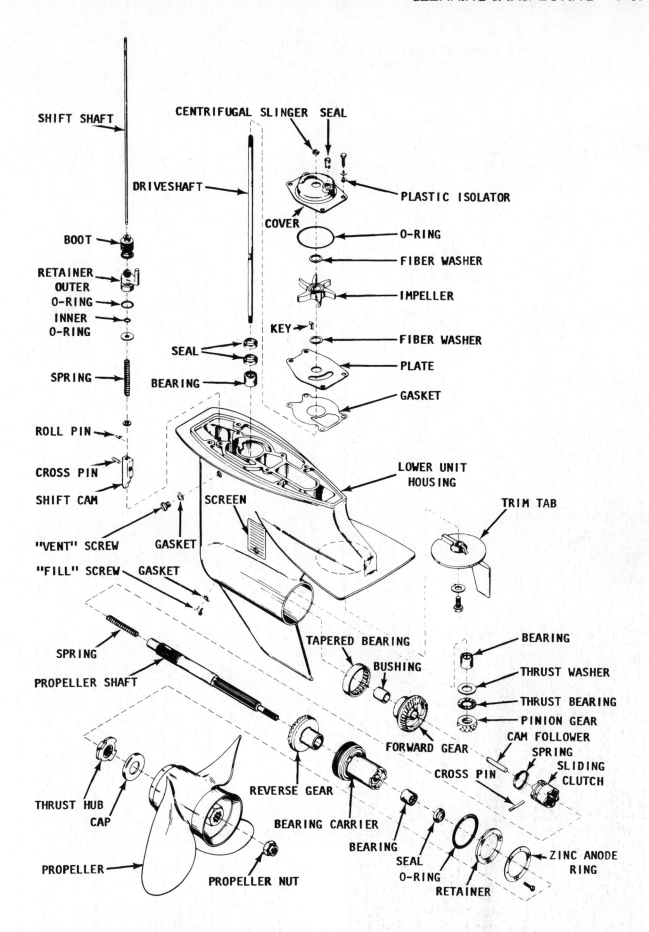

Exploded drawing of an "older" lower unit (manufactured prior to Mid 1987), with major parts identified.

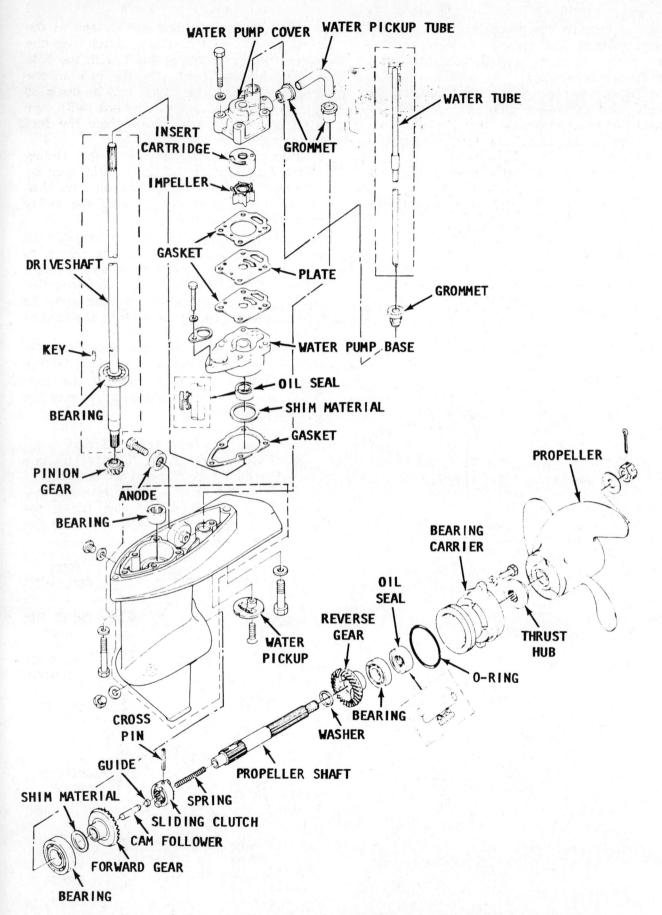

Exploded drawing of a "newer" lower unit (manufactured after mid 1987), with major parts identified.

tact surface on the propeller shaft for pitting, grooves, scoring, uneven wear, embedded metal particles, and discoloration caused from overheating.

Inspect the propeller shaft splines for wear and corrosion damage. Check the propeller shaft for straightness.

Inspect the shift cam for wear, corrosion, or other signs of damage.

Clean all parts with solvent, and then dry them with compressed air.

Inspect all bearing bores for loose fitting bearings.

Check the lower unit housing for impact damage.

Inspect the lower unit housing threads for cross-threading and corrosion damage.

Determine the condition of the labyrinth seal.

ASSEMBLING TYPE "C" UNIT

FIRST, THESE WORDS

As explained in the Description portion of this section, separate steps will be given for **"older"** and **"newer"** units. Models manufactured before mid 1987 are identified as **"older"** units. Units manufacturered since mid 1987 are identified as **"newer"** units.

Sliding Clutch Assembling

1- Slide the spring down into the propeller shaft. Insert a narrow screwdriver into the slot and compress the spring until approximately 1/2" (12mm) is obtained between the top of the slot and the screwdriver.

Hold the compressed spring, and at the same time, slide the sliding clutch over the splines of the propeller shaft with the hole in the clutch aligned with the hole in the propeller shaft. The clutch may be installed either way, preferably the side with the least amount of wear should face the forward gear.

Insert the cross pin into the sliding clutch and through the space held open by the screwdriver. Center the pin and then remove the screwdriver allowing the spring to pop back into place.

2- Older Units: Fit the cross pin ring into the groove around the sliding clutch, to retain the cross pin in place. Insert the flat end of the cam follower into the propeller shaft, with the rounded end protruding to permit the plunger to slide along the cam of the shift rod.

Newer Units: Install the guide into the end of the propeller shaft, followed by the cam follower. As both ends of the cam follower are equally rounded, it may be installed either way.

Forward Gear and Bearing Installation

3- Older units: Install the bearing race using a suitable size mandrel and driver.

Newer units: Install the ball bearing assembly with the numbered side facing toward the installation tool P/N 91-8453M and driver P/N 91-8429M.

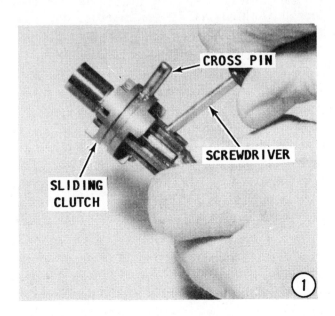

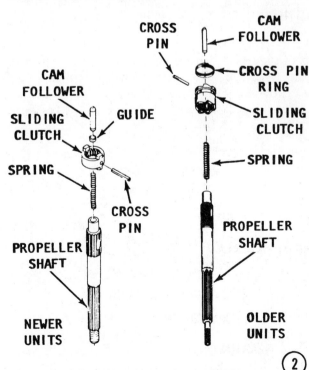

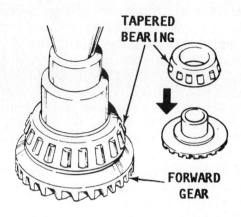

On some "older" lower units, the tapered bearing must be pressed onto the forward gear. Always press on the inner race, never on the cage or on the rollers.

Older units: If the tapered roller bearing was separated from the forward gear in step 17, the bearing and gear must be pressed together.

Position the forward gear tapered bearing over the forward gear. Use a suitable mandrel and press the bearing flush against the shoulder of the gear. **ALWAYS** press on the **INNER** race, never on the cage or the rollers.

All units: Thoroughly lubricate the bearing and gear with Mercury Super Duty Gear Lubricant. Insert the shim material saved during disassembling, Step 17, into the lower unit cavity. The shim material should give the same amount of backlash between the pinion gear and the forward gear as before disassembling. Insert the lubricated forward gear and bearing into the lower unit with the taper of the bearing going in first and the teeth of the gear facing outward, as shown.

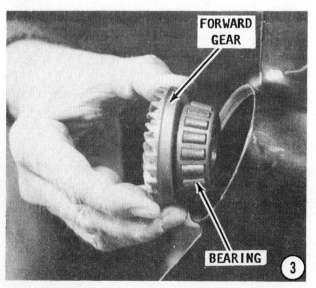

Driveshaft Needle Bearing Installation

The next step applies **ONLY** if the driveshaft needle bearing was removed. If the bearing was not disturbed, proceed directly to the following step.

4- Older units: Lower the needle bearing, the numbered side of the bearing facing **UPWARD**, into the driveshaft cavity. Obtain a suitable size driving socket and long extension. Drive the bearing squarely down until the lower surface of the bearing is flush with the lower unit cavity.

Newer units: Obtain bearing installation/removal tool P/N 91-17351. Slide the new needle bearing onto the end of the tool with the embossed numbers facing **UPWARD**. Slide the tool and the bearing into the top of the driveshaft cavity, making sure the bearing does not tilt. Drive the bearing into place. The bearing will seat itself at the correct location when the head of the driver rod seats against the guide bushing.

Driveshaft and Pinion Gear Installation

5- Hold the thrust washer, thrust bearing, and pinion gear in one hand, as shown. Insert your hand with these parts into the lower unit. Now, with the other hand slide the driveshaft down into the lower unit with the end with the short splines going in first. Slowly rotate the driveshaft until you feel the splines of the pinion gear index with the splines of the driveshaft. Slide any shim material over the driveshaft, which was saved from step 13 of disassembling. The shim material should give the same amount of backlash between the pinion gear and the other two gears as before disassembling.

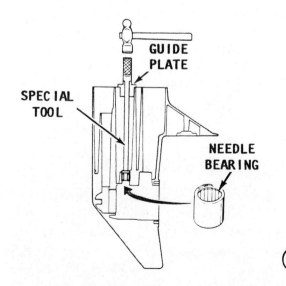

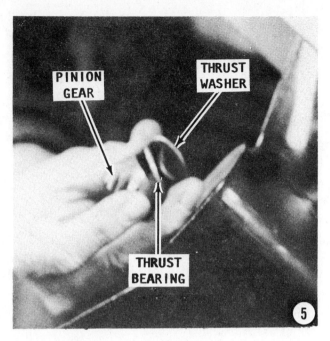

PINION GEAR

THRUST WASHER

THRUST BEARING

5

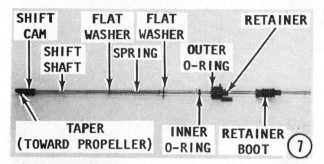

SHIFT CAM

FLAT WASHER

FLAT WASHER

RETAINER

SHIFT SHAFT

SPRING

OUTER O-RING

TAPER (TOWARD PROPELLER)

INNER O-RING

RETAINER BOOT

7

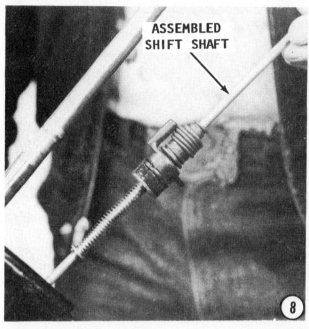

ASSEMBLED SHIFT SHAFT

8

Water Pump Plate Installation
Newer units:

The first part of this step applies **ONLY** if the water pump plate oil seal was removed. If the seal was not disturbed, simply skip the seal installation instructions.

6- Obtain driver P/N 91-84530M and the appropriate driver handle. Place the water pump plate on the workbench as it is to be installed in the lower unit. Coat the seal lip with a good grade of water resistant lubricant. Install the oil seal using the driver with the oil seal lip facing **UPWARD**. Install a new gasket over the driveshaft and lower the plate into place aligning the bolt holes.

Shift Shaft Installation

7- **Older units:** Assemble the shift shaft, if it was disassembled, in the order of parts, as shown. The inner **O**-ring must be installed with the utmost **CARE** to prevent damage to the ring.

Newer units: Coat the surface of a new **O**-ring with a good grade of water resistant lubricant and install the **O**-ring into the water pump plate recess. Slide the bushing over the shift shaft with the shoulder of the

bushing closest to the cam at the end of the shaft.

8- **All units:** Slide the assembled shift shaft into the lower unit with the cam end going in first.

9- Check to be sure the cam follower is on the end of the propeller shaft. This is a loose piece and easily forgotten. Insert the

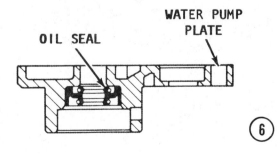

OIL SEAL

WATER PUMP PLATE

6

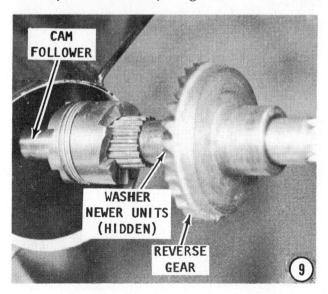

CAM FOLLOWER

WASHER NEWER UNITS (HIDDEN)

REVERSE GEAR

9

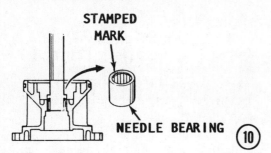

propeller shaft into the lower unit until the end of the shaft indexes into the forward gear center.

SPECIAL WORDS

If neither the needle bearing nor the oil seal were disturbed in steps 11 or 12, simply skip the installation instructions given in the beginning of the following step.

10- Place the bearing carrier on the workbench with the propeller end facing **DOWNWARD**.

Newer units: Obtain installation tool P/N 91-83174M.

Older units: Obtain a suitably sized driver.

All units: Coat the lip of the seal with a good grade water resistant lubricant and install the seal with the lip facing **DOWNWARD**.

Older units: Keep the bearing carrier in the same position on the workbench, and install the needle bearing with the embossed numbers on one face of the bearing facing **UPWARD**.

11- **Newer units:** Obtain bearing installation tool P/N 91-84536M. Move the assembled bearing carrier to an arbor press. Place the bearing squarely over the carrier with the numbered side of the bearing facing **UP-WARD**. Press the bearing into the carrier until seated. Install the washer and then the reverse gear onto the propeller shaft.

Older units: The reverse gear ball bearing is pressed into the bearing carrier on

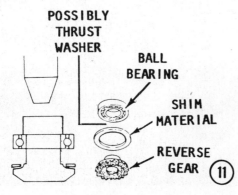

some models. On other models the reverse gear is pressed onto the reverse bear ball bearing.

All units: **DO NOT** forget to install any shim material removed from behind the reverse gear during disassembling. This shim material and, in some cases a thrust washer, **MUST** be installed between the reverse gear and the ball bearing assembly in order to ensure proper mesh between the reverse gear and the pinion gear.

12- **Newer units:** Install a new O-ring around the bearing carrier. Coat the O-ring with water resistant lubricant.

Older units: Install the O-ring, O-ring retainer plate, and then the zinc anode ring in that order. Start the four screws into the bearing carrier. These screws will align all the parts. Coat the threads of the bearing carrier with Mercury Perfect Seal Sealing Compound or equivalent non-hardening gasket sealer. Thread the bearing carrier into

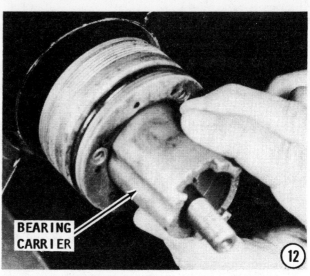

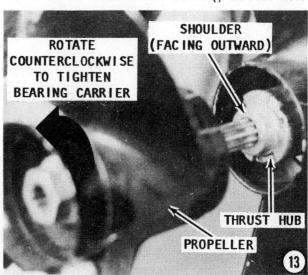

the lower unit **COUNTERCLOCKWISE** by hand. Remember, the threads are **LEFT-HAND**.

Older units: Next Two Steps

13- Slide the splined thrust hub onto the propeller shaft with the shoulder facing outward, as shown. Engage the ears of the thrust hub into the recesses of the bearing carrier. Slide the propeller onto the propeller shaft until the splines of the propeller are fully engaged with the splines of the propeller shaft.

CRITICAL WORDS

The threads on the bearing carrier are **LEFT-HAND**. Therefore, the carrier must be rotated **COUNTERCLOCKWISE** to install and tighten.

Using the proper size socket or end wrench on the propeller, rotate the propeller **COUNTERCLOCKWISE**. Continue rotating the propeller until the bearing carrier is tightened securely.

14- Tighten the four bearing carrier screws securely.

Newer units:

15- Slide the bearing carrier over the installed propeller shaft. Using a soft headed mallet, lightly tap around the circumference of the carrier to seat it. Make sure the two bolt holes are aligned with the lower unit holes. Install the two washers and bolts. Tighten the bolts to a torque value of 5.8 ft lbs (8Nm).

GOOD WORDS

The manufacturer gives no specific instructions for setting up the backlash on

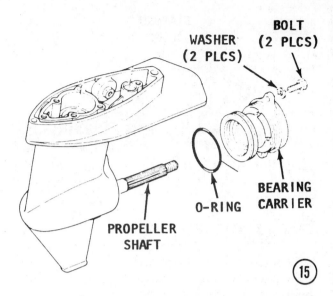

(15)

these units. As long as the shim material found during disassembly was placed in their original locations, the backlash should be acceptable. The manufacturer states: "The amount of play between the gears is not critical, but **NO** play is unacceptable". Therefore, if after the assembly work is complete, the gears are indeed locked, the unit must be disassembled and shim material placed behind all three gears, using a "trial and error" method. If the lower unit is allowed to operate without "some" backlash, very heavy wear on the three gears will take place almost immediately.

WATER PUMP INSTALLATION

16- Slide the gasket down the driveshaft and into place on the lower unit. The screw holes are offset, therefore the gasket holes will only line up properly one way. Slide the plate down the driveshaft and into place.

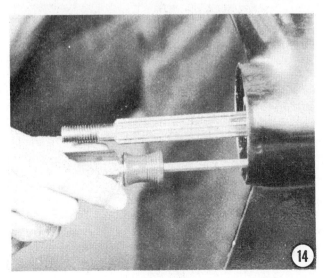

(14)

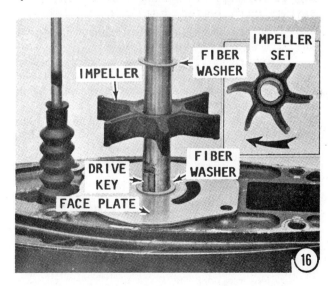

(16)

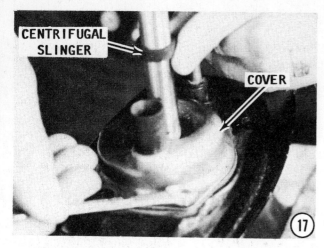

Newer units: Slide another gasket down over the driveshaft.

Older units: Slide the fiber washer down the driveshaft and onto the plate. Apply just a dab of petroleum jelly or grease onto the stainless steel key to hold it in place. Stick the key to the "flat" of the driveshaft. Slide a **NEW** impeller down the driveshaft with the slot in the impeller indexing over the stainless steel key.

GOOD WORDS

If an old impeller is installed be **SURE** the impeller is installed in the same manner from which it was removed. **NEVER** turn the impeller over thinking it will extend its life. On the contrary, the blades would crack and break after just a short time of operation.

Older units: Slide the second fiber washer onto the driveshaft.

17- Newer units: Install the water pump cartridge into the water pump cover, with the locating tab indexing into the hole in the cover.

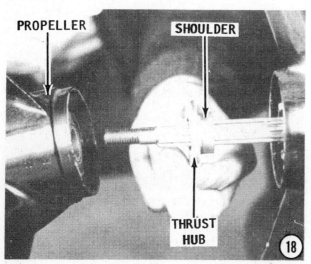

All units: Lubricate the inside surface of the water pump cover with a light coating of petroleum jelly or equivalent grease. Insert a **NEW** O-ring into the groove in the underside surface of the water pump cover. Slide the water pump cover down the driveshaft until it makes contact with the impeller. Apply a small amount of downward pressure on the water pump cover and at the same time rotate the driveshaft **CLOCKWISE** until the cover seats on the plate.

GOOD WORDS FOR OLDER UNITS

The design of the water pump cover allows it to serve as a retainer for the shift shaft and the driveshaft.

All units: Secure the cover in place with the four attaching screws.

Older units: Slide a **NEW** centrifugal slinger down the driveshaft until it seats on the cover.

All units: Push a **NEW** water tube seal onto the water pump cover.

Newer units: Push one end of the water pickup tube into the grommet on the water pump cover. Push the other end of the tube into the grommet at the aft end of the lower unit housing.

18- Install the splined thrust hub onto the propeller shaft with the shoulder facing toward the lower unit, as shown. Coat the propeller shaft with Mercury Perfect Seal Sealing Compound or equivalent waterproof lubricant. Slide the propeller onto the pro-

Every boat owner should have a can of Perfect Seal Sealing Compound ready for use when installing the propeller. This product will prevent the propeller from "freezing" on the shaft -- the propeller splines seizing with the propeller shaft splines.

peller shaft until it seats against the thrust hub. Thread the propeller nut onto the propeller shaft and tighten it securely.

Filling Lower Unit

19- **ALWAYS** fill the lower unit with lubricant and check for leaks **BEFORE** installing the unit to the powerhead.

Lower Unit Installation

20- **Older units:** Swing the exhaust housing outward until the tilt lock lever can be actuated, and then engage the tilt lock. Move the shift lever to the **FORWARD** gear position.

Cover the driveshaft and shift shaft splines with a light coating of Multipurpose Lubricant. **TAKE CARE** not to use an excessive amount of lubricant on the driveshaft splines on any Mercury Outboard.

BAD NEWS

An excessive amount of lubricant on top of the driveshaft to crankshaft splines will be trapped in the clearance space. This trapped lubricant will not allow the driveshaft to fully engage with the crankshaft.

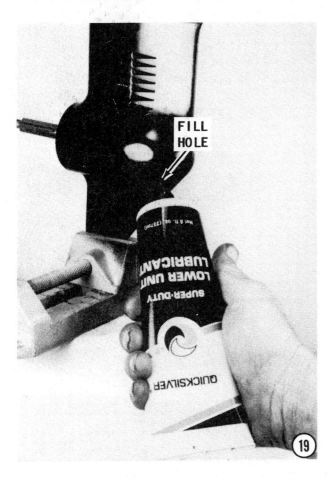

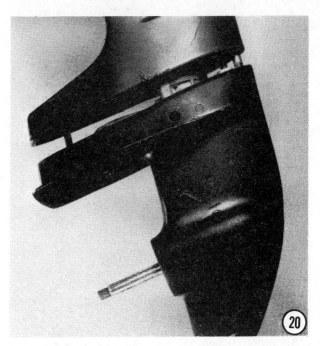

Position the driveshaft into the exhaust housing and at the same time align the water tube with the water pump cover outlet. Maintain the lower unit mating surface parallel with the exhaust housing mating surface. Push the lower unit toward the exhaust housing and at the same time rotate the flywheel to permit the crankshaft splines to index with the driveshaft splines.

Secure the lower unit in place with the attaching bolts. Install the shift lever connector in the bottom cowling behind the carburetor.

Check the completed work for proper shifting.

Manufacturer recommended lubricants and additives will not only keep the unit within the limits of the warranty, but will be a major contributing factor to dependable performance and reduced maintenance work and costs.

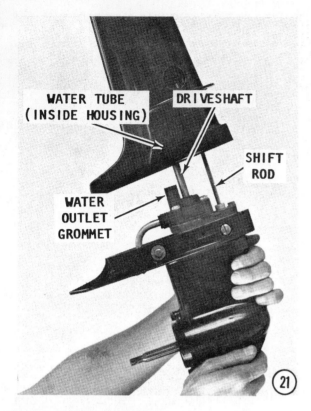

WATER TUBE (INSIDE HOUSING)

DRIVESHAFT

SHIFT ROD

WATER OUTLET GROMMET

(21)

Lower Unit Installation

21- Newer units: Swing the exhaust housing outward until the tilt lock lever can be actuated, and then engage the tilt lock.

Both the lower unit and the shift lever **MUST** be in the full reverse position to enable the upper shift shaft to align with the clamp on the lower shift shaft. Move the shift lever to the **REVERSE** gear position.

Rotate the propeller shaft while pushing **DOWN** on the shift shaft to shift the lower unit into reverse gear. When in reverse, the propeller shaft will not rotate more than a few degrees in either direction.

Cover the driveshaft and shift shaft splines with a light coating of Multipurpose

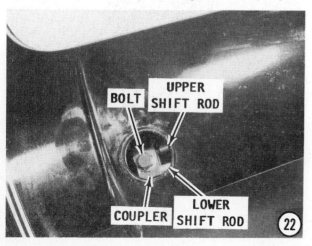

BOLT

UPPER SHIFT ROD

COUPLER   LOWER SHIFT ROD

(22)

Lubricant. **TAKE CARE** not to use an excessive amount of lubricant on the driveshaft splines on any Mercury Outboard.

BAD NEWS

An excessive amount of lubricant on top of the driveshaft to crankshaft splines will be trapped in the clearance space. This trapped lubricant will not allow the driveshaft to fully engage with the crankshaft.

Position the driveshaft into the exhaust housing and at the same time align the water tube with the water pump cover outlet. Feed the lower shift rod into the coupler on the upper shift rod. Maintain the lower unit mating surface parallel with the exhaust housing mating surface. Push the lower unit toward the exhaust housing and at the same time rotate the flywheel to permit the crankshaft splines to index with the driveshaft splines.

22- Secure the lower unit in place with the attaching bolts. Tighten the bolts to a torque value of 5.8 ft lbs (8Nm). Tighten the coupler bolt to the same torque valve, and install the plug into the intermediate housing to cover the coupler bolt. Check the completed work for proper shifting.

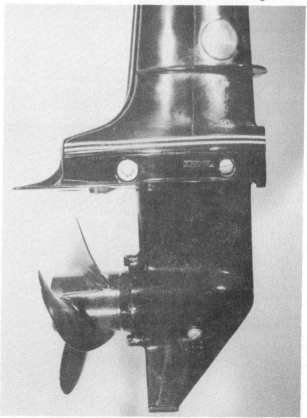

A "newer" (manufacturered since Mid 1987), lower unit, serviced, installed, and ready to serve the owner.

9-7 LOWER UNIT TYPE D
NO REVERSE CAPABILITY
WATER PUMP INSTALLED
ON PROPELLER SHAFT
(See Lower Unit Table in Appendix)

FIRST, THESE WORDS

One of the features making this lower unit unique from all others covered in this manual is the location of the water pump. Instead of the pump being installed on the driveshaft, as on most units, the water pump is installed on the propeller shaft.

In this location, the pump impeller may be replaced without disassembling the lower unit. In fact the propeller shaft does not have to be disassembled. The only work required is to remove the propeller and a couple other simple tasks performed to replace an impeller.

THEREFORE, if the only service work necessary is to replace the water pump impeller, perform Steps 1 thru 5 in the Disassembling portion of this section, and then Steps 11 thru 18 in the Assembling procedures.

The following procedures are provided for a complete overhaul of the lower unit.

DISASSEMBLING

1- Pull the cotter pin from the propeller shaft. A spare cotter pin is provided to a new owner and may be found in the spark plug access cover, courtesy of the manufacturer.

2- Slide the propeller rearward and free of the propeller shaft. Remove the shear pin. A spare shear pin may also be found in the spark plug access cover. Again, compliments of the manufacturer.

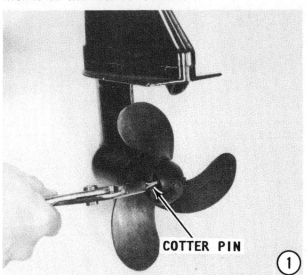

COTTER PIN

①

SHEAR PIN

②

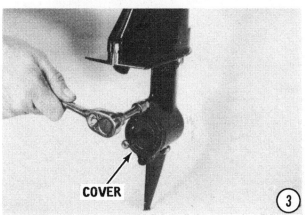

COVER

③

3- Remove the two bolts and washers securing the water pump cover to the lower unit.

4- Jar the water pump cover free of the lower unit using a soft head mallet. Remove the cover from the propeller shaft. Check the inside surface of the cover for signs of wear indicating foreign particles had entered the water pump.

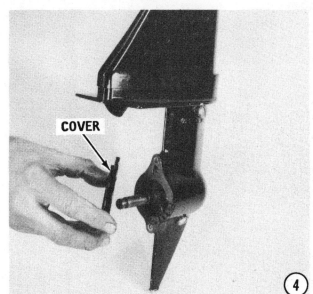

COVER

④

WATER PUMP IMPELLER

⑤

5- Pry the water pump impeller from the lower unit recess. Check the condition of the impeller carefully and replace with a new one if there is any question as to its condition for satisfactory service. **NEVER** turn the impeller over in an attempt to gain further life from the impeller.

SPECIAL WORDS

If the only service on the lower unit is to replace the water pump impeller proceed directly to Step 11 in the Assembling procedures. Actually, assembling entails simply installing a new impeller and then replacing the cover and the propeller with the attaching hardware.

6- Position a suitable container under the lower unit, and then remove the **OIL** screw and the **OIL LEVEL** screw. Allow the gear lubricant to drain into the container. As the lubricant drains, catch some with your fingers from time to time, and rub it between your thumb and finger to determine if any metal particles are present. If metal is detected in the lubricant, the unit must be completely disassembled, inspected, and the damaged parts replaced. Check the color of the lubricant as it drains. A whitish or creamy color indicates the presence of water in the lubricant. Check the drain pan for signs of water separation from the lubricant. The presence of any water in the gear lubricant is bad news. The unit must be completely disassembled, inspected, the cause of the problem determined, and then corrected.

7- Remove the six bolts securing the powerhead to the driveshaft housing.

8- CAREFULLY pry the powerhead free of the driveshaft housing. It may be necessary to tap on the joint with a soft head mallet to break the powerhead loose. Lift the powerhead straight up and clear of the driveshaft.

BAD NEWS

If the unit is several years old, or if it has been operated in salt water, or has not had proper maintenance, or shelter, or any number of other factors, then separating the powerhead from the driveshaft housing may not be a simple task. An air hammer may be required on the studs to shake the corro-

OIL LEVEL SCREW

OIL SCREW

⑥

DRIVESHAFT HOUSING

POWERHEAD

⑦

sion loose; heat may have to be applied to the casting to expand it slightly; or other devices employed in order to remove the powerhead.

One very serious condition would be the driveshaft "frozen" with the crankshaft. In this case, a circular plug-type hole must be drilled and a torch used to cut the drive-shaft. Let's assume the powerhead will come free on the first attempt.

Thoroughly clean any gasket material adhering to the mating surfaces of the powerhead and the driveshaft housing.

9- Remove the anode if it appears to be corroded or "eaten" away. The anode may deteriorate fairly rapidly if the outboard unit is operated in salt water. The electrolysis acting on the anode protects expensive parts.

SPECIAL WORDS

The recess in which the water pump impeller rotates is not perfectly round. Therefore the propeller shaft may be rotated to allow the drive pin to be withdrawn on the "larger" side.

10- Rotate the propeller shaft and pull the drive pin free of the shaft.

11- Withdraw the water pump housing. Rotate the bearing and check for any "rough" spots. If the bearing appears to be rusted, it should be replaced. The bearing may be removed using a suitable bearing puller. Remove and **DISCARD** the O-ring from the base of the water pump housing.

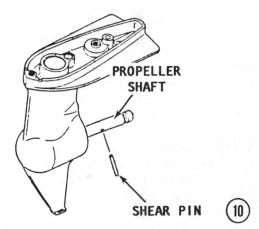

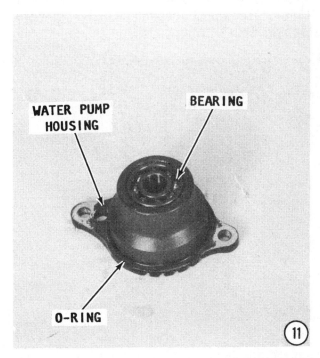

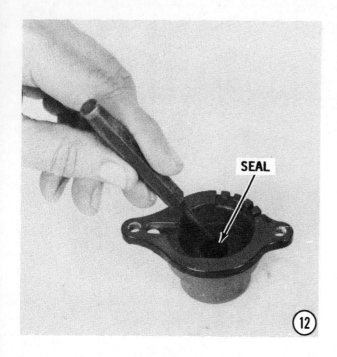

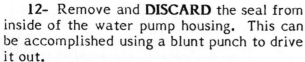

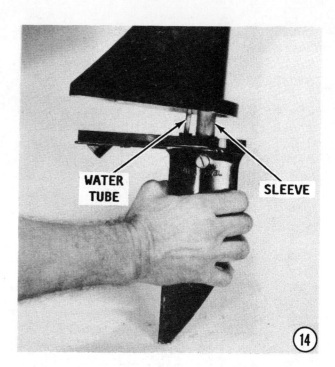

12- Remove and **DISCARD** the seal from inside of the water pump housing. This can be accomplished using a blunt punch to drive it out.

13- Remove the two bolts securing the lower unit gear housing to the mid-section. One bolt is located on the front side of the mid-section and the other goes into the mid-section from the underneath side of the lower gear housing, as shown.

14- Separate the lower gear housing from the mid-section. It may be necessary to tap the seam around the lower gear housing with a soft-head mallet to jar it loose. The upper rectangular driveshaft and sleeve, together with the water tube, will

usually remain in the mid-section when the two units are separated. The lower drive-shaft will remain in the lower gear housing.

15- Using a screwdriver, "pop" the Cir-clip out of the groove in the lower end of the lower driveshaft. This clip holds the pinion gear onto the driveshaft. The clip may not come free on the first try, but have patience and it will come free.

16- With one hand, pull the lower drive shaft straight up and out of the lower gear housing and at the same time catch the pinion gear with the other hand.

17- Pull the propeller shaft out of the lower gear housing. The drive gear **AND** the

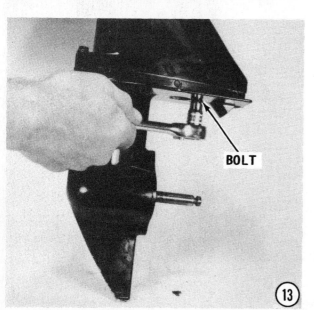

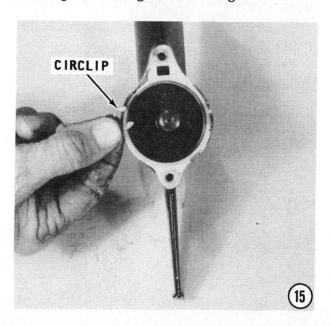

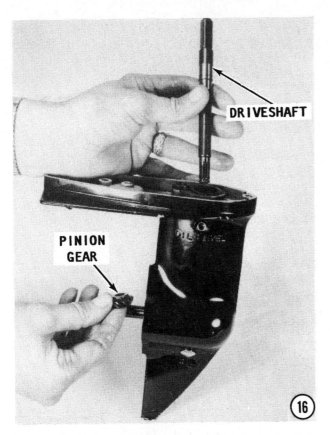

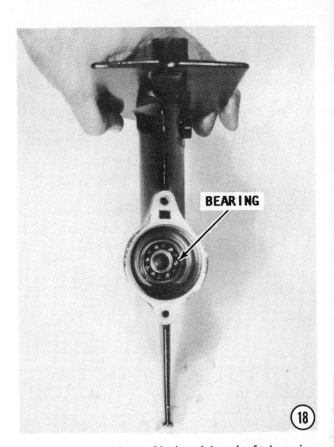

shim material on the shaft will come with it. **SAVE** the shim material.

18- Check the condition of the forward propeller shaft bearing. This bearing need not be removed unless it appears to be unfit for further service. If the bearing does not rotate freely or is rusted, it should be replaced. Remove the bearing using a suitable bearing puller. If a puller is not available, perform Steps 20 and 21 after Step 19.

19- Remove the water tube seal and the driveshaft oil seal from the upper portion of

the lower housing. If the driveshaft bearing is no longer fit for service, use a pair of Truarc pliers and remove the Tru-arc ring securing the driveshaft bearing in place. Use a suitable bearing puller to remove the bearing. If the bearing puller is not available, see the "Good Words" following and it may be possible to remove this bearing in a similar manner.

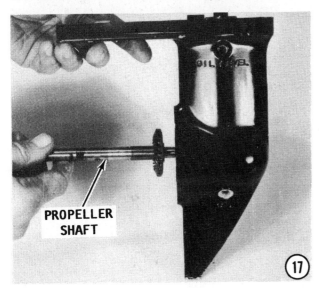

GOOD WORDS

The following two steps are to be performed only if the forward propeller shaft bearing must be removed and a suitable bearing puller is not available. The procedure will change the temperature between the bearing retainer and the housing substantially, hopefully about 80°F (50°C). This change will contract one metal -- the bearing retainer, and expand the other metal --the housing, giving perhaps as much as .003" (.08mm) clearance to allow the bearing to fall free. Read the complete steps before commencing the work because three things are necessary: a freezer, refrigerator, or ice chest; some ice cubes or crushed ice; and a container large enough to immerse about 1-1/2" (3mm) of the forward part of the lower gear housing in boiling water.

After all parts, including all seals, grommets, etc., have been removed from the lower gear housing, place the gear housing in a freezer, preferably overnight. If a freezer is not available try an electric refrigerator or ice chest. The next morning, obtain a container of suitable size to hold about 1-1/2" (3mm) of the forward part of the lower gear housing. Fill the container with water and bring to a rapid boil. While the water is coming to a boil, place a folded towel on a flat surface for padding.

20- After the water is boiling, remove the lower gear housing from its cold storage area. Fill the propeller shaft cavity with ice cubes or crushed ice. Hold the lower gear housing by the trim tab end and the lower end of the housing. Now, immerse the lower unit in the boiling water for about 20 or 30 seconds.

21- Quickly remove the housing from the boiling water; dump the ice; and with the open end of the housing facing downward, **SLAM** the housing onto the padded surface. **PRESTO**, the bearing should fall out.

If the bearing fails to come free, try the complete procedure a second time. Failure on the second attempt will require the use of a bearing puller.

CLEANING AND INSPECTING

Clean all water pump parts with solvent, and then dry them with compressed air. Inspect the water pump cover and base for cracks and distortion. If possible, **ALWAYS** install a new water pump impeller while the lower unit is disassembled. A new impeller will ensure extended satisfactory service and give "peace of mind" to the owner. If the old impeller must be returned to service, **NEVER** install it in reverse to the original direction of rotation. Installation in reverse will cause premature impeller failure.

Inspect the ends of the impeller blades for cracks, tears, and wear. Check for a glazed or melted appearance, caused from operating without sufficient water. If any question exists, as previously stated, install a **NEW** impeller if at all possible.

Inspect the bearing surface of the propeller shaft. Check the shaft surface for pitting, scoring, grooving, imbedded particles, uneven wear and discoloration.

Check the straightness of the propeller shaft with a set of **V**-blocks. Rotate the propeller on the blocks

Good shop practice dictates installation of new **O**-rings and oil seals **REGARDLESS** of their appearance.

Clean the pinion gear and the propeller shaft with solvent. Dry the cleaned parts with compressed air.

Check the pinion gear and the drive gear for abnormal wear.

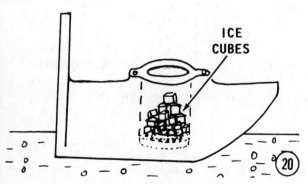

ICE CUBES

20

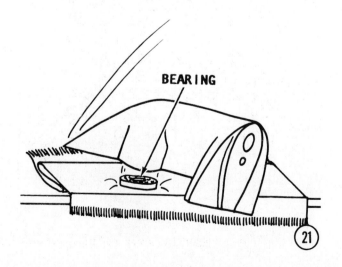

BEARING

21

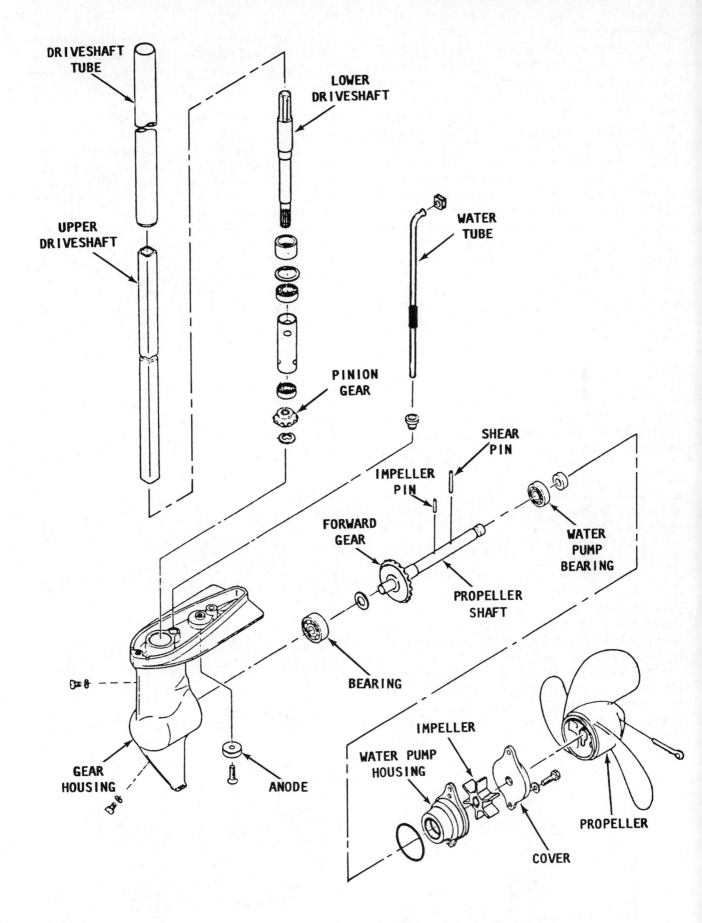

Exploded drawing of the lower unit Type "D" with major parts identified.

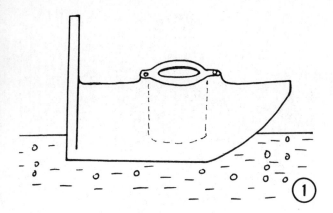

ASSEMBLING LOWER UNIT "D"

FIRST, THESE WORDS

The first two assembling steps apply **ONLY** if the forward propeller shaft bearing was removed from the housing. Therefore, the assumption is made disassembling Steps 20 and 21 were followed in order to remove the bearing. The "Good Words" prior to Step 20 explains in detail the theory behind the procedure. If this bearing was not removed, proceed directly to Step 3.

1- Place a **NEW** forward propeller shaft bearing in a freezer, refrigerator, or ice chest, preferably overnight. The next morning, boil water in a container of sufficient size to allow about 1-1/2" (3.8cm) of the forward part of the lower gear housing to be immersed. While the water is coming to a boil, place a folded towel on a flat surface for padding. Immerse the forward part of the gear housing in the boiling water for about a minute.

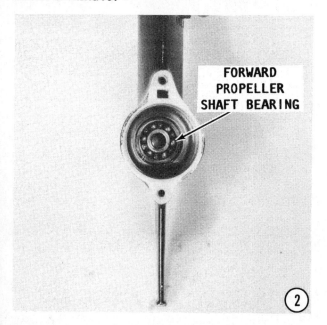

FORWARD PROPELLER SHAFT BEARING

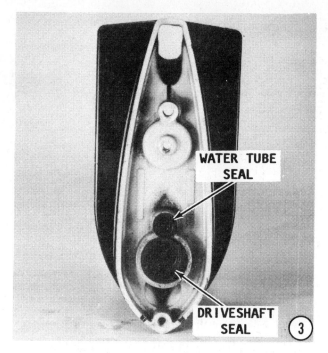

WATER TUBE SEAL

DRIVESHAFT SEAL

2- Quickly remove the lower gear housing from the boiling water and place it on the padded surface with the open end facing upward. At the same time, have an assistant bring the bearing from the cold storage area. Continue working rapidly. Carefully place the bearing squarely into the housing as far as possible. Push the bearing into place. Obtain a blunt punch or piece of tubing to bear on the complete circumference of the retainer. **CAREFULLY** tap the bearing retainer all the way into its forward position -- until it "bottoms-out". Tap evenly around the outer perimeter of the retainer, shifting from one side to the other to **ENSURE** the bearing is going squarely into place. The bearing must be properly installed to receive the forward end of the propeller shaft.

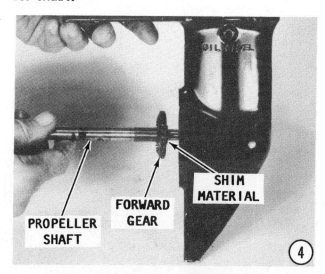

PROPELLER SHAFT

FORWARD GEAR

SHIM MATERIAL

3- If the upper driveshaft bearing was removed, position a new bearing in place and drive it in until it "bottoms-out" in the recess. Use a suitable driver or piece of tubing to ensure the bearing goes in squarely. The bearing must be installed properly to receive the lower driveshaft. Install the Truarc snap ring to secure the bearing in place. Install a **NEW** driveshaft oil seal and water tube seal.

4- Place the shim material saved during disassembling, Step 17, onto the propeller shaft ahead of the drive gear. The shim material should give the same backlash between the pinion and drive gear as before disassembling. Insert the propeller shaft into the lower gear housing. Push the propeller shaft into the forward propeller shaft bearing as far as possible.

5- Slide the lower driveshaft down through the lower gear housing with the splined end going in first -- the square end up, as shown. Hold the driveshaft with one hand and slide the pinion gear up onto the lower end of the driveshaft with the other hand. The splines of the pinion gear will index with the splines of the driveshaft if the gear is rotated slightly as it is moved upward.

6- Now, comes the hard part. Snap the Circlip in the groove on the end of the driveshaft to secure the pinion gear in place. If the first attempt is not successful, try again. Take a break, have a cup of coffee, tea, whatever, then give it another go.

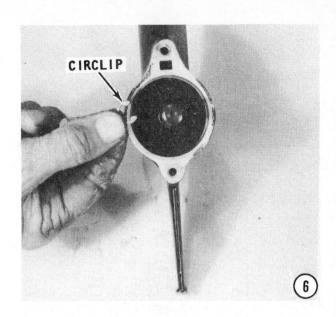

(6)

Checking Gear Backlash

Backlash between two mating gears is the amount of movement one gear will make before turning its mating gear when one gear is rotated back and forth.

7- Pull upward on the lower driveshaft with one hand and at the same time push inward and rotate the propeller shaft left and right with the other hand . The manufacturer recommends there should be a small amount of backlash "play" between the two gears, but does not specify exactly how much. No backlash is **NOT** acceptable.

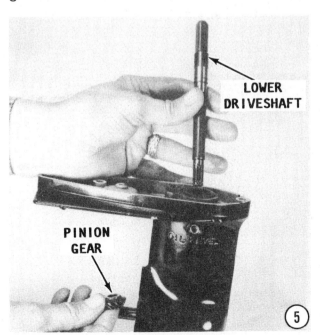

LOWER DRIVESHAFT

PINION GEAR

(5)

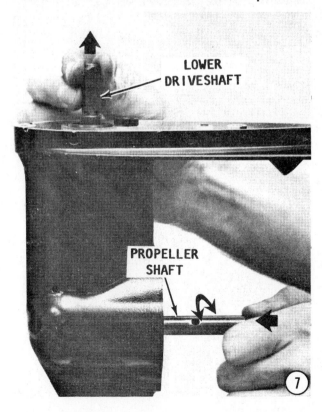

LOWER DRIVESHAFT

PROPELLER SHAFT

(7)

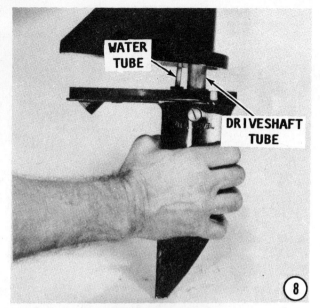

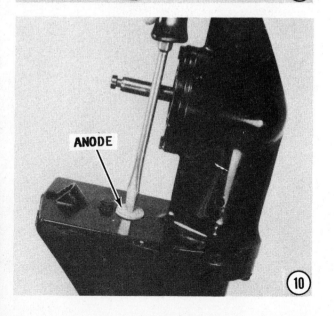

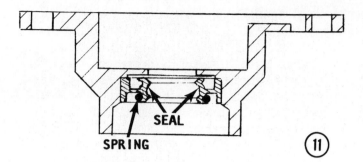

GOOD WORDS

If there is no backlash between the pinion gear and the drive gear, either one or both of two conditions may exist. The propeller shaft may not be indexed completely into the forward propeller shaft bearing or there is too much shim material ahead of the drive gear. Therefore, first try moving the propeller shaft further into the bearing. Let's hope the "no backlash" condition is corrected. If not, the driveshaft and propeller shaft will have to be removed, shim material removed, and the parts assembled, then the backlash checked again.

8- Apply a coating of Multi-purpose lubricant onto the outside diameter of the long tube. Insert the long tube into the midsection housing. Apply a coating of Multi-purpose lubricant to the outside diameter of the rectangular driveshaft. Insert the driveshaft into the tube. Apply a thin coating of Multi-purpose lubricant to the inside diameter of the seals in the lower gear housing. Position a **NEW** gasket in place on the lower gear housing. Begin to bring the midsection housing and the lower gear housing together. As the two units come closer, rotate the propeller shaft slightly to index the upper end of the lower driveshaft with the upper rectangular driveshaft. At the same time, feed the water tube into the water tube seal. Push the lower gear housing and the mid-section together.

9- Secure the lower gear housing and the mid-section together with the two bolts, lockwasher, and regular washer. The lock-

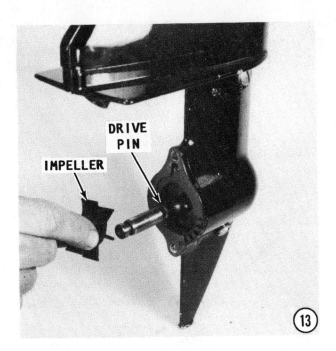

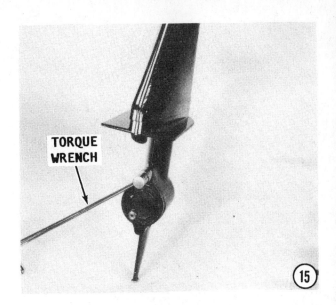

washer should be installed between the bolt head and the regular washer. Tighten the bolts alternately and evenly to a torque value of 25 in lbs (3 Nm).

10- Install a new anode to the underside of the trim bracket of the lower gear housing. Tighten the screw securely.

11- Install a **NEW** seal into the water pump housing, using the correct size socket. The accompanying cross-section illustration indicates the proper location for the spring and seal in the water pump housing.

12- If the bearing was removed, press a new bearing into the water pump housing until the bearing race is fully seated in the recess. Position a **NEW O**-ring in place around the water pump base. Install the water pump housing into place on the lower gear housing.

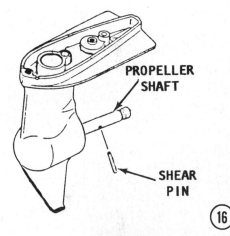

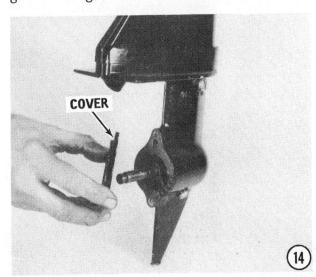

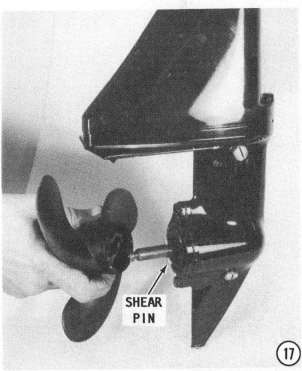

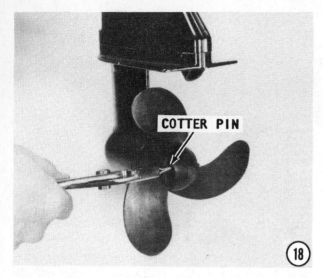

COTTER PIN

⑱

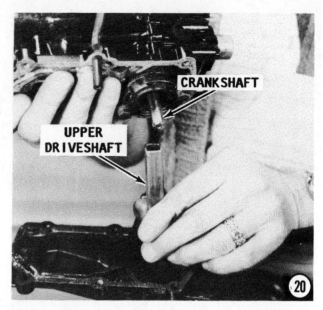

CRANKSHAFT

UPPER
DRIVESHAFT

⑳

13- Rotate the propeller shaft until the hole in the shaft is facing the largest part of the water pump cavity in the gear housing. (This opening is not perfectly round.) Insert the drive pin through the hole in the propeller shaft. Slide a **NEW** water pump impeller onto the propeller shaft. As the impeller begins to enter the pump housing, rotate the impeller **CLOCKWISE** to permit the impeller vanes to curl in the proper direction. Continue to move the impeller into the housing with the cutout in the impeller indexed over the drive pin. As the name implies, the drive pin "drives" the impeller.

14- Install the water pump cover onto the water pump housing. Apply Quicksilver Perfect Seal, or equivalent to the threads of the attaching bolts.

15- Secure the cover in place with the two bolts with a washer on each bolt. Tighten the bolts alternately and evenly to a torque value of 25 in lbs (2.8 Nm).

16- Insert a shear pin through the propeller shaft. To ease the task of removing

the propeller the next time, coat the propeller shaft with Mercury Special Lubricant 101, 2-4-C Multi-Lube, Perfect Seal, or equivalent.

17- Slide the propeller onto the propeller shaft and over the shear pin.

18- Insert a cotter pin through the propeller and propeller shaft. Bend the ends of the cotter pin over to secure it in place.

19- Place the lower unit in a vertical position. With the **OIL** and **OIL LEVEL** screws removed, fill the lower unit with Quicksilver Super Duty Lower Unit Lubricant until the lubricant oozes out the upper hole and no air bubbles are visible. Install both upper and lower screws with a gasket on each.

20- Slide a **NEW** gasket down the driveshaft into position on the driveshaft housing. Now, lower the powerhead onto the driveshaft housing with the lower end of the crankshaft indexed into the hollow core of the driveshaft.

21- Install and tighten the six bolts securing the powerhead to the driveshaft housing.

OIL LEVEL
SCREW

OIL
SCREW

⑲

BOLT
6-PLACES

㉑

10
TRIM/TILT

10-1 INTRODUCTION

All outboard engine installations are equipped with some means of raising or lowering the lower unit for efficient operation under various load, boat design, and water conditions. The most simple form is a mechanical tilt adjustment consisting of a series of holes in the transom mounting bracket through which an adjustment pin passes through to secure the engine at the desired angle. A second and more modern method, especially for the larger units, is a hydraulically operated system controlled from the helmsperson's position.

The system installed on the outboard units covered in this manual consists of two small trim/tilt cylinders. The cylinders extended very slowly through the trim degree range, and then accelerated to move the outboard to the tilt position for trailering or shallow water operation.

CHAPTER ORGANIZATION

All information, including description, operation, special instructions, bleeding the system, troubleshooting, and service of the various components is covered in this chapter as follows:

10-2 Mechanical Pin
10-3 System Description and Operation
10-4 Special Instructions
10-5 Hydraulic Bleeding
10-6 Troubleshooting
10-7 Hydraulic Trim/Tilt Service
10-8 Servicing Hydraulic Pump
10-9 Servicing Electric Motor

INCORRECT
Bow too high -- trim engine down.

INCORRECT
Bow too low -- trim engine up.

CORRECT
Boat and engine properly trimmed.

The trim position of the outboard unit directly affects the bow position and thus the boat performance.

TILT PIN

The tilt position is adjusted by inserting the tilt pin through one of a series of holes in the transom bracket.

10-2 MECHANICAL TILT PIN
ALL UNITS

The mechanical tilt arrangement is found on most outboard units. A change in the tilt angle of the engine is accomplished by inserting the tilt adjustment pin through one of a series of holes in the transom mounting bracket. These holes allow the operator to obtain the desired boat trim under various speeds and loading conditions.

The tilt angle of a lower unit is properly set when the anti-cavitation plate is approximately parallel with the bottom of the boat. The boat trim is corrected by stopping the boat, removing the adjustment pin tilting the engine upward or downward, as desired, and then installing the pin through the new hole exposed in the transom mounting bracket.

To raise the bow of the boat, the engine is raised one hole at-a-time until the operator is satisfied with the boat's performance. If the bow is to be lowered, the lower unit is lowered one hole at-a-time.

Performance will generally be improved if the bow is lowered during operation in rough water. The boat should **NEVER** be operated with the lower unit set at an excessive raised position. Such a tilt angle will cause the boat to "porpoise", which is very dangerous in rough water. Under such conditions the helmsperson does not have complete control at all times.

Instead of making extreme changes in the lower unit angle, it is far better to shift passengers and/or the load to obtain proper performance.

In order to obtain maximum efficiency and safety from the boat and engine, the tilt pin must be installed in the proper position. The wide range of boat designs with their various transom angles, requires a determination be made for each engine installation.

Actually, the tilt pin is only required if the boat handles improperly in the full trimmed "in" position at wide open throttle (WOT). Usually this occurs when the transom "angle" is too large.

This section provides detailed procedures to properly install the tilt pin and also the

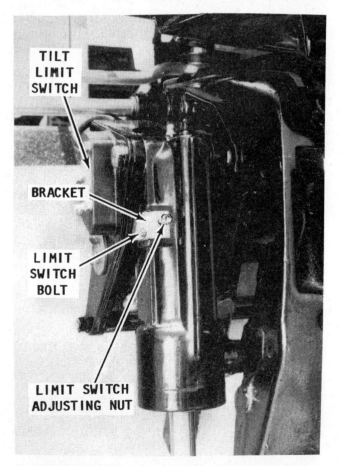

When the outboard is at the full trim out position, the swivel bracket should still be engaged with the clamp bracket flanges by 1/2" (1.75 cm) or more.

Adjusting the trim limit switch on the port side of the outboard.

necessary instructions to adjust the tilt limit switch.

Tilt Pin Installation

Refer to the accompanying illustration while performing the following installation steps.

1- Operate the Power Trim to move the engine inward or outward until the anti-cavitation plate is parallel to the boat bottom. With the engine in this position, notice the position of the swivel bracket in relation to the clamp bracket tilt pin holes. Now, install the tilt pin into the first full pin hole closer to the transom .

2- Install the washer and cotter pin into the tilt pin and open the pin end to secure it in place.

Tilt Limit Switch Adjustment

SAFETY WORDS: The tilt trim limit switch **MUST** be properly adjusted to prevent the swivel bracket from being trimmed outward beyond the clamp bracket tilt pin flanges. Proper tilt limit switch adjustment will also ensure continuous side support for the swivel bracket.

Refer to the two accompanying illustrations while performing the following six steps to adjust the tilt limit switch.

1- Depress the **IN** button and hold it depressed until the engine has reached the bottom of its travel.

2- Depress the **UP/OUT** button and hold it depressed until the pump motor stops. With the engine in this position, the swivel

bracket should still be engaged with the clamp bracket flanges by 1/2" (12.7 mm) or more.

3- Pull out on the lower unit to remove any slack in the hydraulic cylinders while the trim limit tilt position is checked.

GOOD WORDS: If the piston rods retract into the hydraulic cylinders more than 1/8" (3.17 mm) while the lower unit was being pulled out, the hydraulic system **MUST** be bled of air. See Section 10-5.

4- If the engine swivel bracket tilts out further than the limit given in Step 2, the trim limit switch needs adjustment. To adjust the switch, loosen the limit switch bolt on the side of the wrap around bracket. If the engine tilts out too far, turn the limit switch adjustment nut on top of the bracket **COUNTERCLOCKWISE**. If the engine fails to tilt out far enough, turn the adjustment nut **CLOCKWISE**.

5- Repeat Steps 1 thru 4, as necessary until the proper trim limit switch adjustment is attained.

6- Tighten the limit switch bolt securely to lock in the adjustment.

10-3 DESCRIPTION AND OPERATION TRIM/TILT SYSTEM

The hydraulically powered tilt system permits changing the tilt angle of the engine from the helmsperson's position. Controls and indicators for the system are located on the control panel.

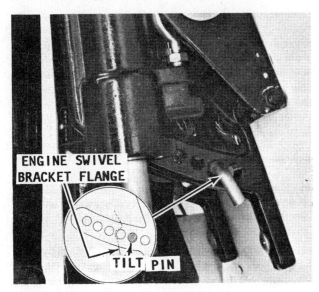

Installation of the tilt pin through the first hole aligned with the engine swivel bracket.

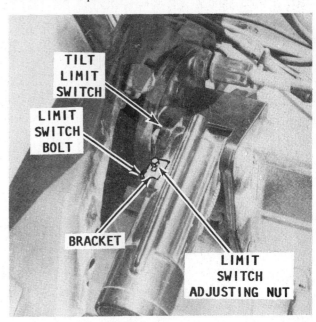

Adjusting the trim limit switch on the starboard side of the outboard.

The angle of the lower unit is properly set when the boat is operating to give maximum performance, including comfort and safety.

The powered tilt system consists of a hydraulic pump, two trim/tilt cylinders, a reverse lock valve, trim indicator sender, controls, an indicator gauge, and associated hoses and fittings.

The hydraulic pump includes a valve body-and-gear assembly, control valve, pump motor, and a reservoir. The controls consist of an **UP** button, an **UP/OUT** button, and an **IN** button. The indicator gauge is installed on the control panel next to the control buttons.

OPERATION

When the **UP** trim switch is operated, the **UP** pump solenoid is actuated and the electric motor circuit is closed. The electric motor drives the oil pump and oil is forced into the **UP** side of the trim cylinders. The engine is trimmed upward until the trim switch is released, or until the trim limit cutout switch opens the circuit and stops the engine swivel bracket within the limits of the clamp bracket supporting flanges.

When the **DOWN** trim switch is operated, the **DOWN** solenoid is actuated and the electric motor circuit is closed, but the motor will run in the opposite direction. Again, the motor drives the oil pump and oil is forced into the **DOWN** side of the trim cylinders and the engine is trimmed down to the desired position.

When the **TRAILERING** switch is operated, the **UP** pump solenoid is actuated and the pump motor circuit is closed. The pump motor will drive the oil pump and force oil into the **UP** side of the trim cylinders. The trailering circuit bypasses the trim limit switch, enabling the engine to be tilted upward for trailering, docking or shallow water operation.

TRAILERING OR LAUNCHING

Two control buttons on the control panel must be pushed at the same time to raise the stern drive to the full up position for trailering or launching the boat -- the normal **UP** button and the middle **UP/OUT** button. By pushing the middle **UP/OUT** button, current is passed to the top **UP** switch. The current passing through the **UP** switch while the button is depressed, will by-pass the trim limit switch so the **UP** circuit will be able to raise the lower unit to the full up position. If the middle **UP/OUT** button is depressed during normal boat operation, a trim limit switch will keep the drive unit from moving out beyond the transom mounting bracket.

When the hydraulic cylinders reach their full extent of travel, the pump motor will labor if the control buttons are not released. Therefore, to prevent damage to the system, a bimetal switch will open the circuit

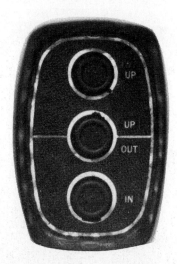

Three buttons are used to control the trim/tilt adjustment. One button is used to raise the engine for trim, another to lower the unit, and the third is used simultaneously with the UP button to position the outboard for trailering or during launch.

The trim gauge advised the helmsperson of the relative position of the outboard at all times.

to stop the pump motor and prevent the motor from overheating. The switch contacts will close automatically after the motor has cooled and the motor can again be operated.

ONE MORE WORD

Whenever the boat is being trailered, a trailer bracket should **ALWAYS** be used to mechanically lock the lower unit in the up position. Such a bracket may be purchased at modest cost from the local marine store and will give "peace of mind" to the owner as he moves the boat from place to place.

SAFETY WORDS

Two precautions **MUST** be followed if the engine is to be operated in shallow water with the engine trimmed beyond the trim limit cut-out:

1- **NEVER** operate the engine above idle rpm. When the engine is trimmed beyond trim limit cut-out, the swivel bracket does not have support.

2- Check to be sure the water intake ports remain submerged. If the intake ports should rise above the water surface, even for a very short time, severe damage could be caused to the water pump or to other expensive parts from overheating.

CAUTION: Water must circulate through the lower unit to the engine any time the engine is run to prevent damage to the water pump in the lower unit. Just five seconds without water will damage the water pump.

ALWAYS release the **DOWN** or **TRAILER** switch as soon as the engine reaches the end of its travel. If the switch is not released,

an overload cut-out switch will open and the pump motor will stop. If the cut-out switch should open, **DO NOT** depress either switch for approximately one minute. During this time, the cut-out switch will reset itself, the switch closes, and the pump may be operated again.

10-4 SPECIAL INSTRUCTIONS TRIM/TILT SYSTEM

1- Use clean **MS, SD, SE, SF 20W,** or **5W-30,** or **10W-40** or equivalent oil to fill the hydraulic reservoir. Use one of these oils to flush the system.

2- **REMEMBER** to remove the vent seal when filling the pump reservoir to prevent overfilling. **TOO MUCH** fluid in the system can cause pump body failure or motor failure. **DO NOT** close the vent screw because air in the reservoir **MUST** be able to escape.

3- The engine unit **MUST** be in the **FULL UP** position with the hydraulic cylinders extended when the reservoir is filled. The correct fluid level is **EVEN** with the bottom of the oil filler hole.

4- If the pump stops during long use, allow the pump motor to cool at least one minute before starting it again. An internal thermal circuit breaker protects the pump motor. If the pump will not operate, the circuit breaker may be tripped in the off-position.

5- Keep the work area **CLEAN** when servicing disassembled parts. The smallest

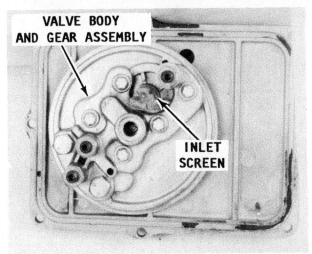

The gear assembly secured in the valve body cannot be serviced. If defective, it must be replaced as a unit.

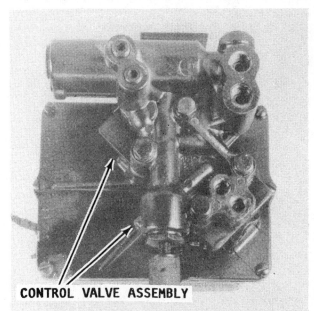

The control valves cannot be serviced. If they are found to be defective, the entire unit must be replaced.

amount of dirt or lint can cause failure of the pump to operate.

6- The valve body-and-gear cage assembly, or the control valve assembly **MUST** be replaced as a unit because of the precision fitting of the valves and gears.

7- The can over the inlet screen of the valve body-and-gear assembly **MUST** be installed securely or the oil will foam out the vent.

8- **TAKE CARE** not to allow Liquid Neoprene to get into the valve body-and-gear assembly. Use the neoprene sparingly.

9- The hydraulic pump motor must be moisture-proof. To moisture-proof a motor after service, apply Liquid Neoprene around the pump motor commutator plate edges, around the motor frame where it contacts the reservoir, and in the area where the wire leads enter the motor. Any sign of oil in the pump motor indicates either the reservoir seal is damaged or the vent screw was closed. If the vent was closed, air could not escape and oil was forced into the pump motor.

10- **DO NOT** allow the trim cylinders to hang by their hydraulic hoses. Such practice may cause damage to the hoses. Use **CARE** when handling trim cylinders during removal/installation of the engine. Rough treatment of the hoses could result in a weakened hose, partial separation at the fitting, or bending of the metal tubing. Such damage could certainly restrict the flow of oil to the trim cylinders.

11- **ALWAYS** hold the metal ferrule on the hose with pliers when tightening the

fitting. The indicator line on the hose **MUST** follow the hose bend without a twist. A twisted hose will cause severe loads resulting in over-stress on the hose and the tubing will be bent.

12- Following service on the system, move the engine from hard-over to hard-over, port and starboard, and from the full **DOWN** to the full **UP** position. During engine movement, check for possible kinks, twists, or severe bends in the hoses or at the fitting ends.

10-5 HYDRUALIC BLEEDING TRIM/TILT SYSTEM

Air in the system, even a small amount, can lead to poor performance or to a malfunction. Any time the system is opened for service, the following bleeding procedures should be closely followed.

1- Move the outboard unit until the cylinders are fully collapsed (retracted).

2- Remove the **FILL** screw and fill the pump with fluid.

3- Operate the trim system completely through the tilt range several times to bleed

Bleed screw location for the **DOWN** side of the cylinder.

Hydraulic pump reservoir installation for the power trim/tilt system. The vent screw should remain open.

TROUBLESHOOTING

PROBLEM	POSSIBLE CAUSES	CORRECTIVE ACTION
Engine fails to trim up or down	Release valve control knob not completely closed.	Turn valve fully clockwise.
	Oil level low or air in system.	Fill reservoir and bleed system. Check for leaks.
	Determine if hydraulic pump motor operates when trim button is depressed.	If motor operates, service the hydraulic system. If motor fails to operate, refer to electrical system troubleshooting.
Engine trims up, but will not trim down.	Determine if hydraulic pump motor operates when UP trim button is depressed.	If motor operates, service hydraulic system. If motor fails to operate, refer to electrical system troubleshooting.
Engine trims down, but will not trim up.	Determine if hydraulic pump motor operates when UP trim button is depressed.	If motor operates, perform hydraulic system UP pressure test. If motor does not operate, refer to elctrical troubleshooting.
Engine will not return completely to DOWN position, or returns part way with jerky motion.	Air in the system.	Fill reservoir and bleed system. Check for leaks.
	Internal cylinder/s leaks.	Service the hydraulic system.
Engine thumps when shifted.	Air in the system.	Fill reservoir and bleed system. Check for leaks.
	Internal cylinder leaks.	Service the hydraulic system.
	Control valve assembly leaks.	Service the hydraulic system.
Engine trails out when backing off throttle at high speed.	Air in the system.	Fill reservoir and bleed system. Check for leaks.
	Defective control valve assembly.	Install new control valve assembly.
	Internal cylinder leaks.	Service the hydraulic system.
Engine fails to hold a trimmed position or will not remain tilted for extended period.	External leak -- fittings or parts.	Tighten fittings, or replace defective parts.
	Internal leaks.	Service hydraulic system.
	Pump check valve leak, high pressure.	Clean check valve by operating system up and down several times to flush system. If problem not corrected, replace pump valve body assembly.
Engine fails to hold in reverse.	Defective reverse lock valve solenoid.	Refer to electrical troubleshooting.
	Internal cylinder leaks.	Service the hydraulic system.
	Worn reverse lock valve seat.	Service the hydraulic system.

air from the system. Check the fluid level in the pump when the cylinders are fully collapsed (retracted). Thread the **FILL** screw into place and tighten it securely.

10-6 TROUBLESHOOTING TRIM TILT SYSTEM

Always attempt to proceed with the troubleshooting in an orderly manner. The "shot-in-the-dark" approach will only result in wasted time, incorrect diagnosis, replacement of unnecessary parts, and frustration.

To be successful, the troubleshooting process must begin with an accurate determination of the problem, followed by a logical reasoning of what caused the problem, and finally intelligently concluding the required corrective action to be taken. The final phase, of course, is the work involved in returning the unit to satisfactory performance.

Any problem with the Power Trim system must first be classified as a malfunction in the electrical system or in the hydraulic system.

The following troubleshooting table is designed to provide a logical approach and sequence of work necessary to arrive at the proper solution.

TESTING TRIM SWITCHES EARLY AND LATE MODELS

Test Panel Control Switch

a- Disconnect the power trim harness from the trim pump.

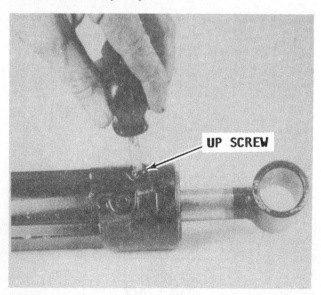

*Bleed screw location for the **UP** side of the cylinder.*

b- Connect a continuity meter or a test lamp between the switch terminals on the back of the switch. The meter should indicate **NO** continuity or the test lamp should not light with the button in the free position. Depress the button. The meter should indicate continuity or the test lamp should light.

Early Model Trim Switches Externally Mounted On Trim Cylinder Bracket Trim Switch Removal

c- Disconnect the power trim harness from the trim pump.

d- Remove the switch control panel from the mounting hole.

e- Remove the four attaching screws, and then remove the switch retainer.

f- Release the soldered connections from the switch.

Trim Switch Installation

g- Solder the wires on the new switch to the terminals. After soldering, coat the terminals with Liquid Neoprene.

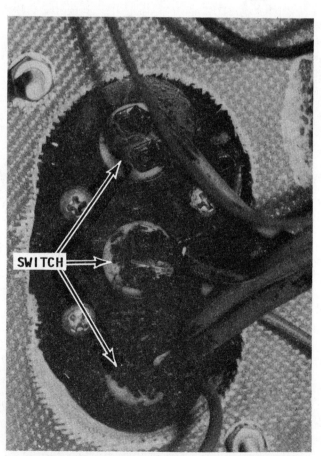

*The control switch, as viewed from the **BACK** side, is accessible for service.*

h- Insert the switch into the panel and secure it in place with the switch retainer and the four attaching screws.

i- Install the switch control panel into the mounting hole.

j- Connect the power trim harness to the trim pump.

Late Model Trim Switch
Mounted On Inside Of Starboard Trim
Cylinder Mounting Bracket

1- Remove the two bolts securing the starboard trim cylinder mounting bracket to the transom. Remove the hex bolt and nut securing the trim cylinder mounting bracket to the outboard. Back-out the bolt securing the lower end of the trim cylinder to the outboard. Lift off the trim cylinder and mounting bracket free of the outboard.

2- Remove the five screws, and then lift off the cover and gasket from the mounting bracket. Lift out and **SAVE** the spring from the actuating plunger. Remove the two screws securing the trim limit switch to the mounting bracket. Lift out the trim limit switch and pull the wires free from the mounting bracket. Verify the actuating plunger operates freely without binding.

Two trim limit switches are externally mounted, one on either side of the trim cylinder bracket.

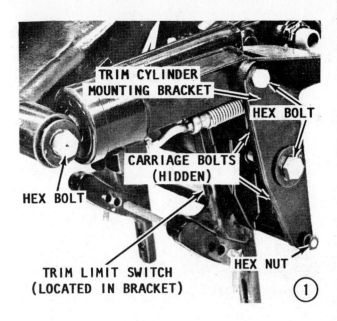

CLEANING AND INSPECTING

Clean the plunger and check the spring to be sure it is not distorted. If trouble-shooting indicates the switch is defective, it must be replaced. Individual parts are not available to "rebuild" the switch.

INSTALLATION

1- Install the actuating plunger. Position the trim limit switch into the mounting bracket. Route the wire harness over the hydraulic hoses and through the hole in the bracket. Coat the trim limit switch and wires with Anti-Corrosion Lubricant or a Silicone Compound. If corrosion was a problem, fill the entire switch cavity with lubricant or the Silicone Compound. Slide the spring into the actuating plunger. Place the gasket and cover onto the mounting bracket with the spring extending through the hole in the gasket. Secure the cover in place

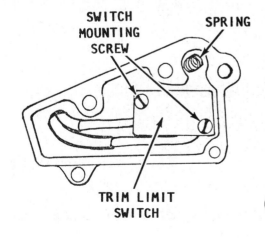

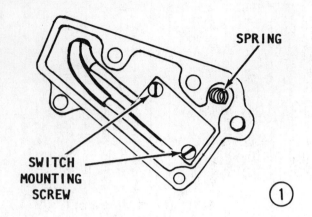

SPRING

SWITCH
MOUNTING
SCREW

(1)

with the five screws. Check the switch by depressing the plunger. The plunger should operate without binding and an audible "click" sound should be heard as the switch turns on and off.

2- Place the trim cylinder and mounting bracket into position on the outboard unit. Install the bolt securing the lower end of the trim cylinder to the outboard. Secure the trim cylinder mounting bracket to the outboard unit with the hex bolt and nut. Coat the shank, not the threads, of the two bolts securing the starboard trim cylinder mounting bracket to the transom with sealing compound. The compound will ensure a watertight installation. Install the bolts and tighten the nuts securely.

SOLENOID TESTING
EARLY AND LATE MODELS

FIRST, THESE WORDS

Some units covered in this manual have and **UP** solenoid and a **DOWN** solenoid. Other units have only one solenoid.

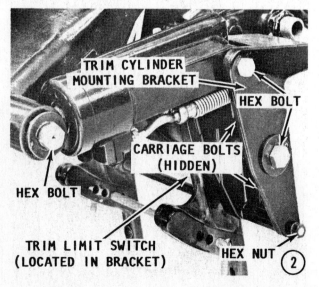

TRIM CYLINDER
MOUNTING BRACKET

HEX BOLT

CARRIAGE BOLTS
(HIDDEN)

HEX BOLT

TRIM LIMIT SWITCH
(LOCATED IN BRACKET)

HEX NUT

(2)

In addition to this fact, two different styles of solenoids were used on the System "A" power trim/tilt units. The accompanying illustration shows how each solenoid may be bench tested.

Disconnect all leads from the solenoid terminals. Use the RX1 scale on an ohmmeter. Connect each meter lead to the large solenoid terminals. Connect the positive lead of a 12-volt battery to one of the solenoid small terminals. Conmnect one end of a piece of wire to the other small solenoid terminal. Now, for just a split-second, make contact with the other end of the wire to the ground terminal of the battery. A loud click sound should be heard from the solenoid and the meter reading should indicate zero ohms. If a audible "click" is not heard and the meter indicates a reading other than zero ohms, the solenoid is defective and must be replaced.

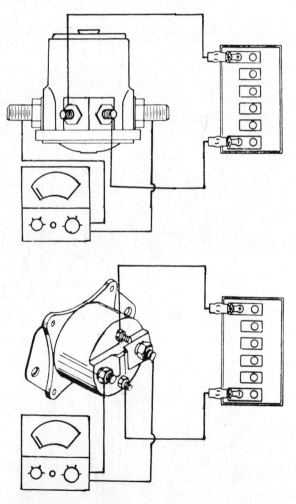

Early model solenoid (above), and late model solenoid (below), hooked up for testing. An audible "click" should be heard from the solenoid and the ohmmeter should indicate zero ohms.

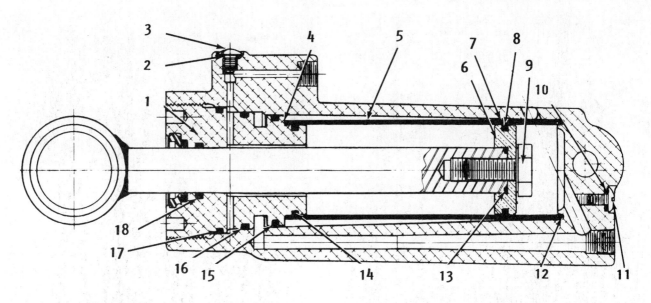

1- RETAINER	10- O-RING - 7/32x11/32x1/16
2- O-RING - 7/32x11/32x1/16	11- DOWN BLEED SCREW
3- UP BLEED SCREW	12- O-RING - 1-1/2x1-5/8x1/16
4- CHAMFERED END	13- O-RING - 5/8x3/4x1/16
5- CYLINDER LINER	14- O-RING - 1-5/16x1-1/2x3/32
6- PISTON	15- O-RING - 1-5/8x1-13/16x3/32
7- SLIPPER SEAL	16- O-RING - 1-3/4x1-15/16x3/32
8- O-RING - 1-1/4x1-11/32x3/32	17- O-RING - 1-7/8x2-1/16x3/32
9- BOLT	18- O-RING - 3/4x27/32x3/32

Cross-section of a 3/4" diameter piston rod cylinder with the bleed screw and O-ring sizes identified.

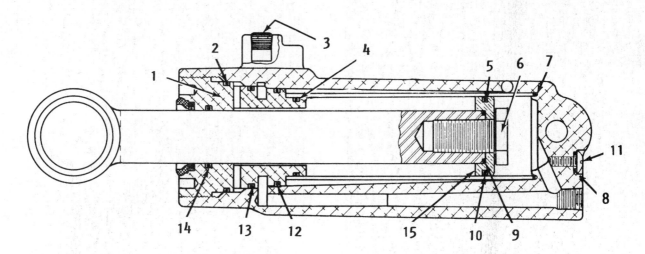

1- RETAINER	8- O-RING - 1-5/16x1-1/2x3/32
2- O-RING - 1-7/8x2-1/16x3/32	9- O-RING - 7/8x1x1/16
3- UP BLEED SCREW	10- SLIPPER SEAL
4- O-RING - 1-5/16x1-1/2x3/32	11- DOWN BLEED SCREW
5- O-RING - 1-1/4x1-7/16x3/32	12- O-RING - 1-5/8x1-13/16x3/32
6- BOLT	13- O-RING - 1-3/4x1-15/16x3/32
7- O-RING - 1-1/2x1-5/8x1/16	14- O-RING - 1x1-3/16x3/32
	15- PISTON

Cross-section of a 1" diameter piston rod cylinder with the bleed screw and O-ring sizes identified.

Typical trim/tilt cylinder and hose installation for the trim/tilt system covered in this manual.

10-7 HYDRAULIC TRIM SERVICE TRIM/TILT CYLINDERS

Trim Cylinder Removal

WARNING: Exercise care when working with the trim cylinders. Take extra precautions to prevent the engine from falling, or moving down rapidly. Such action could cause **BODILY INJURY**.

Disconnect the two trim hoses from the cylinder to be removed. Cap the hoses and plug the holes in the cylinders as a precaution against contamination, including air, from entering the system. The caps and plugs will also prevent an unnecessary loss of oil. Remove the mounting bolts at each end of the cylinder, and then remove the cylinder.

ALWAYS keep the work area clean to prevent contamination. The smallest amount of foreign material can lead to a malfunction in the system.

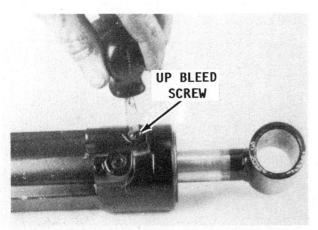

Removing the bleed screws from the cylinder. The UP screw is at the bottom of the cylinder and the DOWN screw is at the top.

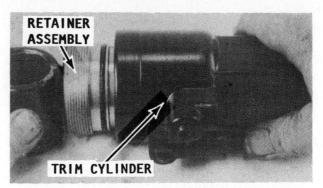

Removing the retainer assembly and liner from the cylinder housing.

Trim Cylinder Disassembling

Remove the **UP** and **DOWN** bleed screws, identified as **3** and **11** in the illustration, previous page. Use a 1-3/4" spanner wrench and unscrew the retainer assembly from the cylinder assembly. After the retainer assembly is loose, pull the rod and piston assembly, retainer assembly, and the cylinder liner from the cylinder housing.

Use a long wire pick to dislodge the O-ring from the cylinder housing, and then remove the ring. Clamp the rod assembly in a vise equipped with soft jaws, and then slide the cylinder liner off the retainer and piston, using a twisting motion. Remove the bolt and piston from the piston rod. Remove the four O-rings from the outside retainer and one O-ring from inside the retainer behind the seal and scraper. **DO NOT** remove or damage the seal or scraper.

CLEANING AND INSPECTING

Make an effort to keep the work area clean, because any contamination in the system could lead to a malfunction.

Inspect the cylinder liner and piston for wear. Check to be sure the piston is not bent or distorted.

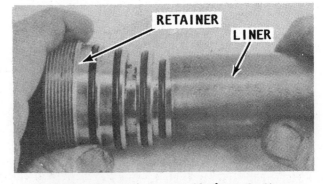

Removing the retainer assembly from the liner.

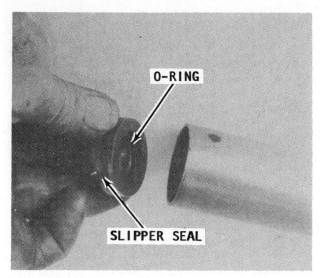

Removing the O-ring and slipper seal from the liner.

Remove all seals and O-rings.

Clean all parts in a solvent, and then blow them dry with compressed air.

Trim Cylinder Assembling

Lubricate all O-rings and O-ring contact surfaces with Multipurpose Lubricant, as protection against damaging the ring during installation. Install the proper O-ring into the cylinder housing. Use Multipurpose Lubricant to hold the O-ring in the recessed groove. Press the O-ring firmly into the bottom of the groove, using the cylinder liner as a tool.

Install the O-ring into the groove of the new brass piston. **CAREFULLY** stretch the slipper seal over the O-ring. **USE CARE** when stretching the slipper seal. If the slipper is overstretched, installation of the piston into the liner will be more difficult. Position the proper O-ring into the recessed end of the piston.

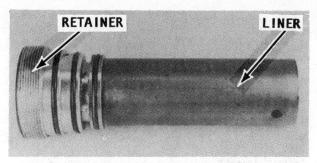

Installing the liner onto the piston.

Install the proper size O-ring into the grooves on the retainer. **CAREFULLY** insert the O-ring into the groove inside the retainer, without damaging the seal or the scraper. Slide the retainer assembly onto the piston rod. Coat the internal surface of the cylinder liner, piston, slipper seal assembly and Installation Tool C-91-69626 with Multipurpose Lubricant. Clamp the rod eye in a vise, as shown.

Position the cylinder liner onto the retainer with the chamfered side **TOWARD** the retainer. Now, position installation tool C-91-69626 on the cylinder liner and place the piston, with the slipper seal installed, into the tool, with the O-ring **TOWARD** the end of the rod.

Push the slipper seal into the tool by hand until the retainer bolt can be started. Remove the retainer bolt and apply a small amount of Blue Loctite to the threads. Install the bolt and use the bolt to draw the piston into the liner. Tighten the bolt to a torque value of 35 ft lbs. (47.6 Nm) on 3/4" diameter rods or to 60 ft lbs. (81.6 Nm) on 1" diameter rods.

Apply Multipurpose Lubricant to the end of the liner which seats against the O-ring in the housing.

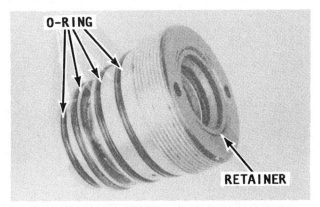

The retainer should be carefully inspected to ensure the threads have not been damaged and the O-ring seats are clean.

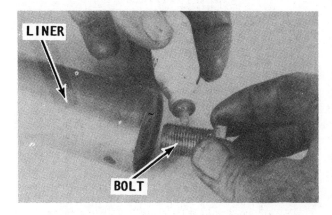

Apply Loctite to the threads of the bolt securing the slipper seal to the end of the cylinder shaft.

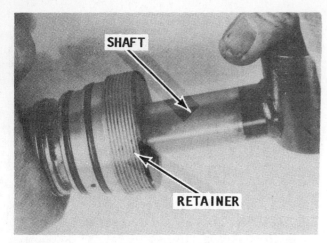

Working the cylinder shaft through the retainer. After the shaft is in place, it should move smoothly through the retainer.

Insert the complete piston, piston rod, cylinder liner, and retainer assemblies into the cylinder housing.

Coat the threads of the retainer with Perfect Seal, or equivalent. Thread the retainer assembly into the cylinder housing and tighten it to a torque value of 35-40 ft-lbs. (47.6-54.4 Nm).

Check the cylinder movement by moving the piston rod completely in and out of the cylinder several times. The movement should be free with no feel of binding.

Install the bleed screws into the cylinder housing. Check to be sure the O-rings under the bleed screws remain in place.

As a protection against corrosion, touch-up any scratches, nicks, etc., by sanding and painting.

Fill and bleed the cylinders according to the procedures outlined in Section 10-5.

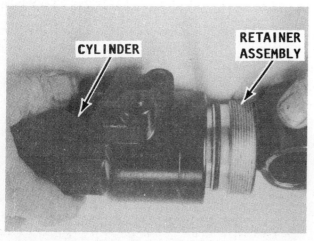

Installing the piston and liner into the pump housing.

Trim Cylinder Installation

WARNING: Exercise care when working with the trim cylinders. Take extra precautions to prevent the engine from falling, or moving down rapidly. Such action could cause **BODILY INJURY**.

Position the trim cylinder in place, as shown. Install the top mounting bolt and tighten it to a torque value of 50 ft lbs. (67.8 Nm).

The starboard cylinder has a washer installed between the outer side of the cylinder and the mounting bracket. Install the lower mounting bolts, with a spacer and rubber bushing. Tighten the bolt to a torque value of 50 ft lbs. (67.8 Nm).

Remove the caps from the hoses and the plugs from the cylinders. Connect the hoses to the cylinders. **TAKE CARE** not to cross-thread the hoses in the cylinders. Tighten the hoses securely.

Operate the trim system and check for leaks. See Section 10-5 to bleed any air that may have entered the system during the service work.

10-8 HYDRAULIC PUMP SERVICE

TESTING

Trim Pump Motor Test

This simple procedure will determine if the trim pump motor requires service.

a- Disconnect the black and blue motor wires from the solenoid terminals.

b- Connect a 12-volt supply between the motor terminals where the black and blue wires were disconnected. The motor should run. If the motor fails to operate, it requires service or replacement, see Section 10-9.

The cylinder can be checked prior to installation by using compressed air at each pressure port. The piston should move smoothly in each direction. Exercise care because the piston will move rapidly and with force.

PUMP REMOVAL

Disconnect the large red and black battery wires at the battery to eliminate the possibility of sparks. Tag the wires and terminals at the pump and solenoid, and then disconnect the wires. Remove the mounting nuts, and then remove the pump-and-mounting bracket assembly. If the pump is to be replaced, **TAKE CARE** to prevent loss of hydraulic fluid, and then disconnect the hoses.

DISASSEMBLING

Clean the outside of the pump with solvent, and then blow it dry with compressed air. The pump **MUST** be thoroughly cleaned before the disassembly work begins to prevent contamination entering the system. Even the smallest amount of foreign material could lead to a malfunction. Keep the work area as clean as possible for the same reason.

Remove the reservoir fill screw, and then drain the oil. Remove the four 7/16" hex-head screws from the control valve and the eight screws from the valve body and gear assembly. Separate the reservoir from the valve body and **DISCARD** the seal.

CLEANING AND INSPECTING

If the pump is damaged or shows signs of wear, the entire body and gear assembly must be replaced as a unit because the valves and gears have such a precision matched fit.

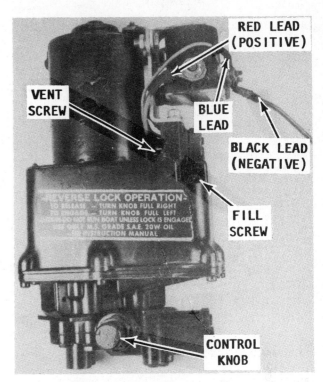

Hydraulic pump for the trim/tilt system with associated parts and wiring prior to removal. Tag the wires as an aid during installation.

ASSEMBLING
TRIM/TILT PUMP

Place a **SMALL** amount of Liquid Neoprene C-92-25711-1 in the seal groove of the valve body to hold the seal in place during assembly. **TAKE CARE** with the Liquid Neoprene because any excess amount may enter the valve body and plug the ports. Result: Failure of the pump. Install the seal on the valve body, and then install the

General view of the trim/tilt installed on the outboard, connected, and ready for operation.

The gear assembly secured in the valve body with Phillips head screws. The assembly is identical to the two at the top of Page 10-17.

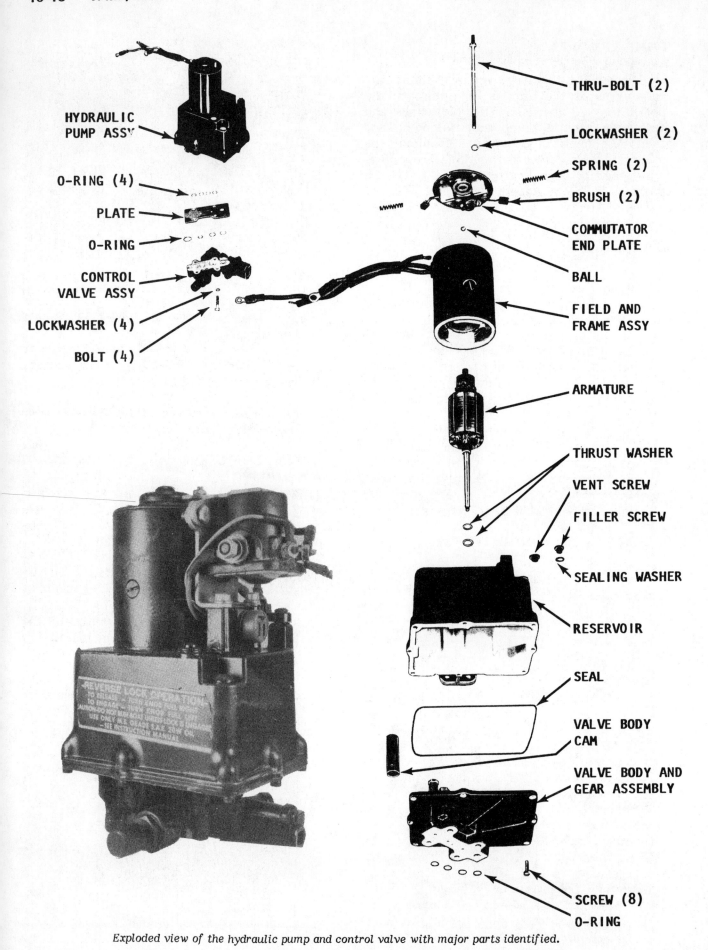

HYDRAULIC PUMP ASSY

O-RING (4)

PLATE

O-RING

CONTROL VALVE ASSY

LOCKWASHER (4)

BOLT (4)

THRU-BOLT (2)

LOCKWASHER (2)

SPRING (2)

BRUSH (2)

COMMUTATOR END PLATE

BALL

FIELD AND FRAME ASSY

ARMATURE

THRUST WASHER

VENT SCREW

FILLER SCREW

SEALING WASHER

RESERVOIR

SEAL

VALVE BODY CAM

VALVE BODY AND GEAR ASSEMBLY

SCREW (8)

O-RING

Exploded view of the hydraulic pump and control valve with major parts identified.

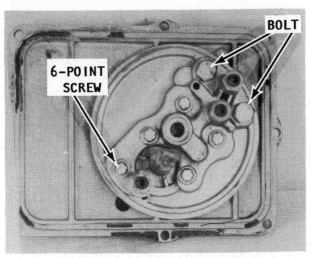

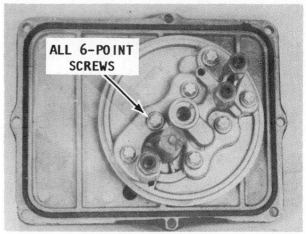

The gear assembly secured in the valve body with 6-point screws and two bolts. The assembly is identical to the one on Page 10-15 and top of next column.

The gear assembly secured in the valve body using 6-point screws. The only difference between this unit, and the one in the left column, and at the bottom of Page 10-15, is the method of attachment to the valve body.

valve body and gear cage assembly onto the reservoir. Secure the parts together with the eight attaching screws. Tighten the opposite screws evenly and securely. Now, apply a coating of Liquid Neoprene, or equivalent, around the edges of the housings to ensure a good seal. After the neoprene has thoroughly dried, paint the parts.

Because the vent screw is left open, the pump is vented, therefore, it is self-bleeding.

A GOOD WORD: Because the vent screw is left open, the pump is vented, therefore, it is self-bleeding.

10-9 ELECTRIC MOTOR SERVICE TRIM/TILT SYSTEM

TROUBLESHOOTING

Before going directly to the electric motor as a source of trouble, check the

INSTALLATION TRIM/TILT PUMP

Attach the pump and mounting bracket assembly to the transom with the mounting nuts. The pump **MUST** be installed vertically.

Connect the hoses to the pump and tighten the fittings securely. Insert the harness adaptor into the receptacle of the hydraulic pump. Connect the wires from the opposite end of the adaptor. **TAKE CARE** to match the color code or the tags made during removal. The wiring diagrams in the Appendix will be helpful if the color code is not apparent or tags were not used during removal. Slide the rubber sleeves over the connections. Place the rubber boot over the large red battery cable, and then connect the wire to the solenoid. Connect the opposite end of the red wire to the positive (+) battery terminal and the large black wire to the solenoid. Connect the opposite end of the black wire to the negative (-) battery terminal.

Typical hydraulic pump with major parts and wiring identified.

battery to be sure it is up to a full charge; inspect the wiring for loose connections, corrosion and the like; and take a good look at the control switches and connections for evidence of trouble.

The control switches may be eliminated as a source of trouble by connecting the pump directly to the battery for testing purposes. This can be accomplished by disconnecting the black and the blue wires from the solenoid, and then connecting the black wire to the negative (-) battery terminal and the blue wire to the positive (+) terminal.

If the pump motor does not run with this direct connection, either the pump motor or the pump and valve assembly is defective. If the pump motor does operate, the problem is in the wiring or the panel switches.

DISASSEMBLING

Remove the two screws from the top of the motor and reservoir assembly. Scribe a mark on the reservoir and a matching mark on the motor housing as an aid to assembling. Separate the reservoir from the motor. **TAKE CARE** not to lose the spacers from the armature. Pull the armature out of the frame. Disconnect the ground wire from the upper end cap. Remove the end cap from the frame and field assembly.

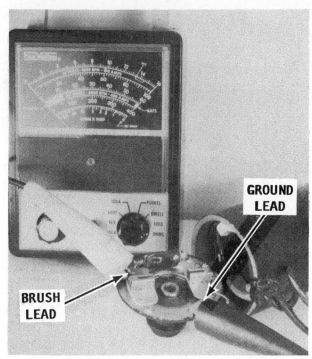

Hookup to check the continuity between the brush holders and the end cap. The meter should read infinity, indicating a lack of continuity.

CLEANING AND TESTING

Any sign of oil in the pump motor indicates either the reservoir seal is damaged or the vent screw was closed. If the vent was closed, air could not escape and oil was forced into the pump motor.

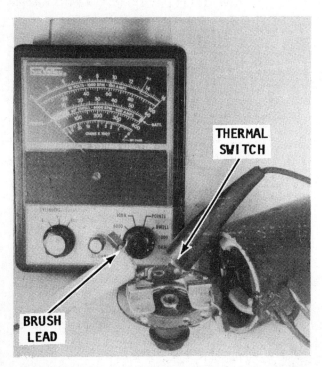

Hookup to check the resistance of the thermal switch. The meter is set on the Rx1 scale.

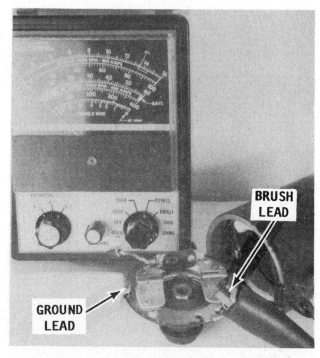

Testing for a short between a brush pigtail and the end cap.

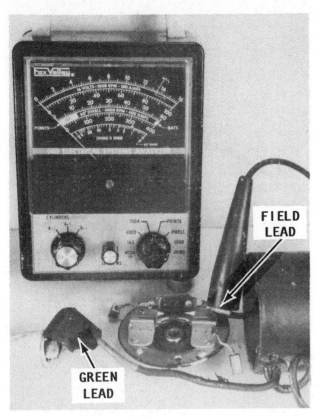

Checking the field coils for a short.

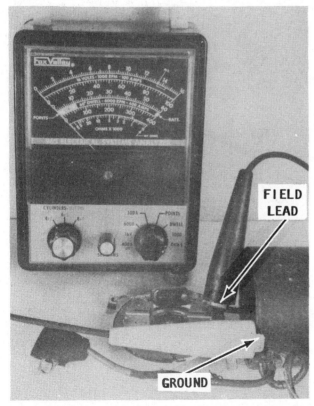

Testing the field coil for continuity. The meter should read infinity, indicating no continuity.

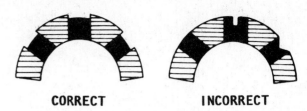

CORRECT   INCORRECT

Armature segments properly cleaned (left), and improperly cleaned (right).

Check the amount of wear to the brushes. If they are worn to half their original length, they should be replaced. If the commutator is worn, true it on a lathe, and undercut the mica.

Check the armature on a growler for shorts, open windings, or shorted windings.

Check the resistance of the thermal switch. If the switch has no continuity or has high resistance, it **MUST** be replaced.

Check between each brush holder and the end cap for a short. There must be no continuity.

GOOD WORDS

If there is any measurable high resistance in any of the tests in this step, the frame and field assembly **MUST** be replaced. Check the field coils for a short, ground, or excessive resistance in the windings. With the tester on Rx1, check for resistance between the green wire and the black jumper wire. Check for resistance between the blue wire and the black jumper wire. Check the black ground wire for resistance.

With the tester still on the Rx1 scale, check for a short between the black jumper

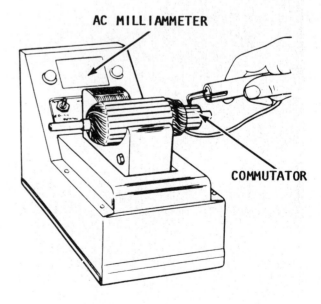

Checking between the commutator segments. Continuity must exist.

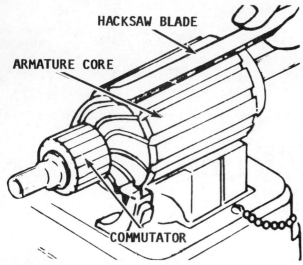

Method of testing the armature for a short circuit using a growler and a hacksaw blade. If the blade vibrates, the mica must be cleaned out or the armature replaced.

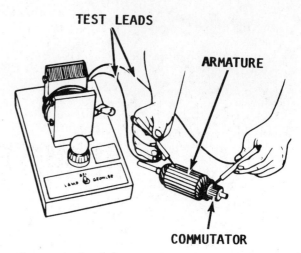

Armature check for a short: One test light lead on each commutator segment alternately, and the other lead on the armature core. No continuity.

wire and ground. A short is indicated if the needle moves to the right. The assembly **MUST** be replaced.

Install the armature into the frame and field assembly. Place the ball bearing on top of the armature shaft, if such a bearing

is used. Depress the brushes into place, and slide them over the commutator bars. **TAKE CARE** not to mar the brushes. Position the assembly onto the reservoir housing and work the shaft into the pump. Install the screws into the cap. Be sure to install the ground wire onto one of the screws securing the cap. Tighten the two bolts to the reservoir housing.

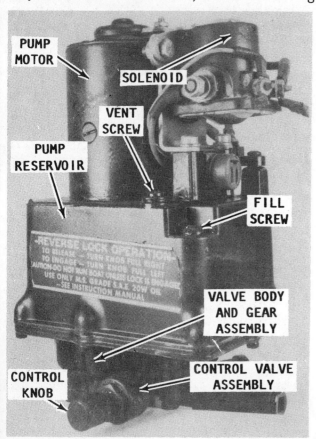

Hydraulic pump with early model solenoid. Other major parts are identified.

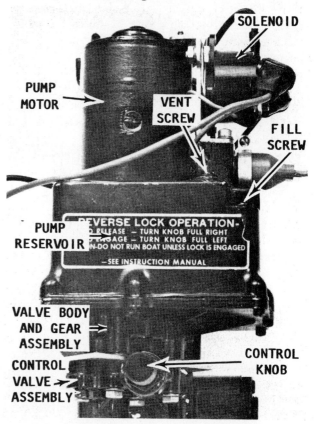

Hydraulic pump identical to the one in the left column except with later model solenoid.

11
REMOTE CONTROLS

11-1 INTRODUCTION

Remote controls are seldom obtained from the original outboard manufacturer. Shift boxes and steering arrangements may be added by the boat manufacturer. Because of the wide assortment, styles, and price ranges of remote controls, the boat manufacturer, or customer, has a wide selection from which to draw, when outfitting the boat.

Therefore, the procedures and suggestions in this chapter are for the "Commander" shift controls widely used with the outboard units covered in this manual. The procedures are specific and in enough detail to allow troubleshooting, repair, and adjustment of the "Commander" shift unit for maximum comfort, performance and safety.

WOULD YOU BELIEVE

Probably 90% of steering cable problems are directly caused by the system not being operated, just sitting idle during the off-season. Without movement, all steering cables have a tendency to "freeze". **Would you also believe:** Service shops report almost 50% of the boat cables replaced every year are due to lack of movement. Therefore, during off-season when the boat is laid up in a yard, or on a trailer alongside the house, take time to go aboard and operate the steering wheel from hard-over to hard-over several times.

These sections provide step-by-step detailed instructions for the complete disassembly, cleaning and inspection, and assembly of the "Commander" shift box. Disassembly may be stopped at any point desired and the assembly process begun at that point. However, for best results and maxi-

mum performance, the entire system should be serviced if any one part is disassembled for repair.

An exploded drawing of the "Commander" shift box is presented between the assembling and disassembling procedures. This diagram will be most helpful in gaining an appreciation of how the shift box functions and the relationship of individual parts to one another.

If at all possible, keep the parts in order as they are removed. Make an effort to keep the work area clean and keep disassembled parts covered with a shop cloth to prevent contamination.

GOOD WORDS

If the control cable has a "Zerk" fitting at the engine end, the cable **MUST** be retracted, then the fitting lubricated with Quicksilver Multi-Purpose lubricant or Quicksilver 2-4C Lubricant.

STEERING CHECKS

The steering system may be checked by performing a few very simple tests. First, move the steering wheel from hard-over to hard-over, port and starboard several times. The outboard unit should move without any sign of stiffness. If binding or stiffness is encountered, the cause may be a defect in the swivel bearing.

Next, remove the steering bolt at the outboard unit, and again turn the steering wheel back-and-forth from hard-over to hard-over, port and starboard several times. If there is any sign of stiffness, it is proof the problem is with the cables. They may be corroded or there may be a defect in the steering mechanism.

11-2 COMMANDER CONTROL SHIFT BOX

REMOVAL AND DISASSEMBLING

The following detailed instructions cover removal and disassembly of the "Commander" control shift box from the mounting panel in the boat.

1- Turn the ignition key to the **OFF** position. Disconnect the high tension leads from the spark plugs, with a twisting motion.

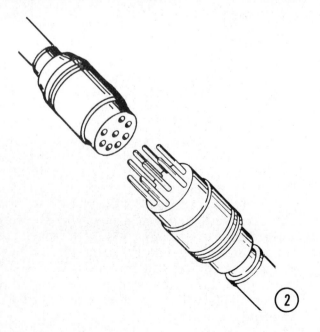

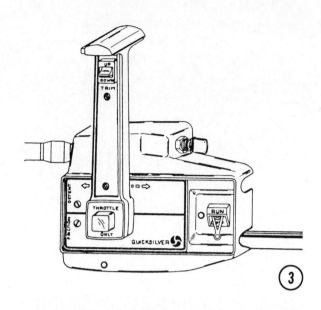

2- Disconnect the remote control wiring harness plug from the outboard trim/tilt motor and pump assembly.

3- Disconnect the tachometer wiring plug from the forward end of the control housing.

4- Remove the three locknuts, flat washers, and bolts securing the control housing to the mounting panel. One is located next to the **RUN** button (the ignition safety stop switch), and the second is beneath the control handle on the lower portion of the plastic case. The third is located behind the control handle when the handle is in the **NEUTRAL** position. Shift the handle into **FORWARD** or **REVERSE** position to remove the bolt, then shift it back into the **NEUTRAL** position for the following steps.

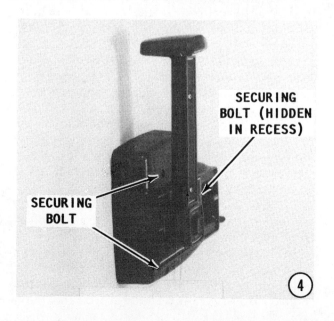

SECURING BOLT (HIDDEN IN RECESS)

SECURING BOLT

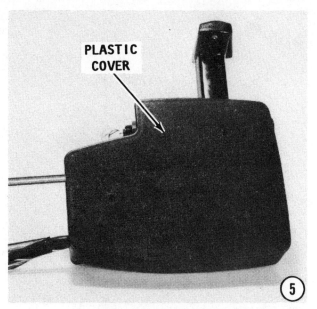

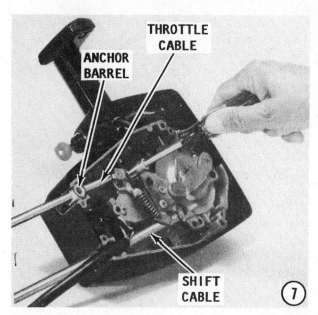

5- Pull the remote control housing away and free of the mounting panel. Remove the plastic cover from the back of the housing. Lift off the access cover from the housing. (Some "Commander" remote control units do not have an access cover.)

6- Remove the two screws securing the cable retainer over the throttle cable, wiring harness, and shift cable. Unscrew the two Phillips-head screws securing the back cover to the control module, and then lift off the cover.

Throttle Cable Removal

7- Loosen the cable retaining nut and raise the cable fastener enough to free the throttle cable from the pin. Lift the cable from the anchor barrel recess.

8- Remove the grommet.

Shift Cable Removal

9- Shift the outboard unit into **REVERSE** gear by depressing the neutral lock bar on the control handle and moving the control handle into the **REVERSE** position. **LOOSEN**, but do not remove, the shift cable retainer nut with a 3/8" deep socket as far as it will go without removing it. Raise the shift cable fastener enough to free the shift cable from the pin.

DO NOT attempt to shift into **REVERSE** while the cable fastener is loose. An attempt to shift may cause the cable fastener to strike the neutral safety microswitch and cause it damage.

Lift the wiring harness out of the cable anchor barrel recess and remove the shift cable from the control housing.

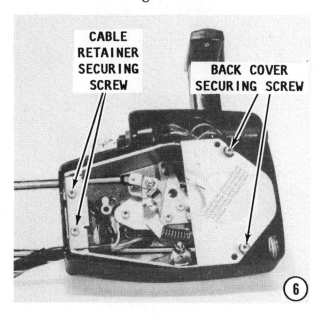

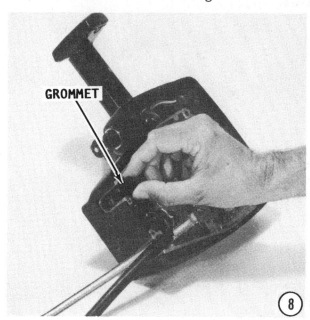

Control Handle Removal
For Power Trim/Tilt
With Toggle Trim Switch
Or Push–Button Trim Switch

GOOD WORDS

For non-power trim/tilt units, it is not necessary to remove the cover of the control handle. If servicing one of these units, proceed directly to Step 13. All others perform Steps 11 and 12.

10- Depress the **NEUTRAL** lock bar on the control handle and shift the control handle back to the **NEUTRAL** position. Remove the two Phillips head screws which secure the cover to the handle, and then lift off the cover. The push button trim switch will come free with the cover, the toggle trim switch will stay in the handle body.

Unsnap and then remove the wire retainer. Carefully unplug the trim wires and straighten them out from the control panel hub for ease of removal later.

11- Back-off the set screw at the base of the control handle to allow the handle to be removed from the splined control shaft.

12- Grasp the "throttle only" button and pull it off the shaft.

SPECIAL WORDS

Take care not to damage the trim wires when removing the control handle, on power trim models.

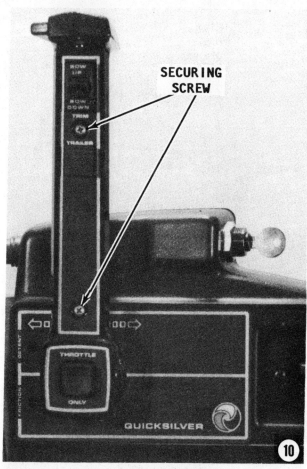

SECURING SCREW

10

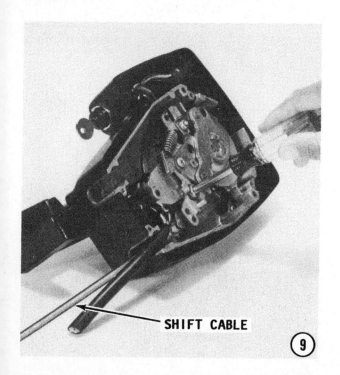

SHIFT CABLE

9

BASE SET SCREW

11

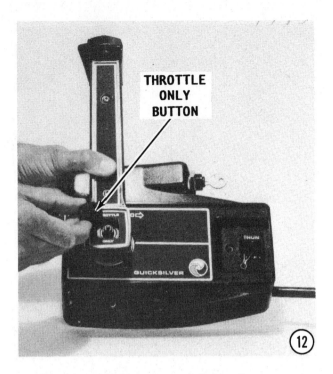

THROTTLE ONLY BUTTON

(12)

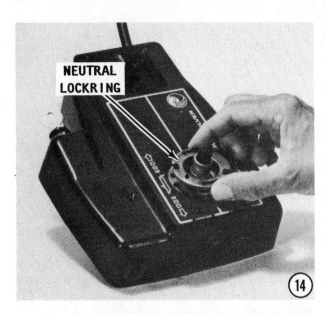

NEUTRAL LOCKRING

(14)

13- Remove the control handle.

14- Lift the neutral lockring from the control housing.

TAKE CARE to support the weight of the control housing to avoid placing any unnecessary stress on the control shaft during the following disassembling steps.

15- Remove the three Phillips-head screws securing the control module to the plastic case. Two are located on either side of the bearing plate and one is in the recess where the throttle cable enters the control housing.

16- Back-out the detent adjustment screw and the control handle friction screw until their heads are flush with the control module casing. This action will reduce the pre-load from the two springs on the detent ball for later removal.

GOOD WORDS

As this next step is performed, count the number of turns for each screw as they are backed-out and record the figure somewhere. This will be a tremendous aid during assembling.

17- Remove the two locknuts securing the neutral safety switch to the plate assembly and lift out the micro-switch from the recess in the assembly.

18- Remove the Phillips-head screw securing the retaining clip to the control module.

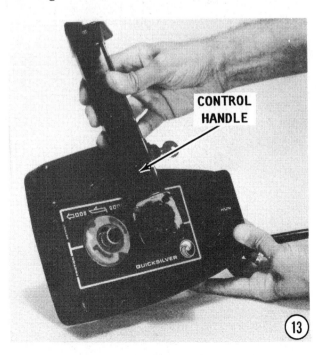

CONTROL HANDLE

(13)

(15)

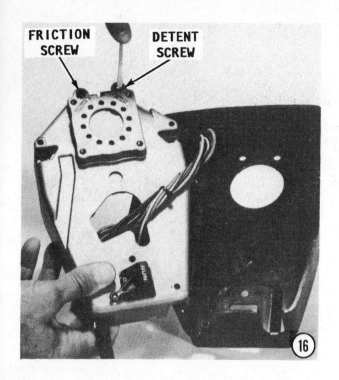

FRICTION SCREW

DETENT SCREW

(16)

SECURING SCREW

RETAINING CLIP

(18)

19- Support the module in your hand and tilt it until the shift gear spring, shift nylon pin (earlier models have a ball), shift gear pin, another ball the shift gear ball (inner), fall out from their recess. If the parts do not fall out into your hand, attach the control handle and ensure the unit is in the **NEUTRAL** position. The parts should come free when the handle is in the **NEUTRAL** position.

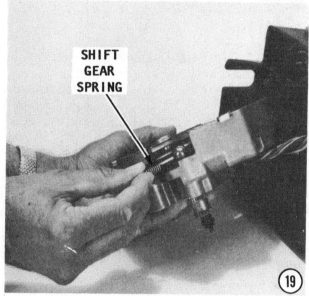

SHIFT GEAR SPRING

(19)

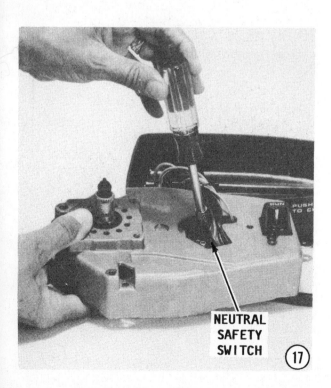

NEUTRAL SAFETY SWITCH

(17)

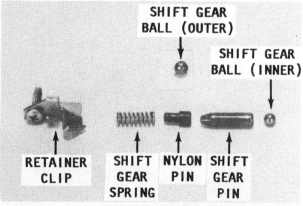

SHIFT GEAR BALL (OUTER)

SHIFT GEAR BALL (INNER)

RETAINER CLIP

SHIFT GEAR SPRING

NYLON PIN

SHIFT GEAR PIN

Arrangement of parts from the control module recess. As the parts are removed and cleaned, keep them in order, ready for installation.

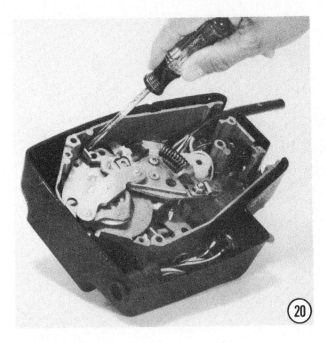

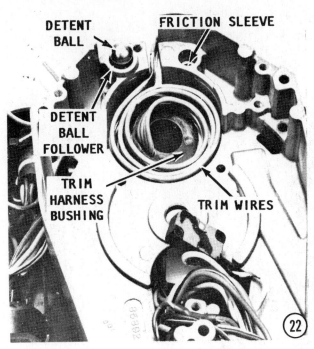

20- Remove the three Phillips-head screws securing the bearing plate assembly to the control module housing.

21- Lift out the bearing plate assembly from the control module housing.

Power Trim/Tilt Units Only

22- Uncoil the trim wires from the recess in the remote control module housing and lift them away with the trim harness bushing attached.

All Units

23- Remove the detent ball, the detent ball follower, and the two compression

springs (located under the follower), from their recess in the control module housing.

24- If it is not part of the friction pad, remove the control handle friction sleeve from the recess in the control module housing.

25- Pull the throttle link assembly from the module. Remove the compression spring from the throttle lever. It is not necessary to remove this spring unless there is cause to replace it. At this point, there is the

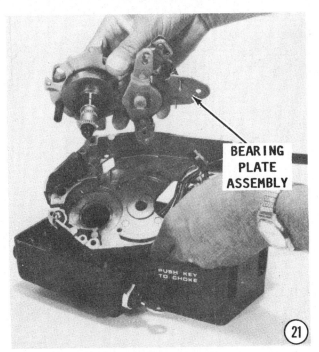

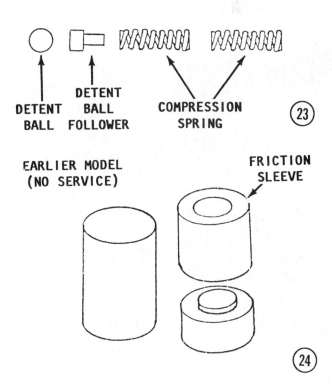

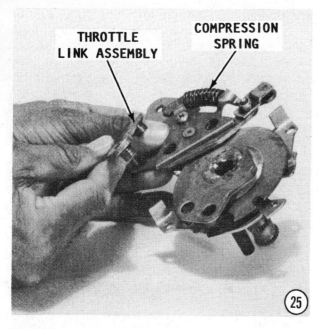

THROTTLE LINK ASSEMBLY   COMPRESSION SPRING

(25)

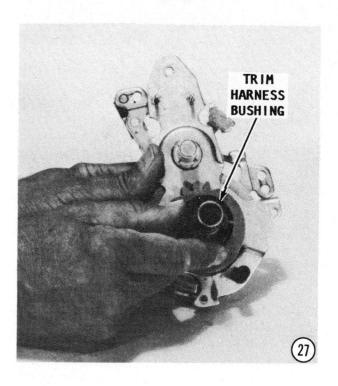

TRIM HARNESS BUSHING

(27)

least amount of tension on the compression spring, therefore, now would be the time to replace it, if required.

26- Lift the shift pinion gear (with attached shift lever), off the pin on the bearing plate. The nylon bushing may come away with the shift lever or stay on the pin. Remove the shift lever and shift pinion gear as an assembly. **DO NOT** attempt to separate them. Both are replaced if one is worn.

Non-Power Trim/Tilt Units Only

27- Remove the trim harness bushing and wiring harness retainer from the control shaft. (On non-power trim/tilt units these two items act as spacers.)

All Units

28- Remove the shift gear retaining ring from its groove with a pair of Circlip pliers.

SPECIAL WORDS

If the Circlip slipped out of its groove, this would allow the shift gear to ride up on the shaft and cause damage to the small parts contained in its recess. The shift gear ball (inner), the shift gear pin, the shift gear ball (outer), or the nylon pin and **PARTICULARLY** the shift gear spring **MUST** be inspected closely.

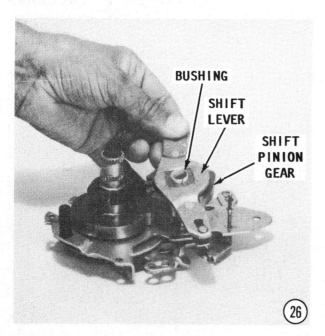

BUSHING

SHIFT LEVER

SHIFT PINION GEAR

(26)

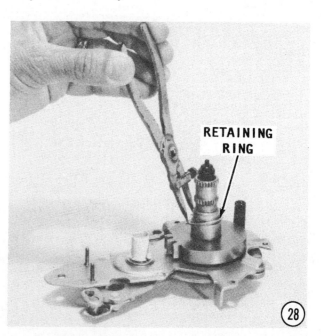

RETAINING RING

(28)

29- Lift the gear from the control shaft.

30- Remove the "throttle only" shaft pin and "throttle only" shaft from the control shaft.

31- Remove the step washer from the base of the bearing plate.

CLEANING AND INSPECTING

Clean all metal parts with solvent, and then blow them dry with compressed air.

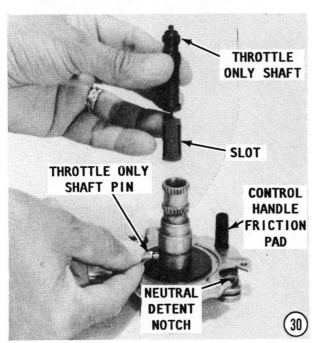

NEVER allow nylon bushings, plastic washers, nylon pins, wiring harness retainers, and the like, to remain submerged in solvent more than just a few moments. The solvent will cause these type parts to expand slightly. They are already considered a "tight fit" and even the slightest amount of expansion would make them very difficult to install. If force is used, the part is most likely to be distorted.

Inspect the control housing plastic case for cracks or other damage allowing moisture to enter and cause problems with the mechanism.

Carefully check the teeth on the shift gear and shift lever for signs of wear. Inspect all ball bearings for nicks or grooves which would CAUSE them to bind and fail to move freely.

Closely inspect the condition of all wires and their protective insulation. Look for exposed wires caused by the insulation rubbing on a moving part, cuts and nicks in the insulation and severe kinking which could cause internal breakage of the wires.

Inspect the surface area above the groove in which the Circlip is positioned for signs of the Circlip rising out of the groove. This would occur if the clip had lost its "spring" or worn away the top surface of the groove as mentioned previously in Step 29. If the Circlip slipped out of its groove, this would allow the shift gear to ride up on the shaft and cause damage to the small parts contained in its recess. The shift gear ball (inner), the shift gear pin, the shift gear ball (outer), or the nylon pin and **PARTICULARLY** the shift gear spring **MUST** be inspected closely.

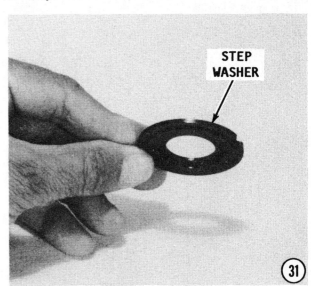

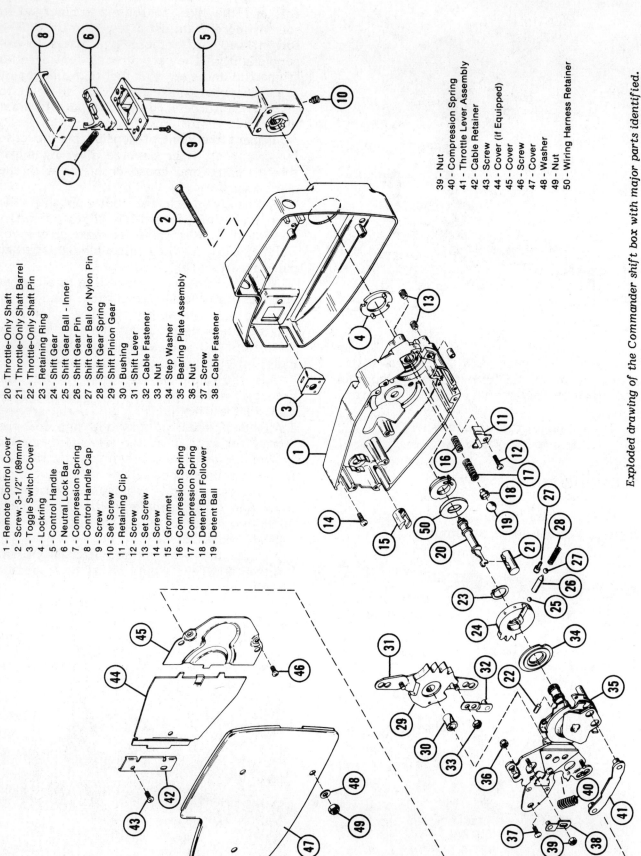

1 - Remote Control Cover
2 - Screw, 3-1/2" (89mm)
3 - Toggle Switch Cover
4 - Locking
5 - Control Handle
6 - Neutral Lock Bar
7 - Compression Spring
8 - Control Handle Cap
9 - Screw
10 - Set Screw
11 - Retaining Clip
12 - Screw
13 - Set Screw
14 - Screw
15 - Grommet
16 - Compression Spring
17 - Compression Spring
18 - Detent Ball Follower
19 - Detent Ball

20 - Throttle-Only Shaft
21 - Throttle-Only Shaft Barrel
22 - Throttle-Only Shaft Pin
23 - Retaining Ring
24 - Shift Gear
25 - Shift Gear Ball - Inner
26 - Shift Gear Pin
27 - Shift Gear Ball or Nylon Pin
28 - Shift Gear Spring
29 - Shift Pinion Gear
30 - Bushing
31 - Shift Lever
32 - Cable Fastener
33 - Nut
34 - Step Washer
35 - Bearing Plate Assembly
36 - Nut
37 - Screw
38 - Cable Fastener

39 - Nut
40 - Compression Spring
41 - Throttle Lever Assembly
42 - Cable Retainer
43 - Screw
44 - Cover (if Equipped)
45 - Cover
46 - Screw
47 - Cover
48 - Washer
49 - Nut
50 - Wiring Harness Retainer

Exploded drawing of the Commander shift box with major parts identified.

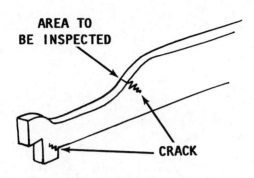

The throttle-only shaft should be inspected for wear along the ramp, as indicated.

Inspect the "throttle only" shaft for wear along the ramp. In early model units, this shaft was made of plastic. Later models have a shaft of stainless steel. Check for excessive wear or cracks on the ramp portion of the shaft, as indicated in the accompanying illustration. Also check the lower "stop" tab to be sure it has not broken away.

SPECIAL WORDS

Good shop practice dictates a thin coat of Multipurpose Lubricant be applied to all moving parts as a precaution against the "enemy" moisture. Of course the lubricant will help to ensure continued satisfactory operation of the mechanism.

ASSEMBLING AND INSTALLATION COMMANDER CONTROL SHIFT BOX

FIRST, THESE WORDS

The Commander control shift box, like others, has a number of small parts that

MUST be assembled in only one order -- the proper order. Therefore, the work should not be "rushed" or attempted if the person assembling the unit is "under pressure". Work slowly, exercise patience, read ahead before performing the task, and follow the steps closely.

1- Place the step washer over the control shaft and ensure the steps of the washer seat onto the base of the bearing plate.
2- Rotate the control shaft until the "throttle only" shaft pin hole is aligned centrally between the neutral detent notch and the control handle friction pad. Lower the "throttle only" shaft into the barrel of the control shaft with the wide slot in the "throttle only" shaft aligned with the line drawn on the accompanying illustration. Secure the shaft in this position with the "throttle only" shaft pin.

SPECIAL WORDS

When the pin is properly installed, it should protrude slightly in line with the plastic bushing, as shown in the accompanying illustration.

Make an attempt to gently pull the "throttle only" shaft out of the control shaft. The attempt should fail, if the shaft and pin are properly installed.
3- Place the shift gear over the control shaft, and check to be sure the "throttle only" shaft pin clears the gear.

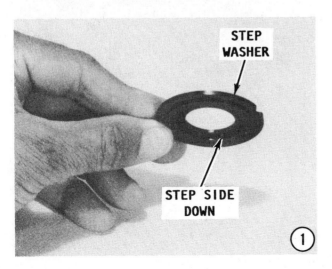

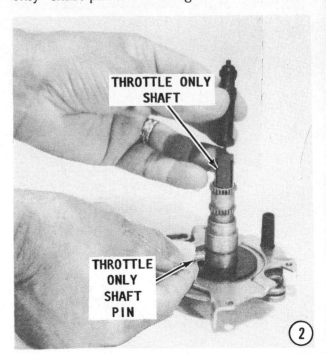

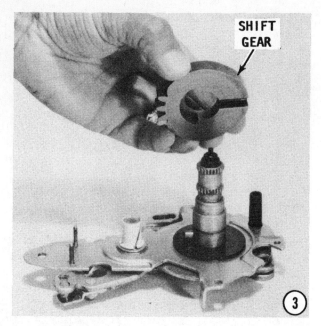

SHIFT GEAR

(3)

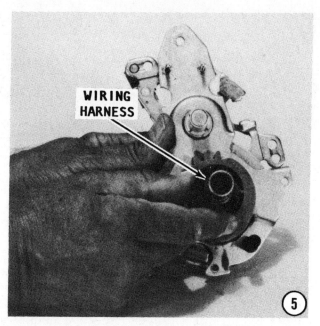

WIRING HARNESS

(5)

4- Install the retaining ring over the control shaft with a pair of Circlip pliers. Check to be sure the ring snaps into place within the groove.

Non-Power Trim/Tilt Units Only

5- Slide the wiring harness retainer and the trim harness bushing over the control shaft. The trim harness bushing is placed "stepped side" **UP** and the notched side toward the forward side of the control housing.

Power Trim/Tilt Units Only

6- Insert the trim harness bushing into the recess of the remote control housing and

carefully coil the wires, as shown in the accompanying illustration. Ensure the black line on the trim harness is positioned at the exact point shown for correct installation. The purpose of the coil is to allow slack in the wiring harness when the control handle is shifted through a full cycle. The bushing and wires move with the handle.

All Units

7- Position the bushing, shift pinion gear and shift lever onto the pin on the bearing plate, with the shift gear indexing with the shift pinion gear.

8- Install the two compression springs,

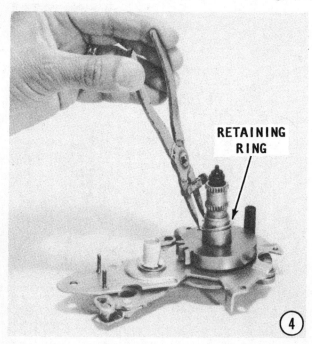

RETAINING RING

(4)

POSITION OF BLACK LINE ON TRIM HARNESS

TRIM HARNESS SLOT

TRIM HARNESS BUSHING

(6)

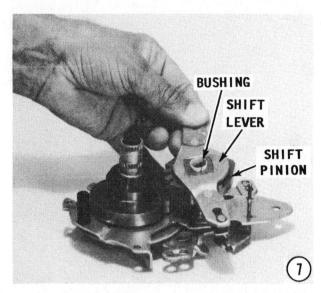

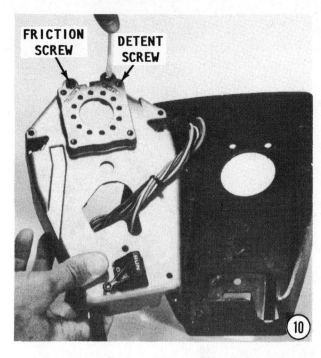

the detent ball follower and the detent ball into their recess in the control module housing.

9- If the friction sleeve is not a part of the friction pad, then place the control handle friction sleeve into its recess in the control module housing.

SPECIAL WORDS

In Step 16 of the disassembling procedures, instructions were given to count the number of turns required to remove the detent adjustment screw and the control handle friction screw. The number of turns is now necessary for ease in performing the next step.

10- Thread the detent adjustment screw and the control handle friction screw the exact number of turns as recorded during Step 16 of the disassembling procedures. A fine adjustment may be necessary after the unit is completely assembled.

11- Place the compression spring (if removed) in position on the shift lever and shift pinion gear assembly against the bearing plate, as shown. Use a rubber band to secure the shift pinion gear to the bearing plate. Lower the complete bearing plate assembly into the control module housing.

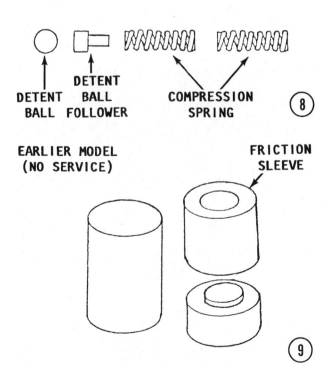

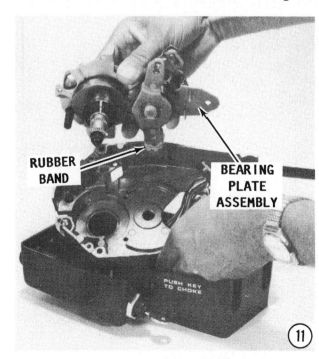

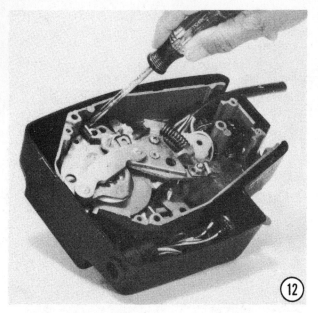

(12)

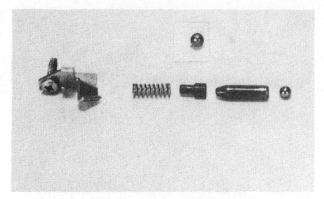

Arrangement of parts, cleaned and ready for installation into the control module recess.

12- Secure the bearing plate assembly to the control module housing with the three Phillips head screws, remove the rubber band.

13- Insert the gear shift ball (inner) into the recess of the shift gear and hole in the "throttle only" shaft barrel. Now, insert the shift gear pin into the recess with the rounded end of the pin away from the control shaft. Insert the nylon pin or shift gear ball (outer) into the same recess. Next, insert the gear shift spring.

14- Hold these small parts in place and at the same time secure them with the retaining clip and the Phillip head screw. (On power trim/tilt units, this retaining clip secures the trim wire to the control module also.

15- Insert the neutral safety microswitch into the recess of the plate assembly and secure it with the two locknuts.

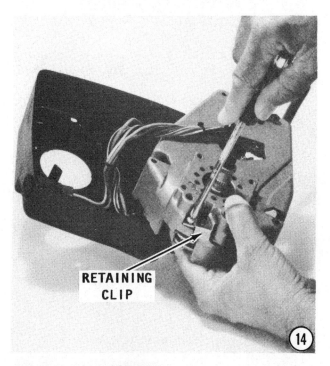

RETAINING CLIP

(14)

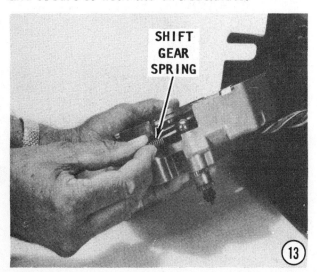

SHIFT GEAR SPRING

(13)

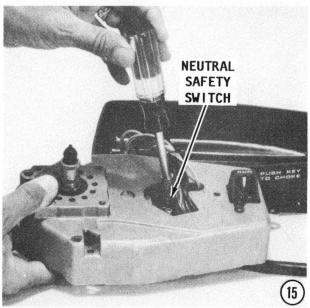

NEUTRAL SAFETY SWITCH

(15)

16- Secure the control module to the plastic control housing case with the three Phillips head screws. Two are located on either side of the bearing plate and the third in the recess where the throttle cable enters the control housing.

17- Temporarily install the control handle onto the control shaft. Shift the unit into forward detent **ONLY**, not full forward, to align the holes for installation of the throttle link. After the holes are aligned, remove the handle. Install the throttle link.

18- Again, temporarily install the control handle onto the control shaft. This time shift the unit into the **NEUTRAL** position, and then remove the handle. Place the neutral lockring over the control shaft, with the index mark directly beneath the small boot on the front face of the cover.

19- Install the control onto the splines of the control shaft. **TAKE CARE** not to cut, pinch, or damage the trim wires on the power trim/tilt unit.

CRITICAL WORDS

When positioning the control handle, ensure the trim wire bushing is aligned with its locating pin against the corresponding slot in the control handle. On a Power Trim/Tilt unit only: if this bushing is **NOT** installed correctly, it will not move with the control handle as it is designed to move -- when shifted. This may pinch or cut the trim wires, causing serious problems.

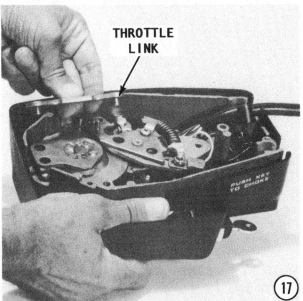

THROTTLE LINK

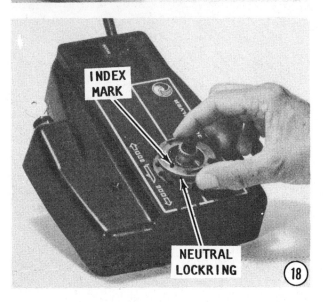

INDEX MARK

NEUTRAL LOCKRING

NEUTRAL LOCKRING

On a Non-Power Trim/Tilt unit, mis-placement of this bushing (it is possible to install this bushing upside down) will not allow the control handle to seat properly against the lockring and housing. This situation will lead to the Allen screw at the base of the control handle to be incorrectly tightened to seat against the splines on the control shaft, instead of gripping the smooth portion of the shaft. Subsequently the control handle will feel "sloppy" and could cause the neutral lock to be ineffective.

WARNING
IF THIS HANDLE IS NOT SEATED PROPERLY, A SLIGHT PRESSURE ON THE HANDLE COULD THROW THE LOWER UNIT INTO GEAR, CAUSING SERIOUS INJURY TO CREW, PASSENGERS, AND THE BOAT.

20- Push the "throttle only" button in place on the control shaft.

21- Ensure the control handle has seated properly, and then tighten the set screw at the base of the handle to a torque value of 70 in. lbs (7.9Nm).

SAFETY WORDS
FAILURE to tighten the set screw to the required torque value, could allow the handle to disengage with a loss of throttle and shift control. An extremely **DANGEROUS** condition.

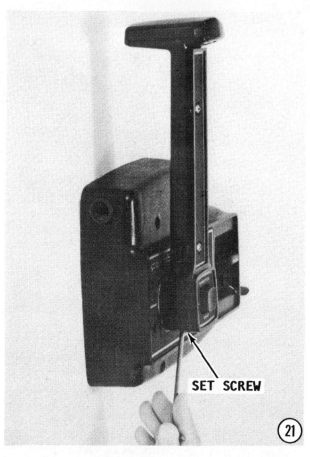

SET SCREW

㉑

**Power Trim/Tilt Models
or
Non-Power Models
If Handle Cover Was Removed**

22- Slide the hooked end of the neutral lock rod into the slot in the neutral lock release. Route the trim wires in the control handle in their original locations. Connect them with the wires remaining in the handle and secure the connections with the wire retainer. Install the handle cover and tighten the two Phillips head screws.

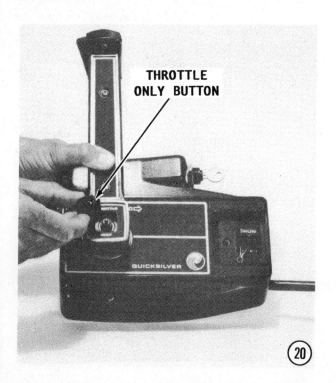

THROTTLE ONLY BUTTON

㉒

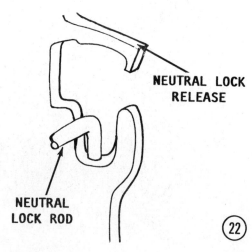

NEUTRAL LOCK RELEASE

NEUTRAL LOCK ROD

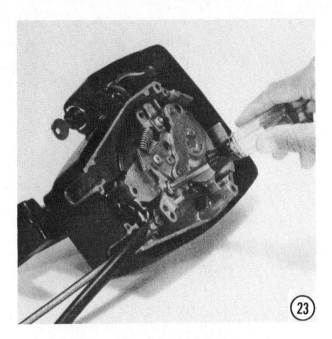

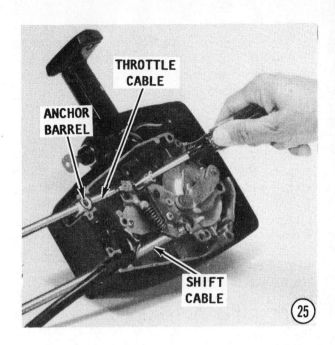

23- Move the wiring harness clear of the barrel recess. Thread the shift cable anchor barrel to the end of the threads, away from the cable converter, and place it into the recess. Hook the pin on the end of the cable fastener through the outer hole in the shift lever. Depress the **NEUTRAL** lock bar on the control handle and shift the handle into the **REVERSE** position. **TAKE CARE** to ensure the cable fastener will clear the neutral safety micro-switch. The access hole is now aligned with the locknut.

STOP

Check to be sure the pin on the cable fastener is all the way through the cable end and the shift lever. A pin partially engaging the cable and the shift lever may cause the cable fastener to **BEND** when the nut is tightened.

Tighten the locknut with a 3/8" deep socket to a torque value of 20 to 25 in. lbs (2.26 to 2.82 Nm). Position the wiring harness over the installed shift cable.

24- Install the grommet into the throttle cable recess.

25- Thread the throttle cable anchor barrel to the end of the threads, away from the cable connector, and then place it into the recess over the grommet. Hook the pin on the end of the cable fastener through the outer hole in the shift lever.

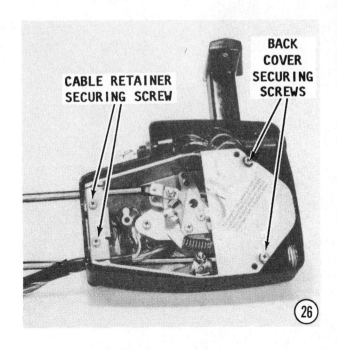

STOP AGAIN

Check to be sure the pin on the cable fastener is all the way through the cable end and the throttle lever. A pin partially engaging the cable and throttle lever may cause the cable fastener to **BEND** when the nut is tightened.

Tighten the locknut to a torque value of 20 to 25 in. lbs (2.26 to 2.82 Nm).

26- Position the control module back cover in place and secure it with the two Phillips-head screws. Tighten the screws to a torque value of 60 in. lbs (6.78 Nm). Install the cable retainer plate over the two cables and secure it in place with the two Phillip-head screws.

27- Place the plastic access cover over the control housing.

28- Position the control housing in place on the mounting panel and secure it with the three long (3-1/2") bolts, flat washers, and locknuts. One is located next to the **RUN** button (the ignition safety stop switch). The second is beneath the control handle on the power portion of the plastic case. The third bolt goes in behind the control handle when the handle is in the **NEUTRAL** position.

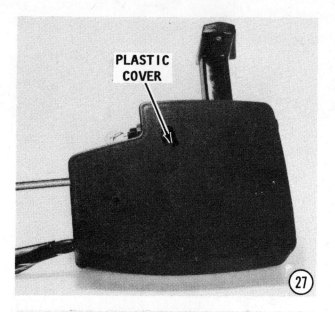

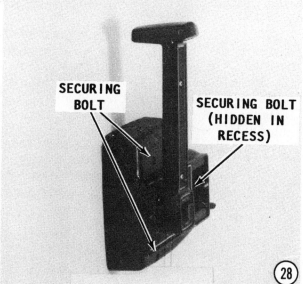

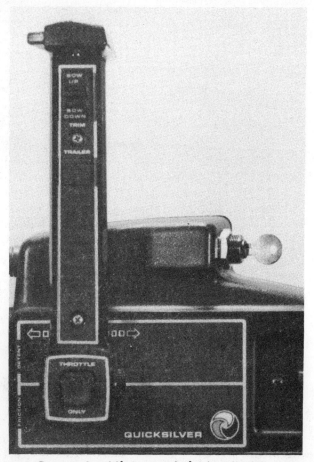

Commander shift box ready for installation.

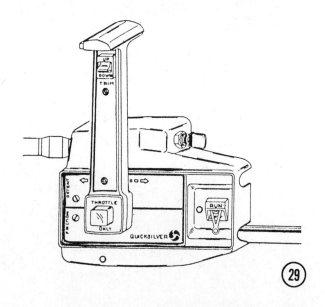

(30)

(31)

Therefore, in order to install this bolt, shift the handle into the **FORWARD** or the **RE-VERSE** position, and then install the bolt. After the bolt is secure, shift the handle back to the **NEUTRAL** position for the next few steps.

29- Connect the tachometer wiring plug to the forward end of the control housing.

GOOD WORDS

Clean the prongs of the connector with crocus cloth to ensure the best connection possible. Exercise care while cleaning to prevent bending the prongs.

30- Connect the remote control wiring harness plug from the outboard trim/tilt motor and pump assembly.

31- Install the high-tension leads to their respective spark plugs.

32- Route the wiring harness alongside the boat and fasten with the "Sta-Straps". Check to be sure the wiring will not be pinched or chafe on any moving part and will not come in contact with water in the bilge. Route the shift and throttle cables the best possible way to make large bends and as few as possible. Secure the cables approximately every three feet (one meter).

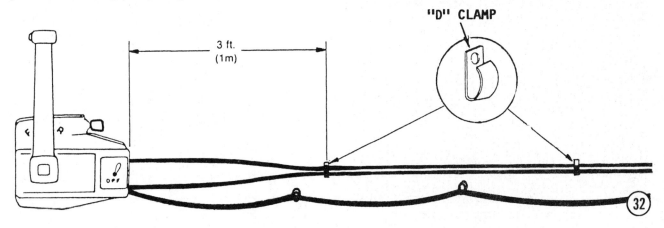

"D" CLAMP

3 ft. (1m)

(32)

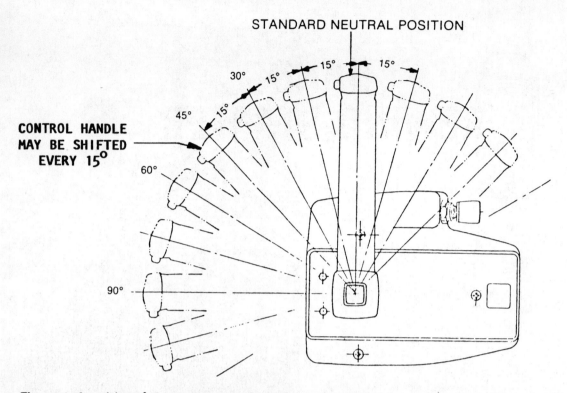

The neutral position of the remote control handle may be changed to any one of a number of convenient angles to meet the owner's preference. The change is accomplished by shifting the handle one spline on the shaft at a time. Each spline equals 15° of arc, as shown. The procedures on Page 11-15 explain the positioning in detail.

12
HAND
REWIND STARTER

12-1 INTRODUCTION

Eight, yes eight, hand rewind starters have been used over the years on the 1- and 2-cylinder powerheads covered in this manual.

Because the number of outboard models having so many different starters is quite lengthy, it would not be practical to list the model and effective years for each unit.

Therefore, starters have been classified as Type **"A"** through Type **"H"** with complete and detailed procedures provided to disassemble, clean and inspect, then assemble and install each type.

If service is required on a hand starter, first, remove the cowling. At this point you may be able to identify the starter with one of the illustrations presented on this page and the page following. If so, proceed directly to the appropriate section.

If identification is not possible, remove the rope retainer (handle), untie the "stop" knot in the rope and allow the rewind spring to unwind within the recess of the sheave. If a shift interlock cable is used, disconnect the cable. Next, remove the attaching hardware, lift the hand starter from the powerhead, turn it over and check the underneath side. Positive identification can now be made by comparing the starter to be serviced with the illustrations. A determination may thus be made as to the procedures to be followed.

Procedures for the eight hand rewind starters begin on the following pages:

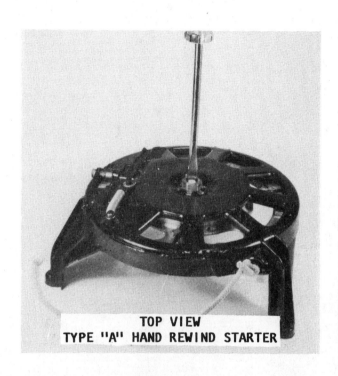

TOP VIEW
TYPE "A" HAND REWIND STARTER

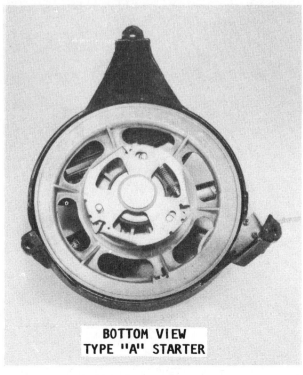

BOTTOM VIEW
TYPE "A" STARTER

TOP VIEW
TYPE "B" HAND REWIND STARTER

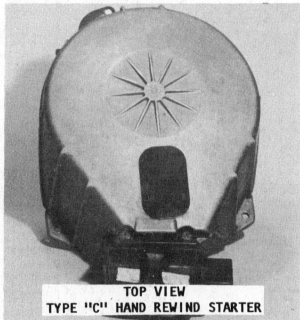

TOP VIEW
TYPE "C" HAND REWIND STARTER

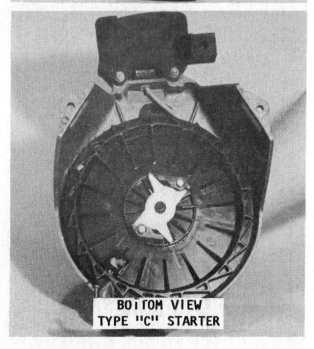

BOTTOM VIEW
TYPE "C" STARTER

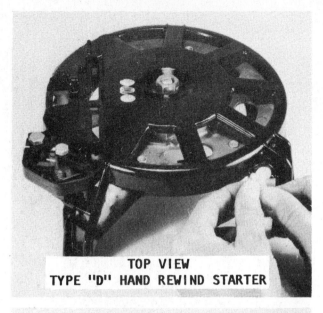

TOP VIEW
TYPE "D" HAND REWIND STARTER

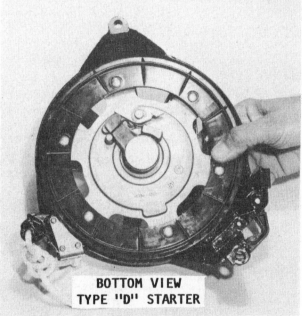

BOTTOM VIEW
TYPE "D" STARTER

TOP VIEW
TYPE "E" HAND REWIND STARTER

BOTTOM VIEW
TYPE "E" STARTER

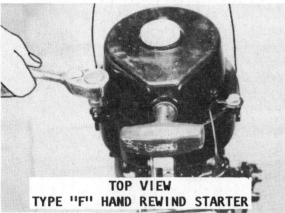

TOP VIEW
TYPE "F" HAND REWIND STARTER

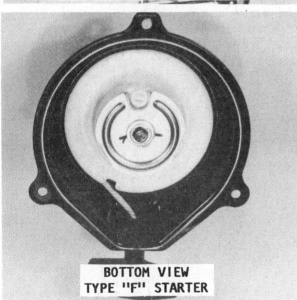

BOTTOM VIEW
TYPE "F" STARTER

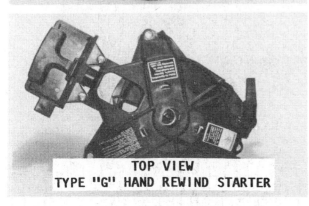

TOP VIEW
TYPE "G" HAND REWIND STARTER

BOTTOM VIEW
TYPE "G" STARTER

TOP VIEW
TYPE "H" STARTER

(Bottom view of the Type "H" hand rewind starter is shown on Page 12-52, under Step 4.)

12-2 TYPE "A"
HAND REWIND STARTER

REMOVAL AND DISASSEMBLING

PRELIMINARY TASKS

First, remove the wrap around cowling. Next, remove the rope retainer (handle) from the rope. This is accomplished by simply pushing the inner portion through the outer shell of the handle. Untie the "stop" knot in the rope and allow the rewind spring to unwind within the recess of the sheave.

If the choke rod passes down through the top of the cowl, disconnect the upper portion of the shaft from the lower portion by pulling the cotter pin securing the two together. Pull the upper portion of the choke shaft up through the cowling.

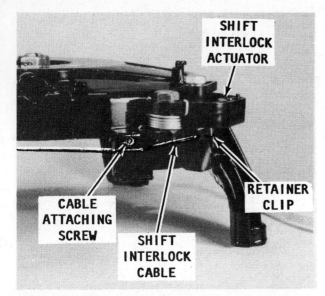

Later model rewind starters are equipped with a shift interlock cable mounted on the side of the starter housing. Early models have the cable mounted on the top of the housing. The interlock cable of both models is secured in the same manner.

If the unit being serviced is equipped with a shift interlock actuator (cable), remove the cable attaching screw and the retainer clip, as shown in the accompanying illustration.

Now, remove the cowling. Remove the screws securing the three legs of the hand rewind starter mechanism to the powerhead.

The following steps pickup the work after these few preliminary tasks have been completed.

DISASSEMBLING

1- Place the hand starter on a suitable work surface. Pry out the one tab bent into the recess in the rewind housing. Use a hammer and flat punch to push aside the two tabs of the tab washer. This washer secures the sheave shaft retaining nut.

CRITICAL WORDS

The sheave shaft retaining nut has **LEFT-HAND** threads.

2- Remove the nut with a 3/4" end wrench by rotating the nut **CLOCKWISE** because it is a **LEFT-HAND** nut.

3- Carefully lift the starter housing from the sheave assembly so as not to disturb the rewind spring encased in the sheave. Place the sheave face down on a flat surface.

4- Lift off the spring retainer plate, and then remove the shaft bushing and the spring bushing from the sheave.

5- Remove the sheave shaft, the fiber washer, and the pawl retainer plate from the sheave.

GOOD WORDS

The return spring will come away with the pawl retainer plate.

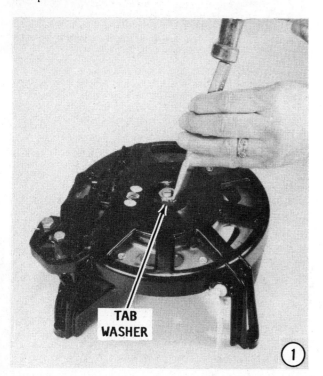

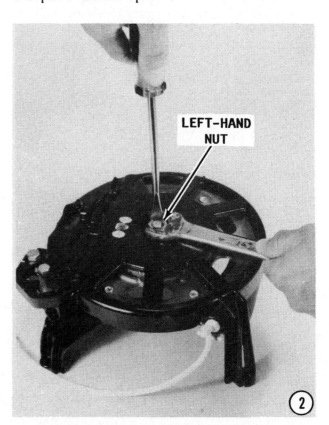

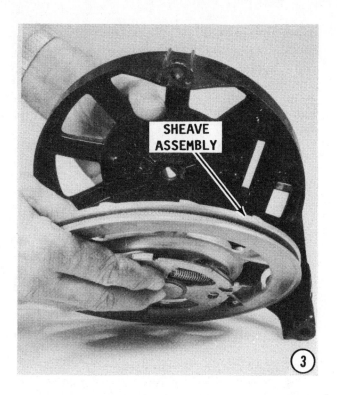

SHEAVE ASSEMBLY

③

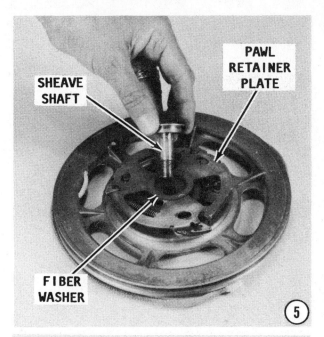

SHEAVE SHAFT

PAWL RETAINER PLATE

FIBER WASHER

⑤

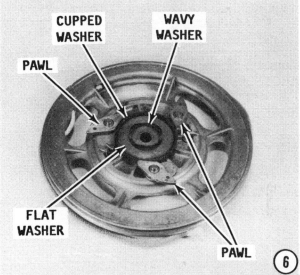

CUPPED WASHER   WAVY WASHER

PAWL

FLAT WASHER

PAWL

⑥

6- Remove and **DISCARD** the wavy washer. It cannot be used a second time. Remove the flat washer and the cupped washer. Lift off the three pawls from their pins. Remove and **DISCARD** the wavy washer under each pawl.

7- Obtain two pieces of wood, a short 2" x 4" (5cm x 10cm) will work fine. Place the two pieces of wood approximately 8" (20 cm) apart on the floor. Center the sheave on top of the wood with the spring side facing **DOWN**. Check to be sure the wood is not touching the spring.

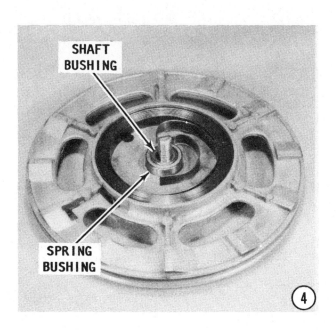

SHAFT BUSHING

SPRING BUSHING

④

WOODEN BLOCK

8" (20 cm)

⑦

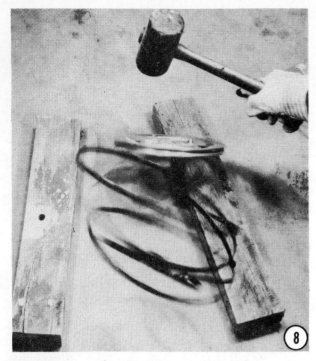

WARNING

THE REWIND SPRING IS A POTENTIAL HAZARD. The spring is under tremendous tension when it is wound -- a real **"tiger"** in a cage! If the spring should accidentally be released, severe personal injury could result from being struck by the spring with force. Therefore, the following step **MUST** be performed with care to pervent personal injury to self and others in the area.

8- Stand behind the wood, keeping away from the openings as the spring unwinds with considerable force. Tap the sheave with a soft mallet. The spring retainer plate will drop down releasing the spring. The spring will fall and unwind almost instantly and with **FORCE.**

9- Unwind the rope from around the sheave and feed it through the holes. Turn the anchor pin 90° in either direction to release the rope.

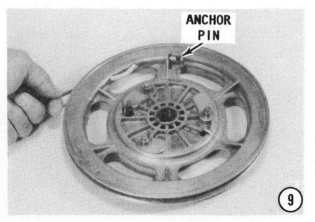

CLEANING AND INSPECTING

Wash all parts except the rope and the handle in solvent, and then blow them dry with compressed air.

Remove any trace of corrosion and wipe all metal parts with an oil dampened cloth.

Inspect the rope. Replace the rope if it appears to be weak or frayed. If the rope is frayed, check the holes through which the rope passes for rough edges or burrs. Remove the rough edges or burrs with a file and polish the surface until it is smooth. Inspect the starter spring end hooks. Replace the spring if it is weak, corroded or cracked. Inspect the tab on the spring retainer plate. This tab is inserted into the inner loop of the spring. Therefore, be sure it is straight and solid. Inspect the inside surface of the sheave rewind recess for grooves or roughness. Grooves may cause erratic rewinding of the starter rope.

Check the condition of the pawl pivot pins for excessive wear. These pins are a part of the sheave casting. They **CANNOT** be serviced. Therefore, if they are not acceptable, a new sheave **MUST** be purchased and installed.

Inspect the pawls for wear around the pivot holes and for rounded outer edges. Replace the pawls in sets of **THREE ONLY.** Check the pawl retainer plate for wear, especially the center hole and the areas where the sides of the pawls contact the retainer plate.

Inspect the pawl retainer plate return spring (if equipped). The manufacturer recommends this spring be replaced each time the rewind starter is disassembled for service. A weak spring could allow the pawls

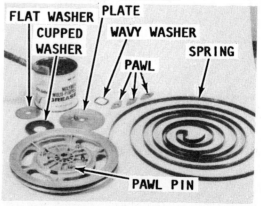

After the parts shown have been cleaned and are ready for assembling, apply a coating of low-temperature lubricant to ensure long time service.

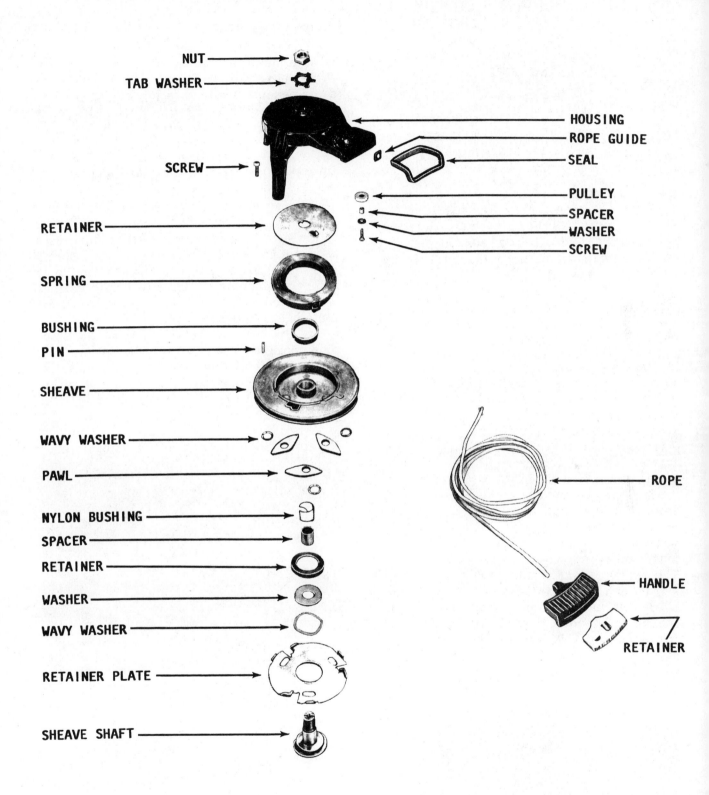

NUT

TAB WASHER

HOUSING

ROPE GUIDE

SEAL

SCREW

PULLEY

SPACER

WASHER

SCREW

RETAINER

SPRING

BUSHING

PIN

SHEAVE

WAVY WASHER

PAWL

NYLON BUSHING

SPACER

RETAINER

WASHER

WAVY WASHER

RETAINER PLATE

SHEAVE SHAFT

ROPE

HANDLE

RETAINER

Arrangement of parts, in order, for the hand rewind starter covered in this section.

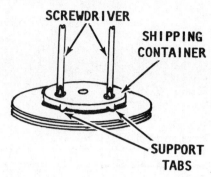

A new rewind spring is installed into the sheave using the shipping container as a helpful tool. Two screwdrivers are used to tap the coiled spring out of the container and into the sheave.

to contact the flywheel while the powerhead is operating and cause a very unpleasant noise and even damage to the edge of the flywheel. A weak spring will also cause excessive wear to the pawls.

It is **STRONGLY** recommended the wavy washer under the pawl retainer plate, and the three wavy washers under the pawls be **REPLACED**. New wavy washers will ensure smooth operation of the rewind mechanism.

Coat the following parts with low-temperature lubricant: the entire length of the used rewind spring (a new spring will be coated with lubricant from the package), the pawl pins, both sides of the cupped washer, the flat washer, the shoulder on the sheave shaft, and the tab on the spring retainer plate.

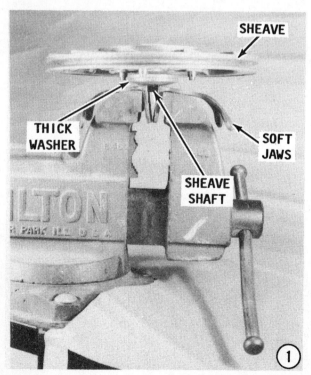

ASSEMBLING AND INSTALLATION TYPE "A" REWIND HAND STARTER

SPECIAL WORDS

The rewind starter may be assembled with a new rewind spring or a used one. Procedures for assembling are **NOT** the same because the new spring will arrive in a shipping container already wound and ready for installation. The used spring must be manually wound into its recess.

NEW REWIND SPRING INSTALLATION

The situation may arise when it is only necessary to replace a broken spring. The following few procedures outline the tasks required to replace the spring. A new spring is already properly wound and will arrive in a special shipping container. This container is designed to be used as an aid to installing the new spring.

a- Remove the retainers from the shipping container used to keep the spring from accidentally falling out of the container. Place the shipping container over the spring recess with the tabs resting on the outer edge of the recess.

b- Align the hook on the outer end of the spring with the sheave anchor or notch.

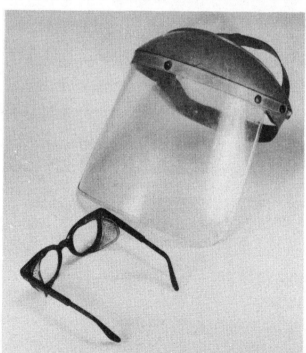

Protect your eyes with a face mask or safety glasses while working with the rewind spring, especially a used one. The spring is a "tiger in a cage", almost 13 feet (4 meters), of spring steel wound and confined into a space less than 4" (about 10 cm) in diameter.

c- Place two large blade screwdrivers into the holes of the shipping container over the tensioned spring. Push on both screwdrivers at the same time to press the spring out of the shipping container and into the spring recess of the sheave.

USED REWIND SPRING INSTALLATION

A used spring naturally will not be wound. Therefore, special instructions are necessary for installation.

SAFETY WORDS

Wear a good pair of gloves while winding and installing the spring. The spring will develop tension and the edges of the spring steel are extremely sharp. The gloves will prevent cuts to the hands and fingers.

1- Insert the sheave shaft through the sheave with the spring recess facing **UP-WARD**. Obtain two large washers and place them over the sheave shaft. Hold the washers in place and at the same time clamp the sheave shaft in a vise equipped with soft jaws. Shift the shaft in the vise to allow for "up and down" clearance and permit the shaft to rotate freely while the rewind spring is being installed.

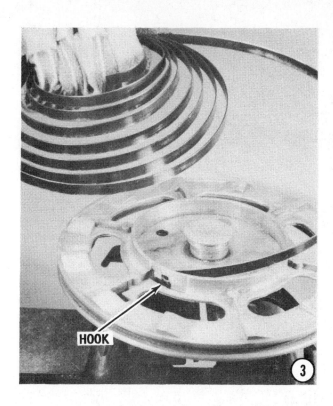

HOOK

③

SAFETY WORDS

It is **STRONGLY** recommended a pair of safety goggles or a face shield be worn while the spring is being installed. As the work progresses a "tiger" is being forced into a cage. If the spring is accidentally released, it will lash out with tremendous ferocity and very likely could cause personal injury to the installer or other persons nearby.

②

FEEDING IN COUNTERCLOCKWISE

④

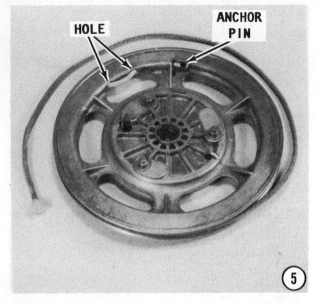

HOLE   ANCHOR PIN

5

PAWL

WAVY WASHER

6

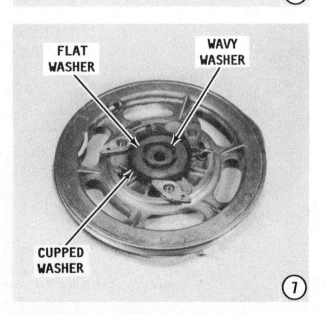

FLAT WASHER   WAVY WASHER

CUPPED WASHER

7

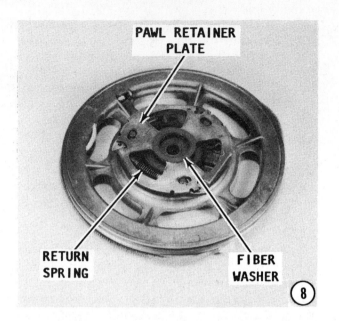

PAWL RETAINER PLATE

RETURN SPRING   FIBER WASHER

8

2- Loop the spring loosely into a coil to enable it to be fed into its recess.

3- Insert the hook on the end of the spring into the notch of the recess.

4- Feed the spring around the inner edge of the recess and at the same time rotate the sheave **CLOCKWISE.** Proceed **WITH GREAT CARE.** Guide the spring into place. The spring will be slippery with lubrication. **DO NOT** lose control of the spring. Carefully remove the sheave from the vise and the sheave shaft and washers from the sheave.

5- Thread the starter rope thru the holes in the sheave and secure it with the anchor pin. Wind the rope **COUNTERCLOCKWISE**

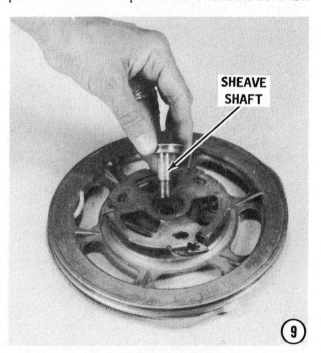

SHEAVE SHAFT

9

around the outer rim of the sheave (when viewed from the pawl side of the sheave).

6- Place one **NEW** wavy washer on each pawl pin. Slide the pawls over the pins and wavy washers. Check to be sure the "dot" or depression on the pawl is facing **UP-WARD** and angled **OUTWARD**, as shown.

7- Place the cupped washer, with the cupped side facing **UPWARD** over the center hole of the sheave. Place the flat washer over the cupped washer with the chamfered side of the flat washer facing **OUTWARD** (away from the sheave). Install another **NEW** wavy washer onto the flat washer.

8- Install a **NEW** return spring between the sheave and the pawl retainer plate, and then position the plate over the pawls. Place the fiber washer over the pawl retainer plate.

9- Insert the sheave shaft into the pawl retainer plate. The large diameter of the shoulder on the sheave shaft **MUST** be flush against the pawl retainer plate. If the wavy washer is not centered beneath the pawl retainer plate, the sheave shaft will appear misaligned. To correct this condition, remove the shaft and the retainer plate, then center the wavy washer and again install the plate and shaft. Again, check the alignment of the sheave shaft.

10- Lift the sheave assembly off the work bench, grasp the threaded end of the sheave shaft and turn the assembly over. Place the unit on the work bench with the spring side **UPWARD**.

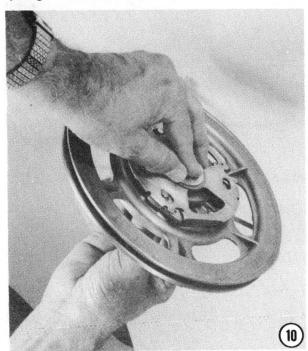

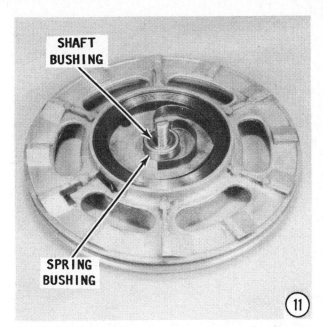

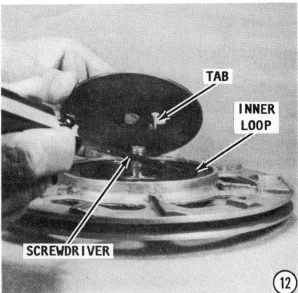

Assembly of the sheave, with the spring retainer plate installed, is now ready for installation into the rewind starter housing.

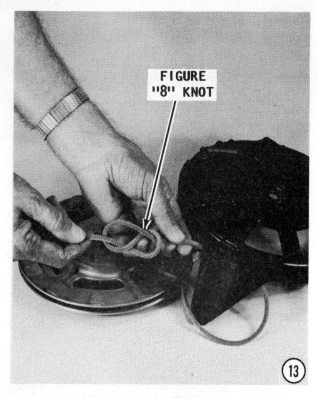

FIGURE "8" KNOT

13

11- Slide the shaft bushing and the spring bushing over the sheave shaft.

12- Lower the spring retainer plate over the shaft. Insert the tab on the plate with the inner loop of the rewind spring. It may be necessary to hook a thin long screwdriver into the loop as an assist to aligning the loop and the tab.

NEW TAB WASHER

14

13- Feed the rope through the rope guide. Tie a figure "8" knot in the rope about 1 foot (30 cm) from the end. Place the sheave assembly into the rewind housing.

14- Support the sheave assembly with one hand. Slide a **NEW** tab washer over the threaded portion of the sheave shaft with the cupped side of the washer facing **DOWN**. Thread the left-hand sheave shaft retaining nut onto the shaft in a **LEFT-HAND (COUNTERCLOCKWISE)** direction until it is just **FINGER-TIGHT**, at this time.

ADJUSTING REWIND SPRING TENSION

15- Insert a large blade screwdriver into the slot of the sheave shaft. Hold the retaining nut with the proper size wrench and at the same time rotate the shaft with the screwdriver **COUNTERCLOCKWISE** until the figure "8" knot in the rope rests against the rope guide. Continue to rotate the shaft through **TWO** full turns after the knot is against the rope guide. The correct tension has now been placed on the rewind spring.

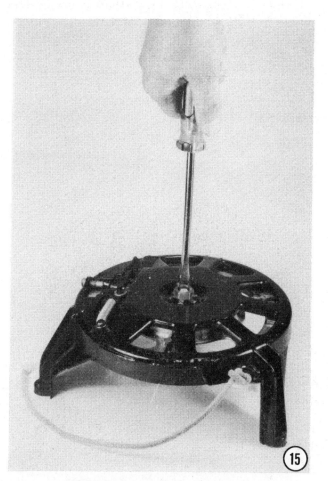

15

16- Hold the tension on the spring with the screwdriver in the shaft slot and at the same time tighten the retaining nut securely. **REMEMBER** -- left-hand thread --tighten **COUNTERCLOCKWISE**.

GOOD WORDS
DO NOT bend the tabs on the washer up against the retaining nut until **AFTER** the rewind operation has been checked.

CHECK REWIND OPERATION
SLOWLY pull the starter rope outward. The pawls must move to the engaged position as the pawl retainer plate begins to turn. If the pawls fail to engage, check the alignment or replace the wavy washers between the pawls. Again, extend the rope to its full length and allow it to rewind. The rope should rewind smoothly without catching. **NEVER** release the rope from the fully extended position.

If the rewind mechanism catches, but fails to rewind, the sheave shaft and its wavy washer are not correctly aligned. The unit **MUST** be disassembled to correct the condition.

Once the rewind operation is satisfactory, proceed to the next step.

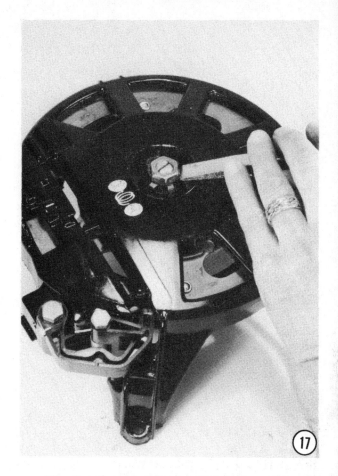

17- Bend two tabs of the tab washer up against the flats of the retaining nut. Use tabs opposite each other. Bend one other tab down into the recess of the rewind housing. Install the rewind starter onto the powerhead and secure the legs with the screws and lockwashers.

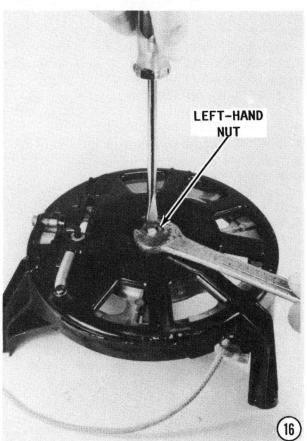

LEFT-HAND NUT

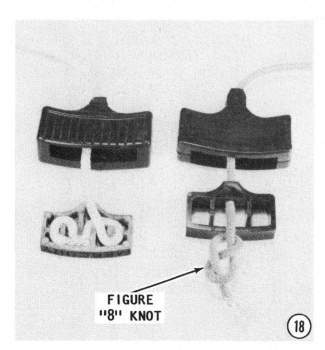

FIGURE "8" KNOT

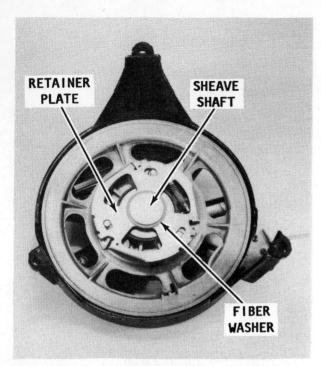

The sheave shaft must be flush with the retainer plate for correct alignment during installation.

SPECIAL WORDS

If the unit being serviced is equipped with a shift interlock actuator (cable), attach the cable securing screw and the retainer clip, as shown in the accompanying illustration.

If the choke handle shaft passes through the top of the cowling, feed it through and make the connection to the lower portion of the shaft with the cotter pin.

18- Feed the rope through the top cowling and install the rope retainer. Two styles of retainer are used on this model rewind starter. Secure the one handle with a figure "8" knot in the end of the rope, as shown.

SHIFT INTERLOCK ADJUSTMENT

Early Model -- Prior to 1977

Secure the cable attaching screw, but **DO NOT** tighten it at this time. Final adjustment will be made later. Hook the end of the cable over the sliding cam and fasten it with the flat washer and cotter pin.

Shift the control handle into the **NEUTRAL** position and adjust the cable to position the toggle pin in line with the spring peg. Tighten the cable attaching screw to hold this position.

Shift the control handle into the **FORWARD** position and try to pull the handle outward. The attempt should fail.

Shift the control handle back into the **NEUTRAL** position and again pull on the handle. The rope should rewind normally.

Late Model
1977 and On

Secure the cable attaching screw to the side of the rewind housing, but **DO NOT** tighten it at this time. Hook the interlock cable over the peg on the interlock actuator with a flat washer and cotter pin. **BE SURE** the control handle is in the **NEUTRAL** position. Adjust the cable until the interlock actuator is positioned on the rise of the interlock cam. Tighten the cable attaching screw to hold this position.

Shift the control handle into the **FORWARD** position and try to pull the handle outward. The attempt should fail.

Shift the handle back into the **NEUTRAL** position and again pull on the handle. The rope should rewind normally.

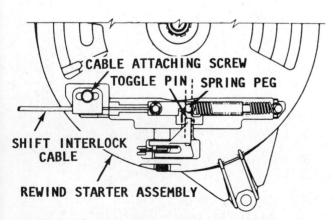

Detailed drawing of an early model rewind housing with the interlock cable installed on top.

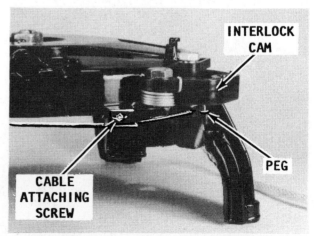

Late model rewind starter with the interlock cable installed on the side of the housing.

12-3 TYPE "B"
SIDE MOUNT
REWIND HAND STARTER

REMOVAL AND DISASSEMBLING

1- Remove the upper cowling from the powerhead. Disconnect the high tension leads from the spark plugs. Pry the cover from the round starter rope handle. Untie the retaining knot in the rope, and then pull the handle free of the rope. Allow the rope to rewind through bottom cowling onto the starter sheave. Remove the three screws securing the starter to the powerhead and lift the starter free.

SPECIAL WORDS
The center bolt of this starter has **LEFT-HAND** threads.

2- Rotate the center bolt **CLOCKWISE**, because it has **LEFT-HAND** threads, and then remove the bolt and the large flat washer from the pinion gear.

3- Lift the pinion spring retainer, pinion spring, and the large flat washer from the pinion gear.

SAFETY WORDS
THE REWIND SPRING IS A POTENTIAL HAZARD. The spring is under tremendous tension when it is wound -- a real **"tiger"** in a cage! If the spring should accidentally be released, severe personal injury could result from being struck by the spring with force. Therefore, the following steps **MUST** be performed with care to prevent personal injury to self and others in the area.

4- With the rope still tightly wound around the pinion gear sheave, hold the sheave with one hand and **CAREFULLY** ro-

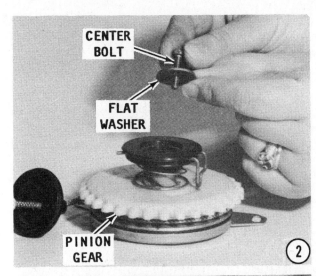

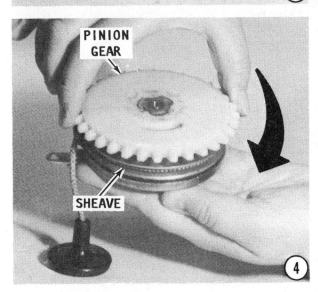

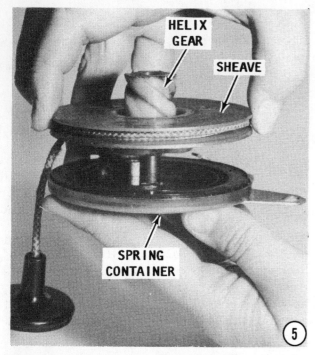

tate the pinion gear **CLOCKWISE** with the other hand. The pinion gear will come free of the helix gear.

5- SLOWLY and **CAREFULLY** separate the sheave from the spring container. Do not allow any part of the spring to come away with the sheave. If the spring does start to separate with the sheave, use a long thin blade screwdriver to confine the spring within the container.

WARNING

If the spring is accidentally released, it will lash out with tremendous ferocity and

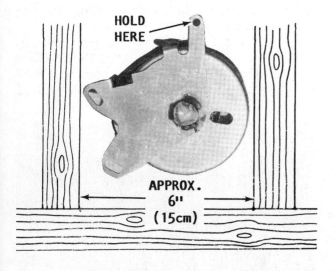

very likely could cause personal injury to the service worker or other persons nearby.

6- Place three pieces of wood on the floor, as shown in the accompanying illustration to confine the spring when it is released. Grasp the spring holder by the mounting flange, the arm without a slotted hole. Strike the top of the container on the floor. The spring will fall free and unwind instantly and with force, but away from you. One end of the spring will remain hooked inside the container.

7- Unwind the pull rope from the sheave. The rope is held in place by two rivets. It may be necessary to drive the rivets out. To accomplish this task, place the sheave between the open jaws of a vise with the helix gear down. Drive the rivets back with a hammer and punch, until the rope is free.

CLEANING AND INSPECTING

Wash all parts except the rope and the handle in solvent, and then blow them dry with compressed air.

Remove any trace of corrosion and wipe all metal parts with an oil dampened cloth.

Inspect the rope. Replace the rope if it appears to be weak or frayed. If the rope is frayed, check the holes through which the rope passes for rough edges or burrs. Remove the rough edges or burrs with a file and polish the surface until it is smooth. Inspect the starter spring end hooks. Replace the spring if it is weak, corroded or cracked. Inspect the inside surface of the sheave rewind recess for grooves or roughness. Grooves may cause erratic rewinding of the starter rope.

Coat the entire length of the used rewind spring (a new spring will be coated with lubricant from the package), with low-temperature lubricant.

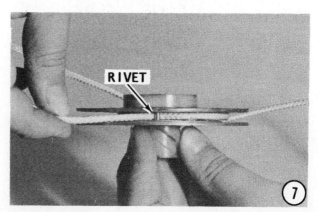

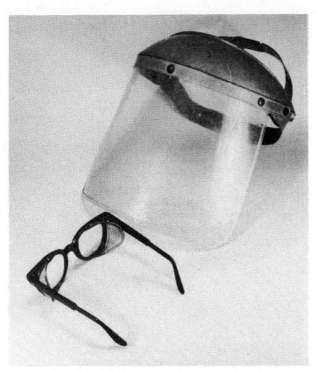

*Protect your eyes with a face mask or safety glasses while working with the rewind spring, especially a used one. The spring is a real **"tiger in a cage"**, almost 13' (4 m), of spring steel wound into less than 4" (about 10cm).*

ASSEMBLING AND INSTALLATION TYPE "B" REWIND STARTER

SAFETY WORDS

Wear a good pair of gloves while winding and installing the spring. The spring will develop tension and the edges of the spring steel are extremely sharp. The gloves will prevent cuts to the hands and fingers.

It is **STRONGLY** recommended a pair of safety goggles or a face shield be worn while the spring is being installed. As the

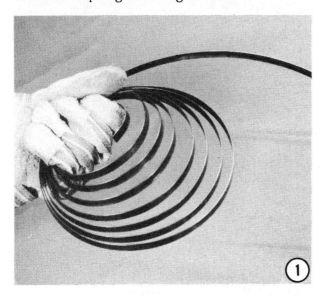

REWIND SPRING

work progresses a "tiger" is being forced into a cage, over 14 ft (4.3m) of spring steel wound into about 4" (10.2cm) circumference. If the spring is accidentally released, it will lash out with tremendous ferocity and very likely could cause personal injury to the installer or other persons nearby.

1- Hook the **"S"** shaped end of the rewind spring into the recess of the container. Feed the spring into the container by holding the spring with one hand and rotating the container **CLOCKWISE** with the other hand.

2- Continue rotating the container until the entire length of the spring is encased within the container.

3- Slide the looped end of the starter rope into the sheave. Hook the rope onto the rivet closest to the helix gear. Drive the rivet into place with a hammer and punch. With the helix gear facing downward, wind the rope **CLOCKWISE** 1-1/2 turns around the sheave. Drive the second rivet into place. Continue winding the remainder of the rope onto the sheave.

4- With the helix gear facing upward, rotate the sheave onto the gear **COUNTERCLOCKWISE**. As the sheave moves downward on the gear, engage the hook end of

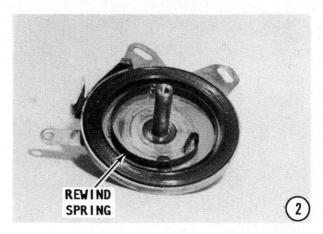

RIVET

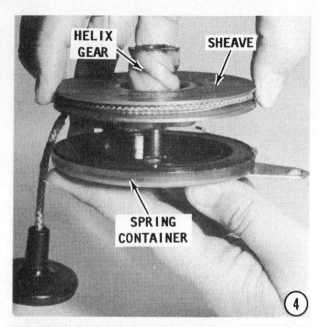

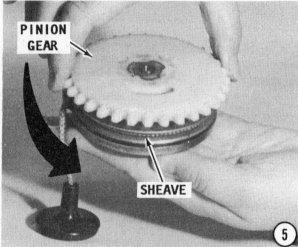

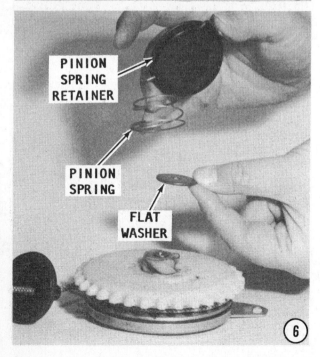

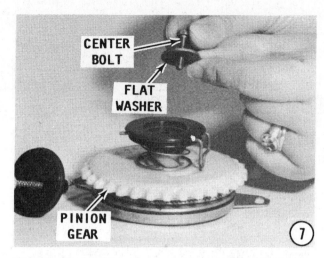

the rewind spring into the slot on the inside of the sheave. A check of the spring engagement with the slot can be made through the slot in the back of the spring container.

5– With the helix gear still facing upward, hold the spring container with one hand and with the other hand rotate the pinion gear onto the helix gear, **COUNTERCLOCKWISE**. The manufacturer recommends **NO** lubricant of any kind be applied to the helix gear.

6– Place the flat washer onto the helix gear, then the pinion spring, and finally the pinion spring retainer into place on the helix gear.

7– Position another flat washer on top of the spring retainer, and then thread the center bolt into the end of the helix gear **COUNTERCLOCKWISE**, because the bolt has **LEFT-HAND** threads. Tighten the bolt to 30in lbs (3Nm).

8– Place the starter assembly on a flat surface with the pinion gear facing upward. Check the distance between the pinion and the curved portion of the pinion gear spring extension, as shown. This measurement should **NOT** be less than 1/16" (1.5mm). Reason: when the starter rope is pulled, there must be sufficient clearance for the pinion gear to move outward to index with the teeth on the flywheel.

9- Install the side mount rewind starter to the powerhead with the three attaching screws. Adjust the starter assembly as far down as the slots in the mounting holes will permit. The screws will then be in the uppermost end of the slots. Thread the end of the pull rope through the bottom cowling. Attach the handle onto the end of the pull rope using a figure **"8"** knot. Pull the rope through the handle and secure it all with the handle cap.

Adjust the return spring tension by looping the rope **COUNTERCLOCKWISE** around the starter sheave inside the rope guide. The rewind spring should rewind the rope handle against the bottom cowling.

Install the upper cowling.

12-4 TYPE "C"
HAND REWIND STARTER
WITH FOUR-ARM CAM

FIRST, THESE WORDS

Be sure to read Section 12-1, and be convinced the starter being serviced is a Type **"C"** as identified from the photographs on Page 12-2.

REMOVAL AND DISASSEMBLING

1- Remove the cowling. Disconnect the high tension leads to the spark plugs. Insert a screwdriver between the rewind starter housing and the fuel filter. A slight downward pressure on the screwdriver will "pop" the fuel filter free. Move the filter and associated fuel lines to one side.

2- Remove the three attaching bolts securing the rewind starter to the powerhead. Lift off the hand starter.

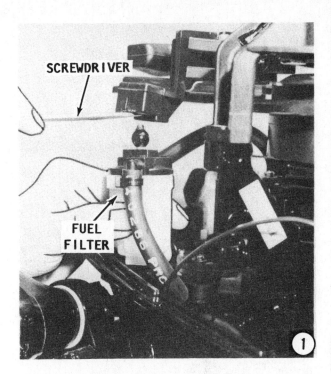

3- Hold the shift interlock lever back with the thumb on one hand, and at the same time, pull approximately 1' (25cm) of the rope out and tie a loose overhand knot in the rope. Do not tighten the knot, because it will be necessary to untie it later with one hand. Allow the spring to rewind. This knot in the rope will permit removing the handle from the end of the rope.

After the handle has been removed, pull the rope further out of the housing. Continue to pull the rope out until the little square window containing the other end of the knoted rope aligns with the rope guide in the starter housing. Exert a good hold on

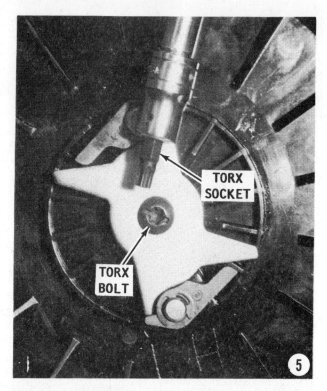

the sheave and at the same time remove the knot from its recess in the sheave. Hold the sheave securely with one hand and untie the knot you made in the other end of the rope earlier in this step. Now, pull the rope back through the rope guide and the sheave until it is free.

4- Carefully allow the spring to unwind by allowing the sheave to rotate in your grasp, until all the tension on the spring is gone.

SPECIAL WORDS

The center bolt on this type rewind starter has standard **RIGHT-HAND** threads.

5- Remove the center Torx bolt from the sheave. If a Torx socket is not available, use the proper size screwdriver. Exercise **CARE** not to damage the bolt head. Using a screwdriver provides very little surface for the blade to grasp, therefore, the

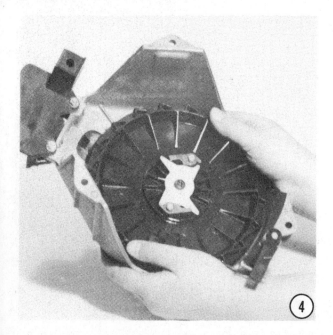

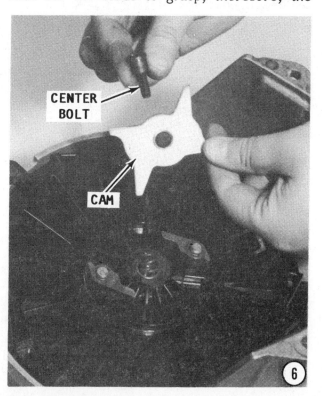

head may be easily damaged. Some **Type "C"** rewind starters may have a regular hex-head bolt -- then no problem.

6- Remove the center bolt, four arm cam and small spring from the sheave.

SAFETY WORDS

THE REWIND SPRING IS A POTENTIAL HAZARD. The spring is under tremendous tension when it is wound -- a real **"tiger"** in a cage! If the spring should accidentally be released, it will lash out with tremendous ferocity and very likely cause severe personal injury to the service worker or other persons nearby.

7- **CAREFULLY** Lift the sheave out of the housing. Remove the felt pad from around the shaft in the housing.

8- Turn the sheave over and **EVER SO CAREFULLY** lift the container with the rewind spring out of the sheave.

SPECIAL WORDS

There is no reason **WHATSOEVER** to remove the spring from the container. If the spring is broken or no longer fit for service for any reason, a new spring must be purchased. The new spring will come in a container, lubricated and ready for installation.

If the old spring is used it will be lubricated, still in the container, prior to installation.

9- Turn the sheave over again. Snap the Circlip off each post. Lift each pawl from its post and then **STOP**. Notice the small spring on each post and the position of the springs. One end of the spring snaps over the pawl and the other end indexes into a recess in the sheave. Lift each spring and observe and **REMEMBER** exactly how it is installed, as an aid during installation.

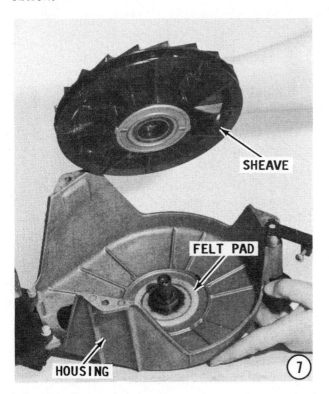

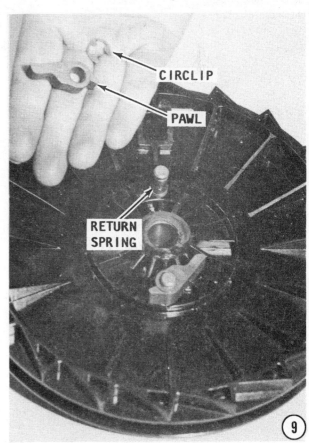

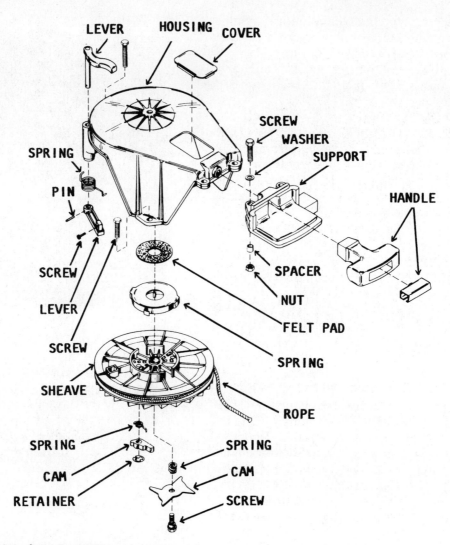

Exploded drawing of the Type "C" hand rewind starter covered in this section. Major parts are identified.

CLEANING AND INSPECTING

Wash all parts except the rope and the handle in solvent, and then blow them dry with compressed air.

Remove any trace of corrosion and wipe all metal parts with an oil dampened cloth.

Inspect the rope. Replace the rope if it appears to be weak or frayed. If the rope is frayed, check the holes through which the rope passes for rough edges or burrs. Remove the rough edges or burrs with a file and polish the surface until it is smooth. Inspect the starter spring end hooks. Replace the spring if it is weak, corroded or cracked. Inspect the inside surface of the sheave rewind recess for grooves or roughness. Grooves may cause erratic rewinding of the starter rope.

Coat the entire length of the used rewind spring (a new spring will be coated with lubricant from the package), with low-temperature lubricant.

ASSEMBLING AND INSTALLATION TYPE "C" REWIND STARTER

1- Slide each small spring down over the pawl posts with one end of the spring indexed into the recess in the sheave as observed in Step 9 of disassembling. Slide each pawl onto the posts with the long end facing inward, as shown. Slip each spring end over the side of the pawl. The illustration shows one pawl correctly installed and the other pawl ready. After installation, the pawls will face in opposite directions with the long flat surface facing outward. Snap a Circlip onto each post to restrain the pawls in place.

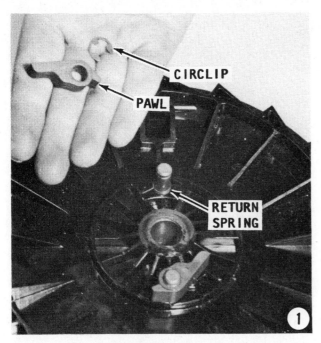

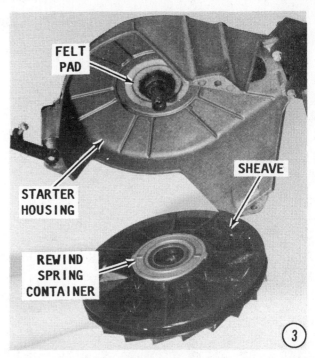

2- Very **CAREFULLY** place the rewind spring container into the sheave with the three tabs indexed into the notches of the sheave.

3- Apply a couple drops of lubricant to one side of the felt pad. Now, place the felt pad over the center shaft with the "lubricated" side going down first to hold the pad in place. Raise and insert the assembled sheave up into the housing from the underneath side, as shown. Rotate the sheave to index the spring loop into the notch on the shaft. Hold the sheave and housing together, and then turn the complete unit over.

4- Slide the spring down the shaft. This spring serves as a spacer between the cam and the sheave and exerts an upward pressure against the cam to keep it indexed over the shoulder on the center bolt. Install the four arm cam onto the shaft on top of the spring.

5- Shift the cam until the arms are positioned against the two pawls, as shown.

SPECIAL WORDS
The center bolt on this type rewind starter has standard **RIGHT-HAND** threads.

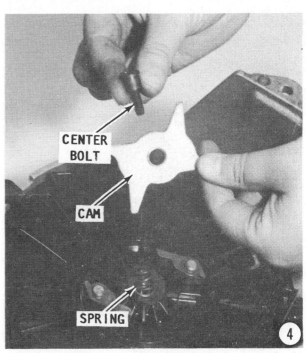

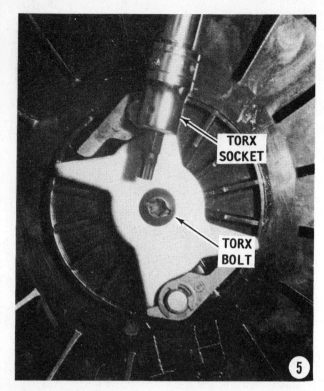

TORX SOCKET

TORX BOLT

5

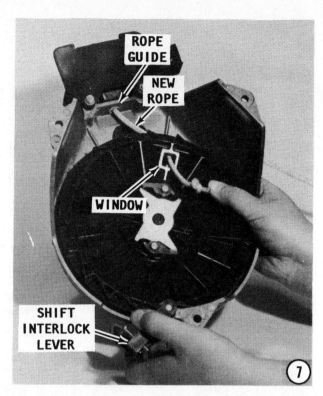

ROPE GUIDE

NEW ROPE

WINDOW

SHIFT INTERLOCK LEVER

7

Secure the cam in place with the Torx bolt. This bolt has a very special shoulder. Tighten the Torx bolt just good and "snug" with a Torx socket or the proper size screwdriver. Exercise **CARE** not to damage the bolt head. Using a screwdriver provides very little surface for the blade to grasp, therefore, the head may be easily damaged. Again, tighten just good and "snug". Some **Type "C"** rewind starters may have a hexhead bolt with a special shoulder -- then no problem.

6- Place tension on the rewind spring by holding the housing and at the same time rotating the sheave **COUNTERCLOCKWISE**. Rotate the sheave as far as possible, thus placing maximum tension on the spring. Hold the tension for the next step.

7- Hold the shift interlock lever clear and allow the sheave to unwind very **SLOWLY** until the rope knot window in the sheave is aligned with the rope guide in the housing. Hold the sheave and housing in this position and at the same time feed a new rope through the window and out the rope guide. Tie an overhand knot in the rope with one hand, then feed the knot into the window. Continue to hold the sheave and housing and tie a knot about 1' (25cm) from the other end of the rope. Check to be sure the knot

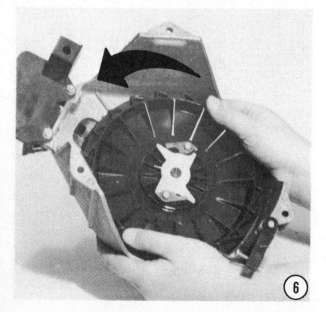

6

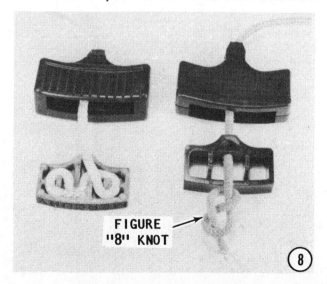

FIGURE "8" KNOT

8

is imbedded in the window, and then slowly release your grip on the sheave and allow the spring to rewind. After the knot in the other end of the rope is against the rope guide, release your grip on the sheave.

8- Install the handle onto the end of the rope. One of two type of handles may be used. After the handle has been installed, untie the knot made about 1 ft (25 cm) from the end and allow the rope to completely rewind into the housing, with the rope guide up against the rope guide of the housing.

9- Install the rewind starter assembly onto the powerhead and secure it in place with the three attaching bolts. Check to be sure the upper and lower lock levers operate properly. Tighten the bolts to the torque value given in the Torque Table in the Appendix.

10- Snap the fuel filter into place on the starter housing bracket. Install the cowling.

12-5 TYPE "D" REWIND HAND STARTER WITH SINGLE PAWL

FIRST, THESE WORDS

Be sure to read Section 12-1, and be convinced the starter being serviced is a Type "D" as identified from the photographs on Page 12-2.

PRELIMINARY TASKS

First, remove the wrap around cowling. Disconnect the high tension leads from the spark plugs. Next, remove the rope retainer (handle) from the rope. This is accomplished by simply pushing the inner portion through the outer shell of the handle. Untie the "stop" knot in the rope and allow the rewind spring to unwind within the recess of the sheave.

If the choke rod passes down through the top of the cowling, disconnect the upper portion of the shaft from the lower portion by pulling the cotter pin securing the two together. Pull the upper portion of the choke shaft up through the cowling.

If the unit being serviced is equipped with a shift interlock actuator (cable), remove the cable attaching screw and the retainer clip, as shown in the accompanying illustration.

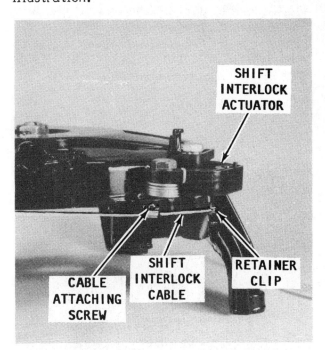

Later model rewind starters are equipped with a shift interlock cable mounted on the side of the starter housing. Early models have the cable mounted on the top of the housing. The interlock cable of both models is secured in the same manner.

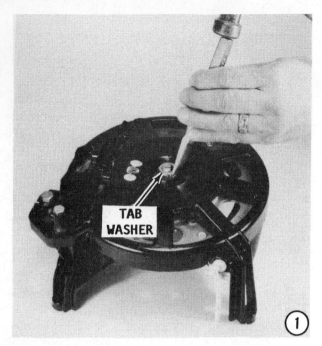

Now, remove the cowling. Remove the screws securing the three legs of the hand rewind starter assembly to the powerhead.

The following steps pickup the work after these few preliminary tasks have been completed.

DISASSEMBLING

1- Place the hand starter on a suitable work surface. Pry out the one tab bent into the recess in the rewind housing. Use a hammer and flat punch to push aside the two tabs of the tab washer. This washer secures the sheave shaft retaining nut.

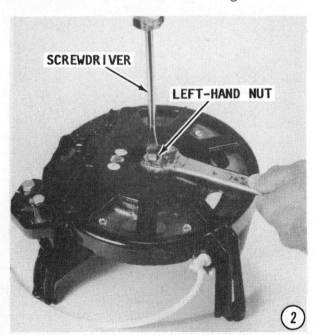

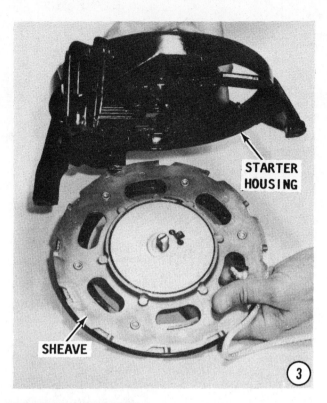

CRITICAL WORDS

The sheave shaft retaining nut has **LEFT-HAND** threads.

2- Remove the nut by rotating the nut **CLOCKWISE** because it has **LEFT-HAND** threads.

SAFETY WORDS

THE REWIND SPRING IS A POTENTIAL HAZARD. The spring is under tremendous tension when it is wound -- a real **"tiger"** in a cage! If the spring should accidentally be

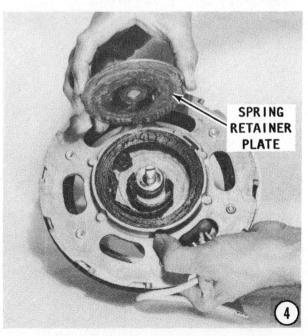

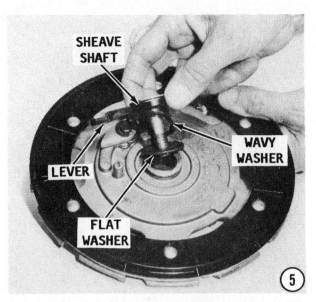

released, severe personal injury could result from being struck by the spring with force. Therefore, the following steps **MUST** be performed with care to prevent personal injury to self and others in the area.

3- Carefully lift the starter housing from the sheave assembly so as not to disturb the rewind spring encased in the sheave. Place the sheave face down on a flat surface.

4- Lift off the spring retainer plate.

5- **CAREFULLY** turn the sheave over. Remove the sheave shaft, lever, wavy washer, and the flat washer from the sheave. **DISCARD** the wavy washer -- it cannot be used a second time. A new one must be obtained.

6- **CAREFULLY** turn the sheave over again. Remove the bushing from the sheave shaft opening.

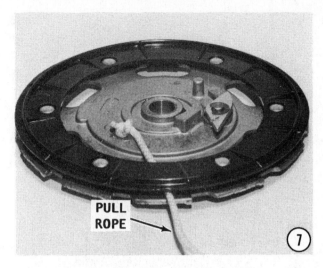

7- Turn the sheave over again. Untie the knot, and then remove the pull rope from the sheave.

8- Obtain two pieces of wood, a short 2" x 4" (5cm x 10cm) will work fine. Place the two pieces of wood approximately 8" (20 cm) apart on the floor. Center the sheave on top of the wood with the spring side facing **DOWN**. Check to be sure the wood is not touching the spring.

WARNING
THE REWIND SPRING IS A POTENTIAL HAZARD. The spring is under tremendous tension when it is wound -- a real tiger in a cage. If the spring should accidentally be released, severe personal injury could result from being struck by the spring with force. Therefore, the following step **MUST** be performed with care to prevent personal injury to self and others in the area.

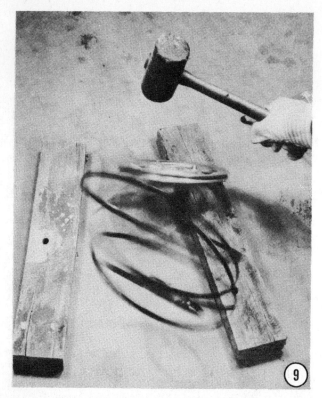

9- Stand behind the wood, keeping away from the openings as the spring unwinds with considerable force. Tap the sheave with a soft mallet. The spring retainer plate will drop down releasing the spring. The spring will fall and unwind almost instantly and with **FORCE**.

10- Snap the Circlip out of the groove on the pawl post.

11- Slide the pawl up and free of the post. The spring may come free with the pawl or it may remain on the post. Observe and REMEMBER how one end of the spring is snapped onto the pawl and the other end of the spring indexes into the recess of the sheave.

CLEANING AND INSPECTING

Wash all parts except the rope and the handle in solvent, and then blow them dry with compressed air.

Remove any trace of corrosion and wipe all metal parts with an oil dampened cloth.

Inspect the rope. Replace the rope if it appears to be weak or frayed. If the rope is frayed, check the holes through which the rope passes for rough edges or burrs. Remove the rough edges or burrs with a file and polish the surface until it is smooth.

Inspect the starter spring end hooks. Replace the spring if it is weak, corroded or cracked. Inspect the tab on the spring retainer plate. This tab is inserted into the inner loop of the spring. Therefore, be sure it is straight and solid. Inspect the inside surface of the sheave rewind recess for grooves or roughness. Grooves may cause erratic rewinding of the starter rope.

Check the condition of the pawl pivot post for excessive wear. The post is a a part of the sheave casting. The post **CANNOT** be serviced. Therefore, if it is not acceptable, a new sheave **MUST** be purchased and installed.

Inspect the pawl for wear around the pivot holes and for rounded outer edges. Check the pawl retainer plate for wear, especially the center hole and the areas where the sides of the pawl contact the retainer plate.

Inspect the pawl retainer plate return spring (if equipped). The manufacturer recommends this spring be replaced each time the rewind starter is disassembled for service. A weak spring could allow the pawl to contact the flywheel while the powerhead is

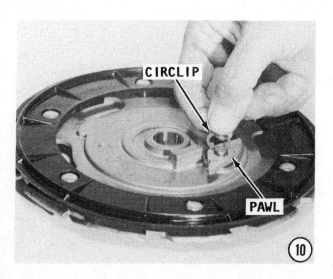

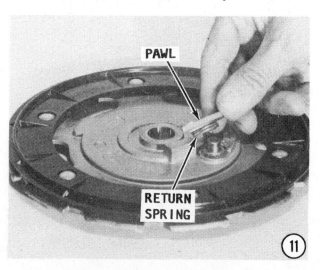

operating and cause a very unpleasant noise and even damage to the edge of the fly-wheel. A weak spring will also cause excessive wear to the pawl.

ASSEMBLING AND INSTALLATION TYPE "D" REWIND STARTER

SAFETY WORDS

Wear a good pair of gloves while winding and installing the spring. The spring will develop tension and the edges of the spring steel are extremely sharp. The gloves will prevent cuts to the hands and fingers.

It is **STRONGLY** recommended a pair of safety goggles or a face shield be worn while the spring is being installed. As the work progresses a "tiger" is being forced into a cage, over 14' (4.3 m) of spring steel wound into about 4" (10.2cm) circumference. If the spring is accidentally released, it will lash out with tremendous ferocity and very likely could cause personal injury to the installer or other persons nearby.

SPECIAL WORDS

The rewind starter may be assembled with a new rewind spring or a used one. Procedures for assembling are **NOT** the

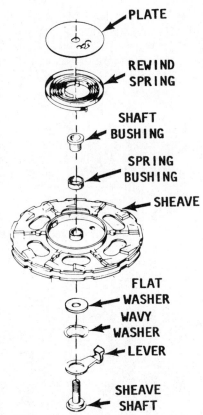

Exploded drawing of the Type "D" hand rewind starter covered in this section.

same because the new spring will arrive in a shipping container already wound, lubricated, and ready for installation. The used spring must be manually wound into its recess.

NEW REWIND SPRING INSTALLATION

The situation may arise when it is only necessary to replace a broken spring. The following few procedures outline the tasks required to replace the spring. A new spring is already properly wound and will arrive in a special shipping container. This container is designed to be used as an aid to installing the new spring.

a- Remove the retainers from the shipping container used to keep the spring from accidentally falling out of the container. Place the shipping container over the spring recess with the tabs resting on the outer edge of the recess.

b- Align the hook on the outer end of the spring with the sheave anchor or notch.

c- Place two large blade screwdrivers into the holes of the shipping container over the tensioned spring. Push on both screwdrivers at the same time to press the spring out of the shipping container and into the spring recess of the sheave.

USED REWIND SPRING INSTALLATION

A used spring naturally will not be wound. Therefore, special instructions are necessary for installation.

SAFETY WORDS

Wear a good pair of gloves while winding and installing the spring. The spring will develop tension and the edges of the spring steel are extremely sharp. The gloves will prevent cuts to the hands and fingers.

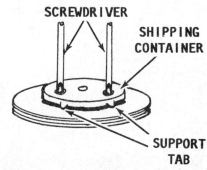

A new rewind spring is installed into the sheave using the shipping container as a helpful tool. Two screwdrivers are used to tap the coiled spring out of the container and into the sheave.

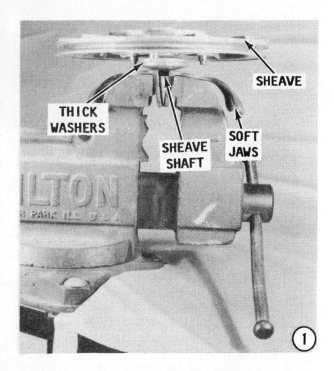

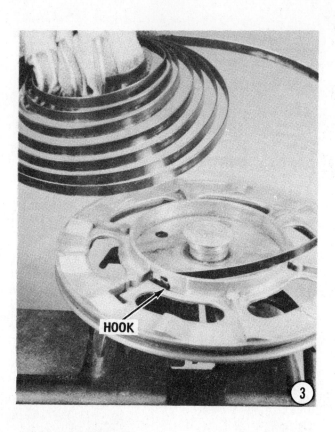

1- Insert the sheave shaft through the sheave with the spring recess facing **UP-WARD**. Obtain two large washers and place them over the sheave shaft. Hold the washers in place and at the same time clamp the sheave shaft in a vise equipped with soft jaws. Shift the shaft in the vise to allow for "up and down" clearance and permit the shaft to rotate freely while the rewind spring is being installed.

SAFETY WORDS

It is **STRONGLY** recommended a pair of safety goggles or a face shield be worn while the spring is being installed. As the work progresses a "tiger" is being forced into a cage. If the spring is accidentally released, it will lash out with tremendous ferocity and very likely could cause personal injury to the installer or other persons nearby.

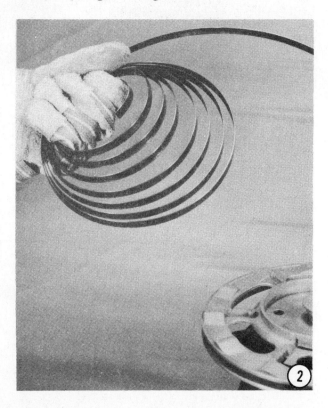

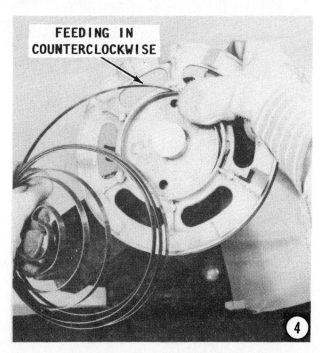

7- Feed a new rope through the hole in the sheave. Tie a figure "8" knot near the rope end. Pull the rope back against the knot, and then wrap it around the sheave **CLOCKWISE**.

8- Lubricate the bushing with Multi-Purpose lubricant. Insert the bushing into the center of the sheave.

9- Turn the sheave over -- with the pawl side facing up. Slide the lever onto the

2- Loop the spring loosely into a coil to enable it to be fed into its recess.

3- Insert the hook on the end of the spring into the notch of the recess.

4- Feed the spring around the inner edge of the recess and at the same time rotate the sheave **CLOCKWISE**. Proceed **WITH GREAT CARE**. Guide the spring into place. The spring will be slippery with lubrication. **DO NOT** lose control of the spring. Carefully remove the sheave from the vise and the sheave shaft and washers from the sheave.

5- Coat the pawl post with a small amount of Multi-Purpose lubricant. Place the spring in position on the underside of the pawl with the hook of the spring over the side of the pawl. Slide the pawl and spring onto the pawl post with the other end of the spring indexed into the spring recess in the sheave.

6- Check to be sure the sheave is in the proper position, as shown. Snap the Circlip onto the post and into the groove to secure the sheave and spring in place.

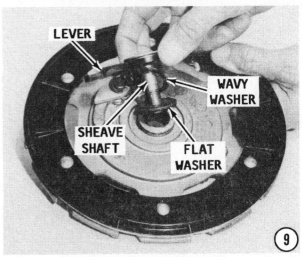

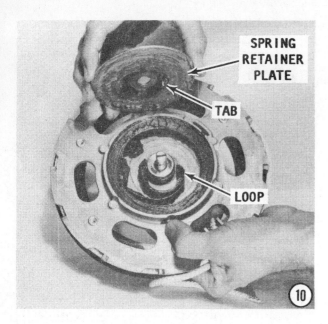

SPRING RETAINER PLATE

TAB

LOOP

10

NEW TAB WASHER

12

sheave shaft, then a **NEW** wavy washer and finally a flat washer. Hold the pieces on the shaft and slide the shaft down through the bushing. The large diameter of the shoulder on the sheave shaft **MUST** be flush against the sheave. Hold the shaft in place and turn the sheave over for the next step.

10- Place the spring retainer plate over the shaft and down into the sheave cavity on top of the spring. The tab on the plate must index into the loop on the end of the spring. If necessary, a small screwdriver may be used to help index the tab into the loop.

11- Hold the starter housing with one hand. With the other hand, hold the sheave assembly together, then lift the assembly up and into the housing with the sheave shaft indexed through the hole in the starter housing. Feed the rope through the rope guide. Tie a figure "8" knot in the rope about 1' (30cm) from the end. Place the sheave assembly into the rewind housing.

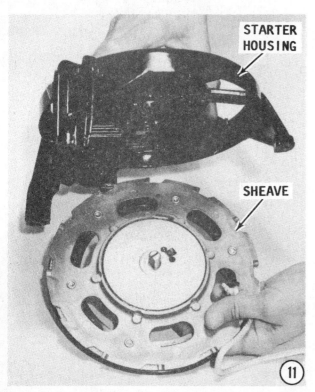

STARTER HOUSING

SHEAVE

11

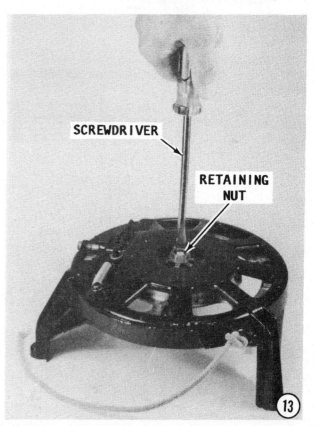

SCREWDRIVER

RETAINING NUT

13

12- Support the sheave assembly with one hand. Slide a **NEW** tab washer over the threaded portion of the sheave shaft with the cupped side of the washer facing **DOWN**. Thread the left-hand sheave shaft retaining nut onto the shaft in a **LEFT-HAND (COUNTERCLOCKWISE)** direction until it is just **FINGER-TIGHT**, at this time.

ADJUSTING REWIND SPRING TENSION

13- Insert a large blade screwdriver into the slot of the sheave shaft. Hold the retaining nut with the proper size wrench and at the same time rotate the shaft with the screwdriver **COUNTERCLOCKWISE** until the figure "8" knot in the rope rests against the rope guide. Continue to rotate the shaft through **TWO** full turns after the knot is against the rope guide. The correct tension has now been placed on the rewind spring.

14- Hold the tension on the spring with the screwdriver in the shaft slot and at the same time tighten the retaining nut securely. **REMEMBER** -- left-hand thread -- tighten **COUNTERCLOCKWISE.**

GOOD WORDS

DO NOT bend the tabs on the washer up against the retaining nut until **AFTER** the rewind operation has been checked.

CHECK REWIND OPERATION

SLOWLY pull the starter rope outward. The pawls must move to the engaged position as the pawl retainer plate begins to turn. If the pawls fail to engage, check the alignment or replace the wavy washers between the pawls. Again, extend the rope to its full length and allow it to rewind. The rope should rewind smoothly without catching. **NEVER** release the rope from the fully extended position.

If the rewind mechanism catches, but fails to rewind, the sheave shaft and its wavy washer are not correctly aligned. The unit **MUST** be disassembled to correct the condition.

Once the rewind operation is satisfactory, proceed to the next step.

15- Bend two tabs of the tab washer up against the flats of the retaining nut. Use tabs opposite each other. Bend one other tab down into the recess of the rewind housing. Install the rewind starter onto the powerhead and secure the legs with the screws and lockwashers.

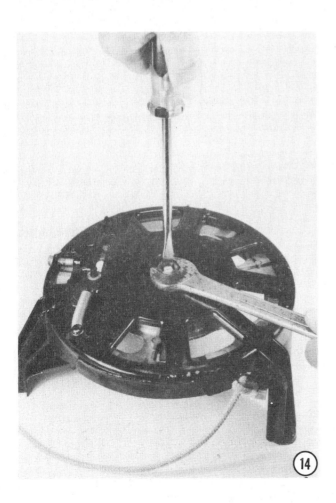

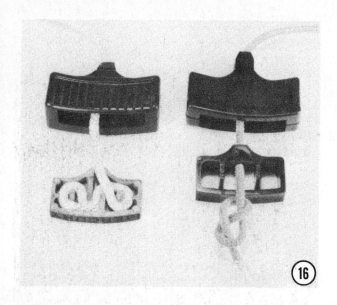

SPECIAL WORDS

If the unit being serviced is equipped with a shift interlock actuator (cable), attach the cable securing screw and the retainer clip, as shown in the accompanying illustration.

If the choke handle shaft passes through the top of the cowl, feed it through and make the connection to the lower portion of the shaft with the cotter pin.

16- Feed the rope through the top cowl and install the rope retainer. Two styles of retainer are used on this model rewind starter. Secure the one handle with a figure "8" knot in the end of the rope, as shown.

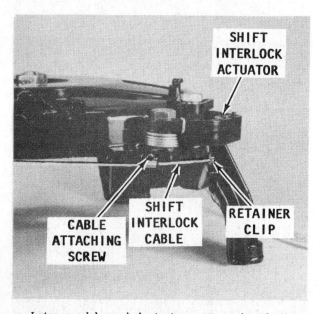

Later model rewind starters are equipped with a shift interlock cable mounted on the side of the starter housing. Early models have the cable mounted on the top of the housing. The interlock cable of both models is secured in the same manner.

SHIFT INTERLOCK ADJUSTMENT

Secure the cable attaching screw to the side of the rewind housing, but **DO NOT** tighten it at this time. Hook the interlock cable over the peg on the interlock actuator with a flat washer and cotter pin. **BE SURE** the control handle is in the **NEUTRAL** position. Adjust the cable until the interlock actuator is positioned on the rise of the interlock cam. Tighten the cable attaching screw to hold this position.

Shift the control handle into the **FORWARD** position and try to pull the handle outward. The attempt should fail.

Shift the handle back into the **NEUTRAL** position and again pull on the handle. The rope should rewind normally.

12-6 TYPE "E"
REWIND HAND STARTER
ONE SMALL PAWL SHOWING
ON UNDERSIDE

FIRST, THESE WORDS

Be sure to read Section 12-1, and be convinced the starter being serviced is a Type **"E"** as identified from the photographs on Page 12-2 and 12-3.

The Type **"E"** starter is a very small plastic unit. Procedures are presented to remove the starter and replace the rope. A new spring and other parts are not available as replacement items. Therefore, if the spring is broken or the starter has suffered other damage, besides a broke pull rope, it must be replaced as a complete assembly.

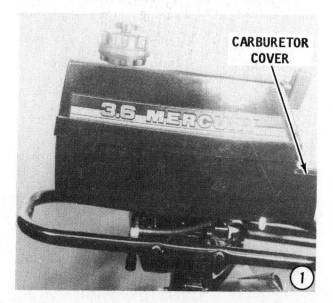

REMOVAL AND DISASSEMBLING

Model 3.6 Only

1- Pull the carburetor front cover free of the cowling. Unscrew the fuel cap; compress the plastic fuel cap retainer; and then remove the fuel cap assembly. Remove the cowling.

Model 3.5 Only

1A- Remove the rubber seal from around the fuel tank filler neck. Lift up on the cowling to disengage the two locating pins. Tilt the cowling forward to clear the starter handle and pull the cowling free.

2- Remove the four nuts securing the rewind hand starter assembly to the power-head.

3- Lift the hand starter and at the same time pull forward to permit the assembly to clear the fuel tank mounting bracket.

4- Untie the knot in the end of the pull rope, and then remove the rope handle. Allow the rewind spring to unwind and the rope to retract inside the starter housing. Unwind the rope from around the sheave. Untie the knot in the other end of the rope. Pull the rope free of the sheave.

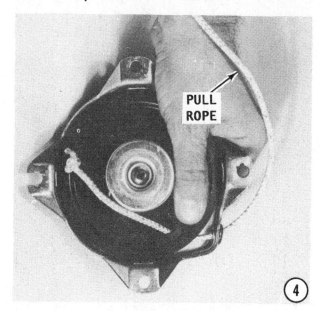

CLEANING AND INSPECTING

As mentioned at the beginning of this section, replacement of the pull rope is about the only service work possible on this type starter. If there is an indication the starter is not functioning properly -- the reason for removal in the first place -- and the indication is something broken inside -- the main spring, pawl spring, etc., then the starter assembly must be replaced. The unit is small, many parts are plastic, and the cost is nominal.

Clean the housing and accessible parts with solvent, then blow them dry with compressed air.

Check the overall condition of the starter housing. Inspect the pawl for what might be termed "excessive" wear. Exert some pressure on the pawl to ensure the spring actuating the pawl has a reasonable amount of tension.

ASSEMBLY AND INSTALLATION
TYPE "E" STARTER

1- Tie an overhand knot about 1' (30cm) from one end of a new rope. Hold the starter assembly with the pawl side facing up. Rotate the sheave **COUNTERCLOCK-WISE** as far as possible to place the rewind spring under maximum tension. Hold tension on the sheave for the next step.

2- While holding tension on the sheave with one hand, allow the sheave to rotate slightly to align the rope hole in the sheave with the rope opening in the starter housing. Continue to hold tension on the sheave with one hand and with the other hand, feed the long end of the   new rope through the

opening and the hole in the sheave. Tie a knot in the end of the rope, and then release a bit of tension on the sheave and allow the rope to rewind into the sheave, until the knot you tied in Step 1 is against the housing.

Feed the short end of the rope through the handle and retainer, then tie a figure "8" knot in the end of the rope. Pull the rope and retainer back into the handle. Untie the knot against the starter housing and allow the rope to rewind until the handle is against the housing.

3- Position the assembled starter on the powerhead with the front of the starter going on first. Lower the back side down with the rear mounting hole of the starter going under the fuel mounting bracket and over the studs.

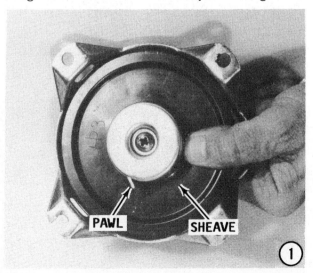

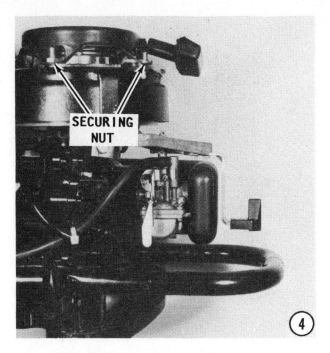

4- Install and tighten the four mounting nuts to secure the starter to the powerhead.

5- Install the cowling and then the rubber seal around the fuel tank filler cap.

Model 3.5 Only

Compress the fuel cap retainer, insert it into the tank, and screw the cap in place.

Model 3.6 Only

5A- Snap the carburetor cover into place.

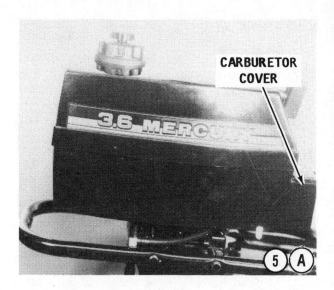

12-7 TYPE "F" REWIND HAND STARTER

FIRST, THESE WORDS

Be sure to read Section 12-1, and be convinced the starter being serviced is a Type **"F"** as identified from the photographs on Page 12-3.

REMOVAL AND DISASSEMBLING

1- Remove the spark plug access cover by pulling down and back on the cover.

2- Five clips and two pair of screws secure the cowling in place. Remove only **ONE** screw from each clip. Leaving one screw in place will hold the clip and assist in replacing the cowling. Two sets of two screws each must be removed. One set on each side of the cowling, located just a

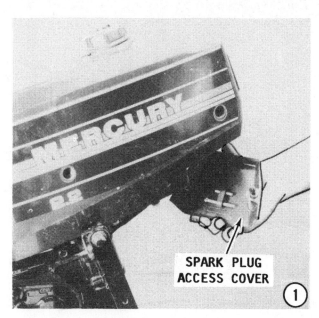

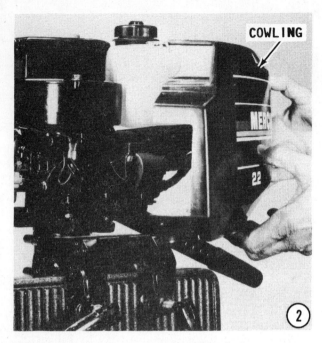

little below the centerline, one towards the rear and the other just forward of the fuel tank shutoff valve.

3- Remove the three bolts securing the rewind starter assembly to the powerhead, then remove the starter assembly.

4- Turn the starter housing over with the sheave facing up. Pull on the handle to gain some slack in the rope; hold the sheave from rewinding; and then remove the handle. Two types of handle arrangement are used.

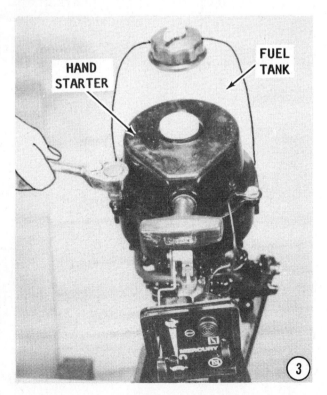

After the handle is removed, ease the grip on the sheave and allow the sheave to completely rewind with the rope inside the sheave.

5- Snap the Circlip out of the groove in the center shaft.

6- Lift the spacer off the friction plate and free of the center shaft.

7- Lift the friction plate slightly and snap the friction spring out of the hole in the center shaft. Remove the friction plate. The return spring will come with the plate. Remove the spring cover, and then the friction spring.

8- Remove the ratchet from the top of the sheave.

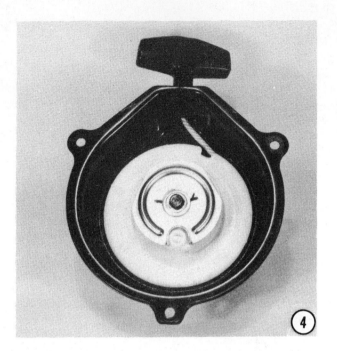

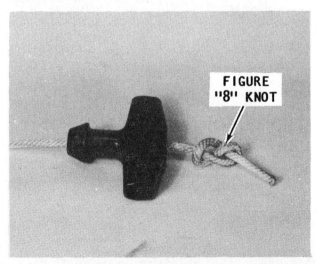

The pull rope is secured in the handle recess with a figure "8" knot tied close to the end. Excess rope beyond the knot can be cut off. Burn the end of a non-fiber rope with a match to prevent it from unraveling.

(5)

SAFETY WORDS

THE REWIND SPRING IS A POTENTIAL HAZARD. The spring is under tremendous tension when it is wound -- a real **"tiger"** in a cage! If the spring should accidentally be released, severe personal injury could result from being struck by the spring with force. Therefore, the following two steps **MUST** be performed with care to prevent personal injury to self and others in the area.

(6)

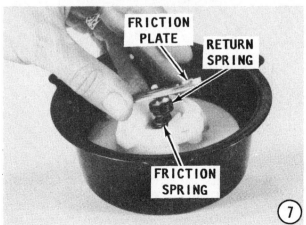

(7)

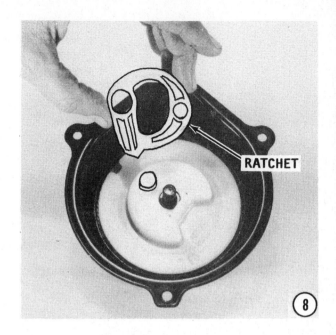

(8)

9- Very **CAREFULLY** "rock" the sheave and at the same time lift the sheave about 1/2" (13mm). The spring will disengage from the sheave and remain in the housing, as shown.

10- Now, **SLOWLY** turn the housing over and **GENTLY** place it on the floor with the spring facing the floor. Tap the top of the housing with a mallet and the spring will fall free of the housing and unwind almost instantly and with **FORCE**, but be contained

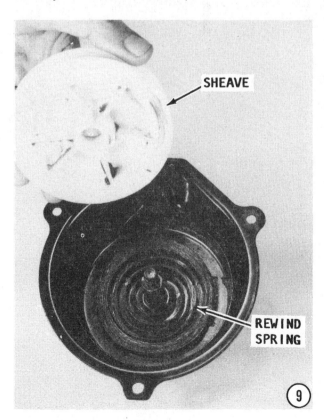

(9)

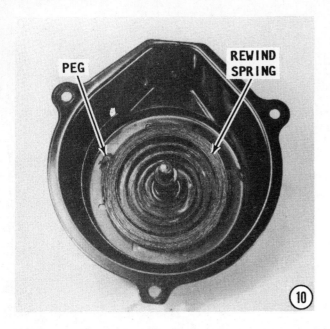

PEG

REWIND
SPRING

10

within the housing. Tilt the container with the opening **AWAY** from you. The spring will be released from the housing and unwind rapidly. Turn the housing over and unhook the end of the spring from the peg in the housing.

11- Untie the knot in the end of the old starter rope and pull the rope free of the sheave.

CLEANING AND INSPECTING

Wash all parts except the rope and the handle in solvent, and then blow them dry with compressed air.

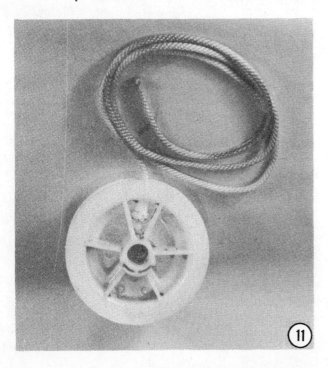

11

Remove any trace of corrosion and wipe all metal parts with an oil dampened cloth.

Inspect the rope. Replace the rope if it appears to be weak or frayed. If the rope is frayed, check the holes through which the rope passes for rough edges or burrs. Remove the rough edges or burrs with a file and polish the surface until it is smooth.

Inspect the starter spring end hooks. Replace the spring if it is weak, corroded or cracked. Inspect the tab on the spring retainer plate. This tab is inserted into the inner loop of the spring. Therefore, be sure it is straight and solid. Inspect the inside surface of the sheave rewind recess for grooves or roughness. Grooves may cause erratic rewinding of the starter rope.

ASSEMBLING AND INSTALLATION TYPE "F" REWIND STARTER

SAFETY WORDS

Wear a good pair of gloves while winding and installing the spring. The spring will develop tension and the edges of the spring steel are extremely sharp. The gloves will prevent cuts to the hands and fingers.

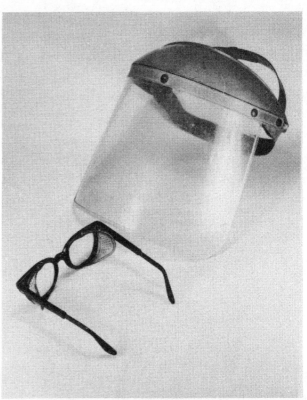

*Protect your eyes with a face mask or safety glasses while working with the rewind spring, especially a used one. The spring is a real **"tiger in a cage"**, almost 13' (4 m), of spring steel wound into less than 4" (about 10cm).*

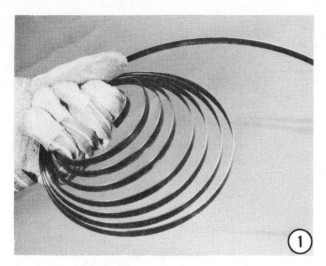

It is **STRONGLY** recommended a pair of safety goggles or a face shield be worn while the spring is being installed. As the work progresses a **"tiger"** is being forced into a cage, over 14' (4.3m) of spring steel wound into about 4" (10.2cm) circumference. If the spring is accidentally released, it will lash out with tremendous ferocity and very likely could cause personal injury to the installer or other persons nearby.

1- Hold the spring in a coil in one gloved hand, as shown.

2- With the other hand, hook the looped end of the spring over the peg in the housing. Feed the spring **CLOCKWISE** into the housing. The easiest way to accomplish this task is to hold the spring in one gloved hand and with the other gloved hand, rotate the housing **COUNTERCLOCKWISE**. Continue working the spring until it is all confined within the housing.

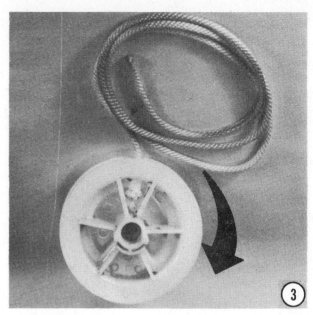

3- Feed one end of a new rope through the hole in the sheave, and then tie a figure "8" knot in the end of the rope. Pull on the rope until the knot is confined inside the sheave recess. Wind the entire pull rope **CLOCKWISE** around the sheave.

4- Lower the sheave down over the center shaft of the housing. As the sheave goes into the housing the hook on the lower face of the sheave **MUST** index into the loop end of the spring. In the illustration, the sheave is turned over to expose the hook to view. Once the hook is indexed into the spring, the sheave may be fully seated in the housing.

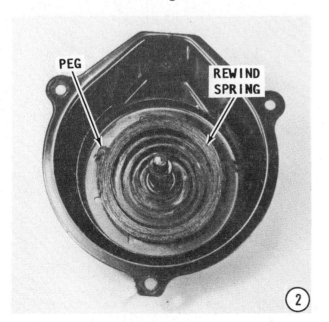

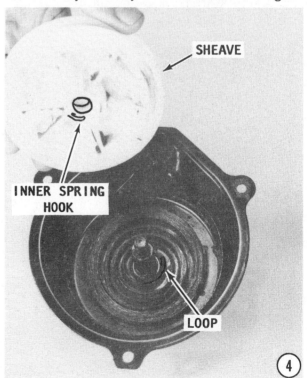

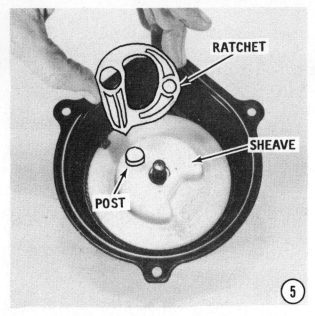

RATCHET

SHEAVE

POST

(5)

FRICTION
PLATE

RETURN
SPRING

FRICTION
SPRING

(6)

SPACER

(7)

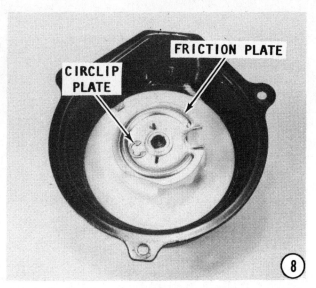

CIRCLIP
PLATE

FRICTION PLATE

(8)

5- Position the ratchet in place on the sheave, with the flat side of the ratchet against the sheave and the round hole indexed over the post.

6- Slide the friction spring down the center shaft. This spring serves as a spacer and exerts an upward pressure on the friction plate. Install the spring cover on top of the spring.

GOOD WORDS

Two holes are located in the sheave on the same side of the center shaft. With the holes on the side of the shaft facing you, one end of the friction spring must index into the right hole.

Hook one end of the return spring into the slot of the friction plate. Now, lower

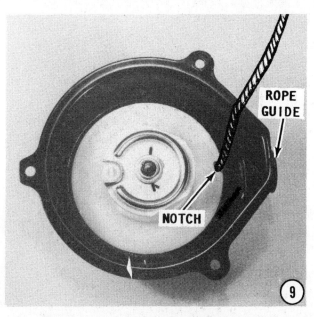

ROPE
GUIDE

NOTCH

(9)

the friction plate onto the center shaft and index the free end of the spring into the right hole, as described in "Good Words", in the sheave.

7- Place the spacer over the hole on the friction plate.

8- Push down on the friction plate, and then snap the Circlip into the goove on the center shaft to secure the plate and associated parts in place.

9- Place the pull rope into the notch in the sheave. With the rope in the notch, rotate the sheave **THREE** complete turns **COUNTERCLOCKWISE**. Hold tension on the rope and at the same time feed the free end of the rope through the rope guide in the starter housing. Continue to hold tension on the sheave for the next step.

10- Feed the free end of the rope through handle and tie a figure "8" knot in the end. Pull the rope back into the handle recess. Relax the tension on the sheave and

allow the sheave to rewind until the handle is against the starter housing.

11- Install the rewind hand starter assembly onto the powerhead and secure it in place with the three attaching bolts. Tighten the bolts securely.

12- Place the two halves of the cowling into place over the powerhead. Secure the cowling with the attaching hardware -- the clips and screws.

13- Connect the high tension leads to the spark plugs. Snap the spark plug access cover into place.

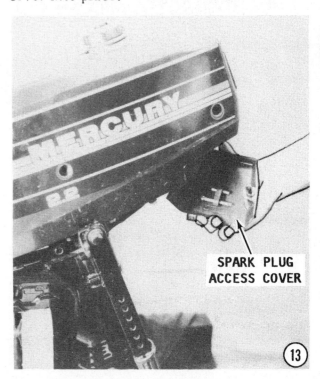

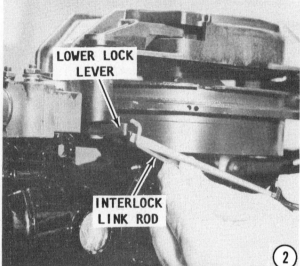

12-8 TYPE "G" REWIND HAND STARTER 1986 AND ON

FIRST, THESE WORDS

Be sure to read Section 12-1, and be convinced the starter being serviced is a Type "G" as identified from the photographs on Page 12-3.

REMOVAL AND DISASSEMBLING

1- Remove the cowling. Insert a narrow blade screwdriver between the fuel filter and the lower edge of the hand starter housing. Exert a slight downward pressure on the screwdiver and at the same pull downward on the filter with the other hand and the fuel filter will "pop" free. Move the filter and connecting hoses to one side out of the way.

2- Unsnap the interlock link rod free of the lower lock lever.

3- Remove the three attaching bolts securing the hand rewind starter to the powerhead.

4- Lift the starter housing up and free of the powerhead.

5- Pull about 1' (30cm) of rope out of the starter housing and tie a loose knot in the rope to prevent the rope from rewinding back into the housing. Do not tighten the

knot because it will be necessary to untie the knot with one hand later.

6- Remove the handle retainer; feed the rope back through the handle; untie the knot; then pull the handle free of the rope. Two type handles are used, as shown.

SPECIAL WORDS

If the only service on the hand starter is to replace a broken rope, the disassembling procedures may be stopped after the next step. Proceed directly to middle of Step 4 in the assembling section.

HOWEVER, if the rewind spring is to be replaced or other service work performed, it is best to leave the rope on the sheave. Therefore, skip Step 7, and continue with Step 8. The illustrations reflect complete disassembling of the starter.

7- Rotate the sheave until the knot in the end of the rope is aligned with the hole in the sheave. Hold the tension on the sheave with one hand, and with the other hand, untie the loose knot in the rope made in Step 5. Pull the rope free of the sheave and housing.

8- Ease your grasp on the sheave and allow the sheave to rotate until all tension on the spring is released. If the rope has not been removed, it will simply wind around the sheave.

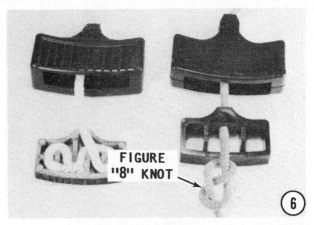

FIGURE "8" KNOT

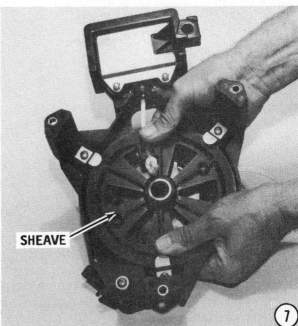

SHEAVE

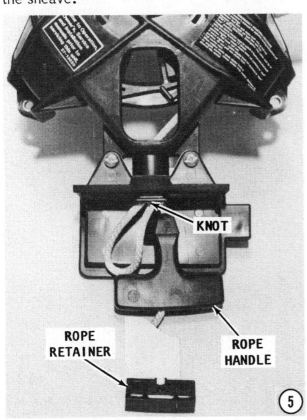

KNOT

ROPE RETAINER

ROPE HANDLE

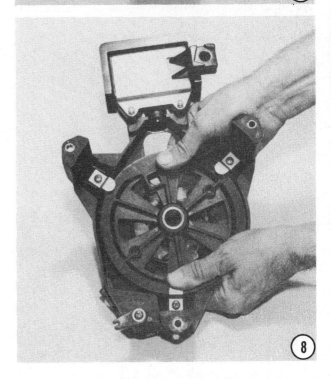

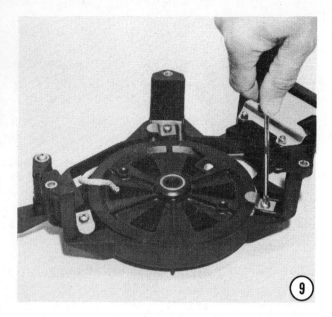

(9)

9- Loosen the three screws securing the tabs holding the sheave in place. It is not necessary to remove the tabs. Rotate the tabs to permit the sheave to clear the housing.

SAFETY WORDS
THE REWIND SPRING IS A POTENTIAL HAZARD. The spring is under tremendous tension when it is wound -- a real **"tiger"** in a cage! If the spring should accidentally be released from its container, severe personal injury could result from being struck by the spring with force. Therefore, the following two steps **MUST** be performed with care to prevent personal injury to self and others in the area.

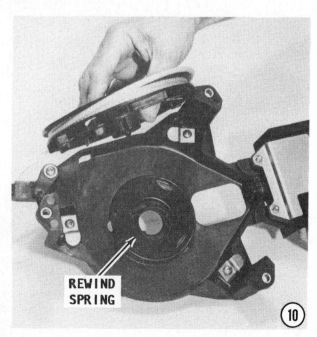

REWIND SPRING

(10)

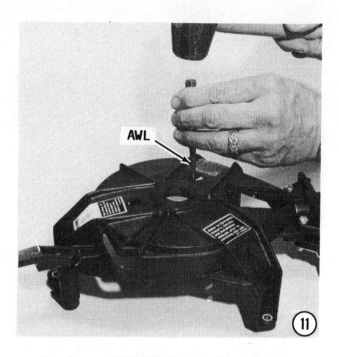

AWL

(11)

10- Very **CAREFULLY** lift and "rock" the sheave. The spring will disengage from the sheave and remain in its container in the housing, as shown.

11- Place the starter housing on a flat surface in the upright position resting on its three legs. Using a hammer and awl, "pop" the rewind spring container out of the starter housing. There is **NO** reason whatsoever to remove the spring from the container. If the spring is in satisfactory condition, it will be lubricated prior to installation. If the spring is broken, a new spring will come in a container, lubricated, and ready for installation.

12- If the rope was not removed in Step 7, untie the knot and remove it from the sheave.

13- Drive the plastic cap free of the sheave using a punch and hammer.

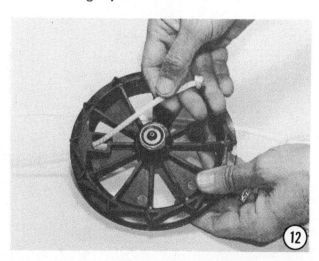

(12)

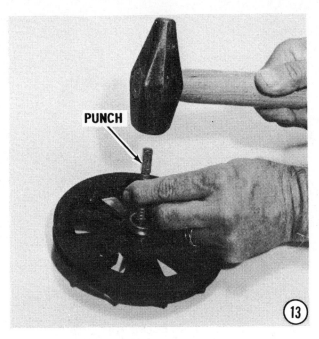

14- Pry the oil seal out of the sheave, with a long thin blade screwdriver.

CLEANING AND INSPECTING

Wash all parts except the rope and the handle in solvent, and then blow them dry with compressed air.

Remove any trace of corrosion and wipe all metal parts with an oil dampened cloth.

Inspect the rope. Replace the rope if it appears to be weak or frayed. If the rope is frayed, check the holes through which the rope passes for rough edges or burrs. Remove the rough edges or burrs with a file and polish the surface until it is smooth.

A new rewind spring will arrive packed in a container, lubricated, and ready for installation.

Inspect the starter spring end hooks. Replace the spring if it is weak, corroded or cracked. Inspect the tab on the spring retainer plate. This tab is inserted into the inner loop of the spring. Therefore, be sure it is straight and solid. Inspect the inside surface of the sheave rewind recess for grooves or roughness. Grooves may cause erratic rewinding of the starter rope.

SPECIAL WORDS

If starter operation was erratic or excessively noisy, check the starter clutch for damage from lack of lubrication. If necessary, replace the complete sheave assembly with a pre-lubricated starter clutch installed.

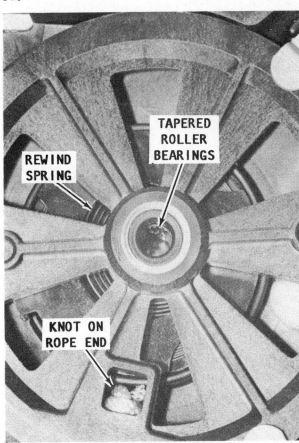

The sheave of the Type "G" starter relies on a clutch arrangement containing tapered roller bearings located in the center of the sheave. If the clutch is defective, the sheave assembly must be replaced.

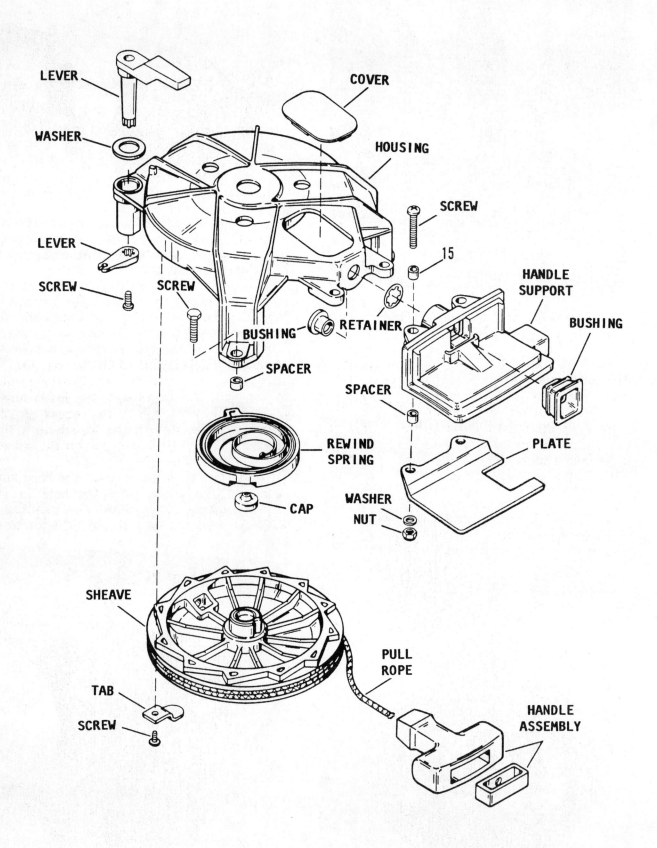

Exploded drawing of the Type "G" hand rewind starter covered in this section. Major parts are identified.

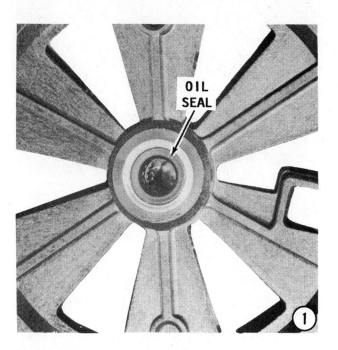

ASSEMBLING AND INSTALLATION TYPE "G" REWIND STARTER

1- Install a new oil seal into the center of the sheave with metal part facing **UP.** This task may be accomplished using the proper size socket and a hammer.

2- Tap the plastic cap into the center of the sheave using a soft head mallet. The cap secures the seal in place.

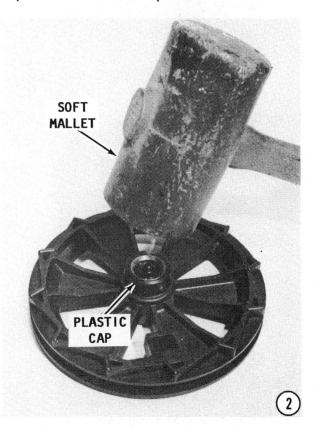

3- Snap the rewind spring container, with the spring inside, into the housing. A new spring will come with the container ready for installation.

4- Insert the sheave into the starter housing. Swing the three tabs around to lock the sheave in place. Tighten the tab screws securely. Hold the sheave to maintain tension and at the same time rotate the sheave **COUNTERCLOCKWISE** as far as possible to wind the spring. Continue holding pressure on the sheave, but let it slip a little until the rope knot recess in the sheave is aligned with the rope hole in the starter housing. Hold tension on the sheave for the next step.

5-   Feed the rope through the rope knot recess of the sheave and the hole in the starter housing. Hold tension on the sheave with one hand and tie a figure "8" knot close

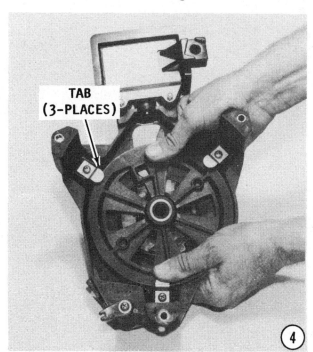

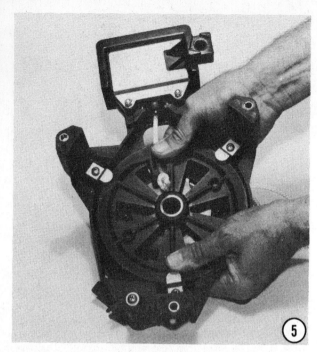

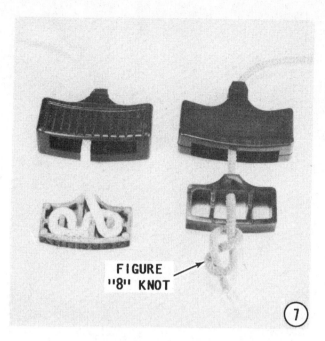

FIGURE "8" KNOT

to the end of the rope. Pull the rope through until the knot is up tight in the rope recess of the sheave. Continue holding tension on the sheave for the next step.

6- Still holding tension on the sheave and at the same time holding the sheave in place within the starter housing, tie a knot in the rope as close as possible to the housing to prevent the sheave from rewind-

ing. Now, release a bit of pressure on the sheave. The spring will rewind just a bit and pull the knot in the rope up tight against the housing.

7- Feed the free end of the rope through the handle and secure it with the retainer, or with a figure "8" knot and the retainer.

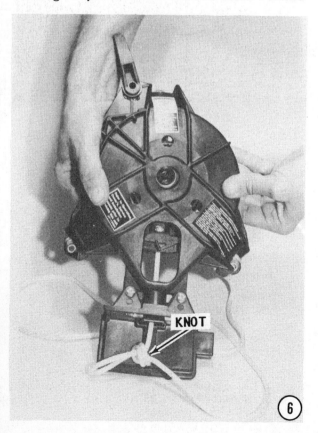

KNOT

Two type handles are used, as shown. After the handle is secured, untie the knot next to the housing and allow the sheave to rewind pulling the rope around the sheave.

8- Slide the starter assembly down over the crankshaft and into position on the powerhead.

9- Secure the starter with the attaching hardware. Tighten the bolts alternately and evenly to the torque value given in the Torque Table in the Appendix.

10- Snap the interlock link rod into the lower lock lever.

11- Snap the clear plastic fuel filter into place on the bracket on the starter housing.

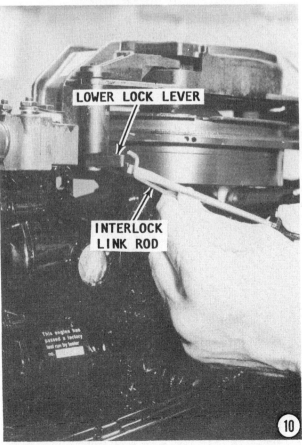

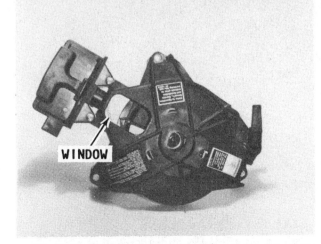

The Type "G" starter has a window on the top of the housing. This window permits a broken pull rope to be replaced without removing the starter from the powerhead.

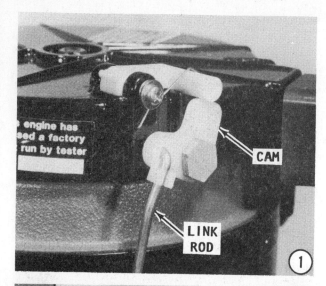

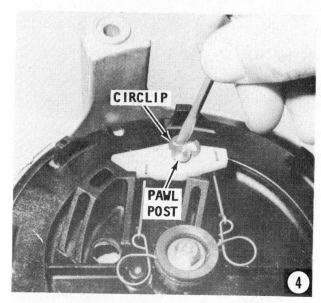

12-9 TYPE "H"
REWIND HAND STARTER

FIRST, THESE WORDS

Be sure to read Section 12-1, and be convinced the starter being serviced is a Type "H" as identified from the photographs on Page 12-3.

REMOVAL AND DISASSEMBLING

1- Pry the shift interlock link rod from the plastic cam on the starter.

2- Remove the three bolts securing the starter legs to the powerhead.

3- Remove the starter and place it upside down on a suitable work surface.

4- Pry the circlip from the pawl post using a narrow slotted screwdriver.

5- Lift the pawl, with the spring attached free of the sheave.

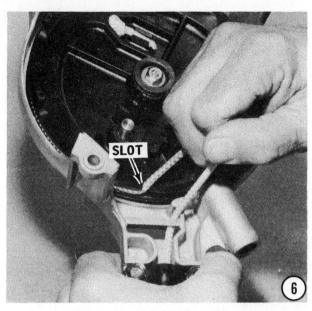

take **EXTRA** care **NOT** to disturb the sheave and spring beneath the sheave.

Hold the sheave against the starter housing to prevent the spring from disengaging from the sheave and **CAREFULLY** rotate the sheave to allow the rope hole to align with the starter handle. Pull the knotted end out of the sheave until all of the rope is free.

If either the sheave or the starter rewind spring is to be replaced the rope may be left in place until the sheave is removed from the starter housing.

SAFETY WORDS

Wear a good pair of heavy gloves and safety glasses while performing the following tasks.

6- Rotate the sheave to align the slot in the sheave with the starter handle, as shown. Lift out a portion of rope, feed the rope into the slot and with a **CONTROLLED MOTION** allow the sheave to rotate in a **CLOCKWISE** direction until the tension on the rewind spring is completely released. **DO NOT** allow the sheave to spin without control.

7- Pry the seal from the handle and push out the knot in the end of the rope. Untie the knot and pull the handle free of the rope.

8- Remove the bolt and washer from the center of the sheave.

9- Remove the sheave bushing and starter housing shaft from the sheave.

SPECIAL WORDS

If the only work to be performed on the hand rewind starter is to replace the rope,

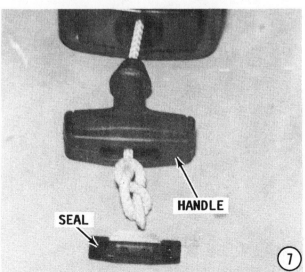

WARNING

THE REWIND SPRING IS A POTENTIAL HAZARD. The spring is under tremendous tension when it is wound -- a real **"TIGER"** in a cage! If the spring should accidentally be released, severe personal injury could result from being struck by the spring with force. Therefore, the following steps **MUST** be performed with care to prevent personal injury to self and others in the area.

DO NOT attempt to remove the spring unless it is unfit for service and a new spring is to be installed.

SPECIAL WORDS

There are two design variations in the housing of the spring. For ease of identification and to ensure the proper procedures are performed, the designs are designated **"a"** and **"b"**. One design starter has the spring encased in a removeable housing and is identified here as Design **"a"**. The other starter has the spring encased inside a recess of the sheave housing and is identified as Design **"b"**. .

When necessary, separate instructions are presented for both starter designs.

Design "a" Hand Rewind Starter
With Removeable Housing

Pry out the two retainer plugs from the sheave. Lift out the tabbed guide plate from the center of the spring housing. Ro-

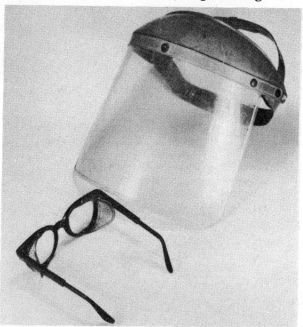

Protect your eyes with a face mask or safety glasses while working with the rewind spring, especially a used one. The spring is a "tiger in a cage", almost 13 feet (4 meters), of spring steel wound and confined into a space less than 4 inches (about 10 cm) in diameter.

tate the sheave in a **CLOCKWISE** direction, while holding the spring housing stationary until the two locking tabs disengage.

Very **CAREFULLY** lift out the spring housing containing the coiled spring from the sheave and remove the backing plate.

Design "b" Hand Rewind Starter
With Spring Encased Inside Sheave

Insert a screwdriver into the hole in the sheave, push down on the section of spring visible through the hole. At the same time gently lift up on the sheave and hold the spring down to confine it in the housing and prevent it from escaping uncontrolled. If the rope has not been removed from the sheave, remove it at this time.

AUTHORS' WORD

The accompanying illustration shows the spring being released from the same type but different model rewind spring. **THEREFORE**, the principle is exactly the same. The procedure outlined in the next step may be followed with safety.

10- Obtain two pieces of wood, a short 2" x 4" (5cm x 10cm) will work fine. Place the two pieces of wood approximately 8" (20 cm) apart on the floor. Center the housing on top of the wood with the spring side facing **DOWN**. Check to be sure the wood is not touching the spring.

Stand behind the wood, keeping away from the openings as the spring unwinds

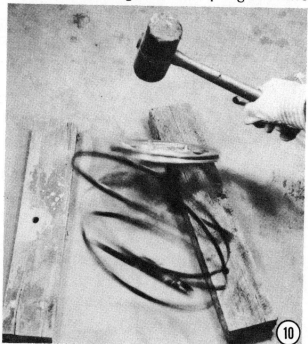

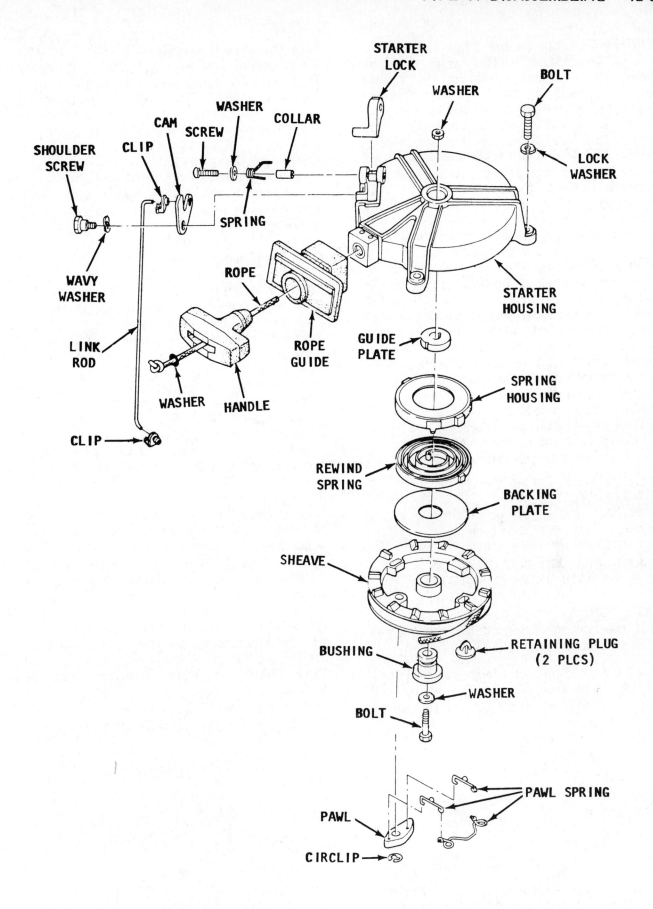

Exploded drawing of hand rewind starter "H", with major parts identified.

with considerable force. Tap the sheave with a soft mallet. The spring retainer plate will drop down releasing the spring. The spring will fall and unwind almost instantly and with **FORCE**.

CLEANING AND INSPECTING

Wash all parts except the rope and the handle in solvent, and then blow them dry with compressed air.

Remove any trace of corrosion and wipe all metal parts with an oil dampened cloth.

Inspect the rope. Replace the rope if it appears to be weak or frayed. If the rope is frayed, check the holes through which the rope passes for rough edges or burrs. Remove the rough edges or burrs with a file and polish the surface until it is smooth. Inspect the starter spring end hooks. Replace the spring if it is weak, corroded or cracked. Inspect the inside surface of the sheave rewind recess for grooves or roughness. Grooves may cause erratic rewinding of the starter rope.

Coat the entire length of the used rewind spring (a new spring will be coated with lubricant from the package), with low-temperature lubricant.

ASSEMBLING AND INSTALLATION TYPE "H" HAND REWIND STARTER

GOOD WORDS

If the rewind spring or the sheave was not removed, proceed directly to Step 6.

Wear a good pair of gloves while winding and installing the spring. The spring will develop tension and the edges of the spring steel are extremely sharp. The gloves will prevent cuts to the hands and fingers.

New Spring Installation

1- Apply a light coating of Quicksilver Mulit-purpose lubricant or equivalent anti-seize lubricant to the inside surface of the starter housing.

A new spring will be wound and held in a steel hoop. Hook the outer end of the new spring onto the starter housing post, and then place the spring inside the housing. **CAREFULLY** remove the steel hoop. The spring should unwind slightly and seat itself in the housing.

Old Spring Installation

SAFETY WORDS

The authors, the manufacturers, and almost anyone else who has handled the spring from this type rewind starter **STRONGLY** recommend a pair of safety goggles or a face shield be worn while the spring is being installed. As the work progresses a **"TIGER"** is being forced into a cage -- over 14' (4.3 m) of spring steel wound into about 4" (10.2cm) circumference. If the spring is accidentally released, it will lash out with tremendous ferocity and very likely could cause personal injury to the installer or other persons nearby.

2- Apply a light coating of Quicksilver Multi-purpose Lubricant or equivalent anti-seize lubricant to the inside surface of the starter housing. Wind the old spring loosely in one hand in a clockwise direction, as shown.

3- Hook the outer end of the spring onto the slot in the removeable housing for Design "a" starters, or the starter housing post for Design "b" starters. Rotate the sheave **CLOCKWISE** and at the same time feed the

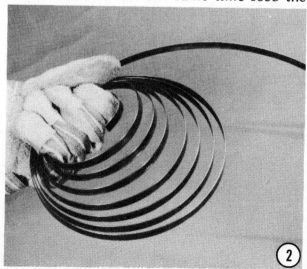

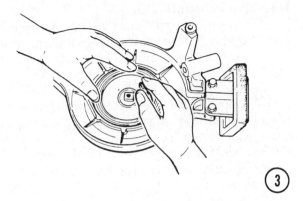

(3)

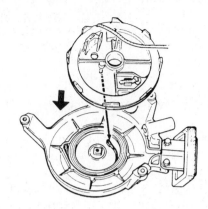

(5)

spring into the housing in a **COUNTER-CLOCKWISE** direction. Continue working the spring into the housing until the entire length has been confined.

Design "a" Starters Continues

Install the backing plate over the sheave hub. Position the spring housing over the sheave, with the two tabs on the housing aligned with the two grooves in the sheave. Rotate spring housing **COUNTERCLOCK-WISE** to engage the two locking tabs against the grooves. Insert the guide plate into the sheave hub with the tab of the plate entering the inner loop of the spring. Insert the retainers into the slots next to the locking tabs of the spring housing. Turn the sheave over and using a pair of needle nose pliers, pull the two ends of the retainers through until seated.

4- Melt the tip of the rope to prevent fraying. Insert the melted end of the rope through the hole in the starter sheave. Tie a figure "8" knot in the end of the rope leaving about one inch (2.5cm) beyond the knot. Tuck the end of the rope beyond the knot into the groove, if so equipped, next to the knot.

5- Wind the rope in a **CLOCKWISE** direction two turns around the sheave, ending

at the slot in the sheave. Lower the sheave into the starter housing. At the same time, for Design **"a"** starters, make sure the tab on the guide plate slides into the hole of the starter housing. For Design **"b"** starters: Use a small screwdriver through the hole to guide the inner loop of the spring onto the post on the underneath side of the sheave.

Units With Spring and Sheave Undisturbed

6- Align the hole in the edge of the sheave with the starter handle. Thread the rope through the hole and up through the top side. Tie a figure "8" knot in the end which was just brought through, leaving about one inch (25cm). Tuck the short free end into the groove next to the hole.

WITHOUT ROTATING THE SHEAVE, feed the rope between the sheave and the edge of the starter housing in a **CLOCK-WISE** direction. Push the rope into place with a narrow screwdriver. Continue feed-

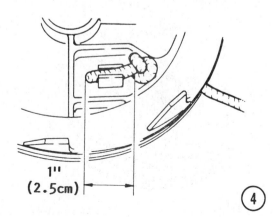

1"
(2.5cm)

(4)

SHEAVE BUSHING

STARTER HOUSING SHAFT

(6)

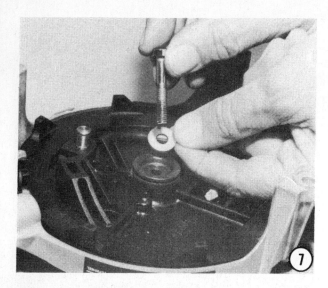

ing and tucking the rope for two turns, ending with the rope at the slot of the sheave.

All Units

Slide the sheave bushing into the starter housing shaft. Insert the shaft and bushing into the center of the sheave.

7- Coat the threads of the center bolt with Loctite. Install the washer and bolt. Tighten the bolt to a torque value of 5.8 ft lbs (8Nm).

8- Thread the rope through the starter handle housing and through the handle. Tie a figure "8" knot in the rope as close to the end as practical. Pull the knot back into the handle recess, and then install the seal in the handle to hide the rope knot.

9- Lift up a portion of rope, and then hook it into the slot of the sheave. Hold the handle tightly and at the same time rotate the sheave **COUNTERCLOCKWISE** until the

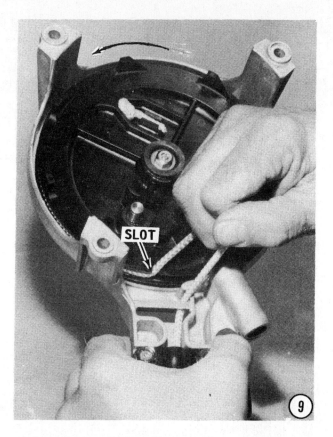

spring beneath is wound tight. This will take about three complete turns of the sheave. Slowly release the tension on the sheave and allow it to rewind **CLOCKWISE** while the rope is taken up as it feeds around the sheave.

10- With the bevelled end of the pawl facing to the left, hook each end of the pawl

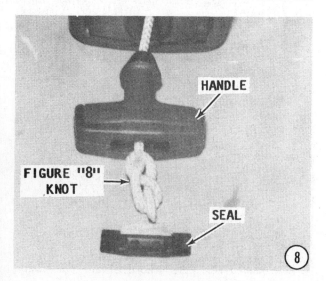

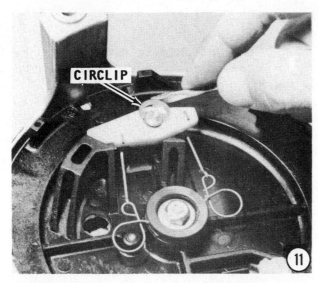

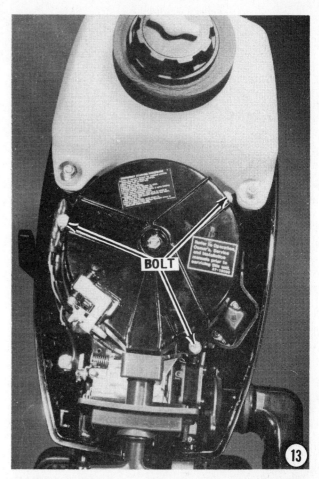

spring into the two small holes in the pawl, from the underneath side of the pawl, and with the pattern of the spring, as shown. The short ends of the spring will then be on the upper surface of the pawl. Move the spring up against the center of the sheave shaft, and then slide the center of the pawl onto the pawl post, as indicated in the accompanying illustration.

11- Snap the circlip into place over the pawl post to secure the pawl in place.

12- Check the action of the rewind starter before further installation work pro-

ceeds. Pull out the starter rope with the handle, then allow the spring to slowly rewind the rope. The starter should rewind smoothly and take up all the rope to lightly seat the handle against the starter housing.

Position the rewind starter in place on the powerhead.

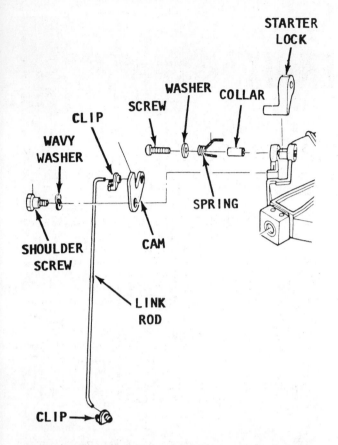

The no-start-in-gear protection system cannot be adjusted. If the system fails to prevent the sheave from rotating in any shift position except NEUTRAL, each component should be inspected. The part should be replaced if there is any evidence of excessive wear or distortion.

13- Apply Loctite to the threads of the three attaching bolts. Secure the starter legs to the powerhead with the bolts, and tighten them to a torque value of 5.8 ft lbs (8Nm).

14- Snap the shift interlock link rod into the plastic cam on the starter housing.

SPECIAL WORDS

If the link rod adjustment at the shift lever and inside the starter housing was undisturbed, the no-start-in-gear protection system should perform satisfactorily. When the unit is **NOT** in **NEUTRAL**, the starter lock should drop down to block the sheave and prevent it from rotating. This means an attempt to pull on the rope with the lower unit in any gear except **NEUTRAL** should **FAIL**.

If the no-start-in-gear protection system fails to function properly, first remove the rewind hand starter from the powerhead. Inspect the blocking surface of the starter lock, the condition of the cam and the return action of the spring. Because no adjustment of the length of the link rod is possible, make sure the lower end of the rod is indeed connected to the shift mechanism below the bottom cowling and the rod itself is not binding or bent. Install the starter on the powerhead and again check the no-start-in-gear system.

APPENDIX

METRIC CONVERSION CHART

LINEAR

inches	X 25.4	= millimetres (mm)
feet	X 0.3048	= metres (m)
yards	X 0.9144	= metres (m)
miles	X 1.6093	= kilometres (km)
inches	X 2.54	= centimetres (cm)

AREA

inches^2	X 645.16	= millimetres^2 (mm^2)
inches^2	X 6.452	= centimetres^2 (cm^2)
feet^2	X 0.0929	= metres^2 (m^2)
yards^2	X 0.8361	= metres^2 (m^2)
acres	X 0.4047	= hectares (10^4 m^2) (ha)
miles^2	X 2.590	= kilometres^2 (km^2)

VOLUME

inches^3	X 16387	= millimetres^3 (mm^3)
inches^3	X 16.387	= centimetres^3 (cm^3)
inches^3	X 0.01639	= litres (l)
quarts	X 0.94635	= litres (l)
gallons	X 3.7854	= litres (l)
feet^3	X 28.317	= litres (l)
feet^3	X 0.02832	= metres^3 (m^3)
fluid oz	X 29.60	= millilitres (ml)
yards^3	X 0.7646	= metres^3 (m^3)

MASS

ounces (av)	X 28.35	= grams (g)
pounds (av)	X 0.4536	= kilograms (kg)
tons (2000 lb)	X 907.18	= kilograms (kg)
tons (2000 lb)	X 0.90718	= metric tons (t)

FORCE

ounces - f (av)	X 0.278	= newtons (N)
pounds - f (av)	X 4.448	= newtons (N)
kilograms - f	X 9.807	= newtons (N)

ACCELERATION

feet/sec^2	X 0.3048	= metres/sec^2 (m/S^2)
inches/sec^2	X 0.0254	= metres/sec^2 (m/s^2)

ENERGY OR WORK (watt-second - joule - newton-metre)

foot-pounds	X 1.3558	= joules (j)
calories	X 4.187	= joules (j)
Btu	X 1055	= joules (j)
watt-hours	X 3500	= joules (j)
kilowatt - hrs	X 3.600	= megajoules (MJ)

FUEL ECONOMY AND FUEL CONSUMPTION

miles/gal	X 0.42514	= kilometres/litre (km/l)

Note:
235.2/(mi/gal) = litres/100km
235.2/(litres/100 km) = mi/gal

LIGHT

footcandles	X 10.76	= lumens/metre^2 (lm/m^2)

PRESSURE OR STRESS (newton/sq metre - pascal)

inches HG (60 F)	X 3.377	= kilopascals (kPa)
pounds/sq in	X 6.895	= kilopascals (kPa)
inches H$_2$O (60 F)	X 0.2488	= kilopascals (kPa)
bars	X 100	= kilopascals (kPa)
pounds/sq ft	X 47.88	= pascals (Pa)

POWER

horsepower	X 0.746	= kilowatts (kW)
ft-lbf/min	X 0.0226	= watts (W)

TORQUE

pound-inches	X 0.11299	= newton-metres (N·m)
pound-feet	X 1.3558	= newton-metres (N·m)

VELOCITY

miles/hour	X 1.6093	= kilometres/hour (km/h)
feet/sec	X 0.3048	= metres/sec (m/s)
kilometres/hr	X 0.27778	= metres/sec (m/s)
miles/hour	X 0.4470	= metres/sec (m/s)

TEMPERATURE

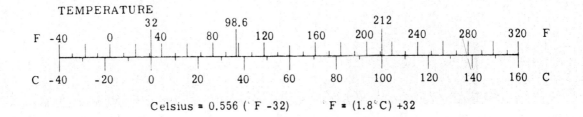

Celsius = 0.556 (°F -32)   °F = (1.8°C) +32

ENGINE SPECIFICATIONS AND

MODEL	BEGIN SERIAL NO.	CYL	HP	CU IN DISPL	W O T	BORE INCHES	STROKE INCHES
1965							
39	1738941	1	3.9	5.5	5000–5400	2.000	1.750
60	1745521	2	6.0	7.2	5000–5400	1.750	1.500
110	1742746	2	9.8	10.9	5000–5400	2.000	1.750
200	1755849	2	20	22	5000–5400	2.562	2.125
350M-350S	1755799	2	35	30	4800–5200	2.875	2.300
1966							
39-39LS	1865633	1	3.9	5.5	5000–5400	2.000	1.750
60-60LS	1958278	2	6.0	7.2	5000–5400	1.750	1.500
110-110LS	1865779	2	9.8	10.9	5000–5400	2.000	1.750
200-200LS	1866496	2	20	22	5000–5400	2.562	2.125
350M, LS, S, SLS	1960827	2	35	32.5	4800–5200	3.000	2.300
1967							
39, 39LS	2098037	1	3.9	5.5	5000–5400	2.000	1.750
60, 60LS	2096171	2	6.0	7.2	5000–5400	1.750	1.500
110, 110LS	2098352	2	9.8	10.9	5000–5400	2.000	1.750
200, 200LS	1997990	2	20	22	5000–5400	2.562	2.125
350M, MLS, S, SLS	2097800	2	35	32.5	4800–5300	3.000	2.300
1968							
39, 39LS	2278851	1	3.9	5.5	5000–5400	2.000	1.750
60, 60LS	2284857	2	6.0	7.2	5000–5400	1.750	1.500
110, 110LS	2290663	2	9.8	10.9	5000–5400	2.000	1.750
200, 200LS	2296267	2	20	22	5000–5400	2.562	2.125
350M, LS, S, SLS	2300972	2	35	32.5	4800–5300	3.000	2.300
1969							
40, 40L	2498136	1	4.0	5.5	4500–5500	2.000	1.750
75, 75L	2529895	2	7.5	10.9	4500–5500	2.000	1.750
110, 110L	2508759	2	9.8	10.9	5000–5400	2.000	1.750
200, 200L	2550065	2	20	22	5000–5400	2.562	1.750
350M, ML, S, SL	2487588	2	35	32.5	4800–5300	3.000	2.300
1970							
40M, 40ML	2771662	1	4.0	5.5	4500–5500	2.000	1.750
75M, 40ML	2810637	2	7.5	10.9	4500–5500	2.000	1.750
110M, ML	2798057	2	9.8	16.7	4500–5500	2.000	1.750
200M, ML	2827677	2	20	21.9	4800–5500	2.562	2.125
400M, ML, E, EL	2874704	2	40	33.3	4800–5300	2.875	2.562
1971							
40, 40L	3028361	1	4.0	5.5	4500–5500	2.000	1.250
75, 75L	3016581	2	7.5	10.9	4500–5500	2.000	1.750
110, 110L	3002814	2	9.8	16.7	4500–5500	2.00	1.750
200, 200L	2991679	2	20	21.9	4800–5500	2.562	2.125
400M, ML, E, EL	3043031	2	40	33.3	4800–5300	2.875	2.562

NOTE: See Appendix Page A-12 for Special Notes, ignition system identification,

SPARK PLUG CHAMP	PLUG GAP	IGN TYPE	PRIMARY PICKUP	MAX TIMING	CARB TYPE	POINT GAP	POWERHEAD TYPE
1965							
J8J	.025	I	Not Adj	Not Adj	A	.020	B
J7J	.025	I	----	----	A	.020	B
J7J	.025	I	----	----	A	.020	A
J6J	.025	I	----	.275BTDC	A	.020	A
J6J	.025	I	----	.222BTDC	A	.020	A
1966							
J8J	.025	I	----	----	A	.020	B
J7J	.025	I	----	----	A	.020	B
J7J	.025	I	----	----	A	.020	A
J6J	.025	I	----	.275BTDC	A	.020	A
J6J	.025	I	----	.300BTDC	A	.020	A
1967							
L9J	.030	I	----	----	A	.020	B
L7J	.030	I	----	----	A	.020	B
L4J	.030	I	----	----	A	.020	A
L4J	.030	I	----	Note 1	A	.020	A
L4J	.030	I	----	.300BTDC	A	.020	A
1968			**1968**				
L9J	.030	I	----	----	A	.020	B
L7J	.030	I	----	----	A	.020	B
L4J	.030	I	----	----	A	.020	A
L4J	.030	I	----	Note 1	A	.020	A
L4J	.030	I	----	.3000BTDC	A	.020	A
1969							
L9J	.030	I	----	----	A	.020	A
L7J	.030	I	----	----	A	.020	A
L4J	.030	I	----	----	A	.020	A
L4J	.030	I	----	Note 1	A	.020	A
L4J	.030	I	----	.300BTDC	A	.020	A
1970							
L77V	Set	II	.005ATDC	Not Adj	A	.020	A
L77V	Set	II	.002ATDC	.193BTDC	A	.020	A
L77V	Set	II	.002ATDC	.193BTDC	A	.020	A
L77V	Set	II	1°BTDC to 4°ATDC	.196BTDC	A	.020	A
L77V	Set	III	5-7°BTDC	27°BTDC		----	A
1971							
L78V	Set	II	.005ATDC	Not Adj	A	.020	A
L78V	Set	II	.002ATDC	.193BTDC	A	.020	A
L78V	Set	II	.002ATDC	.193BTDC	A	.020	A
L78V	Set	II	1°BTDC to 4°ATDC	.196BTDC	A	.020	A
L78V	Set	III	5-7°BTDC 2°BTDC	27°BTDC	A	----	A

carburetor identification, and timing notes called out in this table.

ENGINE SPECIFICATIONS AND

MODEL	BEGIN SERIAL NO.	CYL	HP	CU IN DISPL	W O T	BORE INCHES	STROKE INCHES
1972				**1972**			
40, 40L	3296137	1	4.0	5.5	4500-5500	2.000	1.750
75, 75L	3274633	2	7.5	10.9	4500-5500	2.000	1.750
110, 110L	3263263	2	9.8	10.9	4500-5500	2.000	1.750
200, 200L	3226958	2	20	21.9	4800-5500	2.562	2.125
402M, ML, E, EL	3219678	2	40	33.3	4800-5300	2.875	2.562
1973				**1973**			
40M, 40ML	3529446	1	4.0	5.5	4500-5500	2.000	1.750
75M, 75ML	3488153	2	7.5	10.9	4500-5500	2.000	1.750
110M, MI	3482753	2	9.8	10.9	4500-5500	2.000	1.750
200M, ML	3537531	2	20	21.9	4800-5500	2.562	2.125
402M, ML, E, EL	3474578	2	40	33.3	4800-5300	2.875	2.562
1974				**1974**			
40, 40L	3764825	1	4.0	5.5	4500-5500	2.000	1.875
75, 75L	3801458	2	7.5	10.9	4500-5500	2.000	1.875
110, 110L	3795658	2	9.8	10.9	4500-5500	2.000	1.875
200, 200L	3755885	2	20	21.9	4800-5500	2.562	2.125
402M, ML, E, EL	3859435	2	40	33.3	4800-5300	2.875	2.562
1975				**1975**			
45	4108120	1	4.5	5.5	4500-5500	2.000	1.875
45L	4107220	1	4.5	5.5	4500-5500	2.000	1.875
75	4087015	2	7.5	10.9	4500-5500	2.000	1.875
75L	4085865	2	7.5	10.9	4500-5500	2.000	1.875
110	4079675	2	9.8	10.9	4500-5500	2.000	1.875
110L	4079000	2	9.8	10.9	4500-5500	2.000	1.875
200	4103290	2	20	21.9	4800-5500	2.562	2.125
200L	4102790	2	20	21.9	4800-5500	2.562	2.125
402M	4118385	2	40	33.3	4800-5300	2.875	2.562
402ML	4119010	2	40	33.3	4800-5300	2.875	2.562
402E	4117710	2	40	33.3	4800-5300	2.875	2.562
402EL	4119360	2	40	33.3	4800-5300	2.875	2.562
1976				**1976**			
40M	9075839	2	4.0	5.5	4000-5000	1.562	1.438
45M, 45ML	4377557	1	4.5	5.5	4500-5500	2.000	1.750
75M, ML, E, EL	4314385	2	7.5	10.9	4500-5500	2.000	1.750
110M, ML, E, EL	4304785	2	9.8	10.9	4500-5500	2.000	1.750
200M, ML, E, EL	4293920	2	20	21.9	4800-5500	2.562	2.125
402M, ML, E, EL	4362215	2	40	33.3	4800-5300	2.875	2.562

NOTE: See Appendix Page A-12 for Special Notes, ignition system identification,

SPARK PLUG CHAMP	PLUG GAP	IGN TYPE	PRIMARY PICKUP	MAX TIMING	CARB TYPE	POINT GAP	POWERHEAD TYPE
1972							
L78V	Set	II	5°ATDC	.198BTDC	A	.020	A
L78V	Set	II	2°ATDC	.193BTDC	B	.020	A
L78V	Set	II	2°ATDC	.193BTDC	A	.020	A
L78V	Set	II	1°BTDC-4°ATDC	.196BTDC	A	.020	A
L77V	Set	III	5-7°BTDC 2°BTDC	27°BTDC	A	-----	A
1973							
L78V	Set	II	5°ATDC	.198BTDC	A	.020	A
L78V	Set	II	None	.193BTDC	B	.020	A
L78V	Set	II	None	.193BTDC	A	.020	A
L78V	Set	IV	2°BTDC-2°ATDC	33°BTDC	B	-----	A
L77V	Set	II	5-7°BTDC	27°BTDC	A	-----	A
1974							
L78V	Set	II	5°ATDC	.198BTDC	A	.020	A
L77V	Set	IV	None	34.5°BTDC	B	-----	A
L77V	Set	IV	None	34.5°BTDC	A	-----	A
L78V	Set	IV	2°BTDC-2°ATDC	33°BTDC	B	-----	A
L77V	Set	II	5-7°BTDC	27°BTDC	A	-----	A
1975							
L78V	Set	II	5°ATDC	.198BTDC	A	.020	B
L78V	Set	II	5°ATDC	.198BTDC	A	.020	B
L78V	Set	(IV)	None	35°BTDC	(B)	-----	(A)
L78V	Set	IV	None	35°BTDC	B	-----	A
L78V	Set	IV	None	35°BTDC	B	-----	A
L78V	Set	IV	None	35°BTDC	B	-----	A
L78V	Set	(IV)	3-7°BTDC	33°BTDC	(B)	-----	(A)
L78V	Set	IV	3-7°BTDC	33°BTDC	B	-----	A
L76V	Set	IV	6°BTDC	27°BTDC	A	-----	A
L76V	Set	IV	6°BTDC	27°BTDC	A	-----	A
L76V	Set	IV	6°BTDC	27°BTDC	A	-----	A
L76V	Set	IV	6°BTDC	27°BTDC	A	-----	A
1976							
QL7J5	.050	IV	8°ATDC	24-26°BTDC	C	-----	B
L78V	Set	II	5°ATDC	.198BTDC	A	.020	B
L78V	Set	IV	None	30°BTDC	B	-----	A
L78V	Set	IV	None	35°BTDC	B	-----	A
L78V	Set	IV	3-7°BTDC	33°BTDC	B	-----	A
L76V	Set	IV	6°BTDC	27°BTDC	A	-----	A

carburetor identification, and timing notes called out in this table.

ENGINE SPECIFICATIONS AND

MODEL	BEGIN SERIAL NO.	CYL	HP	CU IN DISPL	W O T	BORE INCHES	STROKE INCHES
1977			**1977**				
40M, 40L	N/A	2	4.0	5.5	4000-5000	1.562	1.438
45M, 45L	4597262	1	4.5	5.5	4500-5500	2.000	1.750
75M, ML, E, EL	4691222	2	7.5	10.9	4500-5500	2.000	1.750
110M, ML, E, EL	4605112	2	9.8	10.9	4500-5500	2.000	1.750
200M, ML, E, EL	4580036	2	20	21.9	4800-5500	2.562	2.125
402M, ML, E, EL	4564002	2	40	33.3	4800-5300	2.875	2.562
1978			**1978**				
40M, ML	9172806	2	4.0	5.5	4300-4700	1.562	1.438
45M, ML	4869518	1	4.5	5.5	4500-5500	2.000	1.750
75M, ML, E, EL	4851693	2	7.5	10.9	4500-5500	2.000	1.750
110M, ML, E, EL	4839254	2	9.8	10.9	4500-5500	2.000	1.750
200M, ML, E, EL	4819249	2	20	21.9	4800-5500	2.562	2.125
400ML, E, EL	4860103	2	40	33.3	5000-5500	2.875	2.562
402M	4860103	2	40	33.3	5000-5500	2.875	2.562
1979			**1979**				
4	9208226	2	4.0	5.5	4300-4700	1.562	1.438
4.5	5270784	1	4.5	5.5	4500-5500	2.000	1.750
7.5	5226935	2	7.5	10.9	4500-5500	2.000	1.750
9.8	5206550	2	9.8	10.9	4500-5500	2.000	1.750
20	5183393	2	20	21.9	4800-5500	2.562	2.125
40	5174089	2	40	33.3	5000-5500	2.875	2.562
1980 -- 1981			**1980 -- 1981**				
3.6	5404657	1	3.6	5.5	4500-5000	2.000	1.750
4	5589202	2	4.0	5.5	4300-4700	1.562	1.437
4.5	5595532	1	4.5	5.5	4500-5500	2.000	1.750
7.5	5524660	2	7.5	10.9	5000-5800	2.000	1.750
9.8	5541280	2	9.8	10.9	5000-5800	2.000	1.750
18	5860332	2	18	24.4	5000-5800	2.562	2.360
20	5606492	2	20	21.9	4800-5800	2.562	2.125
25	5705532	2	25	24.4	5400-6000	2.562	2.360
40	5556230	2	40	33.3	4800-5500	2.875	2.562
1982			**1982**				
3.6	5910968	1	3.6	5.5	4500-5000	2.000	1.750
4.5	6025119	1	4.5	5.5	4500-5500	2.000	1.750
7.5	5936301	2	7.5	10.9	4500-5500	2.000	1.750
9.8	5965263	2	9.8	10.9	4500-5500	2.000	1.750
18	5954911	2	18	24.4	5000-5500	2.562	2.360
25	5942041	2	25	24.4	5400-6000	2.562	2.360
40	5991215	2	40	33.3	5000-5500	2.875	2.562

NOTE: See Appendix Page A-12 for Special Notes, ignition system identification,

TUNE-UP ADJUSTMENTS

SPARK PLUG CHAMP	PLUG GAP	IGN TYPE	PRIMARY PICKUP	MAX TIMING	CARB TYPE	POINT GAP	POWERHEAD TYPE
1977							
QL7J5	.050	IV	8°ATDC	24-26°BTDC	C	----	B
L78V	Set	II	5°ATDC	.198BTDC	A	.020	B
L78V	Set	IV	None	30°BTDC	B	----	A
L78V	Set	IV	None	35°BTDC	B	----	A
L78V	Set	IV	3-7°BTDC	33°BTDC	B	----	A
QL7J5	.050	IV	6°BTDC	27°BTDC	A	----	A
1978							
QL7J5	.050	IV	8°ATDC	24-26°BTDC	C	----	B
L78V	Set	II	5°ATDC	.198BTDC	A	.020	B
L78V	Set	IV	None	30°BTDC	B	----	A
L78V	Set	IV	None	35°BTDC	B	----	A
L78V	Set	IV	3-7°BTDC	33°BTDC	B	----	A
QL7J5	.050	IV	6°BTDC	27°BTDC	A	----	A
QL7J5	.050	IV	6°BTDC	27°BTDC	A	----	A
1979							
QL7J5	.050	IV	8°ATDC	24-26°BTDC	C	----	B
L78V	Set	II	5°ATDC	.198BTDC	A	.020	B
L77J4	.040	IV	None	30°BTDC	B	----	A
L77J4	.040	IV	None	35°BTDC	B	----	A
L78V	Set	IV	3-7°BTDC	33°BTDC	B	----	A
L76V	Set	IV	6°BTDC	27°BTDC	A	----	A
1980 -- 1981							
L81Y	.035	III	At Idle	N/A	D	--	B
L7J	.050	IV	12°-16°ATDC	22-26°BTDC	C	--	B
L78V	Set	II	5°ATDC	.198 BTDCA	A	.020	B
L77J4	.040	IV	At Idle	Note 2	B	--	A
L77J4	.040	IV	At Idle	Note 2	B	--	A
L77J4	.040	IV	12°ATDC	25°BTDC	B	--	B
L78V	Set	IV	Note 2	Note 3	B	--	A
L77J4	.040	IV	Note 4	25°BTDC	B	--	B
L76V	Set	IV	Note 5	Note 2	A	--	A
1982							
L81Y	.035	III	At Idle	None	D	--	B
L78V	Set	II	5°ATDC	.198 BTDC	A	.020	B
L78V	.040	IV	At Idle	Note 2	B	--	A
L78V	.040	IV	At Idle	Note 2	B	--	A
L77J4	.040	IV	12°ATDC	25°BTDC	B	--	B
L77J4	.040	IV	Note 4	25°BTDC	B	--	B
L76V	Set	IV	Note 5	Note 2	A	--	A

carburetor identification, and timing notes called out in this table.

ENGINE SPECIFICATIONS AND

MODEL	BEGIN SERIAL NO.	CYL	H.P.	CU. IN. DISPL.	OPERATING RANGE RPM	BORE INCHES	STROKE INCHES
1983						**1983**	
3.5	6202173	1	3.5	5.5	4500-5000	2.000	1.750
4.5	6232381	1	4.5	5.5	4500-5500	2.000	1.750
7.5	6239857	2	7.5	10.9	4500-5500	2.000	1.750
9.8	6263906	2	9.8	10.9	4500-5500	2.000	1.750
18	6171607	2	18	24.4	5000-5500	2.562	2.360
25	6185159	2	25	24.4	5400-6000	2.562	2.360
40	6228285	2	40	33.3	5000-5500	2.875	2.562
1984						**1984**	
2.2	8075603	1	2.2	4.6	4200-5200	1.850	1.687
3.5	6434771	1	3.5	5.5	4000-4500	2.000	1.750
4.5	6428076	1	4.5	5.5	4500-5500	2.000	1.750
7.5	6430041	2	7.5	10.9	4500-5500	2.000	1.750
(9.8)	6438338	2	9.8	(10.9)	4500-5500	2.000	1.750
18XD	6443973	2	18	24.4	5000-5500	2.562	2.375
25XD	6453343	2	25	24.4	5400-6000	2.562	2.375
35	6445653	2	35	33.3	5400-6000	2.875	2.562
1985						**1985**	
2.2	8087123	1	2.2	4.6	4200-5200	1.850	1.687
3.5	6594744	1	3.5	5.5	3500-4500	2.000	1.750
4.5	6608536	1	4.5	5.5	4500-5500	2.000	1.750
7.5	6611886	2	7.5	10.9	4500-5500	2.000	1.750
9.8	6620196	2	9.8	10.9	5000-6000	2.000	1.750
18XD	6617631	2	18	24.4	4500-5500	2.562	2.375
25XD	6593849	2	25	24.4	5000-6000	2.562	2.375
35	6626696	2	35	33.3	5400-6000	2.875	2.562
1986						**1986**	
2.2	A800761	1	2.2	4.6	4200-5200	1.850	1.687
4.0	A906702	1	4.0	5.5	4500-5500	2.000	1.750
6.0	A910596	2	6.0	12.8	4000-5000	2.125	1.812
8.0	A197112	2	8.0	12.8	4500-5500	2.125	1.812
9.9	A918999	2	9.9	12.8	5000-6000	2.125	1.812
20	A910971	2	20	24.4	4500-5500	2.562	2.375
25	A911346	2	25	24.4	5000-6000	2.562	2.375
35	A922171	2	35	33.3	5400-6000	2.875	2.562
210cc	A918999	2	9.9	12.8	5000-6000	2.125	1.812
90cc	A907702	1	4.0	5.5	4500-5500	2.000	1.750

(handwritten note next to 1984 "9.8" row: 10.99)

NOTE: See Appendix Page A-12 for Special Notes, ignition system identification,

TUNE-UP ADJUSTMENTS

SPARK PLUG CHAMP.	PLUG GAP	IGN. TYPE	PRIMARY PICKUP AT IDLE	MAX. TIMING	CARB. TYPE	POINT GAP	POWERHEAD TYPE
							1983
L81Y	.035	III	At Idle	----	D	--	B
L78V	Set	II	5°ATDC	.198 BTDC	A	.020	B
L78V	.040	IV	At Idle	Note 2	B	--	A
L78V	.040	IV	At Idle	Note 2	B	--	A
L77J4	.040	IV	12°ATDC	25°BTDC	F	--	B
L77J4	.040	IV	Note 4	25°BTDC	F	--	B
L76V	Set	IV	Note 5	Note 2	A	--	A
							1984
RL87YC	.040	I	At Idle	Not Adj	E	.012-.016	C
L87Y	.035	III	At Idle	----	D	----	B
L78V	Set	II	Note 6	.198BTDC	A	.020	B
L77J4	.040	IV	At Idle	Note 2	B	----	A
L77J4	.040	IV	At Idle	Note 2	B	----	A
L76V	Set	IV	Note 7	Note 8	F/G	----	B
L76V	Set	IV	Note 7	Note 8	F/G	----	B
L76V	Set	IV	3°BTDC	Note 10	A	----	A
							1985
RL87YC	.040	I	At Idle	----	E	.012-.016	C
L87Y	.035	III	At Idle	----	D	----	B
L78V	Set	II	Note 6	.198BTDC	A	.020	B
L77J4	.040	IV	At Idle	Note 2	B	----	A
L77J4	.040	IV	At Idle	Note 2	B	----	A
L76V	Set	IV	Note 7	Note 8	G	----	B
L76V	Set	IV	Note 7	Note 8	G	----	B
L76V	Set	IV	3°BTDC	Note 10	A	----	A
							1986
RL87YC	.040	I	At Idle	----	E	.012-.016	C
L78V	Set	II	Note 6	.198BTDC	A	----	B
L82YC	Note 9	IV	6°BTDC	36°BTDC	G	----	B
L82YC	Note 9	IV	6°BTDC	36°BTDC	G	----	B
L82YC	Note 9	IV	6°BTDC	36°BTDC	G	----	B
L76V	Set	IV	Note 7	Note 8	G	----	B
L76V	Set	IV	Note 7	Note 8	G	----	B
L76V	Set	IV	Note 7	Note 8	A	----	A
L82YC	Note 9	IV	6°BTDC	36°BTDC	G	----	B
L78V	Set	II	Note 6	.198TDC	A	----	B

carburetor identification, and timing notes called out in this table.

ENGINE SPECIFICATIONS AND

MODEL	CYL.	H.P.	CU. IN. DISPL.	OPERATING RANGE RPM	BORE INCHES	STROKE INCHES
1987					**1987**	
2.2	1	2.2	4.6	4200–5200	1.850	1.693
4 Early	1	4.0	5.5	4500–5500	2.000	1.750
4 Late	1	4.0	6.2	4500–5500	2.165	1.693
6.0	2	6.0	12.8	4000–5000	2.125	1.812
8.0	2	8.0	12.8	4500–5500	2.125	1.812
9.9	2	9.9	12.8	5000–6000	2.125	1.812
20	2	20	24.4	4500–5500	2.562	2.375
25	2	25	24.4	5000–6000	2.562	2.375
35	2	35	33.3	5400–6000	2.875	2.562
210cc	2	9.9	12.8	5000–6000	2.125	1.812
90cc	1	4.0	5.5	4500–5500	2.000	1.750
1988					**1988**	
2.2	1	2.2	4.6	4200–5200	1.850	1.693
4	1	4.0	6.2	4500–5500	2.165	1.693
5	1	5.0	6.2	4500–5500	2.165	1.693
8	2	8.0	12.8	4500–5500	2.125	1.812
9.9	2	9.9	12.8	5000–6000	2.125	1.812
15	2	15.0	16.0	5000–6000	2.362	1.811
20	2	20	24.4	4500–5500	2.562	2.375
25	2	25	24.4	5000–6000	2.562	2.375
35	2	35	33.3	5400–6000	2.875	2.562
1989					**1989**	
2.2	1	2.2	4.6	4200–5200	1.850	1.693
4	1	4.0	6.2	4500–5500	2.165	1.693
5	1	5.0	6.2	4500–5500	2.165	1.693
8	2	8.0	12.8	4500–5500	2.125	1.812
9.9	2	9.9	12.8	5000–6000	2.125	1.812
20	2	20	24.4	4500–5500	2.562	2.375
25	2	25	24.4	5000–6000	2.562	2.375
35	2	35	33.3	5400–6000	2.875	2.562
1990 & 1991					**1990 & 1991**	
3	1	3.0	4.6	4200–5200	1.850	1.693
4	1	4.0	6.2	4500–5500	2.165	1.693
5	1	5.0	6.2	4500–5500	2.165	1.693
8	2	8.0	12.8	4500–5500	2.125	1.812
9.9	2	9.9	12.8	5000–6000	2.125	1.812
15	2	15.0	16.0	5000–6000	2.362	1.811
20	2	20	24.4	4500–5500	2.562	2.375
25	2	25	24.4	5000–6000	2.562	2.375

NOTE: See Appendix Page A-12 for Special Notes, ignition system identification,

1987

SPARK PLUG CHAMP	PLUG GAP	IGN. TYPE	PRIMARY PICKUP AT IDLE	MAX. TIMING	CARB. TYPE	POINT GAP	PWRHD. TYPE
RL97YC	0.040	I	At Idle	---	E	0.012-0.016	C
L78V	Set	II	Note 6	.198 BTDC	A	---	B
L82YC	0.040	IV	5°BTDC	30°BTDC	H	---	C
L82YC	Note 9	IV	6°BTDC	36°BTDC	G	---	B
L82YC	Note 9	IV	6°BTDC	36°BTDC	G	---	B
L82YC	Note 9	IV	6°BTDC	36°BTDC	G	---	B
L76V	Set	IV	Note 7	Note 8	G	---	B
L76V	Set	IV	Note 7	Note 8	G	---	B
L76V	Set	IV	Note 7	Note 8	A	---	A
L82YC	Note 9	IV	6°BTDC	36°BTDC	G	---	B
L78V	Set	II	Note 6	.198 BTDC	A	---	B

1988

SPARK PLUG CHAMP	PLUG GAP	IGN. TYPE	PRIMARY PICKUP AT IDLE	MAX. TIMING	CARB. TYPE	POINT GAP	PWRHD. TYPE
RL97YC	0.040	I	At Idle	---	E	0.012-0.016	C
L82YC	0.040	IV	5°BTDC	30°BTDC	H	---	C
L82YC	0.040	IV	5°BTDC	30°BTDC	H	---	C
L82YC	Note 9	IV	6°BTDC	36°BTDC	G	---	B
L82YC	Note 9	IV	6°BTDC	36°BTDC	G	---	B
L82YC	0.040	IV	6°BTDC	36°BTDC	G	---	B
L76V	Set	IV	Note 7	Note 8	G	---	B
L76V	Set	IV	Note 7	Note 8	G	---	B
L76V	Set	IV	Note 7	Note 8	A	---	A

1989

SPARK PLUG CHAMP	PLUG GAP	IGN. TYPE	PRIMARY PICKUP AT IDLE	MAX. TIMING	CARB. TYPE	POINT GAP	PWRHD. TYPE
RL97YC	0.040	I	At Idle	---	E	0.012-0.016	C
L82YC	0.040	IV	5°BTDC	30°BTDC	H	---	C
L82YC	0.040	IV	5°BTDC	30°BTDC	H	---	C
L82YC	Note 9	IV	6°BTDC	36°BTDC	G	---	B
L82YC	Note 9	IV	6°BTDC	36°BTDC	G	---	B
L76V	Set	IV	Note 7	Note 8	G	---	B
L76V	Set	IV	Note 7	Note 8	G	---	B
L76V	Set	IV	Note 7	Note 8	A	---	A

1990 & 1991

SPARK PLUG CHAMP	PLUG GAP	IGN. TYPE	PRIMARY PICKUP AT IDLE	MAX. TIMING	CARB. TYPE	POINT GAP	PWRHD. TYPE
RL97YC	0.040	I	At Idle	---	E	0.012-0.016	C
L82YC	0.040	IV	5°BTDC	30°BTDC	H	---	C
L82YC	0.040	IV	5°BTDC	30°BTDC	H	---	C
L82YC	Note 9	IV	6°BTDC	36°BTDC	G	---	B
L82YC	Note 9	IV	6°BTDC	36°BTDC	G	---	B
L82YC	0.040	IV	6°BTDC	36°BTDC	G	---	B
L76V	Set	IV	Note 7	Note 8	G	---	B
L76V	Set	IV	Note 7	Note 8	G	---	B

carburetor identification, and timing notes called out in this table.

SPECIAL NOTES

BTDC equals "Before Top Dead Center"
ATDC equals "After Top Dead Center"
Model 4 1986–Mid '87 has carburetor "A"
Model 4 Mid'87 & On has carburetor "H"

IGNITION TYPE IDENTIFICATION

I   Phelon -- flywheel -- Magneto with points.
II   Thunderbolt -- flywheel -- Phase maker -- with points.
III   Thunderbolt -- flywheel -- C.D. ignition -- pointless.
IV   Thunderbolt -- flywheel -- C.D. ignition -- pointless -- coil per cylinder.

TIMING NOTES

1- Up to serial number 2432535, the maximum spark advance should be set to 0.375" (Gauge No. 91-46707A1) and the full throttle advance should be set to 0.275". (Gauge No. 91-30292A1); after this number, Use Gauge No. C-91-3973A1 (0.300") for the maximum spark advance and Gauge No. 91-26916A1 (0.235") for the full throttle advance.
2- Align straight line on flywheel with timing pointer (or notch on flywheel housing).
3- Align three dots on flywheel with timing pointer (or notch on flywheel housing).
4- Letters "BCIA" stamped on carburetor flange: 12° ATDC (two dots on flywheel). Letters "BCIB" or "BCIC" stamped on carburetor flange: 5° ATDC (four dots on flywheel).
5- Align two dots on flywheel with notch on flywheel housing.
6- 5° ATDC except for models with WMDO-1 (Type G) carburetor, then primary pickup is 10°-12° ATDC.
7- 2° ATDC, one mark to the right of the single dot.
8- 25° BTDC at WOT (wide open throttle), three dots on flywheel.
9- 0.040" for engines equipped with standard ignition coils, not adjustable for engines equipped with high-energy coils.
10- 29° BTDC at cranking speed.

CARBURETOR IDENTIFICATION

A- Side bowl and back drag carburetor.
B- Round bowl -- single float carburetor with integral fuel pump.
C- Center round bowl -- single float side-draft carburetor.
D- Mikuni rectangular bowl -- double float carburetor.
E- Mikuni round bowl -- single float carburetor.
F- Tillotson rectangular bowl -- double float carburetor with fuel pump.
G- Walbro round bowl -- single float carburetor with fuel pump.
H- Round bowl -- single float carburetor stamped "F" with "KEIKHIN" integral fuel pump.

CARBURETOR JET SIZE / ELEVATION CHART

MODEL	YEAR	CYLINDER	H.P.	ZERO TO 2500FT (0–765M)	2501 TO 5000FT (766 to 1525M)	5001 TO 7500FT 1526 to 2285M)	COMMENT
2.2	1984–89	1	2.2	#94	Note 1	Note 1	
3.0	1990 & on	1	2.2	#94	Note 1	Note 1	
3.5	1983–85	1	3.5	#150	#140	#130	
3.6	1980–82	1	3.6	#150	#140	#130	
39	1965–66	1	3.9	.043	.041	.039	
39	1967–68	1	3.9	.036	.034	.032	
40	1969–74	1	4.0	.036	.034	.032	
40	1976–80	2	4.0	.041	.040	.039	
4.0	1986–Mid'87	1	4.0	.040	.039	.041	Carb. Type "A"
4.0	Mid'87 & on	1	4.0	.031	.029	.027	Carb. Type "H"
5	1988 & on	1	5.0	.031	.029	.027	
45	1975–78	1	4.5	.040	.039	.038	
4.5	1979–85	1	4.5	.040	.039	.038	
60	1965–68	2	6.0	.045	.043	.041	
6.0	1986–1987	2	6.0	.046	.044	.042	
75	1969–74	2	7.5	.035	.033	.031	
75	1975*	2	7.5	.034	.033	.032	*4131609 & below
75	1975–76*	2	7.5	.032	.031	.030	*4131610 to 4397536
75	1976–79*	2	7.5	.045	.043	.041	*4397537 & up
7.5	1975–85	2	7.5	.040	.039	.038	
8.0	1986 & on	2	8.0	.046	.044	.042	
110	1965–70	2	9.8	.049	.047	.045	
110	1971–73	2	9.8	.047	.045	.043	
110	1974	2	9.8	.041	.039	.037	
110	1975–79*	2	9.8	.041	.039	.036	*4079000 & up
9.8	1979–85	2	9.8	.039	.038	.037	
9.9	1986 & on	2	9.9	.056	.054	.052	
15	1988 & on	2	15	.056	.054	.052	
18	1981–83	2	18	.052	.050	.048	
18XD	1984–85	2	18	.052	.050	.048	
200	1965–66	2	20	.061	.059	.057	
200	1967–68	2	20	.063	.061	.059	
200	1969–70	2	20	.061	.059	.057	
200	1971–72	2	20	.059	.057	.055	
200	1973–74	2	20	.057	.055	.053	
200	1975*	2	20	.055	.053	.051	*4351589 & below
200	1976–79*	2	20	.057	.055	.053	*4351590 & up
20	1979–80	2	20	.057	.055	.053	
20	1986 & on	2	20	.080	.078	.076	
25	1980–83	2	25	.067	.065	.063	
25	1986 & on	2	25	.080	.078	.076	
25XD	1984–85	2	25	.080	.078	.076	
350	1965–66	2	35	.069	.067	.065	
350	1967–79	2	35	.063	.061	.059	
35	1984–89	2	35	.072	.070	.068	
400	1970–71	2	40	.078	.076	.074	
402	1972–78	2	40	.078	.076	.074	
40	1979–80	2	40	.072	.070	.068	
40	1981–83	2	40	.072	.070	.068	
210cc	1986–87	2	9.9	.056	.054	.052	
90cc	1986–87	1	4.0	.041	.040	.039	

Note: 1 "E" clip on needle: normal position is second groove. Move clip upward (toward end) one groove for each 2500ft (765M) elevation.

PISTON AND CYLINDER SPECIFICATIONS

Model	Year	Cyl.	H.P.	Piston Taper Skirt		Piston Taper Above Rings		Cyl. Block Hone Finish		Oversize Hone Finish	
				Inches	mm	Inches	mm	Inches	mm	Inches	mm
2.2	1984–89	1	2.2	1.846	46.89	1.844	46.84	1.850	46.99	1.869	47.47
3.0	1990 & on	1	2.2	1.846	46.89	1.844	46.84	1.850	46.99	1.869	47.47
3.5	1983–85	1	3.5	1.996	50.70	1.990	50.55	2.000	50.80	2.015	51.18
3.6	1980–82	1	3.6	1.996	50.70	1.990	50.55	2.000	50.80	2.015	51.18
39	1965–68	1	3.9	1.996	50.70	1.990	50.55	2.000	50.80	2.015	51.18
40	1969–74	1	4.0	1.996	50.70	1.990	50.55	2.000	50.80	2.015	51.18
40	1976–80	2	4.0	1.558	39.57	1.552	39.42	1.563	37.70	N/A	N/A
4.0	1986–Mid'87	1	4.0	1.996	50.70	1.990	50.55	2.000	50.80	2.015	51.18
4.0	Mid'87 & on	1	4.0	2.164	54.96	2.158	54.61	2.165	55.00	2.180	55.37
45	1975–78	1	4.5	1.996	50.70	1.990	50.55	2.000	50.80	2.015	51.18
4.5	1979–85	1	4.5	1.996	50.70	1.990	50.55	2.000	50.80	2.015	51.18
5.0	1988 & on	1	4.0	2.164	54.96	2.158	54.61	2.165	55.00	2.180	55.37
60	1968–68	2	6.0	1.756	44.60	1.750	44.45	1.753	44.53	1.768	44.91
6.0	1986–87	2	6.0	2.118	53.80	2.114	53.70	2.125	53.80	2.140	54.36
75	1969–78	2	7.5	1.996	50.70	1.990	50.55	2.000	50.80	2.015	51.18
7.5	1979–85	2	7.5	1.996	50.70	1.990	50.55	2.000	50.80	2.015	51.18
8.0	1986 & on	2	8.0	2.118	53.80	2.114	53.70	2.125	53.80	2.140	54.36
110	1965–78	2	9.8	1.996	50.70	1.990	50.55	2.000	50.80	2.015	51.18
9.8	1979–85	2	9.8	1.996	50.70	1.990	50.55	2.000	50.80	2.015	51.18
9.9	1986 & on	2	9.9	2.118	53.80	2.114	53.70	2.125	53.80	2.140	54.36
15	1988 & on	2	15	2.358	59.89	2.352	59.74	2.362	60.00	2.377	60.37
18	1981–83	2	18	2.558	64.98	2.551	64.80	2.562	65.07	2.577	54.36
18XD	1984–85	2	18	2.558	64.98	2.551	64.80	2.562	65.07	2.577	54.36
200	1968–72*	2	20	2.558	64.98	2.551	64.80	2.563	65.10	2.578	65.48
200	1968–72	2	20	2.558	64.98	2.551	64.80	2.565	65.15	2.580	65.53
200	1973–74	2	20	2.558	64.98	2.566	65.18	2.581	65.56	N/A	N/A
200	1975–79	2	20	2.558	64.98	2.551	64.80	2.565	65.15	2.581	65.56
20	1979–80	2	20	2.558	64.98	2.551	64.80	2.562	65.07	2.577	54.36
20	1986 & on	2	20	2.558	64.98	2.551	64.80	2.562	65.07	2.577	54.36
25	1980–83	2	25	2.558	64.98	2.551	64.80	2.562	65.07	2.577	54.36
25	1986 & on	2	25	2.558	64.98	2.551	64.80	2.562	65.07	2.577	54.36
25XD	1984–85	2	25	2.558	64.98	2.551	64.80	2.562	65.07	2.577	54.36
350	1965	2	35	2.992	76.00	2.857	72.57	2.872	72.95	2.887	73.33
350	1966–69	2	35	2.992	76.00	2.986	75.84	3.000	76.20	3.015	76.58
35	1984–89	2	35	2.872	72.95	2.863	72.72	2.875	73.02	2.890	73.41
400	1970–71	2	40	2.865	72.95	2.857	72.57	2.875	73.02	2.890	73.41
402	1971–74	2	40	2.871	72.92	2.859	72.62	2.875	73.02	2.890	73.41
402	1975–79	2	40	2.872	72.95	2.863	72.72	2.875	73.02	2.890	73.41
40	1979–80	2	40	2.872	72.95	2.863	72.72	2.875	73.02	2.890	73.41
40	1981–83	2	40	2.872	72.95	2.863	72.72	2.875	73.02	2.890	73.41
210cc	1986–87	2	9.9	2.118	53.80	2.114	53.70	2.125	53.80	2.140	54.36
90cc	1986–87	1	4.0	1.996	50.70	1.990	50.55	2.000	50.80	2.015	51.18

* Model 200 with shift.

REED STOP OPENING

MODEL	YEAR	CYL.	H.P.	REED STOP OPENING INCHES	mm	POWERHEAD TYPE
2.2	1984-89	1	2.2	.240	6.096	C
3.0	1990 & on	1	2.2	.240	6.096	C
3.5	1983 & on	1	3.5	.007*	.178	B
3.6	1980-82	1	3.6	.007*	.178	B
39	1965-68	1	3.9	.007*	.178	B
40	1969-74	1	4.0	.154	3.912	A
40	1976-80	2	4.0	.007*	.178	B
4.0	1986-Mid'87	1	4.0	.007*	.178	B
4.0	Mid'87 & on	1	4.0	.240	6.096	C
45	1975-78	1	4.5	.007*	.178	B
4.5	1979-85	1	4.5	.007*	.178	B
5.0	1988 & on	1	5.0	.240	6.096	C
60	1965-68	2	6.0	.109	2.769	A
6.0	1986-87	2	6.0	.007*	.178	B
75	1969-78	2	7.5	.156	3.962	A
7.5	1979-85	2	7.5	.156	3.962	A
8.0	1986 & on	2	8.0	.007*	.178	B
110	1965-78	2	9.8	.156	3.962	A
9.8	1979-85	2	9.8	.156	3.962	A
9.9	1986 & on	2	9.9	.007*	.178	B
15	1988 & on	2	15	.007*	.178	B
18	1981-83	2	18	.007*	.178	B
18XD	1984-85	2	18	.007*	.178	B
200	1965-78	2	20	.187	4.750	A
20	1986 & on	2	20	.007*	.178	B
20	1979-80	2	20	.191	4.851	A
25	1980-83	2	25	.007*	.178	B
25	1986 & on	2	25	.007*	.178	B
25XD	1984-85	2	25	.007*	.178	B
350	1965-69	2	35	.187	4.750	A
35	1984-89	2	35	.162	4.115	A
400	1970-71	2	40	.156	3.962	A
402	1972-74	2	40	.156	3.962	A
402	1975-79	2	40	.162	4.115	A
40	1979-80	2	40	.162	4.115	A
40	1981-83	2	40	.162	4.115	A
210cc	1986-87	2	9.9	.007*	.178	B
90cc	1986-87	1	4.0	.007*	.178	B

* The manufacturer recommends the reeds stand open, but are not more than .007" (.178 mm) off reed block surface.

N/A equals "Not Applicable".

Powerhead Type A Split block without head -- internal reed box around crankshaft.

Powerhead Type B Split block without head -- external reed block.

Powerhead Type C Split block with head -- reed valves beneath crankshaft.

LOWER UNIT OIL CAPACITY AND GEAR CHART

MODEL	YEAR	CYLINDER	H.P.	CAPACITY FLUID OZ.	NO. TEETH FORWARD	NO. TEETH PINION	GEAR RATIO
2.2	1984–89	1	2.2	90cc	27	13	1.85:1
3.0	1990 & on	1	3.0	Info. not available.			2.18:1
3.5	1983–85	1	3.5	2.75	27	13	2.08:1
3.6	1980–82	1	3.6	2.75	27	13	2.08:1
39	1965–68	1	3.9	3.00	26	13	2:1
40	1969–74	1	4	2.00	26	13	2:1
4	1976–80	2	4	2.75	27	13	2.08:1
4	1986–Mid'87	1	4	3.75	26	13	2:1
4	Mid'87 & on	1	4	6.60	28	13	2:151
45	1975–78	1	4.5	3.75	26	13	2:1
4.5	1979–85	1	4.5	3.75	26	13	2:1
5	1988 & on	1	5	6.60	28	13	2:151
60	1965–68	2	6	3.75	26	13	2:1
6	1986–1987	2	6	6.50	27	13	2.08:1
(75)	1969–78	2	7.5	3.75	26	13	2:1
7.5	1979–85	2	7.5	3.75	26	13	2:1
8	1986 & on	2	8.0	6.50	27	13	2.08:1
110	1965–78	2	9.8	3.75	26	13	2:1
(9.8)	1979–85	2	9.8	3.75	26	13	2:1
9.9	1986 & on	2	9.9	6.50	27	13	2.08:1
15	1988 & on	2	15	6.50	27	13	2.08:1
18	1981–83	2	18	5.50	27	12	2.25:1
18XD	1984–85	2	18	7.60	27	12	2.25:1
200	1965–72	2	20	6.00	24	13	1.85:1
200	1973	2	20	6.00	24	14	1.71:1
(200)	1974	2	20	6.50	24	13	1.85:1
(20)	1979–80	2	20	6.50	24	13	1.85:1
20	1986 & on	2	20	7.60	27	12	2.25:1
25	1980–83	2	25	5.50	27	12	2.25:1
25	1986 & on	2	25	7.60	27	12	2.25:1
25XD	1984–85	2	25	7.60	27	12	2.25:1
350	1965–69	2	35	9.00	24	13	1.85:1
35	1984–89	2	35	12.50	26	13	2:1
400	1970–71	2	40	9.00	26	13	2:1
402	1972–78	2	40	12.50	26	13	2:1
40	1979–80	2	40	12.50	26	13	2:1
40	1981–83	2	40	12.50	26	13	2:1
210cc	1986–87	2	9.9	6.50	27	13	2.08:1
90cc	1986–87	1	4.0	3.75	26	13	2:1

LOWER UNIT GEAR BACKLASH TABLE

MODEL	YEAR	CYL.	H.P.	LOWER UNIT TYPE	FORWARD GEAR Inches	mm	REVERSE GEAR Inches	mm
2.2	1984–89	1	2.2	D	Note 1		Note 1	
3.0	1990 & on	1	2.2	D	Note 1		Note 1	
3.5	1983 & on	1	3.5	A	Note 2		N/A	
3.6	1980–82	1	3.6	A	Note 2		N/A	
39	1965–68	1	3.9	B	.003–.005	.076–.127	.003–.005	.076–.127
40	1969–74	1	4.0	A	Note 2		N/A	
40	1976–80	2	4.0	B	.003–.005	.076–.127	.003–.005	.076–.127
4	1986–Mid'87	1	4.0	A	Note 1		Note 1	
4	Mid'87 & on	1	4.0	C	Note 1		Note 1	
45	1975–78	1	4.5	B	.003–.005	.076–.127	.003–.005	.076–.127
4.5	1979–85	1	4.5	B	.003–.005	.076–.127	.003–.005	.076–.127
5	1988 & on	1	5.0	C	Note 1		Note 1	
60	1965–68	2	6.0	B	.003–.005	.076–.127	.003–.005	.076–.127
6	1986–1987	2	6.0	B	Note 1		Note 1	
75	1969–78	2	7.5	B	.003–.005	.076–.127	.003–.005	.076–.127
7.5	1979–85	2	7.5	B	.003–.005	.076–.127	.003–.005	.076–.127
8	1986 & on	2	8.0	B	Note 1		Note 1	
110	1965–78	2	9.8	B	.003–.005	.076–.127	.003–.005	.076–.127
9.8	1979–85	2	9.8	B	.003–.005	.076–.127	.003–.005	.076–.127
9.9	1986 & on	2	9.9	B	Note 1		Note 1	
15	1988 & on	2	15	B	Note 1		Note 1	
18	1981–83	2	18	C	.003–.005	.076–.127	.003–.005	.076–.127
18XD	1984–85	2	18	C	.003–.005	.076–.127	.003–.005	.076–.127
200	1965–78	2	20	B	.003–.005	.076–.127	.003–.005	.076–.127
20	1986 & on	2	20	B	Note 1		Note 1	
20	1979–80	2	20	B	.003–.005	.076–.127	.003–.005	.076–.127
25	1980–83	2	25	C	.003–.005	.076–.127	.003–.005	.076–.127
25	1986 & on	2	25	C	Note 1		Note 1	
25XD	1984–85	2	25	C	.003–.005	.076–.127	.003–.005	.076–.127
350	1965–69	2	35	C	.003–.005	.076–.127	.003–.005	.076–.127
35	1984–89	2	35	C	.007–.010	.178–.254	Note 3	
400	1970–71	2	40	C	.003–.005	.076–.127	.003–.005	.076–.127
402	1972–78	2	40	C	.003–.005	.076–.127	.003–.005	.076–.127
40	1979–80	2	40	B	.007–.010	.178–.254	Note 3	
40	1981–83	2	40	B	.007–.010	.178–.254	Note 3	
210cc	1986–87	2	9.9	B	Note 1		Note 1	
90cc	1986–87	1	4.0	A	Note 1		Note 1	

Note 1 No numerical specification is given by the manufacturer who states: "The amount of play between the gears is not critical, but **NO** play is unacceptable."

Note 2 Gnat lower unit, follow shimming procedure in Section 9-4..

Note 3 Obtain .250 (.64 mm) clearance between shimming tool and pinion gear.

THUNDERBOLT IGNITION STATOR ASSEMBLY
AND STATOR COIL CHECKS

Stator leads **MUST** be disconnected before testing stator assembly.
Stator coil wires **MUST** be disconnected before testing.

MODEL	YEAR	TESTER LEADS TO:	OHM SCALE	SCALE READING
3.5 3.6	1983-85 1980-82	The stator module cannot be tested. If it is suspected of being faulty, proceed to ignition coil tests. If the ignition coil checks OK, replace the stator module.		
40	1970-71	Positive lead to Green stator lead, negative lead to ground.	Rx1000	No Continuity
		Reverse connections	Rx1000	20-50
40 4.0 45 4.5 90cc	1972-74 1986-Mid'87 1975-78 1979-85 1986-87	(High speed) between Yellow and Blue coil leads.	Rx100	6-8
		(Low speed) between Yellow and Red coil leads.	Rx1000	5.3-6.1
4.0 5.0 Note 3	Mid'87 & On 1988 & On	Positive lead to Black/Red lead, negative lead to White lead.	Rx10	9.3-14.2
40 75 110 200	1970-71 1970-71 1970-71 1970-71	Positive lead to Green stator lead, negative lead to Salmon stator lead.	Rx1000	No Continuity
		Reverse connections.	Rx1000	20-50
		Positive lead to Green stator lead, negative lead to ground.	Rx1000	No Continuity
		Reverse connections.	Rx1000	No Continuity
75	1972-73	(High speed) between Yellow and Blue coil leads.	Rx10	18.5-20.5
110 200	1972-73 1972	(Low speed) between Yellow and Red coil leads.	Rx1000	3.95-4.30
200	1973-74	Between Red and Blue coil leads.	Rx10	18.5-20.5
		Between White and Blue coil leads.	Rx1000	6.7
200 Note 1	1975-76	Between Red and Blue coil wires.	Rx1	180-340
		Between White and Blue coil wires.	Rx1000	5.2-7.0
40 75 & 7.5 110 9.8 200 Note 2 20	1976-80 1974-78 1979-85 1974-78 1979-85 1976-78 1979-80	Between Yellow stator wire and ground.	Rx100	15-20
		Between Yellow and White coil wires.	Rx100	7.5-10

THUNDERBOLT IGNITION STATOR ASSEMBLY
AND STATOR COIL CHECKS

Stator leads **MUST** be disconnected before testing stator assembly.
Stator coil wires **MUST** be disconnected before testing.

MODEL	YEAR	TESTER LEADS TO:	OHM SCALE	SCALE READING
6.0 8.0	1986-87 1986 & on	Between Black/White stator leads and ground.	Rx10	12-18
9.9 15.0 210cc	1986 & on 1988 & on 1986-87	Between Black/Yellow stator leads and ground.	Rx1000	3.2-3.8
18 18XD 20 25 25 25XD	1981-83 1984-85 1986 & on 1980-83 1986 & on 1984-85	Between Black/Yellow and Black/White stator leads.	Rx1000	3.1-3.7
400 (Man.)	1970-71	Between Blue coil wire and ground.	Rx1000	21-25
400 (Elec.)	1970-71	Between White coil wire and ground.	Rx10	11.5-13.5
402 40 35 (Man.)	1972-78 1979-83 1984-89	Between Red and Blue coil wires.	Rx100	50-64
402 40 35 (Elec.)	1972-78 1979-83 1984-89	Between Red coil wire and ground.	Rx1	45-60

Note 1 Model 200 Serial No. 4377556 and below.

Note 2 Model 200 Serial No. 4403787 and above.

Note 3 Manufacturer terms this coil "Capacitor Charging Coil".

THUNDERBOLT TRIGGER TEST

Trigger leads **MUST** be disconnected before testing.

MODEL	YEAR	TESTER LEADS TO:	OHM SCALE	SCALE READING
200	1973-74	Between trigger leads.	Rx100	8-10
75 110	1974 1974	To each Brown trigger lead.	Rx10	10.8-16
40 75 7.5 110 9.8 200 Note 1 20	1976-80 1975-78 1979-85 1975-78 1979-85 1976-78 1979-80	Between trigger leads.	Rx10	14-16
200 Note 2	1975-76	Between trigger leads.	Rx100	8-10
4.0 6.0 8.0 9.9 15.0 90cc 210cc 18 18XD 20 25 25 25XD	1986-Mid'87 1986-87 1986 & on 1986 & on 1988 & on 1986-87 1986-87 1981-83 1984-85 1986 & on 1980-83 1986 & on 1984-85	Between trigger leads.	Rx100	6.5-8.5
4.0 5.0	Mid'87 & On 1988 & On	Postive tester lead to Black lead, Negative tester lead to Red/White lead.	Rx10	8-11.5
400 Note 3	1970-71	Between trigger leads.	Rx1	27-41
402 40 35	1972-78 1979-83 1984-89	Between trigger leads.	Rx100	8-10

NOTE 1 Model 200 Serial No. 4403787 and above.

NOTE 2 Model 200 Serial No. 4377556 and below.

NOTE 3 Trigger Part No. A332-4608A2 cannot be tested with an ohmmeter.

IGNITION COIL TESTS -- TYPE I IGNITION SYSTEM

Breaker Points MUST be OPEN when checking coil

PRIMARY IGNITION COIL TEST

MODEL	YEAR	TESTER LEADS TO:	OHM SCALE	SCALE READING
2.2 3.0	1984-89 1990 & on	Postive test lead to Black/White coil lead. Negative test lead to coil ground	Rx1	1.5

SECONDARY IGNITION COIL TEST

MODEL	YEAR	TESTER LEADS TO:	OHM SCALE	SCALE READING
2.2 3.0	1984-89 1990 & on	PRIMARY WINDING TEST Postive test lead to Black/White coil lead. Negative test lead to coil ground.	Rx1	.81 - 1.09
		SECONDARY WINDING TEST Postive test lead to Spark Plug lead terminal. Negative test lead to coil ground.	Rx100	4.25 - 5.75

MODULE TESTS FOR TYPE II IGNITION SYSTEM
1970-85

MODEL	YEAR	TESTER LEADS TO:	OHM SCALE	SCALE READING
All Type II Ignition System	1970-71	Part of stator (See Stator Checks)		
All Type II Ignition System	1972-85	Red test lead to Red terminal and Black test lead to Capacitor terminal	Rx1000	Continuity
		Red test lead to Capacitor terminal and Black test lead to Red terminal	Rx1000	No Continuity
		Red test lead to Blue terminal and Black test lead to Capacitor terminal	Rx1000	Continuity
		Red test lead to Capacitor terminal and Black test lead to Blue terminal	Rx1000	No

THUNDERBOLT IGNITION COIL CHECKS
IGNITION TYPE II, III, AND IV

Disconnect Coil (+) and (-) Leads and Secondary Wire Prior To Testing
Remove High-Tension Leads from Coil Towers

MODEL	YEAR	TEST	TESTER LEADS TO:	OHM SCALE	SCALE READING
3.5 and 3.6	1983-85	Coil Power	Note 1		
	1980-82	Coil Continuity		Note 2	
40 4.0 45 4.5 90cc	1972-75 1986-Mid'87 1975-79 1980-85 1986-87	Primary	(+) and (-) coil terminals.	Rx1	.01-.02
		Secondary	Ground (or pigtail if not mounted) and coil tower.	Rx100	5-6
4.0 5.0	Mid'87 & on 1988 & on	Primary	(+) and (-) coil terminals	Rx1	.02-.38
		Secondary	Positive tester lead to either (+) or (-) coil terminals, negative lead to high-tension lead	Rx1000	3-4.4
40 75 110 200	1970-71 1970-73 1970-73 1970-71	Primary	(+) and (-) coil terminals.	Rx1	.3-.35
		Secondary	Ground (or pigtail if not mounted) and coil tower.	Rx100	5-6
40 (4.0 hp) 6.0 75 7.5 8.0 110 9.8 9.9 15 200 20 35 402 40 (40 hp) 210 cc	1976-80 1986-87 1979-85 1979-85 1986 & on 1974-78 1979-85 1986 & on 1988 & on 1972-78 1979-80 1986-89 1972-78 1979-80 1986-87	Primary	(+) and (-) coil terminals.	Rx1	.02-.04
		Secondary	Ground (or pigtail if not mounted) and coil tower.	Rx100	9-12
18 18XD 20 25 25 XD 25	1981-83 1984-85 1986 & on 1980-83 1984-85 1986 & on	Primary	(+) and (-) coil terminals.	Rx1	0
		Secondary	Between coil tower and either (+) or (-) terminal.	Rx100	8.5-12.0
400 Note 3	1970-71	Primary	(+) Green wire and ground.	Rx1000	0
		Secondary	Ground and coil tower.	Rx1000	24-30
400 Note 4	1970-71	Primary	(+) Green wire and ground	Rx1	.01-.015
		Secondary	Ground and coil tower.	Rx1000	9-10.5

NOTE: See Appendix Page A-23 for special notes called out in this table.

SPECIAL NOTE

Ignition Coil Test

Ohmmeter tests can detect only certain faults in the ignition coils. Replace ignition coil, if ohmmeter readings (listed in the chart) are not as specified. If coil test **OK**, and coil is still suspected of being faulty, use a Quicksilver MultiMeter/DVA Tester (Part No. 91-99750), or a voltmeter (capable of measuring 400 vbolts DC, or higher), and Quicksilver Direct Voltage Adaptor (91-89045) to thoroughly check the coil.

GENERAL NOTES

NOTE 1 Obtain a Merc-O-Tronic Magneto Analyser (Model No. 9800). Connect the small Black test lead to the primary coil Black ground wire. Connect the small Red test lead to the primary coil termnal (the stator wire terminal). Connect the single red test lead to the spark plug lead. Turn the Magneto Analyser current control knob to the extreme left beyond the "LO" position. Turn the selector switch to the No. 1 position (Coil Power Test). Turn the second selector switch to "C.D.1. Now, slowly turn the current control knob clockwise until 2.4 amps is reached. Hold the spark plug lead about 1/4" (6 mm) from the ground wire. A steady spark should occur. If the spark is weak, intermittent or no spark occurs at this reading, the coil is defective and **MUST** be replaced.

NOTE 2 Disconnect the Black (stator module) wire from the coil and the high tension lead from the spark plug. Remove the two screws and lift the coil free. Obtain a Mer-O-tronic Magneto Analyser (Model No. 9800). Turn the Magneto Analyser selector switch to the No. 3 position (Coil Continuity). Clip the small Red and Black test leads together. Calibrate the meter to "Set" position. Connect the small black test lead to the Black ground wire of the ignition coil. Connect the small Red test lead to the high tension lead. A reading between 29 and 39 is acceptable. A reading lower than 29 indicates the secondary wiring is shorted. A reading higher than 39 indicates the secondary winding is open. In either case the coil is defective and **MUST** be replaced.

NOTE 3 For coil Part No. A332-4075A1

NOTE 4 For coil Part No. A336-4592A2

TORQUE SPECIFICATIONS

Fastener Location	1 Cyl. 2.2 1984–89 3.0 1990 & On 4.0 Mid'87 & On 5.0 1988 & On	1 Cyl. 3.5 1983–85	1 Cyl. 3.6 1980–82	1 Cyl. 39 1965–68	1 Cyl. 40 1969–74	2 Cyl. 40 1976–80	1 Cyl. 4.0 1986–Mid'87
Bearing carrier nut	N/A	40 ft lbs 54.3 Nm	40 ft lbs 54.3 Nm	60 ft lbs 81.6 Nm	60 ft lbs 81.6Nm	60 ft lbs 81.6 Nm	60 ft lbs 81.6 Nm
Carburetor adaptor plate bolts	N/A	N/A	N/A	N/A	N/A	N/A	N/A
Carburetor mounting locknut	N/A	80 in lbs 9 Nm	80 in lbs 9 Nm	80 in lbs 9 Nm	80 in lbs 9 Nm	80 in lbs 9 Nm	80 in lbs 9 Nm
Center main bearing lockscrew (3)	N/A	N/A	N/A	N/A	N/A	N/A	N/A
Center main bearing reed stop screw	N/A	10 in lbs 1.1 Nm	10 in lbs 1.1 Nm	10 in lbs 1.1 Nm	10 in lbs 1.1 Nm	10 lbs 1.1 Nm	N/A
Coil terminal nuts	N/A	N/A	N/A	30 in lbs 3.4 Nm	30 in lbs 3.4 Nm	30 in lbs 3.4 Nm	30 in lbs 3.4 Nm
Connecting rod nuts	N/A	15 ft lbs 20.3 Nm	15 ft lbs 20.3 Nm	15 ft lbs 20.3 Nm	15 ft lbs 20.3 Nm	6.3 ft lbs 8.5 Nm	6.3 ft lbs 8.5 Nm
Crankcase to cylinder block	50 in lbs 6 Nm	80 in lbs 9 Nm	80 in lbs 9 Nm	90 in lbs 10.1 Nm	99 in lbs 11.2 Nm	90 in lbs 10.1 Nm	90 in lbs 10.1 Nm
Cylinder block cover	85 in lbs 10 Nm	80 in lbs 9 Nm	80 in lbs 9 Nm	80 in lbs 9 Nm	99 in lbs 11.2 Nm	99 in lbs 11.2 Nm	30 in lbs 3.4 Nm
Exhaust outer cover	N/A	80 in lbs 9 Nm	80 in lbs 9 Nm	70 in lbs 7.9 Nm	88 in lbs 9.9 Nm	88 in lbs 9.9 Nm	30 in lbs 3.4 Nm
Flywheel to crankshaft	30 ft lbs 41 Nm	35 ft lbs 47.6 Nm	35 Ft lbs 47.6 Nm	35 ft lbs 47.6 Nm	34 ft lbs 46.2 Nm	19 ft lbs 25.9 Nm	19 ft lbs 25.9 Nm
Gear housing to driveshaft housing	25 in lbs 3 Nm	N/A	N/A	Note 1	Note 1	Note 1	140 in lbs 15.8 Nm
Ignition trigger cover	N/A	N/A	N/A	N/A	N/A	N/A	N/A
Pinion gear nut	N/A	80 in lbs 9 Nm	80 in lbs 9 Nm	88 in lbs 9.9 Nm	88 in lbs 9.9 Nm	88 in lbs 9.9 Nm	N/A N/A
Powerhead to driveshaft housing	50 in lbs 6 Nm	80 in lbs 9 Nm	80 in lbs 9 Nm	80 in lbs 9 Nm	80 in lbs 9 Nm	80 in lbs 9 Nm	80 in lbs 9 Nm
Propeller nut	N/A	N/A	N/A	N/A	N/A	60 in lbs 6.8 Nm	60 in lbs 6.8 Nm
Reed block clamping screws	N/A	N/A	N/A	N/A	N/A	N/A	N/A
Spark plugs	20 ft lbs 27 Nm	20 ft lbs 27 Nm	20 ft lbs 27 Nm	17 ft lbs 23 Nm	20 ft lbs 27 Nm	20 ft lbs 27 Nm	20 ft lbs 27 Nm
Starter motor to crankcase	N/A	N/A	N/A	N/A	N/A	N/A	N/A
Stator mounting screws	N/A	N/A	N/A	N/A	N/A	N/A	N/A

TORQUE SPECIFICATIONS

Fastener Location	1 Cyl. 2.2 1984–89 3.0 1990 & On 4.0 Mid'87 & On 5.0 1988 & On	1 Cyl. 3.5 1983–85	1 Cyl. 3.6 1980–82	1 Cyl. 39 1965–68	1 Cyl. 40 1969–74	2 Cyl. 40 1976–80	1 Cyl. 4.0 1986–Mid'87
Support stud nuts	N/A	35 in lbs 4 Nm	35 in lbs 4 Nm	N/A	N/A	N/A	N/A
Swivel bracket to driveshaft housing bolts	50 in lbs 5.6 Nm	60 in lbs 6.8 Nm	60 in llbs 6.8 Nm	60 in lbs 6.8 Nm	60 in lbs 6.8 Nm	60 in lbs 6.8 Nm	N/A
Transfer port cover screws	N/A	25 in lbs 2.8 Nm	25 in lbs 2.8 Nm	45 in lbs 5 Nm	25 in lbs 2.8 Nm	25 in lbs 2.8 Nm	N/A
Trim tab bolt	N/A	N/A	N/A	N/A	N/A	N/A	N/A
Upper & lower end caps	50 in lbs 6 Nm	35 in lbs 4 Nm	35 in lbs 4 Nm	40 in lbs 4.5 Nm	41 in lbs 4.6 Nm	41 in lbs 4.6 Nm	N/A
Water pump (plastic)	25 in lbs 2.8 Nm	20 in lbs 2.3 Nm	20 in lbs 2.3 Nm	30 in lbs 3.4 Nm	3- in lbs 3.4 Nm	30 in lbs 3.4 Nm	20 in lbs 2.3 Nm
Water pump base screws	N/A	N/A	N/A	N/A	N/A	20 in lbs 2.3 Nm	N/A

NOTE 1 With 5/16" nut, 30 ft lbs (40.1 Nm); with 3/8" nut, 55 ft lbs (74.8 Nm).

NOTE 2 With 5/16" nut, 35 ft lbs (47.6 Nm); with 3/8" nut, 30 ft lbs (40.1 Nm).

NOTE 3 65 in lbs (7.3 Nm) for Model 6J.

NOTE 4 With 5/16-24 screw, 140 in lbs (15.8 Nm); with 1/4-20 screw, 100 in lbs (11.3 Nm); with 5/16-18 screw, 200 in lbs (22.6 Nm).

NOTE 5 With 3/8" nut, 55 ft lbs (74.8 Nm); with 7/16" nut, 65 ft lbs (88.4 Nm).

NOTE 6 With 3/8" nut, 55 ft lbs (74.8 Nm); with 7/16" nut, 98 ft lbs (133.3 Nm).

NOTE 7 Upper end cap 12.5 ft lbs (17 Nm); lower end cap 8.3 ft lbs (11.3 Nm).

OTHER TORQUE VALUES

The following torque values are for sizes not listed elsewhere.

Size	Torque Value Ft Lbs	Torque Value In Lbs	Torque Value Nm
#6	--	7 - 10	.8
#10	2 - 3	25 - 35	3 - 4
#12	3 - 4	35 - 45	4 - 5
1/4"	5 - 7	60 - 80	7 - 10
5/16"	10 - 12	120 - 140	14 - 19
3/8"	18 - 20	220 - 240	25 - 27

TORQUE SPECIFICATIONS

Fastener Location	1 Cyl. 45 1975-78	1 Cyl. 4.5 1979-85	2 Cyl. 60 1965-68	2 Cyl. 6.0 1986-87	2 Cyl. 75 1969-78	2 Cyl. 7.5 1979-85	2 Cyl. 8.0 1986 & On
Bearing carrier nut	60 ft lbs 81.6 Nm	60 ft lbs 81.6 Nm	60 ft lbs 81.6 Nm	60 ft lbs 81.6 Nm	60 ft lbs 81.6 Nm	60 ft lbs 81.6 Nm	60 ft lbs 81.6 Nm
Carburetor adaptor plate bolts	N/A	N/A	N/A	60 in lbs 6.8 Nm	N/A N/A	N/A N/A	60 in lbs 6.8 Nm
Carburetor mounting locknut	80 in lbs 9 Nm	80 in lbs 9 Nm	80 in lbs 9 Nm	125 in lbs 14.1 Nm	80 in lbs 9 Nm	80 in lbs 9 Nm	125 in lbs 14.1 Nm
Center main bearing lockscrew (3)	N/A	N/A	33 in lbs 3.7 Nm	N/A	42 in lbs 4.7 Nm	15 ft lbs 20.4 Nm	N/A
Center main bearing reed stop screw	10 in lbs 1.1 Nm	10 in lbs 1.1 Nm	22 in lbs 2.5 Nm	N/A	24 in lbs 2.7 Nm	24 in lbs 2.7 Nm	N/A
Coil terminal nuts	30 in lbs 3.4 Nm	30 in lbs 3.4 Nm	30 in lbs 3.4 Nm	35 in lbs 3.9 Nm	30 in lbs 3.4 Nm	30 in lbs 3.4 Nm	35 in lbs 3.9 Nm
Connecting rod nuts	15 ft lbs 20.3 Nm	15 ft lbs 20.3 Nm	15 ft lbs 20.3 Nm	8.3 ft lbs 11.3 Nm	15 ft lbs 20.3 Nm	15 ft lbs 20.3 Nm	8.3 ft lbs 11.3 Nm
Crankcase to cylinder block	100 in lbs 11.3 Nm	100 in lbs 11.3 Nm	90 in lbs 10.1 Nm	16.3 ft bs 22.6 Nm	95 in lbs 10.7 Nm	100 in lbs 11.3 Nm	16.3 ft lbs 22.6 Nm
Cylinder block cover	100 in lbs 11.3 Nm	100 in lbs 11.3 Nm	70 in lbs 7.9 Nm	60 in lbs 6.8 Nm	90 in lbs 10.1 Nm	100 in lbs 11.3 Nm	60 in lbs 6.8 Nm
Exhaust outer cover	90 in lbs 10.1 Nm	90 in lbs 10.1 Nm	70 in lbs 7.9 Nm	60 inlbs 6.8 Nm	70 in lbs 7.9 Nm	90 in lbs 10.1 Nm	60 inlbs 6.8 Nm
Flywheel to crankshaft	35 ft lbs 47.5 Nm	35 ft lbs 47.5 Nm	35 ft lbs 47.5 Nm	50 ft lbs 68 Nm	35 ft lbs 47.5 Nm	35 ft lb 47.5 Nm	50 ft lbs 68 Nm
Gear housing to driveshaft housing	Note 2	Note 2	Note 2	15 ft lbs 20.4 Nm	Note 2	Note 2	15 ft lbs 20.4 Nm
Ignition trigger cover	15 in lbs 1.7 Nm	15 in lbs 1.7 Nm	N/A	30 in lbs 3.4 Nm	15 in lbs 1.7 Nm	15 in lbs 1.7 Nm	30 in lbs 3.4 Nm
Pinion gear nut	90 in lbs 10.2 Nm	90 in lbs 10.2 Nm	N/A	N/A	90 in lbs 10.2 Nm	90 in lbs 10.2 Nm	N/A
Powerhead to driveshaft housing	80 in lbs 9 Nm	80 in lbs 9 Nm	80 in lbs 9 Nm	10 ft lbs 13.6 Nm	80 in lbs 9 Nm	80 in lbs 9 Nm	10 ft lbs 13.6 Nm
Propeller nut	N/A	N/A	N/A	70 in lbs 7.9 Nm	N/A	N/A	70 in lbs 7.9 Nm
Reed block clamping screws	N/A	N/A	30 in lbs 3.4 Nm	20 in lbs 2.3 Nm	45 in lbs 5.1 Nm	45 in lbs 5 Nm	20 in lbs 2.3 Nm
Spark plugs	17 ft lbs 23 Nm	17 ft lbs 23 Nm	17 ft lbs 23 Nm	20 ft lbs 27 Nm	17 ft lbs 23 Nm	20 ft lbs 27 Nm	20 ft lbs 27 Nm
Starter motor to crankcase	N/A	N/A	N/A	N/A	N/A	N/A	N/A
Stator mounting screws	N/A N/A	30 in lbs 3.4 Nm	N/A N/A	30 in lbs 3.4 Nm	30 in lbs 3.4 Nm	30 in lbs 3.4 Nm	30 in lbs 3.4 Nm

TORQUE SPECIFICATIONS

Fastener Location	1 Cyl. 45 1975-78	1 Cyl. 4.5 1979-85	2 Cyl. 60 1965-68	2 Cyl. 6.0 1986-87	2 Cyl. 75 1969-78	2 Cyl. 7.5 1979-85	2 Cyl. 8.0 1986 & On
Support stud nuts	40 in lbs 4.5 Nm	40 in lbs 4.5 Nm	40 in lbs 4.5 nm	60 in lbs 6.8 Nm	40 in lbs 4.5 Nm	40 in lbs 4.5 Nm	60 in lbs 6.8 Nm
Swivel bracket to driveshaft housing bolts	60 in lbs 6.8 Nm	60 in lbs 6.8 Nm	35 in lbs 3.9 Nm	35 in lbs 3.9 Nm	35 in lbs 3.9 Nm	35 in lbs 3.9 Nm	35 in lbs 3.9 Nm
Transfer port cover screws	25 in lbs 2.8 Nm	25 in lbs 2.8 Nm	Note 3	60 in lbs 6.8 Nm	45 in lbs 5.0 Nm	45 in lbs 5.0 Nm	60 in lbs 6.8 Nm
Trim tab bolt	N/A	N/A	N/A	N/A	N/A	N/A	N/A
Upper & lower end caps	41 in lbs 4.6 Nm	41 in lbs 4.6 Nm	41 in lbs 4.6 Nm	N/A	41 in lbs 4.6 Nm	41 in lbs 4.6 Nm	N/A
Water pump (plastic)	30 in lbs 3.4 Nm	30 in lbs 3.4 Nm	30 in lbs 3.4 Nm	N/A	30 in lbs 3.4 Nm	30 in lbs 3.4 Nm	N/A
Water pump base screws	N/A	N/A	N/A	40 in lbs 4.5 Nm	N/A	N/A	40 in lbs 4.5 Nm

NOTE 1 With 5/16" nut, 30 ft lbs (40.1 Nm); with 3/8" nut, 55 ft lbs (74.8 Nm).

NOTE 2 With 5/16" nut, 35 ft lbs (47.6 Nm); with 3/8" nut, 30 ft lbs (40.1 Nm).

NOTE 3 65 in lbs (7.3 Nm) for Model 6J.

NOTE 4 With 5/16-24 screw, 140 in lbs (15.8 Nm); with 1/4-20 screw, 100 in lbs (11.3 Nm); with 5/16-18 screw, 200 in lbs (22.6 Nm).

NOTE 5 With 3/8" nut, 55 ft lbs (74.8 Nm); with 7/16" nut, 65 ft lbs (88.4 Nm).

NOTE 6 With 3/8" nut, 55 ft lbs (74.8 Nm); with 7/16" nut, 98 ft lbs (133.3 Nm).

NOTE 7 Upper end cap 12.5 ft lbs (17 Nm); lower end cap 8.3 ft lbs (11.3 Nm).

OTHER TORQUE VALUES

The following torque values are for sizes not listed elsewhere.

Size	Torque Value Ft Lbs	Torque Value In Lbs	Torque Value Nm
#6	--	7 - 10	.8
#10	2 - 3	25 - 35	3 - 4
#12	3 - 4	35 - 45	4 - 5
1/4"	5 - 7	60 - 80	7 - 10
5/16"	10 - 12	120 - 140	14 - 19
3/8"	18 - 20	220 - 240	25 - 27

TORQUE SPECIFICATIONS

Fastener Location	2 Cyl. 110 1965–78	2 Cyl. 9.8 1979–85	2 Cyl. 9.9 1986 & On 15.0 1988 & On	2 Cyl. 18 1981–83	2 Cyl. 18XD 1984–85	2 Cyl. 200 1965–72	2 Cyl. 20 1973–80
Bearing carrier nut	60 ft lbs 81.6 Nm	60 ft lbs 81.6 Nm	60 ft lbs 81.6 Nm	80 ft lbs 108.8 Nm	80 ft lbs 108.8 Nm	60 ft lbs 81.6 Nm	100 ft lbs 136 Nm
Carburetor adaptor plate bolts	N/A	N/A	60 in lbs 6.8 Nm	30 in lbs 3.4 Nm	30 in lbs 3.4 Nm	N/A	N/A
Carburetor mounting locknut	80 in lbs 9 Nm	80 in lbs 9 Nm	125 in lbs 14.1 Nm	180 in lbs 20.3 Nm	180 in lbs 20.3 Nm	80 in lbs 9 Nm	80 in lbs 9 Nm
Center main bearing lockscrew (3)	42 in lbs 4.7 Nm	42 in lbs 4.7 Nm	N/A	N/A	N/A	12.5 ft lbs 16.9 Nm	15 ft lbs 20.4 Nm
Center main bearing reed stop screw	22 in lbs 2.5 Nm	22 in lbs 2.5 Nm	N/A	25 in lbs 2.8 Nm	25 in lbs 2.8 Nm	37 in lbs 4.2 Nm	40 in lbs 4.5 Nm
Coil terminal nuts	30 in lbs 3.4 Nm	30 in lbs 3.4 Nm	35 in lbs 3.9 Nm	30 in lbs 3.4 Nm	30 in lbs 3.4 Nm	30 in lbs 3.4 Nm	30 in lbs 3.4 Nm
Connecting rod nuts	15 ft lbs 20.3 Nm	15 ft lbs 20.3 Nm	8.5 ft lbs 11.3 Nm	13.3 ft lbs 18 Nm	13.3 ft lbs 18 Nm	15 ft lbs 20.3 Nm	15 ft lbs 20.3 Nm
Crankcase to cylinder block	95 in lbs 10.7 Nm	95 in lbs 10.7 Nm	16.3 ft lbs 22.6 Nm	29 ft lbs 39.4 Nm	29 ft lbs 39.4 Nm	15.4 ft lbs 20.9 Nm	Note 4
Cylinder block cover	90 in lbs 10.1 Nm	90 in lbs 10.1 Nm	60 in lbs 6.8 Nm	90 in lbs 10.1 Nm	90 in lbs 10.1 Nm	90 in lbs 10.1 Nm	100 in lbs 11.3 Nm
Exhaust outer cover	70 in lbs 7.9 Nm	70 in lbs 7.9 Nm	60 in lbs 6.8 Nm	90 in lbs 10.1 Nm	90 in lbs 10.1 Nm	12.5 ft lbs 16.9 Nm	95 in lbs 10.7 Nm
Flywheel to crankshaft	35 ft lbs 47.5 Nm	35 ft lbs 47.5 Nm	50 ft lbs 68 Nm	50 ft lbs 68 Nm	50 ft lbs 68 Nm	65 ft lbs 88.4 Nm	35 ft lbs 47.5 Nm
Gear housing to driveshaft housing	Note 2	Note 2	15 ft lbs 20.4 Nm	25 ft lbs 34 Nm	25 ft lbs 34 Nm	Note 1	Note 1
Ignition trigger cover	15 in lbs 1.7 Nm	15 in lbs 1.7 Nm	30 in lbs 3.4 Nm	N/A	N/A	N/A	N/A
Pinion gear nut	90 in lbs 10.2 Nm	90 in lbs 10.2 Nm	N/A	100 in lbs 11.3 Nm	100 in lbs 11.3 Nm	N/A	9.7 ft lbs 13.2 Nm
Powerhead to driveshaft housing	80 in lbs 9 Nm	80 in lbs 9 Nm	10 ft lbs 13.6 Nm	16.7 ft lbs 22.6 Nm	16.7 ft lbs 22.6 Nm	12.5 ft lbs 16.9 Nm	80 in lbs 9 Nm
Propeller nut	10 ft lbs 13.6 Nm	10 ft lbs 13.6 Nm	5.8 ft lbs 7.9 Nm	10 ft lbs 13.6 Nm	10 ft lbs 13.6 Nm	10 ft lbs 13.6 Nm	N/A
Reed block clamping screws	25 in lbs 2.8 Nm	25 in lbs 2.8 Nm	20 in lbs 2.3 Nm	30 in lbs 3.4 Nm	30 in lbs 3.4 Nm	25 in lbs 2.8 Nm	N/A N/A
Spark plugs	20 ft lbs 27 Nm	20 ft lbs 27 Nm	20 ft lbs 27 Nm	20 ft lbs 27 Nm	20 ft lbs 27 Nm	17 ft lbs 23 Nm	20 ft lbs 27 Nm
Starter motor to crankcase	N/A	N/A	N/A	N/A	N/A	N/A	N/A
Stator mounting screws	30 in lbs 3.4 Nm	30 in lbs 3.4 Nm	30 in lbs 3.4 Nm	30 in lbs 3.4 Nm	30 in lbs 3.4 Nm	60 in lbs 6.8 Nm	60 in lbs 6.8 Nm

TORQUE SPECIFICATIONS

Fastener Location	2 Cyl. 110 1965–78	2 Cyl. 9.8 1979–85	2 Cyl. 9.9 1986 & On 15.0 1988 & On	2 Cyl. 18 1981–83	2 Cyl. 18XD 1984–85	2 Cyl. 200 1965–72	2 Cyl. 20 1973–80
Support stud nuts	40 in lbs 4.5 Nm	40 in lbs 4.5 Nm	60 in lbs 6.8 Nm	N/A	N/A	80 in lbs 9 Nm	80 in lbs 9 Nm
Swivel bracket to driveshaft housing bolts	35 in lbs 4.0 Nm	35 in lbs 4.0 Nm	35 in lbs 4.0 Nm	200 in lbs 22.6 Nm	200 in lbs 22.6 Nm	35 in lbs 4.0 Nm	35 in lbs 4.0 Nm
Transfer port cover screws	45 in lbs 5 Nm	45 in lbs 5 Nm	60 in lbs 6.8 Nm	30 in lbs 3.4 Nm	30 in lbs 3.4 Nm	60 in lbs 6.8 Nm	60 in lbs 6.8 Nm
Trim tab bolt	N/A	N/A	N/A	100 in lbs 11.3 Nm	100 in lbs 11.3 Nm	N/A	N/A
Upper & lower end caps	41 in lbs 4.6 Nm	41 in lbs 4.6 Nm	N/A	N/A	N/A	12.5 ft lbs 16.9 Nm	13.3 ft lbs 18 Nm
Water pump (plastic)	30 in lbs 3.4 Nm	30 in lbs 3.4 Nm	N/A	N/A	N/A	40 in lbs 4.5 Nm	40 in lbs 4.5 Nm
Water pump base screws	N/A	N/A	40 in lbs 4.5 Nm	25 in lbs 2.8 Nm	25 in lbs 2.8 Nm	N/A	40 in lbs 4.5 Nm

NOTE 1 With 5/16" nut, 30 ft lbs (40.1 Nm); with 3/8" nut, 55 ft lbs (74.8 Nm).

NOTE 2 With 5/16" nut, 35 ft lbs (47.6 Nm); with 3/8" nut, 30 ft lbs (40.1 Nm).

NOTE 3 65 in lbs (7.3 Nm) for Model 6J.

NOTE 4 With 5/16-24 screw, 140 in lbs (15.8 Nm); with 1/4-20 screw, 100 in lbs (11.3 Nm); with 5/16-18 screw, 200 in lbs (22.6 Nm).

NOTE 5 With 3/8" nut, 55 ft lbs (74.8 Nm); with 7/16" nut, 65 ft lbs (88.4 Nm).

NOTE 6 With 3/8" nut, 55 ft lbs (74.8 Nm); with 7/16" nut, 98 ft lbs (133.3 Nm).

NOTE 7 Upper end cap 12.5 ft lbs (17 Nm); lower end cap 8.3 ft lbs (11.3 Nm).

OTHER TORQUE VALUES

The following torque values are for sizes not listed elsewhere.

Size	Torque Value Ft Lbs	Torque Value In Lbs	Torque Value Nm
#6	--	7 - 10	.8
#10	2 - 3	25 - 35	3 - 4
#12	3 - 4	35 - 45	4 - 5
1/4"	5 - 7	60 - 80	7 - 10
5/16"	10 - 12	120 - 140	14 - 19
3/8"	18 - 20	220 - 240	25 - 27

TORQUE SPECIFICATIONS

Fastener Location	2 Cyl. 20 1986 & On	2 Cyl. 25 1980–83	2 Cyl. 25 1986 & On	2 Cyl. 25XD 1984–85	2 Cyl. 350 1965–69	2 Cyl. 35 1984–89	2 Cyl. 400 1970–71
Bearing carrier nut	80 ft lbs 108.8 Nm	80 ft lbs 108.8 Nm	80 ft lbs 108.8 Nm	80 ft lbs 108.8 Nm	110 ft lbs 150 Nm	100 ft lbs 136 Nm	110 ft 150 Nm
Carburetor adaptor plate bolts	30 in lbs 3.4 Nm	30 in lbs 3.4 Nm	30 in lbs 3.4 Nm	30 in lbs 3.4 Nm	N/A	N/A	N/A
Carburetor mounting locknut	15 ft lbs 20.3 Nm	15 ft lbs 20.3 Nm	15 ft lbs 20.3 Nm	15 ft lbs 20.3 Nm	12 ft lbs 16.3 Nm	12 ft lbs 16.3 Nm	12 ft lbs 16.3 Nm
Center main bearing lockscrew (3)	N/A	N/A	N/A	N/A	12.5 ft lbs 16.9 Nm	N/A	12.5 ft lbs 16.9 Nm
Center main bearing reed stop screw	25 in lbs 2.8 Nm	25 in lbs 2.8 Nm	25 in lbs 2.8 Nm	25 in lbs 2.8 Nm	37 in lbs 4.2 Nm	35 in lbs 4.0 Nm	22 in lbs 2.5 Nm
Coil terminal nuts	30 in lbs 3.4 Nm	30 in lbs 3.4 Nm	30 in lbs 3.4 Nm	30 in lbs 3.4 Nm	30 in lbs 3.4 Nm	30 in lbs 3.4 Nm	30 in lbs 3.4 Nm
Connecting rod nuts	13.3 ft lbs 18 Nm	13.3 ft lbs 18 Nm	13.3 ft lbs 18 Nm	13.3 ft lbs 18 Nm	15 ft lbs 20.3 Nm	15 ft lbs 20.3 Nm	15 ft lbs 20.3 Nm
Crankcase to cylinder block	29 ft lbs 39.4 Nm	29 ft lbs 39.4 Nm	29 ft lbs 39.4 Nm	29 ft lbs 39.4 Nm	12.5 ft lbs 16.9 Nm	16.6 ft lbs 22.6 Nm	17.5 ft lbs 23.7 Nm
Cylinder block cover	90 in lbs 10.1 Nm	90 in lbs 10.1 Nm	90 in lbs 10.1 Nm	90 in lbs 10.1 Nm	70 in lbs 7.9 Nm	8.3 ft lbs 11.3 Nm	70 in lbs 7.9 Nm
Exhaust outer cover	90 in lbs 10.1 Nm	90 in lbs 10.1 Nm	90 in lbs 10.1 Nm	90 in lbs 10.1 Nm	12.5 ft lbs 16.9 Nm	16.6 ft lbs 22.6 Nm	12.5 ft lbs 16.9 Nm
Flywheel to crankshaft	50 ft lbs 68 Nm	50 ft lbs 68 Nm	50 ft lbs 68 Nm	50 ft lbs 68 Nm	65 ft lbs 88.4 Nm	70 ft lbs 95 Nm	65 ft lbs 88.4 Nm
Gear housing to driveshaft housing	25 ft lbs 34 Nm	25 ft lbs 34 Nm	25 ft lbs 34 Nm	25 ft lbs 34 Nm	65 ft lbs 88.4 Nm	55 ft lbs 74.6 Nm	Note 5
Ignition trigger cover	N/A	N/A	N/A	N/A	N/A	N/A	N/A
Pinion gear nut	100 in lbs 11.3 Nm	100 in lbs 11.3 Nm	100 in lbs 11.3 Nm	100 in lbs 11.3 Nm	115 in lbs 13 Nm	115 in lbs 13 Nm	45 ft lbs 61.2 Nm
Powerhead to driveshaft housing	16.6 ft lbs 22.6 Nm	16.6 ft lbs 22.6 Nm	16.6 ft lbs 22.6 Nm	16.6 ft lbs 22.6 Nm	12.5 ft lbs 16.9 Nm	12.5 ft lbs 16.9 Nm	15 ft lbs 20.3 Nm
Propeller nut	10 ft lbs 11.3 Nm	10 ft lbs 11.3 Nm	10 ft lbs 11.3 Nm	10 ft lbs 11.3 Nm	N/A	N/A	N/A
Reed block clamping screws	30 in lbs 3.4 Nm	25 in lbs 2.8 Nm	30 in lbs 3.4 Nm	25 in lbs 2.8 Nm	30 in lbs 3.4 Nm	30 in lbs 3.4 Nm	30 in lbs 3.4 Nm
Spark plugs	20 ft lbs 27 Nm	20 ft lbs 27 Nm	20 ft lbs 27 Nm	20 ft lbs 27 Nm	17 ft lbs 23 Nm	17 ft lbs 23 Nm	17 ft lbs 23 Nm
Starter motor to crankcase	N/A	N/A	N/A	N/A	10.4 ft lbs 14.1 Nm	13.3 ft lbs 18 Nm	85 in lbs 9.6 Nm
Stator mounting screws	30 in lbs 3.4 Nm	30 in lbs 3.4 Nm	30 in lbs 3.4 Nm	30 in lbs 3.4 Nm	N/A N/A	60 in lbs 6.8 Nm	60 in lbs 6.8 Nm

TORQUE SPECIFICATIONS

Fastener Location	2 Cyl. 20 1986 & On	2 Cyl. 25 1980-83	2 Cyl. 25 1986 & On	2 Cyl. 25XD 1984-85	2 Cyl. 350 1965-69	2 Cyl. 35 1984-89	2 Cyl. 400 1970-71
Support stud nuts	N/A	N/A	N/A	N/A	70 in lbs 7.9 Nm	70 in lbs 7.9 Nm	70 in lbs 7.9 Nm
Swivel bracket to driveshaft housing bolts	N/A	N/A	N/A	N/A	N/A	N/A	N/A
Transfer port cover screws	30 in lbs 3.4 Nm	30 in lbs 3.4 Nm	30 in lbs 3.4 Nm	30 in lbs 3.4 Nm	60 in lbs 6.8 Nm	60 in lbs 6.8 Nm	60 in lbs 6.8 Nm
Trim tab bolt	100 in lbs 11.3 Nm	100 in lbs 11.3 Nm	100 in lbs 11.3 Nm	100 in lbs 11.3 Nm	25 ft lbs 33.9 Nm	25 ft lbs 33.9 Nm	25 ft lbs 33.9 Nm
Upper & lower end caps	N/A	N/A	N/A	N/A	12.5 ft lbs 16.9 Nm	Note 7	12.5 ft lbs 16.9 Nm
Water pump (plastic)	N/A	N/A	N/A	N/A	30 in lbs 3.4 Nm	35 in lbs 4 Nm	35 in lbs 4 Nm
Water pump base screws	25 in lbs 2.8 Nm	25 in lbs 2.8 Nm	25 in lbs 2.8 Nm	25 in lbs 2.8 Nm	N/A	N/A	N/A

NOTE 1 With 5/16" nut, 30 ft lbs (40.1 Nm); with 3/8" nut, 55 ft lbs (74.8 Nm).

NOTE 2 With 5/16" nut, 35 ft lbs (47.6 Nm); with 3/8" nut, 30 ft lbs (40.1 Nm).

NOTE 3 65 in lbs (7.3 Nm) for Model 6J.

NOTE 4 With 5/16-24 screw, 140 in lbs (15.8 Nm); with 1/4-20 screw, 100 in lbs (11.3 Nm); with 5/16-18 screw, 200 in lbs (22.6 Nm).

NOTE 5 With 3/8" nut, 55 ft lbs (74.8 Nm); with 7/16" nut, 65 ft lbs (88.4 Nm).

NOTE 6 With 3/8" nut, 55 ft lbs (74.8 Nm); with 7/16" nut, 98 ft lbs (133.3 Nm).

NOTE 7 Upper end cap 12.5 ft lbs (17 Nm); lower end cap 8.3 ft lbs (11.3 Nm).

OTHER TORQUE VALUES

The following torque values are for sizes not listed elsewhere.

Size	Torque Value Ft Lbs	Torque Value In Lbs	Torque Value Nm
#6	--	7 - 10	.8
#10	2 - 3	25 - 35	3 - 4
#12	3 - 4	35 - 45	4 - 5
1/4"	5 - 7	60 - 80	7 - 10
5/16"	10 - 12	120 - 140	14 - 19
3/8"	18 - 20	220 - 240	25 - 27

TORQUE SPECIFICATIONS

Fastener Location	2 Cyl. 402 1972–78	2 Cyl. 40 1978–80	2 Cyl. 40 1981–83	2 Cyl. 210cc 1986–87	1 Cyl. 90cc 1986–87
Bearing carrier nut	110 ft lbs 150 Nm	100 ft lbs 136 Nm	100 ft lbs 136 Nm	60 ft lbs 81.6 Nm	60 ft lbs 81.6 Nm
Carburetor adaptor plate bolts	M/A	N/A	N/A	60 in lbs 6.8 Nm	N/A
Carburetor mounting locknut	11.6 ft lbs 15.7 Nm	11.6 ft lbs 15.7 Nm	11.7 ft lbs 15.9 Nm	10.4 ft lbs 14.1 Nm	80 in lbs 9 Nm
Center main bearing lockscrew (3)	N/A	N/A	N/A	N/A	N/A
Center main bearing reed stop screw	25 in lbs 2.8 Nm	25 in lbs 2.8 Nm	25 in lbs 2.8 Nm	N/A	N/A
Coil terminal nuts	30 in lbs 3.4 Nm	30 in lbs 3.4 Nm	30 in lbs 3.4 Nm	35 in lbs 4 Nm	30 in lbs 3.4 Nm
Connecting rod nuts	15 ft lbs 20.3 Nm	15 ft lbs 20.3 Nm	15 ft lbs 20.3 Nm	8.3 ft lbs 11.3 Nm	15 ft lbs 20.3 Nm
Crankcase to cylinder block	16.6 ft lbs 22.6 Nm	16.6 ft lbs 22.6 Nm	16.6 ft lbs 22.6 Nm	16.6 ft lbs 22.6 Nm	100 in lbs 11.3 Nm
Cylinder block cover	70 in lbs 7.9 Nm	70 in lbs 7.9 Nm	8.3 ft lbs 11.3 Nm	60 in lbs 6.8 Nm	100 in lbs 11.3 Nm
Exhaust outer cover	16.6 ft lbs 22.6 Nm	16.7 ft lbs 22.6 Nm	16.6 ft lbs 22.6 Nm	90 in lbs 6.8 Nm	30 in lbs 10.1 Nm
Flywheel to crankshaft	75 ft lbs 102 Nm	75 ft lbs 102 Nm	75 ft lbs 102 Nm	50 ft lbs 68 Nm	35 ft lbs 47.5 Nm
Gear housing to driveshaft housing	Note 6	55 ft lbs 74.6 Nm	55 ft lbs 74.6 Nm	15 ft lbs 20.4 Nm	Note 2
Ignition trigger cover	N/A	N/A	N/A	30 in lbs 3.4 Nm	N/A
Pinion gear nut	70 ft lbs 95.2 Nm	45 ft lbs 61 Nm	45 ft lbs 61 Nm	N/A	90 in lbs 10.2 Nm
Powerhead to driveshaft housing	15 ft lbs 20.3 Nm	15 ft lbs 20.3 Nm	12 ft lbs 16.9 Nm	10 ft lbs 13.6 Nm	80 in lbs 9 Nm
Propeller nut	N/A	N/A	N/A	70 in lbs 7.9 Nm	60 in lbs 6.8 Nm
Reed block clamping screws	25 in lbs 2.8 Nm	25 in lbs 2.8 Nm	30 in lbs 3.4 Nm	20 in lbs 2.2 Nm	N/A
Spark plugs	17 ft lbs 23 Nm	20 ft lbs 27 Nm	20 ft lbs 27 Nm	20 ft lbs 27 Nm	17 ft lbs 23 Nm
Starter motor to crankcase	13.3 ft lbs 18.1 Nm	13.3 ft lbs 18.1 Nm	13.3 ft lbs 18.1 Nm	N/A	N/A
Stator mounting screws	60 in lbs 6.8 Nm	60 in lbs 6.8 Nm	60 in lbs 6.8 Nm	30 in lbs 3.4 Nm	30 in lbs 3.4 Nm

TORQUE SPECIFICATIONS

Fastener Location	2 Cyl. 402 1972-78	2 Cyl. 40 1978-80	2 Cyl. 40 1981-83	2 Cyl. 210cc 1986-87	1 Cyl. 90cc 1986-87
Support stud nuts	70 in lbs 7.9 Nm	70 in lbs 7.9 Nm	70 in lbs 7.9 Nm	60 in lbs 6.8 Nm	40 in lbs 4.5 Nm
Swivel bracket to driveshaft housing bolts	N/A	N/A	N/A	35 in lbs 3.9 Nm	60 in lbs 6.8 Nm
Transfer port cover screws	60 in lbs 6.8 Nm	60 in lbs 6.8 Nm	60 in lbs 6.8 Nm	60 in lbs 6.8 Nm	25 in lbs 2.8 Nm
Trim tab bolt	25 ft lbs 34 Nm	25 ft lbs 34 Nm	25 ft lbs 34 Nm	N/A	N/A
Upper & lower end caps	13.3 ft lbs 18 Nm	Note 7	Note 7	N/A	41 in lbs 4.6 Nm
Water pump (plastic)	30 in lbs 3.4 Nm	30 in lbs 3.4 Nm	30 in lbs 3.4 Nm	N/A	30 in lbs 3.4 Nm
Water pump base screws	40 in lbs 4.5 Nm	40 in lbs 4.5 Nm	40 in lbs 4.5 Nm	40 in lbs 4.5 Nm	N/A

NOTE 1 With 5/16" nut, 30 ft lbs (40.1 Nm); with 3/8" nut, 55 ft lbs (74.8 Nm).

NOTE 2 With 5/16" nut, 35 ft lbs (47.6 Nm); with 3/8" nut, 30 ft lbs (40.1 Nm).

NOTE 3 65 in lbs (7.3 Nm) for Model 6J.

NOTE 4 With 5/16-24 screw, 140 in lbs (15.8 Nm); with 1/4-20 screw, 100 in lbs (11.3 Nm); with 5/16-18 screw, 200 in lbs (22.6 Nm).

NOTE 5 With 3/8" nut, 55 ft lbs (74.8 Nm); with 7/16" nut, 65 ft lbs (88.4 Nm).

NOTE 6 With 3/8" nut, 55 ft lbs (74.8 Nm); with 7/16" nut, 98 ft lbs (133.3 Nm).

NOTE 7 Upper end cap 12.5 ft lbs (17 Nm); lower end cap 8.3 ft lbs (11.3 Nm).

OTHER TORQUE VALUES

The following torque values are for sizes not listed elsewhere.

Size	Torque Value Ft Lbs	Torque Value In Lbs	Torque Value Nm
#6	--	7 - 10	.8
#10	2 - 3	25 - 35	3 - 4
#12	3 - 4	35 - 45	4 - 5
1/4"	5 - 7	60 - 80	7 - 10
5/16"	10 - 12	120 - 140	14 - 19
3/8"	18 - 20	220 - 240	25 - 27

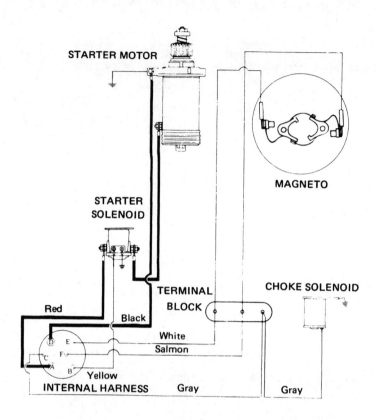

Model 350 Electric Start -- 1965-1969.

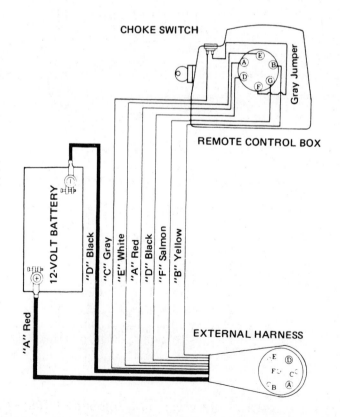

Model 402 Manual -- 1976-1978 and Model 40 Manual -- 1979-1981

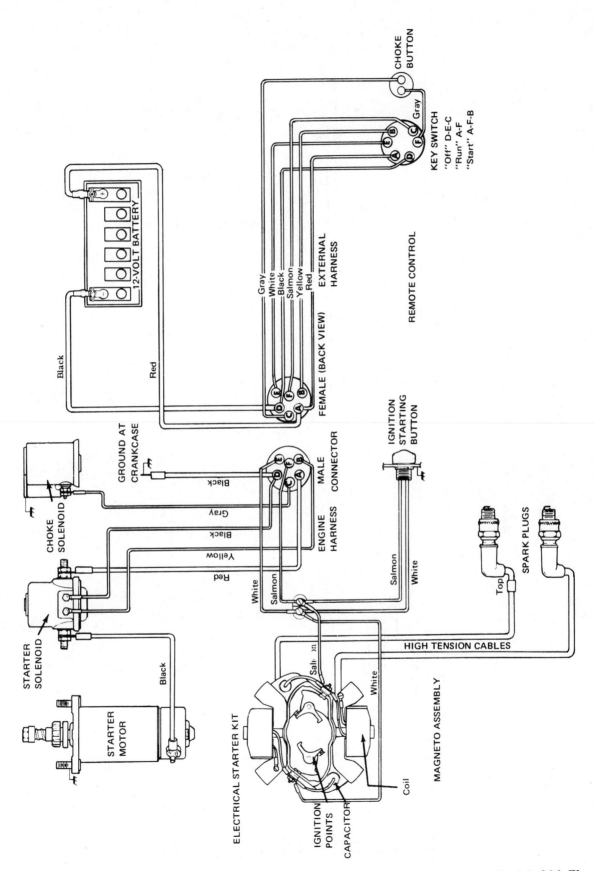

Model 60 Electric Start -- 1965-1968; Model 110 Electric -- 1965-1969; Model 200 Electric -- 1965-1971.

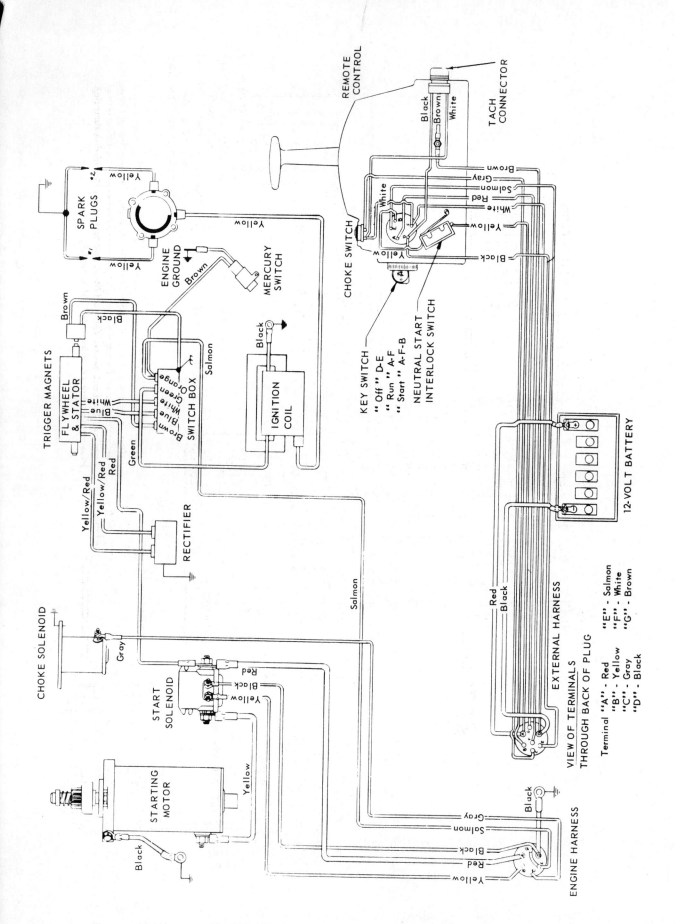

Model 400 Electric Start -- 1970-1971.

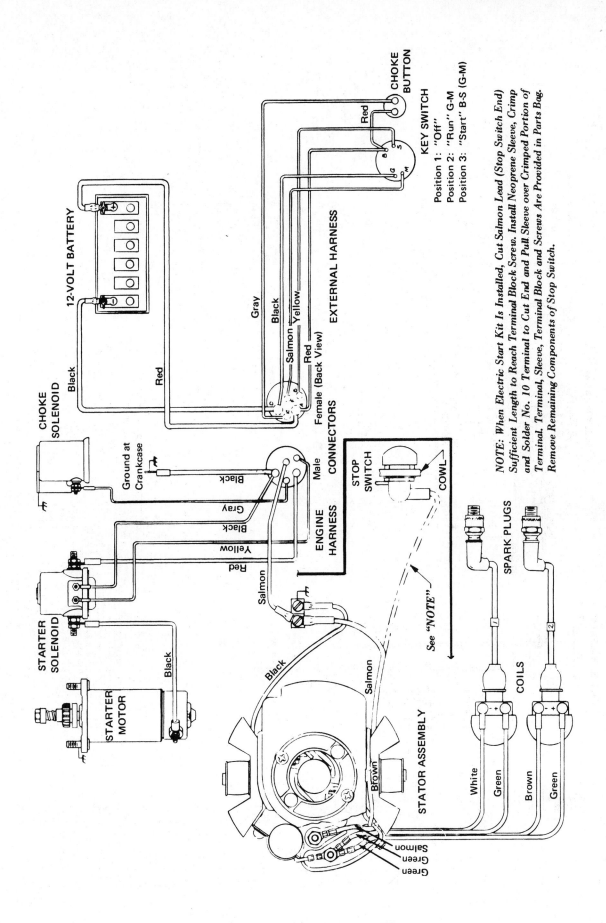

Model 200 Electric Start -- 1972.

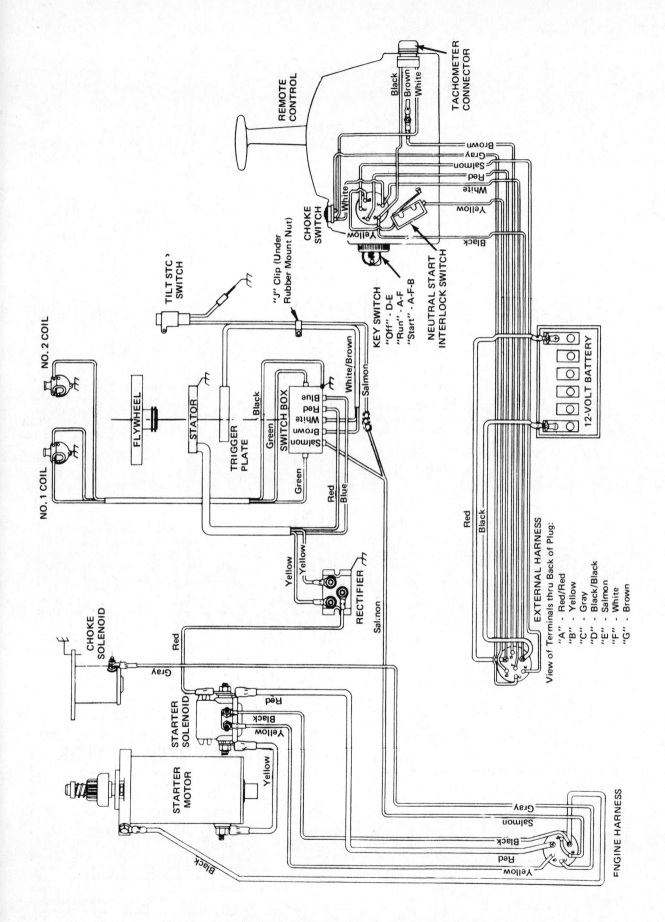

Model 402 Electric Start -- 1972-1974.

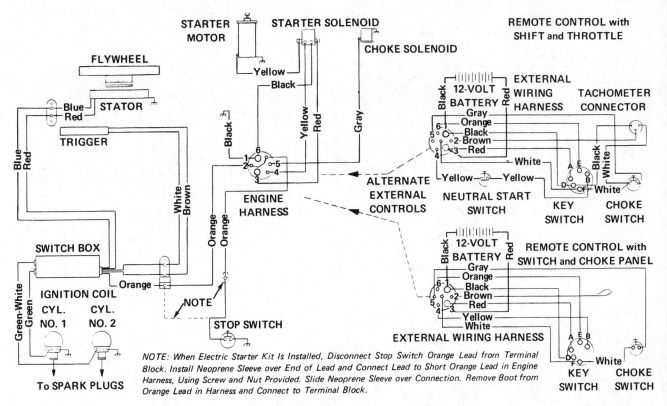

Model 200 Electric Start -- 1973-1975.

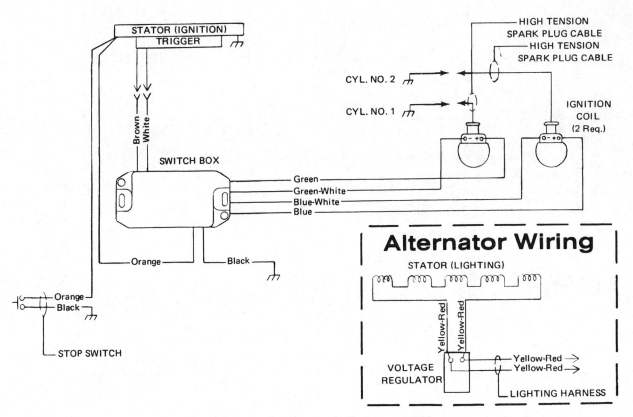

Model 200 with Alternator -- 1973-1978.

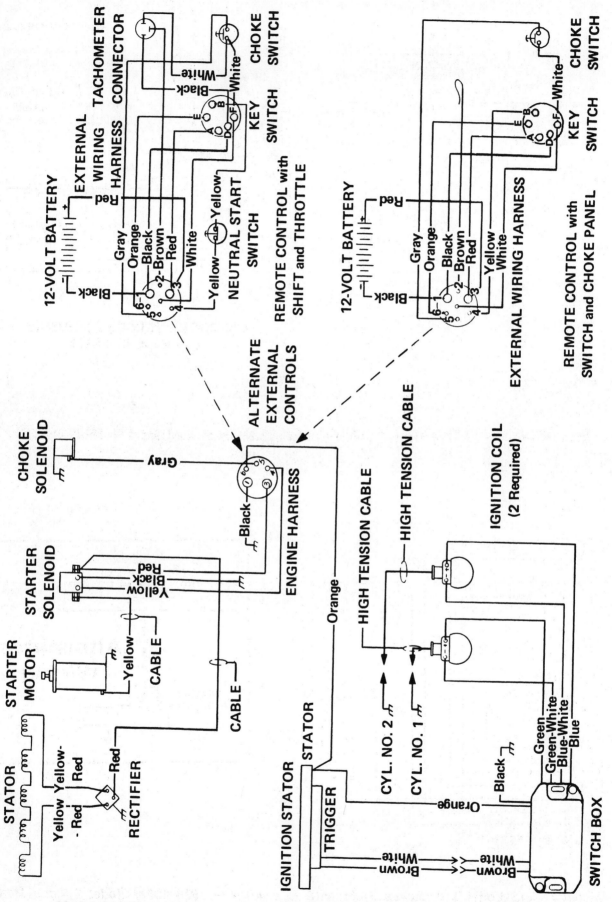

Model 200 Electric Start -- 1976-1978; Model 20 (20 hp) -- 1979-1980.

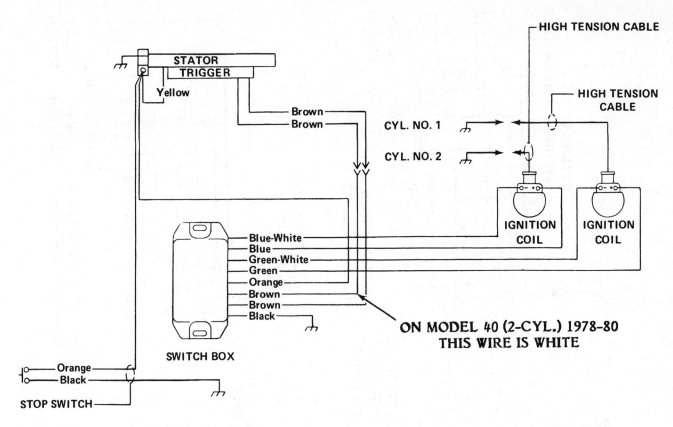

Model 75 and Model 110 Manual Start -- 1975; Model 40 Manual Start -- 1976-1980.

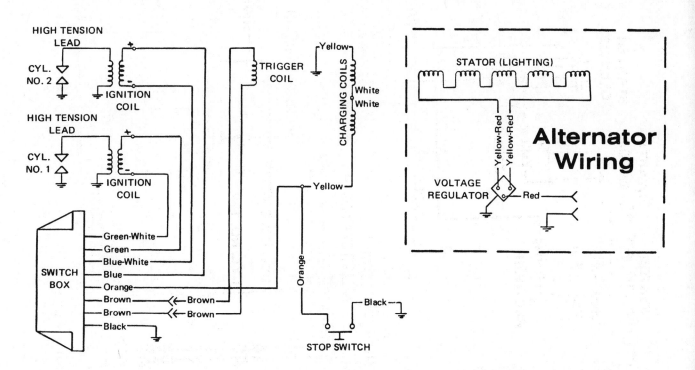

Model 75 and Model 110 Manual Start with Alternator -- 1976-1978; Model 7.5 and Model 9.8 with Alternator --1979-1985.

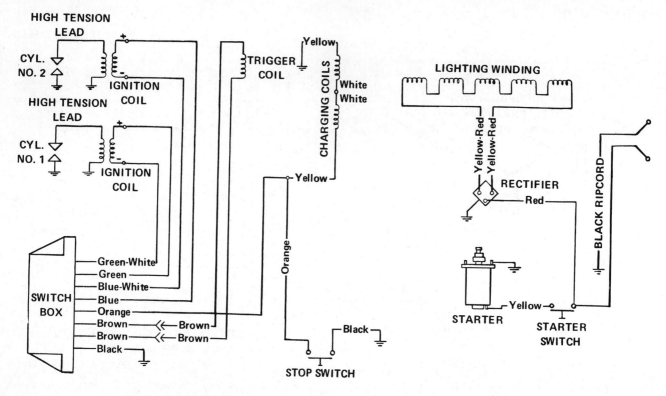

Model 75 and Model 110 -- 1976-78; Model 7.5 and 9.8 -- 1979-1985

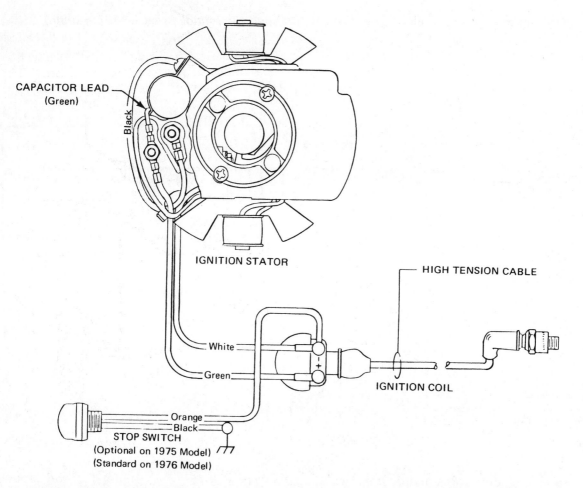

Model 40 Manual Start -- 1970-1974; Model 4.5 -- 1979-1985; Model 4.0 1986-Mid 1987.

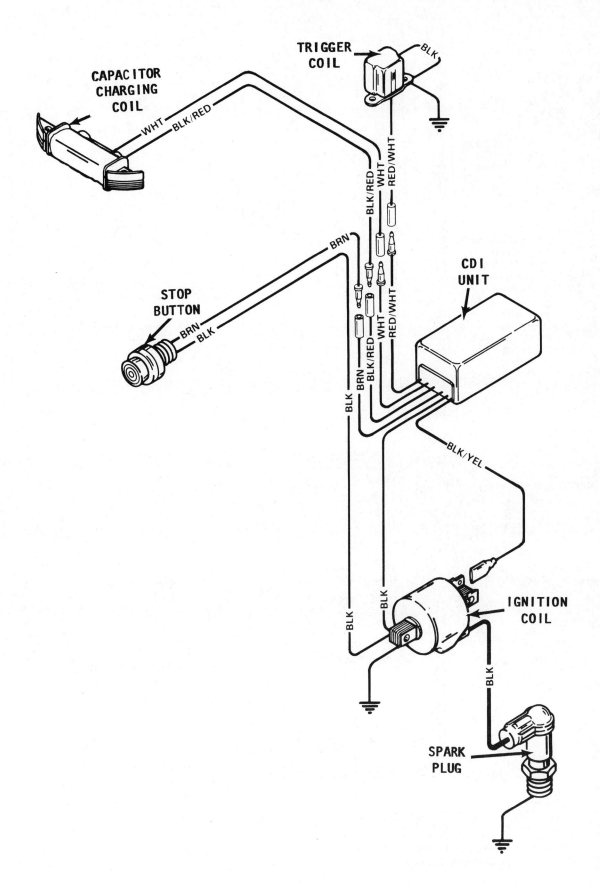

Model 4.0 — Mid 1987 & on; Model 5.0 — 1988 & on; and Model 90cc — 1986-87.

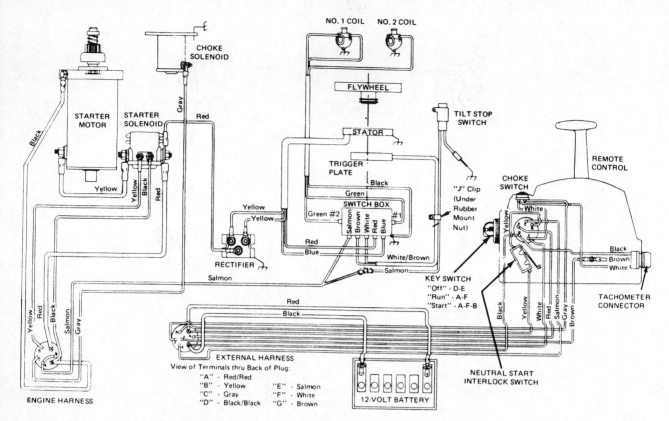

Model 402 Electric Start -- 1975 and Model 40 1979-1981.

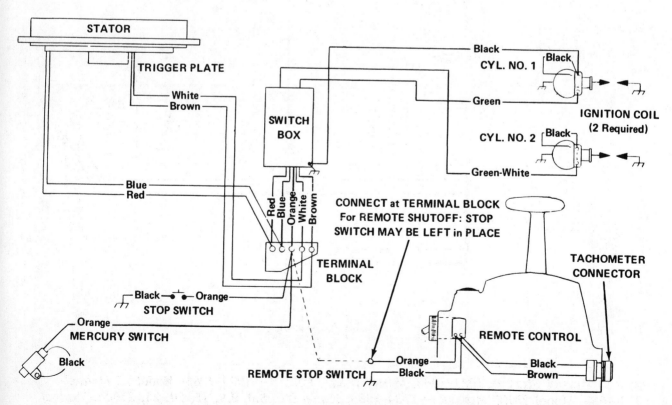

Model 402 Manual Start -- 1976-1978 and Model 40 Manual -- 1979-1981.

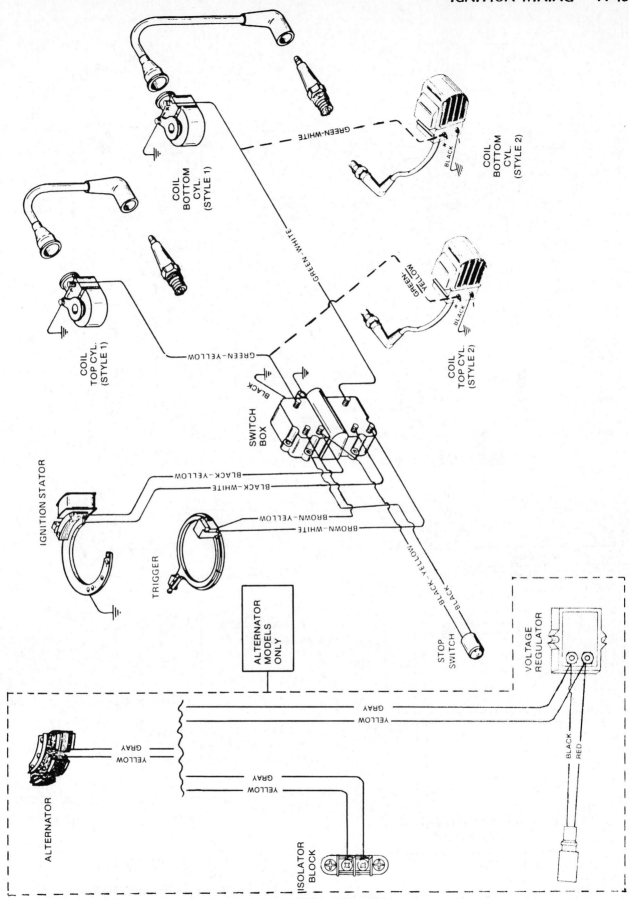

*Model 18 Manual Start -- 1981-1983; Model 18XD Manual -- 1984-1985; Model 25 Manual --1980-1983; Model 25XD Manual -- 1984-1985; Model 6.0, 8.0, 9.9, 15, 20, 25, 210cc, Manual --1986 and On. All models listed **NOT** equipped with remote control.*

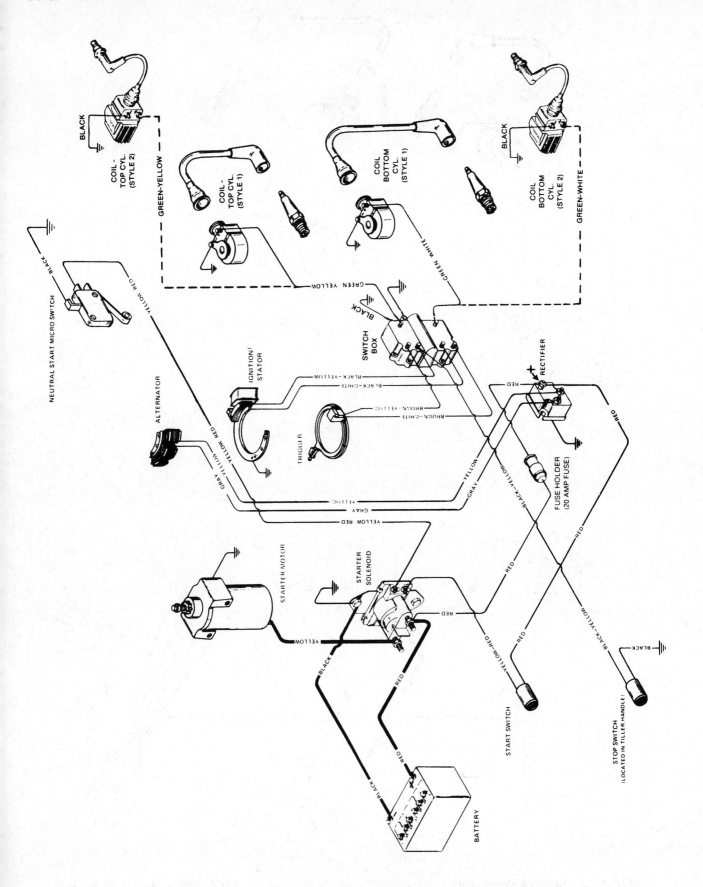

*Model 18 Electric Start 1981-1983; Model 18XD Electric -- 1984-1985; Model 25 Electric --1980-1983; Model 25XD Electric -- 1984-1985; Model 6.0, 8.0, 9.9, 15, 20, 25, 210cc, Electric --1986 and On. All models listed **NOT** equipped with remote control.*

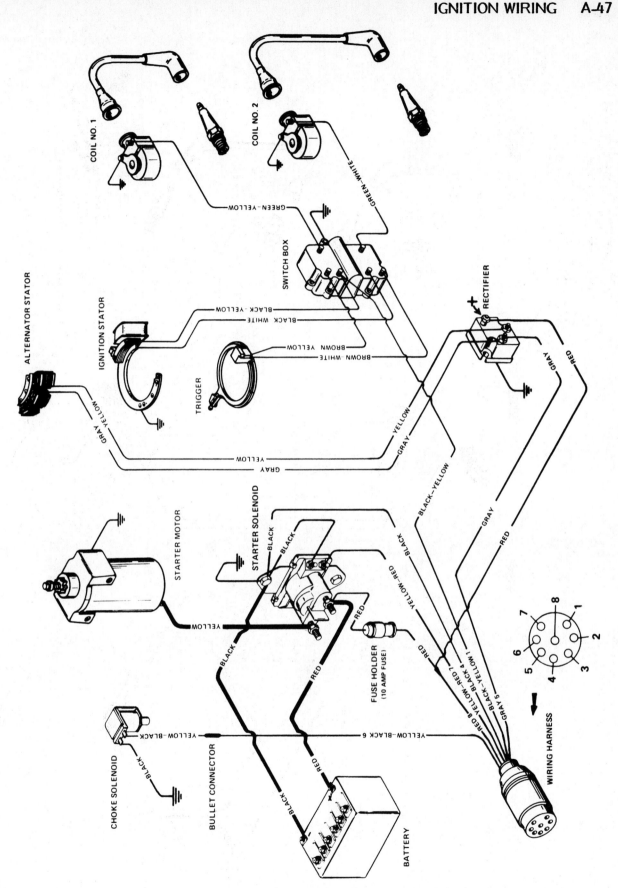

Model 18 Electric Start 1981-1983; Model 18XD Electric — 1984-1985; Model 25 Electric —1980-1983; Model 25XD Electric — 1984-1985; Model 6.0, 8.0, 9.9, 15, 20, 25, 210cc, Electric —1986 and On. All models listed equipped with tiller handle, ignition key/choke panel and remote control.

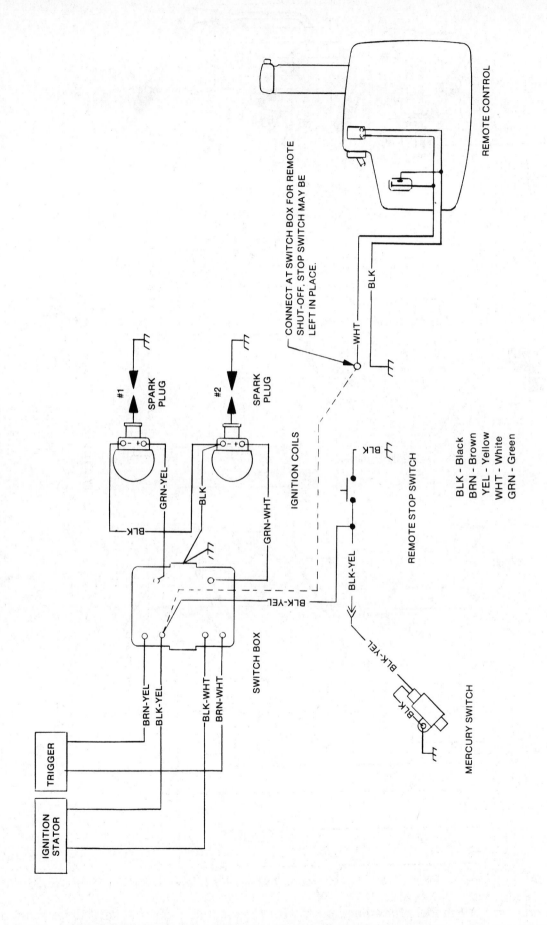

Model 35 Manual -- 1984-89; Model 40 Manual 1982-83.

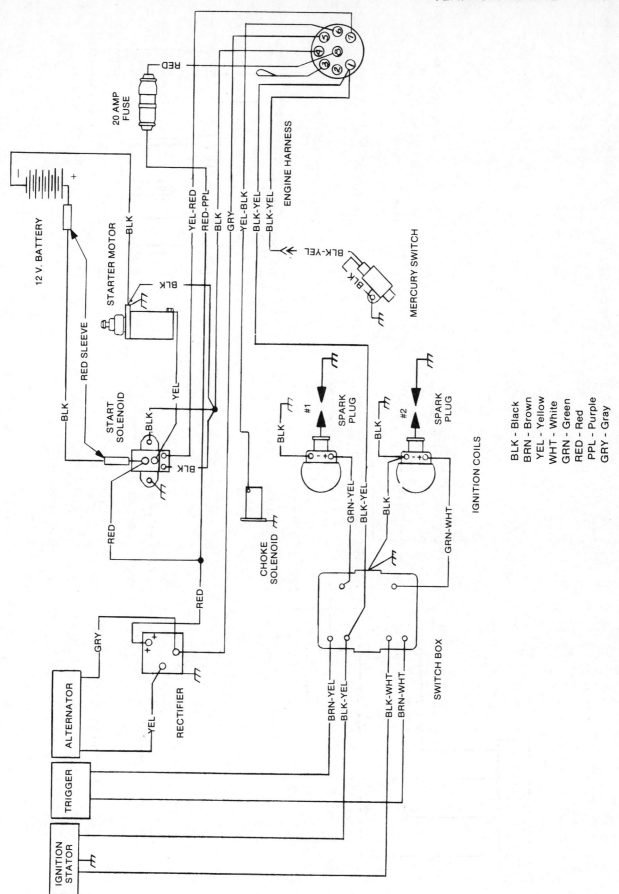

Model 35 Electric Start -- 1984-89 and Model 40 Electric Start -- 1982-83.

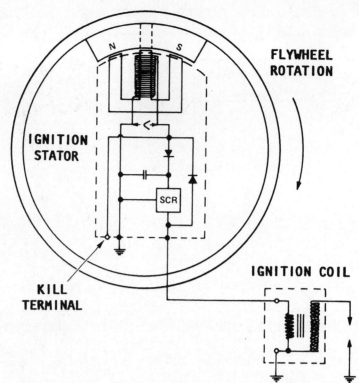

Simplified diagram of the Type III ignition system for the Model 3.5 — 1983-85 and the Model 3.6 — 1980-82.

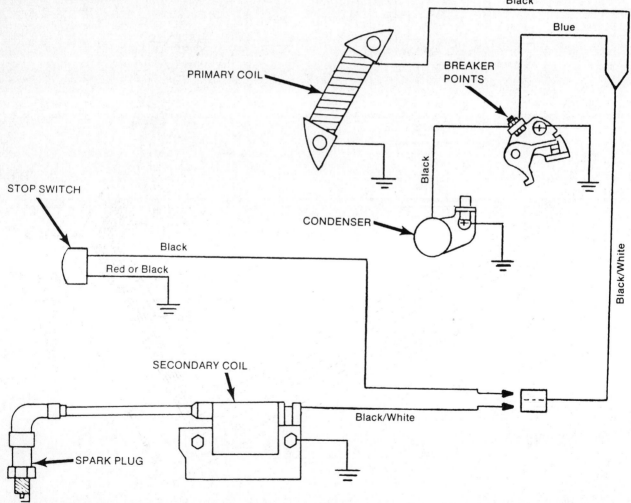

Functional diagram of the unique Type I ignition system on the Model 2.2 — 1984-89 and Model 3.0 — 1990 and on.

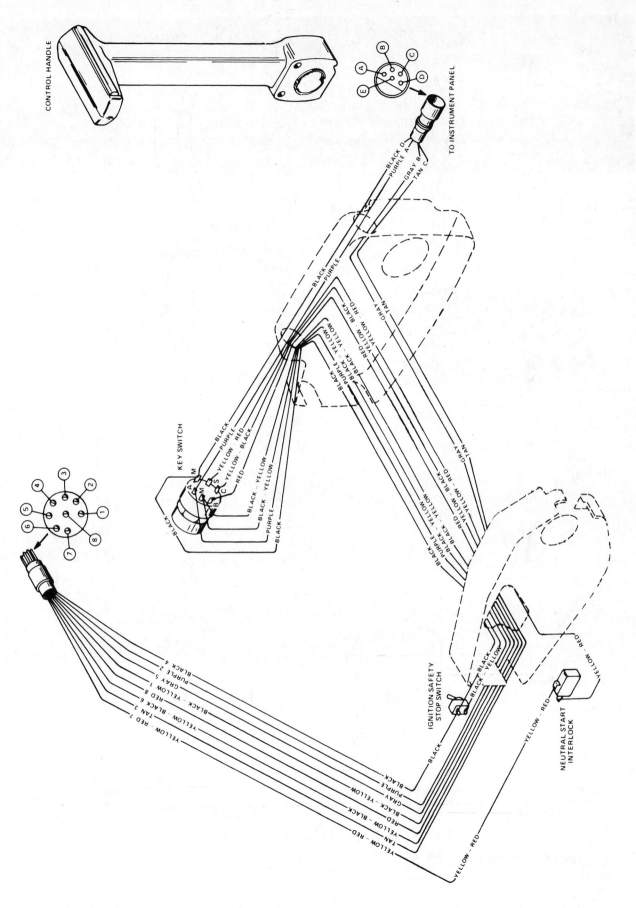

Commander Remote Control.

Chilton and Seloc Marine® offer repair and maintenance manuals for nearly everything with an engine.

Whether you need to care for a car, truck, sport-utility vehicle, boat, motorcycle, ATV, snowmobile or even a lawn mower, we can provide the information you need to get the most from your vehicles.

With over 230 manuals available in more than 6 lines of Do-It-Yourself books, you're bound to find what you want. Here are just a few types of books we have to offer.

TOTAL CAR CARE

With over 140 titles covering cars, trucks and sport utilities from Acura to Volvo, the Total Car Care series contains everything you need to care for or repair your vehicle. Every manual is based on vehicle teardowns and includes tons of step-by-step photographs, procedures and exploded views. Each manual also contains wiring diagrams, vacuum diagrams and covers every topic from simple maintenance to electronic engine

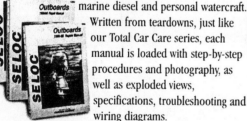

COLLECTOR'S SERIES HARD-COVER MANUALS

Chilton's Collector's Editions are perfect for enthusiasts of vintage or rare cars. These hard-cover manuals contain

repair and maintenance information for all major systems that might not be available elsewhere. Included are repair and overhaul procedures using thousands of illustrations. These manuals offer a range of coverage from as far back as 1940 and as recent as 1997, so you don't need an antique car or truck to be a collector.

SELOC MARINE MANUALS

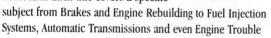

Chilton Marine offers 40 titles covering outboard engines, stern drives, jet drives, marine diesel and personal watercraft. Written from teardowns, just like our Total Car Care series, each manual is loaded with step-by-step procedures and photography, as well as exploded views, specifications, troubleshooting and wiring diagrams.

GENERAL INTEREST / RECREATIONAL BOOKS

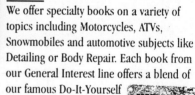

We offer specialty books on a variety of topics including Motorcycles, ATVs, Snowmobiles and automotive subjects like Detailing or Body Repair. Each book from our General Interest line offers a blend of our famous Do-It-Yourself procedures and photography with additional information on enjoying automotive, marine and recreational products. Learn more about the vehicles you use and enjoy while keeping them in top running shape.

TOTAL SERVICE SERIES / SYSTEM SPECIFIC MANUALS

These innovative books offer repair, maintenance and service procedures for automotive related systems. They cover today's complex vehicles in a user-friendly format, which places even the most difficult automotive topic well within the reach of every Do-It-Yourselfer. Each title covers a specific subject from Brakes and Engine Rebuilding to Fuel Injection Systems, Automatic Transmissions and even Engine Trouble

MULTI-VEHICLE SPANISH LANGUAGE MANUALS

Chilton's Spanish language manuals offer some of our most popular titles in Spanish. Each is as complete and easy to use as the English-language counterpart and offers the same maintenance, repair and overhaul information along with specifications charts and tons of illustrations.

Visit your local Chilton® Retailer

For a Catalog, for information, or to order call toll-free: 877-4CHILTON.

1020 Andrew Drive, Suite 200 • West Chester, PA 19380-4291
www.chiltononline.com

1PVerB

HP	SN	YR
20	4103665	75
9.8	6516825	84
7.5	413 4081	75
3.3	OG 968606	2000

20 HP parts motor

cylinder head   '75 - '79

carb   '75   < 4351589 (SN)

reed valves   -'78

lower unit   '74 - '78

ignition   '75 - '74